Studies in Theoretical and Applied Statistics
Selected Papers of the Statistical Societies

For further volumes:
http://www.springer.com/series/10104

Series Editors

Spanish Society of Statistics and Operations Research (SEIO)

Société Française de Statistique (SFdS)

Società Italiana di Statistica (SIS)

Sociedade Portuguesa de Estatística (SPE)

Nicola Torelli • Fortunato Pesarin
Avner Bar-Hen
Editors

Advances in Theoretical and Applied Statistics

Springer

Editors
Nicola Torelli
Department of Statistics
University of Trieste
Trieste
Italy

Fortunato Pesarin
Department of Statistics
University of Padua
Padova
Italy

Avner Bar-Hen
Université Paris Descartes
Paris
France

ISSN 2194-7767 ISSN 2194-7775 (electronic)
ISBN 978-3-642-35587-5 ISBN 978-3-642-35588-2 (eBook)
DOI 10.1007/978-3-642-35588-2
Springer Heidelberg New York Dordrecht London

Library of Congress Control Number: 2013940738

Printed on acid-free paper

Springer is part of Springer Science+Business Media (www.springer.com)

Foreword

Dear reader, On behalf of the four Scientific Statistical Societies – the SEIO, Sociedad de Estadística e Investigación Operativa (Spanish Statistical Society and Operation Research); SFdS, Société Française de Statistique (French Statistical Society); SIS, Società Italiana di Statistica (Italian Statistical Society); and the SPE, Sociedade Portuguesa de Estatística (Portuguese Statistical Society) – we would like to inform you that this is a new book series of Springer entitled "Studies in Theoretical and Applied Statistics," with two lines of books published in the series: "Advanced Studies" and "Selected Papers of the Statistical Societies."

The first line of books offers constant up-to-date information on the most recent developments and methods in the fields of theoretical statistics, applied statistics, and demography. Books in this series are solicited in constant cooperation between the statistical societies and need to show a high-level authorship formed by a team preferably from different groups so as to integrate different research perspectives.

The second line of books presents a fully peer-reviewed selection of papers on specific relevant topics organized by the editors, also on the occasion of conferences, to show their research directions and developments in important topics, quickly and informally, but with a high level of quality. The explicit aim is to summarize and communicate current knowledge in an accessible way. This line of books will not include conference proceedings and will strive to become a premier communication medium in the scientific statistical community by receiving an Impact Factor, as have other book series such as "Lecture Notes in Mathematics."

The volumes of selected papers from the statistical societies will cover a broad range of theoretical, methodological as well as application-oriented articles, surveys and discussions. A major goal is to show the intensive interplay between various, seemingly unrelated domains and to foster the cooperation between scientists in different fields by offering well-founded and innovative solutions to urgent practice-related problems.

On behalf of the founding statistical societies I wish to thank Springer, Heidelberg and in particular Dr. Martina Bihn for the help and constant cooperation in the organization of this new and innovative book series.

Rome, Italy — Maurizio Vichi

Preface

Statistics is the science of data analysis and inference. New theoretical developments and new methods are stimulated by real research problems in many substantive areas. The aim of this volume is to collect significant and innovative contributions on statistical methodology and applications of statistics and data analysis in different subject areas.

The volume is organized into six sections. The first two sections present contributions on statistical theory. The second section is focused more specifically on methods for the analysis of time series data or data with a spatial structure. Most of the papers included in the third section illustrate new methods for modeling complex data, while some of them present tools for data analysis. Section 4 includes papers that are focused on methods for survey design, methods for the analysis of data from complex surveys, and problems of relevance for producing official statistics. Application of statistical methods for social, demographic, health and biostatistic analysis is the common element among the papers included in Sect. 5. The last section focuses on econometrics and new methods and original applications for the analysis of economic data.

The papers included were selected from those presented at the 45th Meeting of the Italian Statistical Society held in Padua in June 2010. More than 400 scientists and experts from approximately 20 countries presented 280 papers that were accepted after a review process. More than 80 extended versions of these papers were subsequently submitted for potential inclusion in this volume. A careful double-blind review process was utilized and more than one hundred reviewers were involved. Selecting from among the many interesting papers presented was no easy task. We are grateful to all of the referees for their conscientious reviews.

Finally we would also like to thank Alice Blanck from Springer-Verlag for her patience and for her valued assistance in preparing this volume.

Nicola Torelli
Fortunato Pesarin
Avner Bar-Hen

Contents

Part III Statistical Modelling and Data Analysis

Part I

Statistical Theory

Default Priors Based on Pseudo-Likelihoods for the Poisson-GPD Model

1

Stefano Cabras

Abstract

Extreme values are usually modeled with the peaks over the threshold approach by means of the Poisson-Generalized Pareto Distribution (Poisson-GPD). This model is governed by three parameters: the Poisson rate, the scale and the shape of the GPD. The quantity of interest, in many applications, is the mean return level which is a function of Poisson-GPD parameters. Moreover, the shape parameter of GPD is itself of interest in order to gain more insights on the underlying extremal process. For a suitable orthogonal parametrization, we derive matching priors for shape, scale and Poisson rate parameters based on an approximate conditional pseudo-likelihood. The formal rule, used here to obtain such priors, in some cases leads to the same priors obtained with Jeffreys' and Reference procedures. Moreover, we can provide a formal proof that each marginal prior for shape and scale parameters, respectively, are second order matching priors. We estimate the coverages of the corresponding posterior credible intervals and apply our approach to an example from hydrology.

Keywords

Approximate conditional likelihood • Bayesian statistics • Extreme values • Matching priors • Orthogonal parameters

S. Cabras (✉)
Department of Mathematics and Informatics, University of Cagliari, Via Ospedale, 72 09124-Cagliari, Italy

Department of Statistics, University Carlos III de Madrid, C/ Madrid, 126 - 28903-Getafe (Madrid), Spain
e-mail: stefano.cabras@uc3m.es; s.cabras@unica.it

N. Torelli et al. (eds.), *Advances in Theoretical and Applied Statistics*, Studies in Theoretical and Applied Statistics, DOI 10.1007/978-3-642-35588-2_1,

1.1 Introduction

In the past two decades there has been an increasing interest for statistical models of extreme events in numerous disciplines such as environmental sciences, finance and insurance, among others (see for instance [11] and [27]).

We consider the class of models arising from extreme value theory (see i.e. [24]) with particular attention to model the exceedances of a random variable $Z \in \mathscr{Z} \subset \mathbb{R}$ over a fixed high threshold, u [15]. Let $Z \sim H(z)$ represent the observable variable (i.e. rainfall level or return of a financial asset) and $Y = Z - u | Z > u$ represent the exceedance over u. Pickands [24] shows that, under regularity conditions on $H(z)$, the limiting distribution of Y, given $Z > u$, when $u \rightarrow z^* = \sup \mathscr{Z}$, is the Generalized Pareto Distribution (GPD) with shape ξ and scale $\sigma > 0$,

$$f\,(y \mid \xi, \sigma) = \begin{cases} \sigma^{-1}\,(1 + y\xi/\sigma)^{-(1+\xi)/\xi}\,, & \text{if } (1 + y\xi/\sigma) > 0 \\ \sigma^{-1}\exp\,(-y/\sigma)\,, & \xi = 0 \end{cases}$$

In most applications the shape parameter is of interest because:
(a) ξ regulates the existence of an upper limit in the Y variation. In fact, when $\xi < 0$, $z^* = u - \sigma/\xi < \infty$, while $\xi \geq 0$ implies $z^* = \infty$.
(b) ξ regulates tail thickness of $H(z)$ and, in particular, $1/\xi$ determines the number of finite moments of Y, i.e. for $\xi = 1$, $E(Y) < \infty$ and $E(Y^2) = \infty$.
Because of (a), model $f(\cdot)$ is irregular: for $\xi < 0$ the support of Y depends on σ and, in general, it does not satisfy usual regularity conditions for the maximum likelihood estimator (MLE) [28]. In particular, the MLE does not exists when $\xi < -1$ and they are not regular for $-1 < \xi < -0.5$, in order to avoid such irregularities, we assume $\xi > -0.5$ in the rest of the paper. In practical data analyses, $\xi > 0$ is usually encountered.

One of the main quantity of interest in extreme value analysis is the k-year mean return level, R_k, that is the value of Y that will be exceeded once every k years. In order to model this quantity we need to resort to a point process approach and use the Poisson-GPD model [27]. Let $\mathbf{y} = (y_1, \ldots, y_n)$ be n exceedances over a fixed threshold u that have been observed during T years. The corresponding likelihood function for the Poisson-GPD model is

$$L(\boldsymbol{\theta}) = p(\mathbf{y}|\boldsymbol{\theta}) \propto \gamma^n \exp\{-\gamma T\} \times \prod_{i=1}^{n} f(y_i|\xi, \sigma), \tag{1.1}$$

where $\boldsymbol{\theta} = (\xi, \sigma, \gamma)$ and

$$R_k = u + (\sigma/\xi)\left[(\gamma k)^{\xi} - 1\right].$$

Several papers deal with point and interval estimation of $\boldsymbol{\theta}$ from a frequentist perspective; see e.g.: [14] and [28]. Frequentist approach, in extreme value analysis, poses difficulties not only because of model irregularities but also because n cannot be large as extreme values are rare events. Such difficulties induced several authors [15, 27] to resort to a Bayesian approach where $L(\boldsymbol{\theta})$ is combined with a suitable prior $\pi(\boldsymbol{\theta})$ that mitigates the odd behavior of $L(\boldsymbol{\theta})$. Bayesian methods however, introduce the problem of eliciting such prior $\pi(\boldsymbol{\theta})$. In particular, elicitation is problematic because of the lack of a physical meaning of $\boldsymbol{\theta}$. In this sense, paper [3] proposes an elicitation procedure on $\boldsymbol{\theta}$, while [16] use a separate elicitation for $\xi < 0$ and $\xi > 0$, leading to a prior which is discontinuous at $\xi = 0$. The use of vague priors is further complicated by the need of a sensitivity analysis of the posterior with respect to the prior and this has to be done for a particular sample $\mathbf{y}$. This is not needed under a default Bayesian perspective, where $\pi(\boldsymbol{\theta})$ is determined by a formal rule equal for all possible $\mathbf{y}$. The default Bayesian perspective is that used here along with other papers such as [1] that proposes a non-informative prior for the regular case of $\xi > 0$ and [8] that discussed the use of the joint improper Jeffreys' prior for (ξ, σ),

$$\pi^J(\xi, \sigma) \propto \sigma^{-1}(1+\xi)^{-1}(1+2\xi)^{-1/2}, \quad \xi > -1/2, \sigma > 0,$$

which leads to a proper posterior distribution [8, Theorem 1]. The Jeffreys' rule is found to produce accurate inference on (ξ, σ) mitigating the odd behavior of $L(\boldsymbol{\theta})$. Finally, a recent work of [20] studies a matching prior for the quantiles of the GPD. However, differently from this work, none of the above papers study matching priors for parameters of the Poisson-GPD model in Eq. (1.1).

The main contribution of this paper is to propose matching priors $\pi^*(\cdot)$, for the parameters of model (1.1), based on the approximate conditional (AC) likelihood function in [12]. This kind of priors have received increasing interest in the literature as illustrated in [13]. The prior $\pi^*(\cdot)$ is derived according to the formal rule developed in [31]. We thus have a formal proof that the matching priors $\pi^*(\xi)$ and $\pi^*(\sigma)$ are second order matching priors, while for the joint prior $\pi^*(\xi, \sigma) \propto \pi^J(\xi, \sigma)$ we have heuristic arguments and numerical evidence suggesting that $\pi^*(\xi, \sigma)$ is jointly matching, according to the definition in [13], only for $\xi > 0$.

The rest of the paper is organized as follows: Sect. 1.2 recalls the use of pseudo-likelihoods for deriving matching priors, with particular focus on the AC likelihood. Derivation of $\pi^*(\cdot)$ along with the estimation of the coverage errors of posterior credible intervals (C.I.) are contained in Sect. 1.3. An application to hydrology data is illustrated in Sect. 1.4, while Sect. 1.5 contains further remarks.

1.2 Matching Priors from Pseudo-Likelihoods Under Orthogonal Parametrization

In this section we summarize, under a more general setup, the procedure in [31] that leads to the formal rule here employed in order to obtain $\pi^*(\cdot)$. We consider a partition of the parameter $\boldsymbol{\theta} \in \Theta$, as $\boldsymbol{\theta} = (\psi, \boldsymbol{\lambda})$, where ψ represents the parameter of interest and $\boldsymbol{\lambda}$ is the nuisance parameter. The elimination of $\boldsymbol{\lambda}$ can be achieved using an appropriate pseudo-likelihood function, $L^*(\psi)$, which is a function of ψ and $\mathbf{y}$, with properties similar to those of a genuine likelihood function. Commonly used pseudo-likelihood functions include: marginal, usually employed in Bayesian analysis, conditional, profile and modified profile likelihoods; see, for instance, [2, 23, 26]. The use of $L^*(\psi)$, as if it were a true likelihood, leads to the posterior distribution for ψ,

$$\pi^*(\psi|\mathbf{y}) \propto \pi(\psi)L^*(\psi),$$

where $\pi(\psi)$ is a suitable prior distribution on ψ only. Under a non-orthodox Bayesian point of view, there are two main advantages in computing $\pi^*(\psi|\mathbf{y})$ instead of the usual marginal posterior

$$\pi(\psi|\mathbf{y}) \propto \int_{\lambda\in\Lambda} L(\boldsymbol{\theta})\pi(\lambda|\psi)\pi(\psi)\mathrm{d}\lambda,$$

based on the full likelihood $L(\boldsymbol{\theta})$. In particular:

- We avoid elicitation of $\pi(\lambda|\psi)$
- Numerical integration on the full parameter space Θ is no longer needed.

Bayesian inference based on $L^*(\psi)$ has been considered, for instance, in [4–6, 9, 17–19, 21, 25, 31, 32].

We define a matching prior $\pi^*(\psi)$ as the prior that satisfies the following equation [13, Chap. 2]

$$P_{\boldsymbol{\theta}}(\psi \le \psi_\alpha(\mathbf{Y})) = P_\pi(\psi \le \psi_\alpha(\mathbf{Y})) + O_p(n^{-1}),$$

where $P_{\boldsymbol{\theta}}(\cdot)$ denotes the frequentist probability under $\boldsymbol{\theta}$, $P_\pi(\cdot)$ is the posterior probability and ψ_α is the α-quantile of $\pi^*(\psi|\mathbf{y})$. If $\boldsymbol{\theta}$ is a vector of orthogonal parameters (i.e. the Fisher information matrix is strictly diagonal), then $\pi^*(\psi)$ is available via the approximate conditional likelihood in [12]

$$L_{\mathrm{AC}}(\psi) = L_p(\psi)|j_{\boldsymbol{\lambda}\boldsymbol{\lambda}}(\psi, \hat{\boldsymbol{\lambda}}_\psi)|^{-1/2},$$

where $\hat{\boldsymbol{\lambda}}_\psi$ is the restricted MLE of $\boldsymbol{\lambda}$ for fixed ψ, $L_p(\psi) = L(\psi, \hat{\boldsymbol{\lambda}}_\psi)$ is the profile likelihood and $j_{\boldsymbol{\lambda}\boldsymbol{\lambda}}(\psi, \boldsymbol{\lambda}) = -\partial^2\ell(\psi, \boldsymbol{\lambda})/\partial\boldsymbol{\lambda}\partial\boldsymbol{\lambda}^{\mathsf{T}}$ is the observed Fisher information for $\boldsymbol{\lambda}$, with $\ell(\psi, \boldsymbol{\lambda}) = \log L(\psi, \boldsymbol{\lambda})$. According to Proposition 2.1 in [31], the matching prior for ψ associated with $L_{\mathrm{AC}}(\psi)$ is given by

$$\pi^*(\psi) \propto i_{\psi\psi}(\psi, \hat{\boldsymbol{\lambda}}_\psi)^{1/2}, \tag{1.2}$$

where $i_{\psi\psi}(\psi, \boldsymbol{\lambda})$ is the (ψ, ψ)-component of the Fisher information matrix $i(\psi, \boldsymbol{\lambda})$ based on $L(\psi, \boldsymbol{\lambda})$. The corresponding pseudo-posterior is

$$\pi^*(\psi|\mathbf{y}) \propto L_{\mathrm{AC}}(\psi)i_{\psi\psi}(\psi, \hat{\boldsymbol{\lambda}}_\psi)^{1/2};$$

see also [29, 30] and [9].

Table 1.1 Partitions of $\boldsymbol{\theta}$ along with corresponding matching priors and posteriors

Case	Partition	Prior $\pi^*(\psi)$	Posterior $\pi^*(\psi\|\mathbf{y})$
(i)	$\psi = \xi, \boldsymbol{\lambda} = (\gamma, \nu)$	$\propto (1+\xi)^{-1}$	numerically
(ii)	$\psi = \nu, \boldsymbol{\lambda} = (\gamma, \xi)$	$\propto \nu^{-1}(1+2\hat{\xi}_\nu)^{-1}$	numerically
(iii)	$\psi = \gamma, \boldsymbol{\lambda} = (\xi, \nu)$	$\propto \gamma^{-1/2}$	*Gamma*$(n+1/2, 1)$
(iv)	$\psi = (\xi, \nu), \lambda = \gamma$	$\propto \pi^J(\xi, \nu) \propto \nu^{-1}(1+\xi)^{-1}(1+2\xi)^{-1/2}$	see [8]

1.3 Matching Priors for Parameters in the Poisson-GPD Model

Let us consider a reparametrization of the GPD density in terms of orthogonal parameters ξ and $\nu = \sigma(1+\xi)$ according to [10]. The GPD density becomes

$$f(y|\xi,\nu) = \begin{cases} \nu^{-1}(1+\xi)(1+y\xi(1+\xi)/\nu)^{-(1+\xi)/\xi}, & (1+y\xi(1+\xi)/\nu) > 0 \\ \nu^{-1}\exp(-y/\nu), & \xi = 0 \end{cases}. \tag{1.3}$$

The corresponding log-likelihood for the Poisson-GPD model, for $\xi \neq 0$, is

$$\ell(\boldsymbol{\theta}) = n\log(\gamma) - \gamma T - n\log(\nu) + n\log(1+\xi) - \frac{1+\xi}{\xi}\sum_{i=1}^{n}\log\left(1 + \frac{\xi(1+\xi)}{\nu}y_i\right),$$

where, with an abuse of notation, $\boldsymbol{\theta} = (\gamma, \xi, \nu)$, and the expected Fisher information matrix, for one observation, is

$$i_{\theta}(\boldsymbol{\theta}) = \begin{pmatrix} \gamma^{-1} & 0 & 0 \\ 0 & (1+\xi)^{-2} & 0 \\ 0 & 0 & \nu^{-2}(1+2\xi)^{-1} \end{pmatrix}.$$

We consider four relevant different partitions of $\boldsymbol{\theta}$: (i) $\psi = \xi$, (ii) $\psi = \nu$, (iii) $\psi = \gamma$ and (iv) $\psi = (\xi, \nu)$. It is worth noting that the conditional MLE of the Poisson rate, $\hat{\gamma}_\psi = n/T$, does not depend on ξ and ν. The corresponding priors (1.2), for each partition, are all improper and illustrated in Table 1.1.

Only for case (ii) we have a data-dependent prior because the conditional MLE $\hat{\xi}_\nu$ enters into the prior.

Prior for case (iii) is the Jeffreys/Reference and the first order matching prior for the Poisson rate. Moreover, because of factorization $L(\boldsymbol{\theta}) = L(\gamma) \times L(\xi, \nu)$, $\pi^*(\gamma)$ is also the unique first order matching prior for the Poison-GPD model according to Theorem 2.5.1 in [13].

For case (iv), we have a vector of parameters of interest and the resulting prior $\pi^*(\psi)$ equals the Jeffreys' prior $\pi^J(\xi, \nu)$. Such prior is the same in [8], $\pi^J(\xi, \sigma)$, under the non-orthogonal parametrization. This can be easily seen by recalling that the Jeffreys' prior is invariant under reparametrization. However, although Jeffreys' and [31] procedures both lead to the same prior, in this paper we only have numerical

Table 1.2 Empirical coverages for case (i) with $\nu = 1$ and (ii) with $\xi = 1$

	n		
ξ	10	20	30
−0.4	0.94	0.96	0.95
−0.0	0.98	0.97	0.96
−1.0	0.97	0.96	0.95
−2.0	0.97	0.97	0.96
ν			
−0.5	0.95	0.95	0.95
−1.0	0.94	0.95	0.94
−2.0	0.95	0.96	0.95
−5.0	0.95	0.96	0.95

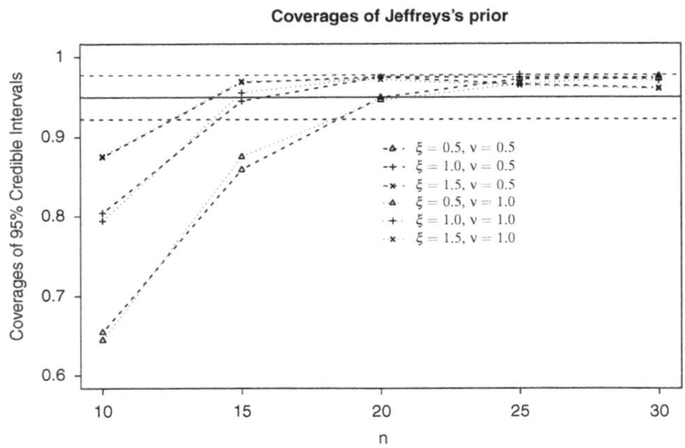

Fig. 1.1 Empirical coverages of $\pi^*(\xi, \sigma|\mathbf{y})$, for the regular case of $\xi > 0$. Horizontal lines indicate coverages compatible with the nominal 95%

evidence that $\pi^*(\xi, \nu|\mathbf{y})$ is a joint matching prior according to the definition in [13] that is,

$$P_\theta(\xi \leq \xi_\alpha, \nu \leq \nu_\alpha) = P_\pi(\xi \leq \xi_\alpha, \nu \leq \nu_\alpha) + O_p(n^{-1}).$$

Such evidence is substantial for the regular case of $\xi > 0$, while coverages of posterior C.I., for $\xi < 0$, are less than the nominal value.

We use Monte Carlo simulations to evaluate frequentist coverages of 95% posterior C.I. for the corresponding priors in Table 1.1 for cases (i), (ii) and (iv). We use, for each combination of ξ, ν and n, 5,000 simulated C.I.. Cases (i) and (ii) are new to the literature and coverages are shown in Table 1.2 with significant digits, while coverages for case (iv) are showed in Fig. 1.1.

We can see from Table 1.2 and Fig. 1.1 that with $n = 30$ coverages are all compatible with the nominal value of 95%.

For the mean return level, we propose to use cases (iii) and (iv) to derive the prior for $R_k(\boldsymbol{\theta})$ by means of

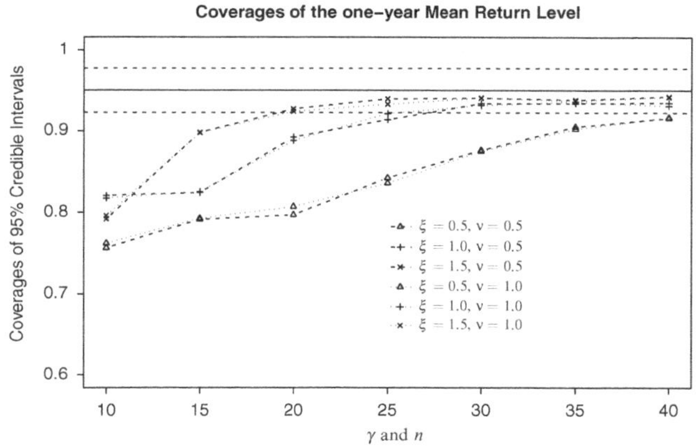

Fig. 1.2 Empirical coverages of $\pi^J(R_k(\boldsymbol{\theta})|\mathbf{y})$, for the regular case of $\xi > 0$. Horizontal lines indicate coverages compatible with the nominal 95%

$$\pi^J(\boldsymbol{\theta}) \propto \pi^*(\gamma)\pi^*(\xi, \nu).$$

Such prior is the Jeffreys' prior for the Poisson-GPD model. The corresponding posterior, $\pi^J(R_k(\boldsymbol{\theta})|\mathbf{y})$, has been also used in [8] and here we study coverages of the corresponding 95% posterior C.I. that are showed in Fig. 1.2 for $\xi > 0$ and $k = 1$ year. From $n = 40$ coverages are compatible with the nominal 95%.

Finally, it is possible to note, from Figs. 1.1 and 1.2, that model irregularities, occurring when $\xi \to -1/2$, have effect on the coverages of posterior C.I.. In fact, coverages tend to be smaller when $\xi = 0.5$ with respect to the other values of ξ considered in the simulation study. However, in real applications, negative values of ξ are not very common, while large positive values of ξ denote the presence of heavy tails in the distribution of Y.

The difference, between the actual coverage and the nominal 95%, may be interpreted also in terms of distance between the actual $\pi^*(\cdot|\mathbf{y})$ and an ideal posterior, that based on the full likelihood and a matching prior, would lead to C.I. with exactly 95% of coverage.

1.4 An Application to Hydrology

We apply the proposed inference procedure to a sample of 154 exceedances over the level 65 m^3/s by the River Nidd at Hunsingore Weir from 1934 to 1969 [22]. This data set has been analyzed by several authors: [8, 15]. These data involve problems of trend, seasonality and threshold selection. Here we use them to demonstrate our procedure. One of the main feature of this sample is the tail bumpiness for thresholds between 100 and 120 m^3/s, because the number of observations over the threshold u does not smoothly decreases as u increases. This feature complicates

Table 1.3 Ninety-five percent credible intervals for ξ and ν at different thresholds for Nidd data

		ξ			ν	
u	n	95% C.I.	Length	$\log BF^*$	95% C.I.	Length
70	138	(0.14,0.60)	0.46	9	(23,36)	13
80	86	(0.09,0.76)	0.68	6	(26,45)	20
90	57	(−0.07,0.79)	0.86	3	(30,58)	29
100	39	(−0.30,0.65)	0.95	1	(37,75)	38
110	31	(−0.37,0.71)	1.08	0	(37,81)	44
120	24	(−0.44,0.58)	1.02	−1	(40,89)	49
130	22	(−0.42,0.87)	1.29	0	(35,88)	54
140	18	(−0.44,0.97)	1.41	0	(34,95)	61

the threshold selection problem, which is not explicitly addressed here. For different thresholds, the C.I. for ξ and ν are showed in Table 1.3, while C.I. for R_k are the same intervals in [8]. The length of ξ and ν C.I. increases with u because of the sample size. Inference is sensitive to u because values of u up to 100 m^3/s provide similar inference (i.e. all C.I. contain $\xi = 0$) and this differs from that obtained with $u > 100$. For these data, it could be of interest whether or not the exceedances of the Nidd River are bounded. This amounts to calculate

$$BF^* = \pi^*(\xi > 0|\mathbf{y})/\pi^*(\xi < 0|\mathbf{y}),$$

that is the pseudo-Bayes Factor on the sign of ξ. Differently from the usual BF, based on $L(\boldsymbol{\theta})$, as that in [8], BF^* is not based on the posterior marginal distribution of ξ, but instead on pseudo-likelihood $L^*(\xi)$. The general use of $L^*(\psi)$ to obtain BF^* is further discussed in [7].

The weight of evidence, $\log BF^*$, is showed in Table 1.3. For $u < 110$, the evidence in favor of $\xi > 0$ is substantial and it is larger than the one obtained in [8].

1.5 Concluding Remarks

For the Poisson-GPD model (1.1) and a reduced set of parameters of interest, we derived priors for inference based on the AC likelihood. Some of these priors are also first order matching priors and some of them coincide with Jeffreys' priors. The main objective of this work is to show the potential of the approach based on pseudo-likelihoods to derive matching priors [31]. In fact, for the Poisson-GPD model, matching priors are derived quite easily.

Finally, one may employ priors $\pi^*(\psi)$ here developed and obtain the posterior $\pi(\psi|\mathbf{y})$ with the usual integration of the marginal likelihood

$$L^m(\psi) \propto \int_{\lambda \in \Lambda} L(\boldsymbol{\theta})\pi(\lambda|\psi)\mathrm{d}\lambda,$$

such that the posterior is

$$\pi(\psi|\mathbf{y}) \propto L^m(\psi)\pi^*(\psi).$$

However, the procedure here proposed avoids elicitation on $\pi(\lambda|\psi)$ as in can be particularly cumbersome for the parameters of the Poisson-GPD model, moreover integration over λ is avoided in our analysis. Therefore, this approach may result to be very useful in the presence of many nuisance parameters as in the context of regression (see [10]). Finally, the matching argument holds for $\pi^*(\psi|\mathbf{y})$, while the marginal posterior $\pi(\psi|\mathbf{y})$ is not guaranteed to be matching.

Acknowledgements This work has been partially supported by M.I.U.R. of Italy.

References

1. Arnold, B., Press, S.J.: Bayesian estimation and prediction for Pareto data. J. Am. Stat. Assoc. **84**, 1079–1084 (1989)
2. Barndorff-Nielsen, O., Cox, D.: Inference and Asymptotics. Chapman and Hall/CRC Press, London (1994)
3. Behrens, C.N., Lopes, H.F., Gamerman, D.: Bayesian analysis of extreme events with threshold estimation. Stat. Model. **4**, 227–244 (2004)
4. Bertolino, F., Racugno, W.: Analysis of the linear correlation coefficient using pseudo-likelihoods. J. Ital. Stat. Soc. **1**, 33–50 (1992)
5. Bertolino, F., Racugno, W.: Robust Bayesian analysis of variance and the χ^2-test by using marginal likelihoods. Statistician **43**, 191–201 (1994)
6. Cabras, S., Racugno, W., Ventura, L.: Bayesian inference on the scalar skew-normal distribution. In, 17th Compstat Symposium of the IASC (Roma-Italy, 28/8-1/9) (ISBN, 3-7908-1708-2), 1373–1381 (2006)
7. Cabras, S., Racugno, W., Ventura, V.: Pseudo-Bayes factors. In, 19th Compstat Symposium of the IASC (Paris-France, 22/8-27/8) (2010)
8. Castellanos, M., Cabras, S.: A default Bayesian procedure for the generalized Pareto distribution. J. Stat. Plann. Infer. **137**, 473–483 (2007)
9. Chang, H., Mukerjee, R.: Probability matching property of adjusted likelihoods. Stat. Probab. Lett. **76**, 838–842 (2006)
10. Chavez-Demoulin, V., Davison, A.: Generalized additive modelling of sample extremes. Appl. Stat. **54**, 207–222 (2005)
11. Coles, S.: An Introduction to Statistical Modeling of Extreme Values. Springer, New York (2001)
12. Cox, D., Reid, N.: Parameter orthogonality and approximate conditional inference (with discussion). J. Roy. Stat. Soc. B **49**, 1–39 (1987)
13. Datta, G.S., Mukerjee, R.: Probability Matching Priors, Higher Order Asymptotics, Lecture Notes in Statistics, vol. 178. Spriner, New York (2004)
14. Davison, A.: Statistical extremes and Applications, Reidel, Dordrecht, Chp. Modelling excesses over high thresholds, with an application, 461–482 (1984)
15. Davison, A.C., Smith, R.L.: Models for exceedance over high threshold. J. Roy. Stat. Soc. B **52**, 393–442 (1990)
16. de Zea Bermudez, P., Turkman, M.A.: Bayesian approach to parameter estimation of the generalized Pareto distribution. Test **12**, 259–277 (2003)
17. Efron, B.: Bayes and likelihood calculations from confidence intervals. Biometrika **80**, 3–26 (1993)

18. Fraser, D.A.S., Reid, N., Wong, A., Yun, Yi.G.: Direct Bayes for interest parameters. Valencia **7**, 529–533 (2003)
19. Greco, L., Racugno, W., Ventura, L.: Robust likelihood functions in Bayesian inference. J. Stat. Plann. Infer. **138**, 1258–1270 (2008)
20. Ho, K.W.: A matching prior for extreme quantile estimation of the generalized Pareto distribution. J. Stat. Plann. Infer. **140**, 1513–1518 (2010)
21. Lin, L.: Quasi Bayesian likelihood. Stat. Meth. **3**, 444–455 (2006)
22. Natural Environment Research Council: Flood Studies Report. Natural Environment Research Council (1975)
23. Pace, L., Salvan, A.: Principles of Statistical Inference. World Scientific, Singapore (1997)
24. Pickands, J.: Statistical inference using extreme order statistics. Ann. Stat. **3**, 119–131 (1975)
25. Racugno, W., Salvan, A., Ventura, L.: Bayesian analysis in regression models using pseudo-likelihoods. Comm. Stat. Theor Meth. **39**, 3444–3455 (2010)
26. Severini, T.A.: Likelihood Methods in Statistics. Oxford University Press, Oxford (2000)
27. Smith, R.: Extreme Values in Finance, Telecommunications and the Environment. Chapman and Hall/CRC Press, London (2003)
28. Smith, R.L.: Statistical Extremes and Applications. Reidel, Dordrecht (1984)
29. Sweeting, T.J.: On the implementation of local probability matching priors for interest parameters. Biometrika **92**, 47–57 (2005)
30. Tibshirani, R.: Noninformative priors for one parameter of many. Biometrika **76**, 604–608 (1989)
31. Ventura, L., Cabras, S., Racugno, W.: Prior distributions from pseudo-likelihoods in the presence of nuisance parameters. J. Am. Stat. Assoc. **104**, 768–774 (2009)
32. Ventura, L., Cabras, S., Racugno, W.: Default prior distributions from quasi- and quasi-profile likelihoods. J. Stat. Plann. Inference **140**(11), 2937–2942 (2010)

2 Robustness, Dispersion, and Local Functions in Data Depth

Mario Romanazzi and Claudio Agostinelli

Abstract
Data depth is a rapidly growing area in nonparametric statistics, especially suited for the analysis of multidimensional data. This chapter covers influence functions and robustness, depth-based dispersion measures, and a generalization of the basic notion of depth function, called local depth, able to deal with multimodal data.

Keywords
Influence function • Local depth • Multivariate dispersion • Nonparametric statistical methods • Simplicial depth

2.1 Introduction

Data depth provides a coherent framework for nonparametric analysis of multivariate data and is currently registering new methodological achievements. Let $X = (X_1, \ldots, X_p)^{\mathrm{T}}$ be a random vector with distribution F and let $F_{A,b}$ denote the distribution of $AX + b$. A depth function $d(\cdot; F) : \mathbb{R}^p \rightarrow \mathbb{R}^+ \cup \{0_p\}$ should meet several requirements [8, 17]:

1. Affine invariance, i.e., $d(x; F) = d(Ax + b; F_{A,b})$.
2. If F is centro-symmetric about $c \in \mathbb{R}^p$, then $d(c; F) \geq d(x; F)$ for all $x \in \mathbb{R}^p$.
3. Ray-monotonicity: if $\theta \in \mathbb{R}^p$ satisfies $d(\theta; F) \geq d(x; F)$ for all $x \in \mathbb{R}^p$, then $d(\theta + \lambda_1 u; F) \geq d(\theta + \lambda_2 u; F)$ for any p-vector u and $0 \leq \lambda_1 < \lambda_2$.
4. When the norm $\| x \|$ is large, the rank of x is negligible, i.e., $\lim_{\|x\| \rightarrow \infty} d(x; F) = 0$.

M. Romanazzi (✉) · C. Agostinelli
Department of Environmental Sciences, Informatics and Statistics, Ca' Foscari University, Venice, Italy
e-mail: romanaz@unive.it; claudio@unive.it

N. Torelli et al. (eds.), *Advances in Theoretical and Applied Statistics*, Studies in Theoretical and Applied Statistics, DOI 10.1007/978-3-642-35588-2_2,

The main result of data depth is a ranking of the points with respect to F which describes the degree of centrality of x, highest values corresponding to the center of the distribution, lowest values corresponding to the tails. Several depth-based functionals summarizing the features of F have been suggested. The most important one is the deepest point $\theta(F) = \arg\max_x d(x; F)$ describing the location of F. In general, for a given depth value $d \geq 0$, one can consider the depth regions $D(d; F) = \{x \in \mathbb{R}^p : d(x; F) \geq d\}$ whose boundary is the d-contour of the depth function. When properties (1)–(4) hold, the depth regions form a family of nested affine equivariant subsets of $\mathbb{R}^p$ conveying valuable information about F. For example, the spread of F can be described by the scale curve [9] $\gamma(d) = \{(p(d), vol(D(d; F))), d \geq 0\}$, where $p(d)$ is the probability mass enclosed by $D(d; F)$. In a similar way, the correlation curve [14] summarizes the degree of interdependence of the components of F. The D–D plot is another graphical display for pairwise comparison of observed versus theoretical depth distribution, and it can be used for informal inferences, similarly to a Q–Q plot. The reader is referred to [9] for a comprehensive coverage of the methodology. A more recent development is functional depth [10] which opens new perspectives in applications involving infinite dimensional distributions.

In this work more recent developments in the classical finite dimensional setting are presented. The influence function and breakdown point for the simplicial depth and its maximizer are presented in Sect. 2.2. The results allow a more precise comparison of depth functions according to robustness, complementing the work by [17]. Depth-based dispersion functionals are discussed in Sect. 2.3. Here the result is a general scalar measure of dispersion for multivariate distributions. The final topic, illustrated in Sect. 2.4, is local depth, a generalization of data depth to multimodal distributions and clustered data sets. Results are available for several geometrical depth functions based on simple structures, like Tukey's halfspace depth, Liu's simplicial depth and Oja's simplicial volume depth, but here we concentrate on simplicial depth whose definition for a probability distribution F is recalled below.

Definition 1. The simplicial depth function [8] $d_S(\cdot; F) \equiv d_S(\cdot)$ is the probability that a random simplex $S_{p+1} = S(x_1, \ldots, x_{p+1})$ covers x,

$$d_S(x) = P_F(S_{p+1} : x \in S_{p+1}).$$

2.2 Robustness

The Influence Function (IF) [7] is the main tool to investigate local sensitivity of a statistical functional to an infinitesimal perturbation of F. In this section the IF of depth values $d(x; F)$, $x \in \mathbb{R}^p$, and depth maximizer $\theta(F)$ are reviewed. Proofs and illustrations of the results can be found in [3, 13, 15] and [4].

We write δ_z for the point mass distribution at $z \in \mathbb{R}^p$. For $0 \leq k \leq p+1$, let A_k be the event that a random sample of $p+1$ elements from $F(\epsilon, z) = (1-\epsilon)F + \epsilon\delta_z$ includes exactly k observations from δ_z and let $p_k(\epsilon)$ be the corresponding probability

$$p_k(\epsilon) \equiv P(A_k) = \binom{p+1}{k} \epsilon^k (1-\epsilon)^{p+1-k}.$$

Moreover, let $d_S(x|A_k, z)$ be the simplicial depth of x conditional on A_k, that is, evaluated from random simplices with $0 \leq k \leq p+1$ vertices coincident with z. The perturbed simplicial depth value turns out to be $d_S(x; F(\epsilon, z)) = \sum_{k=0}^{p+1} p_k(\epsilon) d_S(x|A_k, z)$ and the IF is [15]

$$IF(z; d_S(x; F)) = (p+1)(d_S(x|A_1, z) - d_S(x; F)). \tag{2.1}$$

The IF of simplicial depth is a bounded function of $x \in \mathbb{R}^p$, linearly dependent on the dimension of the space. This suggests that the simplicial ranks could be exposed to the risk of distortion by outliers in high dimensions. However, the rank of the center of a symmetric distribution seems to be remarkably stable.

Example 1. Let F be an absolutely continuous distribution centro-symmetric about the origin 0_p. The origin is the deepest point and $d_S(0_p) = 2^{-p}$. Moreover,

$$IF(z; d_S(0_p; F)) = \begin{cases} (p+1)(1-2^{-p}), & z = 0_p, \text{closed simplices} \\ -(p+1)/2^p, & z = 0_p, \text{open simplices} \\ 0, & z \neq 0_p. \end{cases}$$

Therefore, the simplicial rank of the center should be highly resistant to perturbations, for all p.

Let us now pass to depth maximizers. Suppose F to be absolutely continuous and centro-symmetric about the origin. Under the same hypotheses, the perturbed simplicial median is the maximizer of

$$d_s(x; F(\epsilon, z)) = \begin{cases} (1-\epsilon)^{p+1}(d_S(x; F) + (p+1)\frac{\epsilon}{1-\epsilon} d_s(x \mid A_1, z)), & x \neq z, \\ (1-\epsilon)^{p+1}(d_S(x; F) + \frac{1-(1-\epsilon)^{p+1}}{(1-\epsilon)^{p+1}}), & x = z. \end{cases}$$

When F is absolutely continuous and spherically symmetric about the origin, the perturbed simplicial median can be represented as $\theta_S(\epsilon, z) = t(\epsilon, z)u_z$, with $0 \leq t(\epsilon, z) \leq \| z \|$, which shows that it belongs to the closed segment joining the origin and the perturbing point z. However, a complete characterization of $t(\epsilon, z)$ and IF are still lacking. We end with a result on the breakdown point. The (augmented) finite sample breakdown point ε^* [5] of the empirical simplicial median satisfies

$$\lim_{n\to\infty} \epsilon^*(\theta_S(\hat{F}_n)) \leq \frac{2^p \sup_x d_S(x;F)}{p+1+2^p \sup_x d_S(x;F)} \leq \frac{1}{p+2}, \tag{2.2}$$

almost surely [3, Proposition 5]. These results (see also (2.1)) illustrate a general property of simplicial depth functionals, the dependence on dimension p. For example, when p grows higher, the breakdown point (2.2) becomes negligible. This behavior appears remarkably different from halfspace depth, e.g., the corresponding upper limit in (2.2) is $1/3$.

2.3 Multivariate Dispersion

The information on F provided by a depth function is by no means confined to location but also includes dispersion, as confirmed by the scale-curve [9] and the generalized quantile curves [16]. Hereafter, we will show that some well-known multivariate dispersion coefficients are the Lebesgue integral of a depth function. For any simplex $S_{p+1} = S(x_1, \ldots, x_{p+1})$, we write V for the $(p+1) \times p$ matrix whose rows are the vertices $x_1, \ldots, x_{p+1}$ of the simplex and we write 1_{p+1} for the $(p+1)$-vector with unit elements. We recall that the volume of the simplex is a function of the determinant of the augmented matrix $(1_{p+1} \vdots V)$, i.e.

$$vol(S_{p+1}) = (p!)^{-1} \mid \det(1_{p+1} \vdots V) \mid .$$

Moreover, for any nonsingular affine transformation $x \to Ax + b$

$$vol(S(Ax_1 + b, \ldots, Ax_{p+1} + b)) = \mid \det A \mid \; vol(S(x_1, \ldots, x_{p+1})). \tag{2.3}$$

Standard dispersion orderings in the univariate case are based on the comparison of interquantile intervals. For random numbers X and Y with cumulative distribution functions F and G, X is said to be less dispersed than Y if $F^{-1}(\beta) - F^{-1}(\alpha) \leq G^{-1}(\beta) - G^{-1}(\alpha)$ for all $0 < \alpha < \beta < 1$. A multivariate generalization of the above formula to the multivariate setting based on the volume of simplices was suggested by [11] and is reproduced below.

Definition 2. Suppose there is a function $h : \mathbb{R}^p \to \mathbb{R}^p$ such that $Y = h(X)$. The p-variate random vector $X \sim F$ is less dispersed than the p-variate random vector $Y \sim G$ if for any set of $p+1$ points $x_1, \ldots, x_{p+1}$ of $\mathbb{R}^p$

$$vol(S(x_1, \ldots, x_{p+1})) \leq vol(S(h(x_1), \ldots, h(x_{p+1}))).$$

We write $X \leq_\Delta Y$ (or $F \leq_\Delta G$) to mean that Definition 2 holds. Let $\mathcal{F}$ be a suitable family of probability distributions on $\mathbb{R}^p$ including the absolutely continuous distributions. A sensible dispersion measure is required to be a Δ-monotone function, and it must be translation invariant and satisfy a type of scale equivariance with respect to linear transformations [11, Definition 4.2].

Definition 3. A function $\psi : \mathscr{F} \to \mathbb{R}^+ \cup \{0\}$ is a dispersion measure if
i. for $F, G \in \mathscr{F}$, $F \leq_\Delta G$ implies $\psi(F) \leq \psi(G)$;
ii. a nonsingular affine transformation $Y = AX + b$ with distribution $F_{A,b}$ satisfies $\psi(F_{A,b}) = \mid \det A \mid \psi(F)$.

Two well-known examples of dispersion measures based on the volume of random simplices are Gini's multivariate coefficient $\psi^{(1)}(F)$ and Wilks' generalized variance $\psi^{(2)}(F)$

$$\psi^{(1)}(F) = E_{X_1,\dots,X_{p+1}} \{vol(S_{p+1})\},$$
$$\psi^{(2)}(F) = (p+1)^{-1} E_{X_1,\dots,X_{p+1}} \{(vol(S_{p+1})^2)\} = \det(Cov(X)).$$

We now illustrate the relations between the previous coefficients and the depth functions.

Proposition 1. *For $F \in \mathscr{F}$,*

$$\psi^{(1)}(F) = \int_{\mathbb{R}^p} d_S(x; F)\mathrm{d}x.$$

Proof. *The result follows from Robbins' theorem [12].*

Recall that for continuous univariate distributions $d_S(x; F) = 2F(x)(1 - F(x))$ and $2\int_{\mathbb{R}} F(x)(1 - F(x))\mathrm{d}x$ is a well-known computational formula of Gini's mean difference. The next example deals with the multivariate normal distribution.

Example 2. Let $X \sim N(0_p, I_p)$. According to Proposition 1,

$$\psi^{(1)}(X) = \int_{\mathbb{R}^p} d_S(x; F)\mathrm{d}x = 4p^{-1}(p+1)\frac{\pi^{p/2}}{\Gamma(p/2)} \int_{-\infty}^{\infty} \phi^{p+1}(t)\mathrm{d}t,$$

where $\Gamma(\cdot)$ is the gamma function and $\phi(\cdot)$ denotes the probability density function of the standard normal distribution [see 6, p. 343, for further details]. Hence $\psi^{(1)}(X)$ only depends on dimension p. The results for Wilks' generalized variance are similar.

2.4 Local Depth

The monotonicity property is the reason why a depth function always exhibits a unique center, even for data showing multiple centers. To obtain a ranking system able to recognize multimodality, we constrain the depth function to use information from the nearby region of any given point, instead of using information from the whole space. The ranks of different points will be comparable provided the neighborhood size is kept constant. More theoretical background can be found in

[2]. We illustrate the general idea with the local version of simplicial depth. For any simplex S_{p+1} we denote with $t(S_{p+1})$ a scalar measure of its size, like diameter, perimeter, or volume.

Definition 4. For any $\tau > 0$, the local simplicial depth function is given by

$$lds(x;\tau) = P_F(S_{p+1} : x \in S_{p+1} \cap t(S_{p+1}) \leq \tau). \tag{2.4}$$

A discussion of the computational problems of (local) simplicial depth with large samples and/or high dimensions can be found in [1]. An R package `localdepth` for computing local and global depth is available at CRAN.

Example 3. In the scalar case, (2.4) is equivalent to

$$lds(x;\tau) = P_F\left((X_{(1)} \leq x \leq X_{(2)}) \cap (X_{(2)} - X_{(1)} \leq \tau)\right),$$

($X_{(1)} = \min\{X_1, X_2\}$, $X_{(2)} = \max\{X_1, X_2\}$) that is, the probability that a random interval with length not greater than τ covers x. In the case of absolutely continuous distributions on the real line the expression becomes

$$lds(x;\tau) = \int_{x-\tau}^{x} (F(y+\tau) - F(x))f(y)\mathrm{d}y + \int_{x}^{x+\tau} (F(x) - F(y-\tau))f(y)\mathrm{d}y. \tag{2.5}$$

Using (2.5), when X has a uniform distribution on the interval (a, b) and $0 < \tau < (b-a)/2$, we obtain

$$lds(x;\tau) = \begin{cases} 0, & x < a \text{ or } x > b, \\ 2F(x)(\tau(b-a) - F(x)/2), & a \leq x \leq a+\tau, \\ (\tau/(b-a))^2, & a+\tau < x \leq b-\tau, \\ 2(1-F(x))(\tau/(b-a) - (1-F(x))/2), & b-\tau < x \leq b, \end{cases} \tag{2.6}$$

with $F(\cdot)$ denoting cumulative distribution function. The simplicial depth function is monotone increasing in $(a, (a+b)/2)$ and monotone decreasing in $((a+b)/2, b)$. On the other hand, the local version (2.6) is constant in $(a+\tau, b-\tau)$, monotone increasing in the left tail $(a, a+\tau)$, and monotone decreasing in the right tail $(b-\tau, b)$.

Example 4. A sample of 200 observations was simulated from the mixture $F = (F_1 + F_2)/2$, where $F_i \sim N_2(\mu_i, I_2)$, $i = 1, 2$, and $\mu_1 = (0, 0)^{\mathrm{T}}$, $\mu_2 = (2.5, 2.5)^{\mathrm{T}}$. Figure 2.1 shows the estimated density contours and simplicial depth contours. As expected, simplicial depth has an overall maximum close to the total mean $\mu = (1.25, 1.25)^{\mathrm{T}}$ and monotonically decreases along any ray from this point. On the other hand, the local simplicial depth contours (see Fig. 2.2; left panel: $\tau = 0.198$, corresponding to the 10% quantile of the volume of the simplices; right panel: $\tau =$

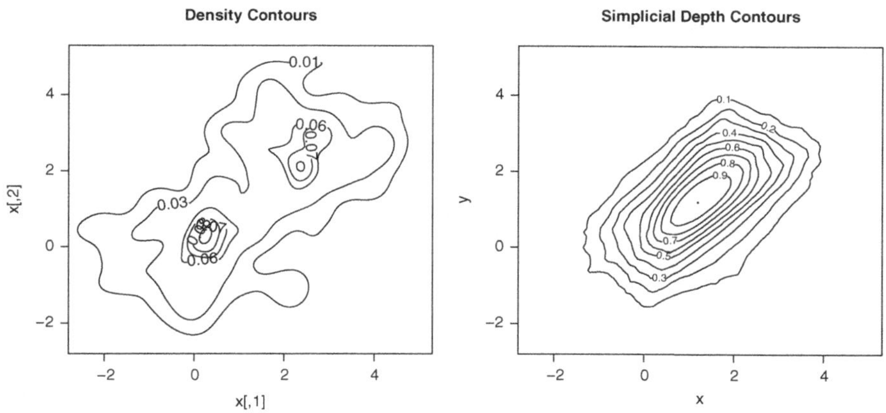

Fig. 2.1 Normal mixture distribution. Estimated density contours and simplicial depth contours

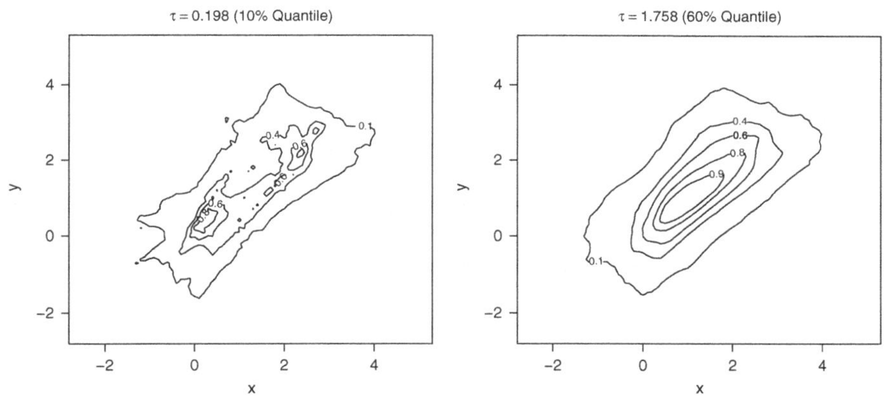

Fig. 2.2 Normal mixture distribution. Estimated local simplicial depth contours

1.758, corresponding to the 60% quantile) are no longer nested and they show the partial centers, provided that τ is sufficiently low.

Local depth functions should meet the general properties of depth functions mentioned in Sect. 2.1, except monotonicity property. In particular, when the reference distribution is centro-symmetric and unimodal about $c \in \mathbb{R}^p$, local depth should agree with global depth.

Proposition 2. *Fix $x \in \mathbb{R}^p$ and let $P_F(\{x\})$ be the probability of the singleton x according to the distribution function F.*

i. If $t(S_{p+1}) = vol(S_{p+1})$, $ld_S(\cdot;\tau)$ is affine invariant whereas, if $t(S_{p+1}) = diam(S_{p+1})$, $ld_S(\cdot;\tau)$ is translation and rotation invariant;

ii. $0 < \tau_1 < \tau_2$ *implies* $ld_S(x;\tau_1) \leq ld_S(x;\tau_2)$;
iii. *for* $\tau > 0$, $P_F(\{x\}) \leq ld_S(x;\tau) \leq d_S(x)$;
iv. $lim_{\tau\to\infty} ld_S(x;\tau) = d_S(x)$;
v. *if* $t(S_{p+1}) = diam(S_{p+1})$, $lim_{\tau\to 0^+} ld_S(x;\tau) = P_F(\{x\})$.

Proof. Proofs are straightforward and are omitted.

Local depth $ld_S(\cdot;\tau)$ is a family of functions indexed by the values of τ, the parameter that defines the neighborhood size and the degree of smoothing. To be effective, the choice of τ must avoid the opposite risks of over-smoothing—potentially important information filtered out—and under-smoothing—uninteresting point-to-point variation shown up. We suggest to set τ equal to a low-order percentile of the distribution of simplex sizes (e.g., diameters or volumes). Percentile orders from 10% to 30% prove effective in most situations.

Local depth, like global depth, implicitly defines a (vector) location parameter

$$\theta_\tau(F) = \arg\max_x ld_S(x;\tau) \tag{2.7}$$

whose components are the partial centers of F, for a given τ. The following proposition describes the behavior of (2.7) for local simplicial depth in the scalar case, when F is unimodal.

Proposition 3. *Let X have an absolutely continuous distribution with cumulative distribution function F and density function f strictly positive for all $x \in \mathbb{R}$. Suppose the distribution to be unimodal about $c \in \mathbb{R}$ with median $x_{0.5}$.*
i. *If the distribution is symmetric,* $\theta_\tau(F) = \theta(F) = c$ *for all* $\tau > 0$;
ii. *if the distribution is asymmetric, local simplicial depth admits a unique maximizer* $\theta_\tau(F)$ *and* $c < \theta_\tau(F) < \min\{c+\tau, x_{0.5}\}$ *or* $\max\{c-\tau, x_{0.5}\} < \theta_\tau(F) < c$, *according to whether the distribution is positive or negative asymmetric;*
iii. *if* $0 < \tau_1 < \tau_2$ *and the distribution is asymmetric,* $\theta_{\tau_1}(F) < \theta_{\tau_2}(F)$ *or* $\theta_{\tau_1}(F) > \theta_{\tau_2}(F)$ *according to whether the distribution is positive or negative asymmetric.*

Proof. [2].

According to Proposition 3, in the unimodal case the maximizer of local simplicial depth is intermediate between the mode and the median, and it converges to the mode or to the median according to whether $\tau \to 0^+$ or $\tau \to \infty$.

Example 5. Let $X \sim 0.5N(\mu_1 = -2, \sigma_1 = 3/2) + 0.5N(\mu_2 = 2, \sigma_1 = 1)$. The simplicial depth function is increasing for $x < x_{0.5} = 0.4$ and decreasing for $x > x_{0.5}$. The local version with τ equal to 10% quantile of interval lengths has two local maxima $\theta_\tau^{(1)} \simeq -1.994$ and $\theta_\tau^{(2)} \simeq 1.962$, near to the modes. The plot of the (normalized) depth functions is shown in the left panel of Fig. 2.3. While the

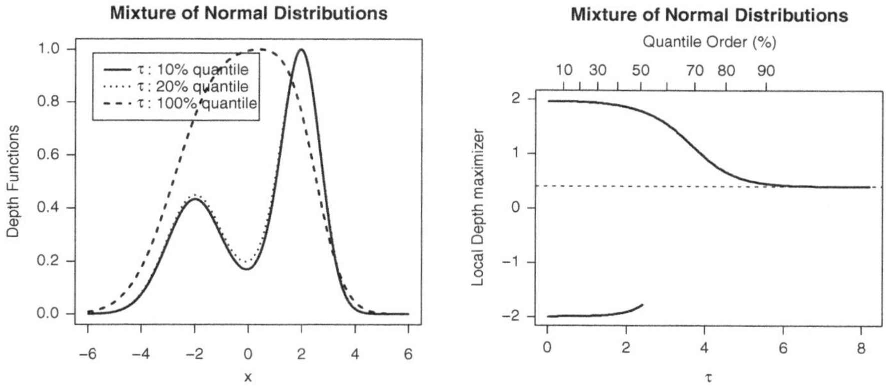

Fig. 2.3 Mixture of normal distributions. *Left*: global and local simplicial depth (functions are normalized with respect to maximum value). *Right*: local maxima versus τ parameter; horizontal line marks the median

results do depend on the value of τ, both the number and the locations of the local maxima stay almost constant for a wide range of the smoothing parameter, up to 52% quantile order (see the right panel of Fig. 2.3). Convergence to the maximum value of the global depth, i.e., the median of the distribution, is observable starting from 90% quantile order.

In the multivariate case, the behavior of local depth is more complex and the details can be found in [1].

2.5 Discussion

Data depth extends the classical nonparametric domain to the analysis of multivariate data. Geometrical depth functions, like halfspace and simplicial depth, are of prominent interest because their information is based on simple geometrical structures. This makes depth methodology particularly suited to initial stages of statistical investigations, preliminary to more precise model-based analyses. Data depth basic ideas can easily fit more general data structures than just units-by-variables data matrices. A notable example is functional data depth [10].

Robustness is a key property of sample statistics and from this point of view halfspace depth performs better than simplicial depth. The IF of depth values is dimension-free for halfspace depth whereas it is linearly dependent on dimension for simplicial depth. This points out a possible weakness of simplicial depth because the depth values are more exposed to distortion by contaminating observations in high dimensions. Halfspace median is a highly robust location estimator, according to both IF and breakdown point, able to resist at least $100(p+1)^{-1}\%$ contamination fractions. The available results show that simplicial median has a lower breakdown point.

Together with a location estimator, a depth function provides an overall measure of dispersion given by its Lebesgue integral. This measure is location free, has nice geometrical interpretations for the usual depth functions, and coincides with well-known dispersion coefficients in the scalar case.

Local depth is a recent development aiming to properly deal with multimodal data. The rationale is similar to the nearest neighborhood method, plus affine invariance of depth ranks and more general types of neighborhoods. The main tools are the local depth values, to be used as ranks of multivariate observations according to the degree of centrality in equal-size neighborhoods, and the depth maximizers, to be used as estimators of the partial centers. A nice feature of simplicial local depth is that when the neighborhood size tends to infinity, the usual (global) depth is recovered. Promising applications of local depth are mode estimation and clustering of the units.

References

1. Agostinelli, C., Romanazzi, M.: Multivariate local depth. Technical Report (2008)
2. Agostinelli, C., Romanazzi, M.: Local depth. J. Stat. Plann. Infer. **141**, 817–830 (2011)
3. Chen, Z.: Bounds for the breakdown point of the simplicial median. J. Multivariate Anal. **55**, 1–13 (1995)
4. Chen, Z., Tyler, D.E.: The influence function and maximum bias of Tukey median. Ann. Stat. **30**, 1737–1759 (2002)
5. Donoho, D., Gasko, M.: Breakdown properties of location estimates based on halfspace depth and projected outlyingness. Ann. Stat. **20**, 1803–1827 (1992)
6. Efron, B.: The convex hull of a random set of points. Biometrika **52**, 331–343 (1965)
7. Hampel, F.: The influence curve and its role in robust estimation. J. Am. Stat. Assoc. **69**, 383–393 (1974)
8. Liu, R.Y.: On a notion of data depth based on random simplices. Ann. Stat. **18**, 405–414 (1990)
9. Liu, R.Y., Parelius, J.M., Singh, K.: Multivariate analysis by data depth: descriptive statistics, graphics and inference. Ann. Stat. **27**, 783–858 (1999)
10. López-Pintado, S., Romo, J.: On the concept of depth for functional data. J. Am. Stat. Assoc. **104**, 718–734 (2009)
11. Oja, H.: Descriptive statistics for multivariate distributions. Stat. Probab. Lett. **1**, 327–332 (1983)
12. Robbins, H.E.: On the measure of a random set. I. Ann. Math. Stat. **15**, 70–74 (1944)
13. Romanazzi, M.: Influence function of halfspace depth. J. Multivariate Anal. **77**, 138–161 (2001)
14. Romanazzi, M.: Data depth and correlation. Allgemeines Statistisches Archiv **88**, 191–214 (2004)
15. Romanazzi, M.: A note on simplicial depth function. Adv. Stat. Anal. **92**, 235–253 (2008)
16. Serfling, R.: Generalized quantile processes based on multivariate depth functions, with applications in nonparametric multivariate analysis. J. Multivariate Anal. **83**, 232–247 (2002)
17. Zuo, Y., Serfling, R.: General notions of statistical depth function. Ann. Stat. **28**, 461–482 (2000)

3 New Distribution-Free Plotting Position Through an Approximation to the Beta Median

Pasquale Erto and Antonio Lepore

Abstract

Even in the modern software, graphical techniques are often utilized to visualize the data and to determine the actual underlying distribution. In this chapter, a new and better distribution-free formula is obtained via an axiomatic approach leading to a new approximation to the median of the Beta distribution. A comparative study, carried out also by using extensive Monte Carlo simulation, shows the advantages of the new solution especially in estimating the median return period and for each considered sample dimension ($N = 5, 15, 30, 50$).

Keywords

Beta Distribution • Distribution-free Plotting Positions • Graphics and Data Visualization • Return Period

3.1 Introduction

Let us consider the observations $(x_{(1)}, \ldots, x_{(N)})$ of the order statistics $(x_{(1)}, \ldots, x_{(N)})$ (arranged in non-decreasing order) corresponding to N random variables $(x_1, \ldots, x_N)$ which are mutually independent and identically-distributed. The basic problem of the graphical estimation is how to determine the best estimate $\hat{F}_i$ of the cumulative distribution function (cdf) $F(x_{(i)})$ (i.e. the best plotting position) corresponding to the i-th order observation $x_{(i)}$. In particular, the appropriate plotting position choice becomes crucial when estimating the return period (i.e. inverse of the probability of exceeding a fixed value x_T). If the parent distribution is known it is trivial to observe that exact unbiased plotting positions

P. Erto (✉) · A. Lepore
Department of Aerospace Engineering, University of Naples Federico II, P.le Tecchio 80, 80125 Naples, Italy
e-mail: ertopa@unina.it; antonio.lepore@unina.it

N. Torelli et al. (eds.), *Advances in Theoretical and Applied Statistics*, Studies in Theoretical and Applied Statistics, DOI 10.1007/978-3-642-35588-2_3,

can be obtained via order statistics-theory. Unfortunately, they are usually complex to determine, thus invalidating the practical aim of the graphical estimation.

Harter [1] provided a detailed review of the main plotting positions up to 1984, concluding that the optimum choice of plotting position depends on the distribution of the variable under consideration (i.e. the parent distribution) and on the use that has to be made of the results. However, the crux is that ideally the parent distribution does not necessarily need to be known [2]; in fact, probability papers are often utilized to determine the underlying distribution. On this assumption, Makkonen [3, 4] states that the probability of non-exceedance of the i-th value in N order ranked events equals

$$\hat{F}_i = \frac{i}{N+1} \qquad i = 1, \ldots, N \tag{3.1}$$

the so-called Weibull formula [5, 6]. Therefore, regardless of the parent distribution and the application, the many other suggested plotting formulas and numerical methods to determine them should be abandoned. From such a background, in [7] the question of determining distribution-free plotting position has been reopened and discussed.

3.2 The Proposed Axiomatic Approach

The following two main properties are assumed in order to lead to a new plotting position formula $\hat{F}_i$.

1. The estimate of the cdf corresponding to the median observation should equal exactly 0.5.
2. The estimate of the cdf corresponding to the largest observation should equal exactly the median $\tilde{F}_N$ of the random variable $F_N = F(X_{(N)})$ (which is independent from the parent distribution).

Both properties are desirable for $\hat{F}_i$ to more accurately estimate the median $\tilde{T}_i$ of the return period

$$T_i = 1/[1 - F(X_{(i)})] \qquad i = 1, \ldots, N \tag{3.2}$$

through the estimator

$$\hat{T}_i = 1/(1 - \hat{F}_i) \qquad i = 1, \ldots, N \tag{3.3}$$

The practical form

$$\hat{F}_i = \frac{i - A}{N + 1 - 2A} \qquad i = 1, \ldots, N;\ A \geq 0 \tag{3.4}$$

already proposed by Blom [8] satisfies the first property. Moreover imposing that formula (3.4) satisfies also property 2, the following value for the constant A is deduced [9]

$$A = N + \frac{N-1}{2^{1/N} - 2} \tag{3.5}$$

which results in the new plotting position

$$\hat{F}_i = \frac{N - i - 2^{-1/N}(1 - 2i + N)}{N-1} \qquad i = 1, \ldots, N \tag{3.6}$$

The adopted criteria to compare plotting positions is to test their ability in estimating $\tilde{T}_i$ by evaluating the following Mean Absolute Relative Error (*MARE*)

$$MARE = \frac{1}{N}\sum_{i=1}^{N}\left|\frac{\tilde{T}_i - \frac{1}{1-\hat{F}_i}}{\tilde{T}_i}\right| = \frac{1}{N}\sum_{i=1}^{N}\left|\frac{\tilde{F}_i - \hat{F}_i}{1 - \hat{F}_i}\right| \tag{3.7}$$

where $\tilde{F}_i = 1 - 1/\tilde{T}_i$ is the median of the random variable $F_i = F(X_{(i)})$ and depends only on i and N. To be convinced, it is enough to note that F_i is a Beta random variable apart from the parent distribution of X, with probability density function (pdf)

$$f_{F_i}(t) = \frac{\Gamma(a+b)}{\Gamma(a)\Gamma(b)} t^{a-1}(1-t)^{b-1} \tag{3.8}$$

where $a = i$ and $b = N - i + 1$. Therefore, the new plotting position in Eq. (3.6) can be assumed as a new approximation of the median value of the Beta random variable with integer parameters $a = i$ and $b = N - i + 1$.

Nevertheless, it is worth pointing out the need to refer to $\tilde{T}_i$, rather than to the expected value $E[T_i]$. In fact, for a random sample $X_1, \ldots, X_N$ from a continuous population with cdf $F(x)$ and pdf $f(x)$, we have

$$\begin{aligned} E[T_i] &= i\binom{N}{i}\int_{-\infty}^{\infty}[F(x)]^{i-1}[1-F(x)]^{N-i-1}f(x)\mathrm{d}x \\ &= i\binom{N}{i}\int_{0}^{1}[F]^{i-1}[1-F]^{N-i-1}\,\mathrm{d}F \end{aligned}$$

By using the Beta function identity (e.g. [10]), we can finally write

$$E[T_i] = i\binom{N}{i}\frac{\Gamma(i)\,\Gamma(N-i)}{\Gamma(N)} = \frac{N}{N-i} \tag{3.9}$$

which is infinite if $i = N$.

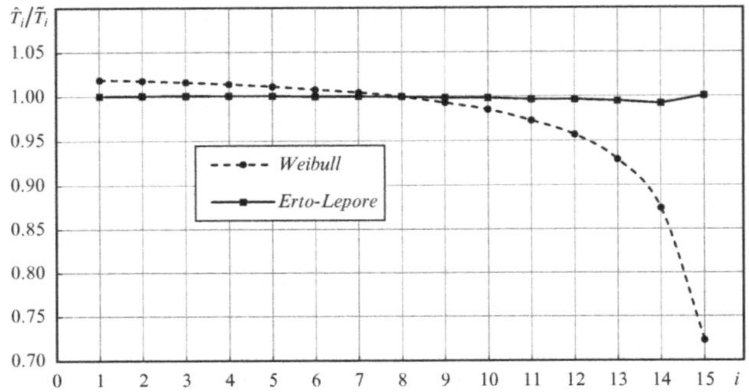

Fig. 3.1 Ratio $\hat{T}_i/\tilde{T}_i$ achieved by the proposed plotting position and the Weibull one ($N = 5$)

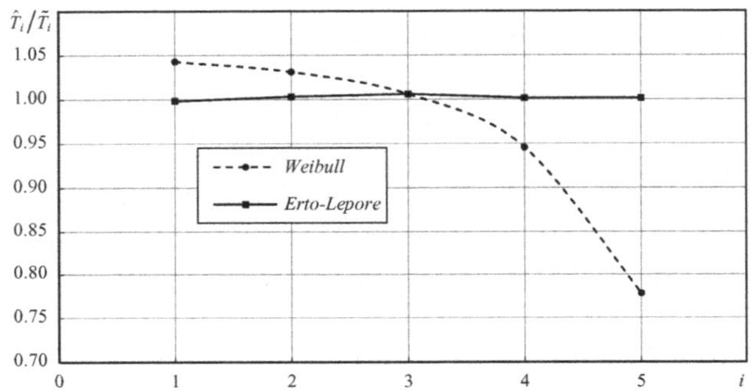

Fig. 3.2 Ratio $\hat{T}_i/\tilde{T}_i$ achieved by the proposed plotting position and the Weibull one ($N = 15$)

3.3 Findings and Conclusion

Starting from $m = 10,000$ pseudo-random samples $(F(x_1), \ldots, F(x_N))$ generated by a Monte Carlo simulation from the standard uniform distribution and arranged in non-decreasing order, m pseudo-random samples from $(T_1, \ldots, T_N)$ are obtained through Eq. (3.2). Then, for each $i = 1, \ldots, N$, the median $\tilde{T}_i$ can be calculated as well as the ratios $\hat{T}_i/\tilde{T}_i$ which are achieved by the proposed plotting position (Eq. (3.6)) and the Weibull one (Eq. (3.1)) upon using Eq. (3.3).

In order to visualize simultaneously the behaviour of both plotting positions, the ratio $\hat{T}_i/\tilde{T}_i$ is plotted in Figs. 3.1 and 3.2 for $N = 5$ and $N = 15$, respectively. It is clear from those figures that the proposed plotting position, whose ratio $\hat{T}_i/\tilde{T}_i$ is much closer to unity, always out-performs the Weibull one in estimating the median return period.

Moreover, the new plotting position advantages are also confirmed by the *MARE* values reported in Table 3.1 for $N = 5, 15, 30, 50$. Those values are obtained from Eq. (3.7) where $\tilde{F}_i$ is numerically evaluated as the median of the Beta random variable with parameters $a = i$ and $b = N - i + 1$.

Table 3.1 *MARE* of the median return period

	Plotting positions	
	Erto-Lepore	Weibull
$N = 5$	0.0024	0.069
$N = 15$	0.0015	0.043
$N = 30$	0.0012	0.028
$N = 50$	0.0010	0.020

References

1. Harter, H.L.: Another look at plotting positions. Comm. Stat. Theory Method **13**(13), 1613–33 (1984)
2. Yu, G.-H., Huang, C.-C.: A distribution free plotting position. Stoch. Environ. Res. Risk Assess. **15**(6), 462–476 (2001). doi:10.1007/s004770100083
3. Makkonen, L.: Bringing closure to the plotting position controversy. Comm. Stat. Theory Method **37**, 460–467 (2008a)
4. Makkonen, L.: Problems in the extreme value analysis. Struct. Saf. **30**(5), 405–419 (2008b). doi: 10.1016/j.strusafe.2006.12.001
5. Gumbel, E.J.: Statistics of Extremes. Columbia University Press, New York (1958)
6. Weibull, W.: A statistical theory of strength of materials. Ing Vet Ak Handl (Stockholm) **151**, 45 (1939)
7. Erto, P., Lepore, A.: A Note on the Plotting Position Controversy and a New Distribution-free Formula. In: Proceedings of the 45th Scientific Meeting of the Italian Statistical Society, University of Padua, 16–18 June, pp. 1–7, (2010) ISBN 9788861295667
8. Blom, G.: Statistical Estimates and Transformed Beta Variables. Wiley, New York (1958)
9. Lepore, A.: An integrated approach to shorten wind potential assessment, PhD Thesis, University of Naples Federico II, Department of Aerospace Engineering, Naples (2010)
10. Davis, P.J.: Gamma Function and Related Functions. In: Abramowitz, M., Stegun, I.A. (eds.) Handbook of Mathematical Functions with Formulas, Graphs, and Mathematical Tables, p. 258. Dover, New York (1972)

On Gaussian Compound Poisson Type Limiting Likelihood Ratio Process 4

Sergueï Dachian and Ilia Negri

Abstract

Different change-point type models encountered in statistical inference for stochastic processes give rise to different limiting likelihood ratio processes. Recently it was established that one of these likelihood ratios, which is an exponential functional of a two-sided Poisson process driven by some parameter, can be approximated (for sufficiently small values of the parameter) by another one, which is an exponential functional of a two-sided Brownian motion. In this chapter we consider yet another likelihood ratio, which is the exponent of a two-sided compound Poisson process driven by some parameter. We establish that the compound Poisson type likelihood ratio can also be approximated by the Brownian type one for sufficiently small values of the parameter. We equally discuss the asymptotics for large values of the parameter.

Keywords

Asymptotic efficiency • Bayesian estimators • Change-point • Limiting distribution • Limiting variance • Maximum likelihood estimator • Non-regularity • Parametric estimation • Probability and stochastic processes

---

S. Dachian
Laboratoire de Mathématiques, Université Blaise Pascal, 63177 Aubière CEDEX, France
e-mail: Serguei.Dachian@math.univ-bpclermont.fr

I. Negri (✉)
Department of Information Technology and Mathematical Methods, University of Bergamo, 24044 Dalmine, Italy
e-mail: ilia.negri@unibg.it

N. Torelli et al. (eds.), *Advances in Theoretical and Applied Statistics*, Studies in Theoretical and Applied Statistics, DOI 10.1007/978-3-642-35588-2_4,

4.1 Introduction

Different change-point type models encountered in statistical inference for stochastic processes give rise to different limiting likelihood ratio processes. In [3] a relation between two of these likelihood ratios was established by one of the authors. More precisely, it was shown that the first one, which is an exponential functional of a two-sided Poisson process driven by some parameter, can be approximated (for sufficiently small values of the parameter) by the second one, defined by

$$Z_0(x) = \exp\left\{ W(x) - \frac{1}{2}|x| \right\}, \quad x \in \mathbb{R}, \tag{4.1}$$

where W is a standard two-sided Brownian motion. In this chapter we consider another limiting likelihood ratio process arising in some change-point type models.

We introduce the random process Z_γ on $\mathbb{R}$ as the exponent of a two-sided compound Poisson process given by

$$\ln Z_\gamma(x) = \begin{cases} \gamma \sum_{k=1}^{\Pi_+(x)} \varepsilon_k^+ - \frac{\gamma^2}{2}\Pi_+(x), & \text{if } x \geq 0, \\ \gamma \sum_{k=1}^{\Pi_-(-x)} \varepsilon_k^- - \frac{\gamma^2}{2}\Pi_-(-x), & \text{if } x \leq 0, \end{cases} \tag{4.2}$$

where $\gamma > 0$, Π_+ and Π_- are two independent Poisson processes of intensity 1 on $\mathbb{R}_+$, the random variables ε_k^+ and ε_k^-, $k \in \mathbb{N}$, are i.i.d. standard Gaussian random variables and are also independent of Π_+ and Π_-, and we use the convention $\sum_{k=1}^{0} \alpha_k = 0$ (for any sequence α_k). We equally introduce the random variables

$$\begin{aligned} \zeta_\gamma &= \frac{\int_{\mathbb{R}} x\, Z_\gamma(x)\, \mathrm{d}x}{\int_{\mathbb{R}} Z_\gamma(x)\, \mathrm{d}x}, \\ \xi_\gamma^- &= \inf\Big\{ z : Z_\gamma(z) = \sup_{x\in\mathbb{R}} Z_\gamma(x) \Big\}, \\ \xi_\gamma^+ &= \sup\Big\{ z : Z_\gamma(z) = \sup_{x\in\mathbb{R}} Z_\gamma(x) \Big\}, \\ \xi_\gamma^\alpha &= \alpha\, \xi_\gamma^- + (1-\alpha)\, \xi_\gamma^+, \quad \alpha \in [0,1], \end{aligned} \tag{4.3}$$

related to this process, as well as their second moments $B_\gamma = \mathbf{E}\zeta_\gamma^2$ and $M_\gamma^\alpha = \mathbf{E}(\xi_\gamma^\alpha)^2$.

The process Z_γ, up to a linear time change, arises in some non-regular, namely change-point type, statistical models (e.g., the threshold autoregressive model studied by Chan and Kutoyants in [2]) as the limiting likelihood ratio process, and the variables ζ_γ and ξ_γ^α as the limiting distributions of the Bayesian estimators and of the appropriately chosen maximum likelihood estimator, respectively. Here the maximum likelihood estimator is not unique, and the appropriate choice is a linear combination with weights α and $1-\alpha$ of its minimal and maximal values. In particular, B_γ and M_γ^α are the limiting variances of these estimators, and the Bayesian estimators being asymptotically efficient, the ratio $E_\gamma^\alpha = B_\gamma / M_\gamma^\alpha$ is the asymptotic efficiency of this maximum likelihood estimator.

On the other hand, many change-point type statistical models encountered in various fields of statistical inference for stochastic processes rather have as limiting likelihood ratio process, up to a linear time change, the process Z_0. In this case, the limiting distributions of the Bayesian estimators and of the maximum likelihood estimator are given by

$$\zeta_0 = \frac{\int_{\mathbb{R}} x\, Z_0(x)\,\mathrm{d}x}{\int_{\mathbb{R}} Z_0(x)\,\mathrm{d}x} \quad \text{and} \quad \xi_0 = \operatorname*{argsup}_{x\in\mathbb{R}} Z_0(x), \tag{4.4}$$

respectively, and the limiting variances of these estimators are $B_0 = \mathbf{E}\zeta_0^2$ and $M_0 = \mathbf{E}\xi_0^2$.

A well-known example is the model of a discontinuous signal in a white Gaussian noise exhaustively studied by Ibragimov and Khasminskii in [10, Chapter 7.2] (see also their previous work [9]), but one can also cite change-point type models of dynamical systems with small noise considered by Kutoyants in [12] and [13, Chapter 5], those of ergodic diffusion processes examined by Kutoyants in [14, Chapter 3], a change-point type model of delay equations analyzed by Küchler and Kutoyants in [11], an i.i.d. change-point type model explored by Deshayes and Picard in [4], a model of a discontinuous periodic signal in a time inhomogeneous diffusion investigated by Höpfner and Kutoyants in [8], and so on.

Let us also note that Terent'yev in [16] determined explicitly the distribution of ξ_0 and calculated the constant $M_0 = 26$. These results were taken up by Ibragimov and Khasminskii in [10, Chapter 7.3], where by means of numerical simulation they equally showed that $B_0 = 19.5 \pm 0.5$, and so $E_0 = 0.73 \pm 0.03$. Later in [7], Golubev expressed B_0 in terms of the second derivative (with respect to a parameter) of an improper integral of a composite function of modified Hankel and Bessel functions. Finally in [15], Rubin and Song obtained the exact values $B_0 = 16\,\zeta(3)$ and $E_0 = 8\,\zeta(3)/13$, where ζ is Riemann's zeta function defined by $\zeta(s) = \sum_{n=1}^\infty 1/n^s$.

In this chapter we establish that the limiting likelihood ratio processes Z_γ and Z_0 are related. More precisely, we show that as $\gamma \to 0$, the process $Z_\gamma(y/\gamma^2)$, $y \in \mathbb{R}$, converges weakly in the space $\mathscr{D}_0(-\infty,+\infty)$ (the Skorohod space of functions on $\mathbb{R}$ without discontinuities of the second kind and vanishing at infinity) to the process Z_0. So, the random variables $\gamma^2\zeta_\gamma$ and $\gamma^2\xi_\gamma^\alpha$ converge weakly to the random variables ζ_0 and ξ_0, respectively. We show equally that the convergence of moments of these random variables holds, that is, $\gamma^4 B_\gamma \to 16\,\zeta(3)$, $\gamma^4 M_\gamma^\alpha \to 26$ and $E_\gamma^\alpha \to 8\,\zeta(3)/13$. Besides their theoretical interest, these results have also some practical implications. For example, they allow to construct tests and confidence intervals on the base of the distributions of ζ_0 and ξ_0 (rather than on the base of those of ζ_γ and ξ_γ^α, which are not known explicitly) in models having the process Z_γ with a small γ as a limiting likelihood ratio. Also, the limiting mean squared errors of the estimators and the asymptotic relative efficiency of the maximum likelihood estimator can be approximated as $B_\gamma \approx 16\,\zeta(3)/\gamma^4$, $M_\gamma^\alpha \approx 26/\gamma^4$ and $E_\gamma^\alpha \approx 8\,\zeta(3)/13$ in such models.

These are the main results of this chapter, and they are presented in Sect. 4.2, where we also briefly discuss the second possible asymptotics $\gamma \to +\infty$. The necessary lemmas are proved in Sect. 4.3 and, finally, in Sect. 4.4 we discuss some directions for future development.

4.2 Asymptotics of the Limiting Likelihood Ratio

Consider the process $X_\gamma(y) = Z_\gamma(y/\gamma^2)$, $y \in \mathbb{R}$, where $\gamma > 0$ and Z_γ is defined by (4.2). Note that $\frac{\int_{\mathbb{R}} y\, X_\gamma(y)\, \mathrm{d}y}{\int_{\mathbb{R}} X_\gamma(y)\, \mathrm{d}y} = \gamma^2 \zeta_\gamma$. Moreover $\inf\{z : X_\gamma(z) = \sup_{y\in\mathbb{R}} X_\gamma(y)\} = \gamma^2 \xi_\gamma^-$ and $\sup\{z : X_\gamma(z) = \sup_{y\in\mathbb{R}} X_\gamma(y)\} = \gamma^2 \xi_\gamma^+$, where the random variables ζ_γ, ξ_γ^+ and ξ_γ^- are defined by (4.3).

Remind also the process Z_0 on $\mathbb{R}$ defined by (4.1) and the random variables ζ_0 and ξ_0 defined by (4.4). Recall finally the quantities $B_\gamma = \mathbf{E}\zeta_\gamma^2$, $M_\gamma^\alpha = \mathbf{E}(\xi_\gamma^\alpha)^2$, $E_\gamma^\alpha = B_\gamma / M_\gamma^\alpha$, $B_0 = \mathbf{E}\zeta_0^2 = 16\,\zeta(3)$, $M_0 = \mathbf{E}\xi_0^2 = 26$ and $E_0 = B_0/M_0 = 8\,\zeta(3)/13$. Now we can state the main result of this chapter.

Theorem 1. *The process X_γ converges weakly in the space $\mathscr{D}_0(-\infty, +\infty)$ to the process Z_0 as $\gamma \to 0$. In particular, the random variable $\gamma^2\zeta_\gamma$ converges weakly to the random variable ζ_0 and, for any $\alpha \in [0,1]$, the random variable $\gamma^2\xi_\gamma^\alpha$ converges weakly to the random variable ξ_0. Moreover, for any $k > 0$ we have*

$$\gamma^{2k}\,\mathbf{E}\zeta_\gamma^k \to \mathbf{E}\zeta_0^k \quad \textit{and} \quad \gamma^{2k}\,\mathbf{E}(\xi_\gamma^\alpha)^k \to \mathbf{E}\xi_0^k.$$

In particular, $\gamma^4 B_\gamma \to 16\,\zeta(3)$, $\gamma^4 M_\gamma^\alpha \to 26$ and $E_\gamma^\alpha \to 8\,\zeta(3)/13$.

The results concerning the random variable ζ_γ are direct consequence of [10, Theorem 1.10.2] and the following three lemmas.

Lemma 1. *The finite-dimensional distributions of the process X_γ converge to those of Z_0 as $\gamma \to 0$.*

Lemma 2. *For all $\gamma > 0$ and all $y_1, y_2 \in \mathbb{R}$ we have*

$$\mathbf{E}\left|X_\gamma^{1/2}(y_1) - X_\gamma^{1/2}(y_2)\right|^2 \leq \frac{1}{4}\,|y_1 - y_2|\,.$$

Lemma 3. *For any $c \in \,]0\,,\,1/8\,[$ we have*

$$\mathbf{E}X_\gamma^{1/2}(y) \leq \exp(-c\,|y|)$$

for all sufficiently small γ and all $y \in \mathbb{R}$.

Note that these lemmas are not sufficient to establish the weak convergence of the process X_γ in the space $\mathscr{D}_0(-\infty,+\infty)$ and the results concerning the random variable ξ_γ^α. However, the increments of the process $\ln X_\gamma$ being independent, the convergence of its restrictions (and hence of those of X_γ) on finite intervals $[A,B]\subset\mathbb{R}$ $\big($i.e., convergence in the Skorohod space $\mathscr{D}[A,B]$ of functions on $[A,B]$ without discontinuities of the second kind$\big)$ follows from [6, Theorem 6.5.5], Lemma 1, and the following lemma.

Lemma 4. *For any $\varepsilon>0$ we have*

$$\lim_{h\to 0}\lim_{\gamma\to 0}\sup_{|y_1-y_2|<h}\mathbf{P}\Big\{\big|\ln X_\gamma(y_1)-\ln X_\gamma(y_2)\big|>\varepsilon\Big\}=0.$$

Now, Theorem 1 follows from the following estimate on the tails of the process X_γ by standard argument $\big($see, for example, [10]$\big)$.

Lemma 5. *For any $b\in\,]0\,,\,1/12\,[$ we have*

$$\mathbf{P}\Big\{\sup_{|y|>A}X_\gamma(y)>\mathrm{e}^{-bA}\Big\}\le 4\,\mathrm{e}^{-bA}$$

for all sufficiently small γ and all $A>0$.

Before giving the proofs of the above lemmas, let us discuss the second possible asymptotics $\gamma\to+\infty$. One can show that in this case, the process Z_γ converges weakly in the space $\mathscr{D}_0(-\infty,+\infty)$ to the process $Z_\infty(x)=\mathbb{1}_{\{-\eta<x<\tau\}}$, $x\in\mathbb{R}$, where η and τ are two independent exponential random variables with parameter 1. So, the random variables ζ_γ, ξ_γ^-, ξ_γ^+, and ξ_γ^α converge weakly to the random variables

$$\zeta_\infty=\frac{\int_{\mathbb{R}}x\,Z_\infty(x)\,\mathrm{d}x}{\int_{\mathbb{R}}Z_\infty(x)\,\mathrm{d}x}=\frac{\tau-\eta}{2},$$

$$\xi_\infty^-=\inf\Big\{z: Z_\infty(z)=\sup_{x\in\mathbb{R}}Z_\infty(x)\Big\}=-\eta,$$

$$\xi_\infty^+=\sup\Big\{z: Z_\infty(z)=\sup_{x\in\mathbb{R}}Z_\infty(x)\Big\}=\tau$$

and

$$\xi_\infty^\alpha=\alpha\,\xi_\infty^-+(1-\alpha)\,\xi_\infty^+=(1-\alpha)\,\tau-\alpha\,\eta,$$

respectively. One can equally show that, moreover, for any $k>0$ we have

$$\mathbf{E}\zeta_\gamma^k\to\mathbf{E}\zeta_\infty^k\quad\text{and}\quad\mathbf{E}(\xi_\gamma^\alpha)^k\to\mathbf{E}(\xi_\infty^\alpha)^k.$$

In particular, denoting $B_\infty = \mathbf{E}\zeta_\infty^2$, $M_\infty^\alpha = \mathbf{E}(\xi_\infty^\alpha)^2$ and $E_\infty^\alpha = B_\infty / M_\infty^\alpha$, we finally have

$$B_\gamma \to B_\infty = \mathbf{E}\Big(\frac{\tau - \eta}{2}\Big)^2 = \frac{1}{2},$$

$$M_\gamma^\alpha \to M_\infty^\alpha = \mathbf{E}\big((1-\alpha)\,\tau - \alpha\,\eta\big)^2 = 6\left(\alpha - \frac{1}{2}\right)^2 + \frac{1}{2} \tag{4.5}$$

and

$$E_\gamma^\alpha \to E_\infty^\alpha = \frac{1}{12\left(\alpha - \frac{1}{2}\right)^2 + 1}. \tag{4.6}$$

Let us note that these convergences are natural, since the process Z_∞ can be considered as a particular case of the process Z_γ with $\gamma = +\infty$ if one admits the convention $+\infty \cdot 0 = 0$.

Note also that Z_∞ is the limiting likelihood ratio process in the problem of estimating the parameter θ by i.i.d. uniform observations on $[\theta, \theta + 1]$. So, in this problem, the variables ζ_∞ and ξ_∞^α are the limiting distributions of the Bayesian estimators and of the maximum likelihood estimator, respectively, B_∞ and M_∞^α are the limiting variances of these estimators and, the Bayesian estimators being asymptotically efficient, E_∞^α is the asymptotic efficiency of the maximum likelihood estimator.

Finally observe that the formulae (4.5) and (4.6) clearly imply that in the latter problem (as well as in any problem having Z_∞ as limiting likelihood ratio) the best choice of the maximum likelihood estimator is $\alpha = 1/2$, and that the so chosen maximum likelihood estimator is asymptotically efficient. This choice was also suggested by Chan and Kutoyants in [2] for problems having Z_γ as limiting likelihood ratio. For large values of γ this suggestion is confirmed by our asymptotic results. However, we see that for small values of γ the choice of α will not be so important, since the limits in Theorem 1 do not depend on α.

4.3 Proofs of the Lemmas

First we prove Lemma 1. Note that the restrictions of the process $\ln X_\gamma$ (as well as those of the process $\ln Z_0$) on $\mathbb{R}_+$ and on $\mathbb{R}_-$ are mutually independent processes with stationary and independent increments. So, to obtain the convergence of all the finite-dimensional distributions, it is sufficient to show the convergence of one-dimensional distributions only, that is,

$$\ln X_\gamma(y) \Rightarrow \ln Z_0(y) = W(y) - \frac{|y|}{2} = \mathcal{N}\Big(-\frac{|y|}{2}, |y|\Big)$$

for all $y \in \mathbb{R}$. Moreover, these processes being symmetric, it is sufficient to consider $y \in \mathbb{R}_+$ only. Here and in the sequel "$\Rightarrow$" denotes the weak convergence of the random variables, and $\mathcal{N}(m, V)$ denotes a "generic" random variable distributed according to the normal law with mean m and variance V.

The characteristic function $\varphi_\gamma(t)$ of $\ln X_\gamma(y)$ is

$$\begin{aligned}\varphi_\gamma(t) &= \mathbf{E}\,\mathrm{e}^{\mathrm{i}t\ln X_\gamma(y)} = \mathbf{E}\,\mathrm{e}^{\mathrm{i}t\gamma\sum_{k=1}^{\Pi_+(y/\gamma^2)}\varepsilon_k^+ - \mathrm{i}t\frac{\gamma^2}{2}\Pi_+(y/\gamma^2)}\\ &= \mathbf{E}\,\mathbf{E}\left(\mathrm{e}^{\mathrm{i}t\gamma\sum_{k=1}^{\Pi_+(y/\gamma^2)}\varepsilon_k^+ - \mathrm{i}t\frac{\gamma^2}{2}\Pi_+(y/\gamma^2)} \;\middle|\; \mathscr{F}_{\Pi_+}\right)\\ &= \mathbf{E}\left(\mathrm{e}^{-\mathrm{i}t\frac{\gamma^2}{2}\Pi_+(y/\gamma^2)}\prod_{k=1}^{\Pi_+(y/\gamma^2)}\mathbf{E}\,\mathrm{e}^{\mathrm{i}t\gamma\varepsilon_k^+}\right) = \mathbf{E}\,\mathrm{e}^{-\frac{\gamma^2}{2}(\mathrm{i}t+t^2)\Pi_+(y/\gamma^2)}\end{aligned}$$

where we have denoted $\mathscr{F}_{\Pi_+}$ the σ-algebra related to the Poisson process Π_+, used the independence of ε_k^+ and Π_+ and recalled that $\mathbf{E}\,\mathrm{e}^{\mathrm{i}t\varepsilon_k^+} = \mathrm{e}^{-t^2/2}$.

Then, noting that $\Pi_+(y/\gamma^2)$ is a Poisson random variable of parameter y/γ^2 with moment generating function $\mathbf{E}\,\mathrm{e}^{t\Pi_+(y/\gamma^2)} = \exp\bigl(\frac{y}{\gamma^2}(\mathrm{e}^t - 1)\bigr)$, we get

$$\begin{aligned}\ln\varphi_\gamma(t) &= \frac{y}{\gamma^2}\left(\mathrm{e}^{-\frac{\gamma^2}{2}(\mathrm{i}t+t^2)} - 1\right) = \frac{y}{\gamma^2}\left(-\frac{\gamma^2}{2}(\mathrm{i}t + t^2) + o(\gamma^2)\right)\\ &= -\frac{y}{2}(\mathrm{i}t + t^2) + o(1) \to -\frac{y}{2}(\mathrm{i}t + t^2) = \ln\mathbf{E}\,\mathrm{e}^{\mathrm{i}t\mathcal{N}(-y/2,\,y)}\end{aligned}$$

as $\gamma \to 0$, and so Lemma 1 is proved.

Now we turn to the proof of Lemma 3 (we will prove Lemma 2 just after). Calculations similar to the above ones show that

$$\mathbf{E}X_\gamma^{1/2}(y) = \exp\left(\frac{|y|}{\gamma^2}\left(\mathrm{e}^{-\frac{\gamma^2}{8}} - 1\right)\right) \tag{4.7}$$

for all $y \in \mathbb{R}$ and, since

$$\frac{1}{\gamma^2}\left(\mathrm{e}^{-\frac{\gamma^2}{8}} - 1\right) = \frac{1}{\gamma^2}\left(-\frac{\gamma^2}{8} + o(\gamma^2)\right) \to -\frac{1}{8}$$

as $\gamma \to 0$, for any $c \in \left]0\,,\,1/8\right[$ we have $\mathbf{E}X_\gamma^{1/2}(y) \le \exp\bigl(-c\,|y|\bigr)$ for all sufficiently small γ and all $y \in \mathbb{R}$. Lemma 3 is proved.

Further we verify Lemma 2. We first consider the case $y_1, y_2 \in \mathbb{R}_+$ (say $y_1 \ge y_2$). Using (4.7) and taking into account the stationarity and the independence of the increments of the process $\ln X_\gamma$ on $\mathbb{R}_+$, we can write

$$\begin{aligned}
\mathbf{E}\left|X_\gamma^{1/2}(y_1)-X_\gamma^{1/2}(y_2)\right|^2 &= \mathbf{E}X_\gamma(y_1)+\mathbf{E}X_\gamma(y_2)-2\,\mathbf{E}X_\gamma^{1/2}(y_1)X_\gamma^{1/2}(y_2)\\
&= 2-2\,\mathbf{E}X_\gamma(y_2)\,\mathbf{E}\frac{X_\gamma^{1/2}(y_1)}{X_\gamma^{1/2}(y_2)}\\
&= 2-2\,\mathbf{E}X_\gamma^{1/2}\big(|y_1-y_2|\big)\\
&= 2-2\exp\!\left(\frac{|y_1-y_2|}{\gamma^2}\Big(\mathrm{e}^{-\frac{\gamma^2}{8}}-1\Big)\right)\\
&\le -2\,\frac{|y_1-y_2|}{\gamma^2}\Big(\mathrm{e}^{-\frac{\gamma^2}{8}}-1\Big)\le\frac{1}{4}\,|y_1-y_2|\,.
\end{aligned}$$

The other cases can be treated similarly, and so, Lemma 2 is proved.

Now let us check Lemma 4. First let $y_1, y_2 \in \mathbb{R}_+$ (say $y_1 \ge y_2$) such that $\Delta = |y_1 - y_2| < h$. Then, noting that conditionally to $\mathscr{F}_{\Pi_+}$ the random variable

$$\ln X_\gamma(\Delta)=\gamma\sum_{k=1}^{\Pi_+(\Delta/\gamma^2)}\varepsilon_k^+-\frac{\gamma^2}{2}\Pi_+(\Delta/\gamma^2)$$

is Gaussian with mean $-\frac{\gamma^2}{2}\Pi_+(\Delta/\gamma^2)$ and variance $\gamma^2\Pi_+(\Delta/\gamma^2)$, we get

$$\begin{aligned}
\mathbf{P}\Big\{\big|\ln X_\gamma(y_1)-\ln X_\gamma(y_2)\big|>\varepsilon\Big\} &\le \frac{1}{\varepsilon^2}\,\mathbf{E}\big|\ln X_\gamma(y_1)-\ln X_\gamma(y_2)\big|^2\\
&= \frac{1}{\varepsilon^2}\,\mathbf{E}\,\mathbf{E}\left(\big(\ln X_\gamma(\Delta)\big)^2\,\Big|\,\mathscr{F}_{\Pi_+}\right)\\
&= \frac{1}{\varepsilon^2}\,\mathbf{E}\left(\gamma^2\Pi_+(\Delta/\gamma^2)+\frac{\gamma^4}{4}\big(\Pi_+(\Delta/\gamma^2)\big)^2\right)\\
&= \frac{1}{\varepsilon^2}\left(\Delta+\frac{\gamma^4}{4}\left(\frac{\Delta}{\gamma^2}+\frac{\Delta^2}{\gamma^4}\right)\right)\\
&= \frac{1}{\varepsilon^2}\,\big((1+\gamma^2/4)\Delta+\Delta^2/4\big)<\frac{1}{\varepsilon^2}\,\big(\beta(\gamma)\,h+h^2/4\big)
\end{aligned}$$

where $\beta(\gamma) = 1 + \gamma^2/4 \to 1$ as $\gamma \to 0$. So, we have

$$\begin{aligned}
\lim_{\gamma\to 0}\sup_{|y_1-y_2|<h}\mathbf{P}\Big\{\big|\ln X_\gamma(y_1)-\ln X_\gamma(y_2)\big|>\varepsilon\Big\} &\le \lim_{\gamma\to 0}\frac{1}{\varepsilon^2}\,\big(\beta(\gamma)\,h+h^2/4\big)\\
&= \frac{1}{\varepsilon^2}\left(h+\frac{h^2}{4}\right),
\end{aligned}$$

and hence

$$\lim_{h\to 0}\lim_{\gamma\to 0}\sup_{|y_1-y_2|<h}\mathbf{P}\Big\{\big|\ln X_\gamma(y_1)-\ln X_\gamma(y_2)\big|>\varepsilon\Big\}=0,$$

where the supremum is taken only over $y_1, y_2 \in \mathbb{R}_+$.

The same conclusion can be obtained similarly in all the other cases. Lemma 4 is proved.

It remains to verify Lemma 5. Taking into account the symmetry of the process $\ln X_\gamma$, as well as the stationarity and the independence of its increments on $\mathbb{R}_+$, we obtain

$$\begin{aligned}\mathbf{P}\Big\{\sup_{|y|>A} X_\gamma(y)>\mathrm{e}^{-bA}\Big\}&\le 2\,\mathbf{P}\Big\{\sup_{y>A} X_\gamma(y)>\mathrm{e}^{-bA}\Big\}\\&\le 2\,\mathrm{e}^{bA/2}\,\mathbf{E}X_\gamma^{1/2}(A)\,\mathbf{E}\sup_{y>A}\frac{X_\gamma^{1/2}(y)}{X_\gamma^{1/2}(A)}\\&=2\,\mathrm{e}^{bA/2}\,\mathbf{E}X_\gamma^{1/2}(A)\,\mathbf{E}\sup_{z>0}X_\gamma^{1/2}(z).\end{aligned}\tag{4.8}$$

In order to estimate the last factor we write

$$\begin{aligned}\mathbf{E}\sup_{z>0}X_\gamma^{1/2}(z)&=\mathbf{E}\exp\left(\frac{1}{2}\sup_{z>0}\left(\gamma\sum_{k=1}^{\Pi_+(z/\gamma^2)}\varepsilon_k^+-\frac{\gamma^2}{2}\Pi_+(z/\gamma^2)\right)\right)\\&=\mathbf{E}\exp\left(\frac{1}{2}\sup_{n>0}\left(\gamma\sum_{k=1}^{n}\varepsilon_k^+-\frac{n\gamma^2}{2}\right)\right).\end{aligned}$$

Now, let us observe that the random process $S_n=\sum_{k=1}^n\varepsilon_k^+$, $n\in\mathbb{N}$, has the same law as the restriction on $\mathbb{N}$ of a standard Brownian motion W. So,

$$\begin{aligned}\mathbf{E}\sup_{z>0}X_\gamma^{1/2}(z)&=\mathbf{E}\exp\left(\frac{1}{2}\sup_{n>0}\big(\gamma W(n)-n\gamma^2/2\big)\right)\\&=\mathbf{E}\exp\left(\frac{1}{2}\sup_{n>0}\big(W(n\gamma^2)-n\gamma^2/2\big)\right)\\&\le\mathbf{E}\exp\left(\frac{1}{2}\sup_{t>0}\big(W(t)-t/2\big)\right)=\mathbf{E}\exp\left(\frac{1}{2}S_0\right)\end{aligned}$$

with an evident notation. It is known that the random variable S_0 is exponential of parameter 1 (see, for example, Borodin and Salminen [1]) and hence, using its moment generating function $\mathbf{E}\,\mathrm{e}^{tS_0}=(1-t)^{-1}$, we get

$$\mathbf{E}\sup_{z>0}X_\gamma^{1/2}(z)\le 2.\tag{4.9}$$

Finally, taking $b \in \,]0\,,\,1/12\,[$ we have $3b/2 \in \,]0\,,\,1/8\,[$ and, combining (4.8), (4.9) and using Lemma 3, we finally obtain

$$\mathbf{P}\Big\{\sup_{|y|>A} X_\gamma(y) > \mathrm{e}^{-bA}\Big\} \le 4\,\mathrm{e}^{bA/2}\exp\Big(-\frac{3b}{2}A\Big) = 4\,\mathrm{e}^{-bA}$$

for all sufficiently small γ and all $A > 0$, which concludes the proof.

4.4 Final Remarks

In conclusion let us mention that in [2], the Gaussianity of the jumps of the limiting likelihood ratio process Z_γ is due to a Gaussianity assumption on the underlying threshold autoregressive model. Without such type assumption, many change-point type statistical models—for example, the multi-phase regression model considered by Fujii in [5] (and probably the threshold autoregressive model itself)—gives raise to a more general compound Poisson type limiting likelihood ratio process $Z_{\gamma,f}$. This process is still an exponent of a two-sided compound Poisson process, but the jumps of the latter are no longer necessarily Gaussian. More precisely, it is given by

$$\ln Z_{\gamma,f}(x) = \begin{cases} \sum_{k=1}^{\Pi_+(x)} \ln \dfrac{f(\varepsilon_k^+ + \gamma)}{f(\varepsilon_k^+)}, & \text{if } x \ge 0, \\ \sum_{k=1}^{\Pi_-(-x)} \ln \dfrac{f(\varepsilon_k^- - \gamma)}{f(\varepsilon_k^-)}, & \text{if } x \le 0, \end{cases}$$

where $\gamma > 0$, Π_+ and Π_- are two independent Poisson processes of intensity 1 on $\mathbb{R}_+$, and the random variables ε_k^+ and ε_k^-, $k \in \mathbb{N}$, are i.i.d. with density f, mean 0 and variance 1 and are also independent of Π_+ and Π_-. The guess is that our results hold in this general situation under some regularity conditions on f.

References

1. Borodin, A.N., Salminen, P.: Handbook of Brownian motion—facts and formulae. Probability and Its Applications. Birkhäuser Verlag, Basel (2002)
2. Chan, N.H., Kutoyants, Y.A.: On parameter estimation of threshold autoregressive models. Stat. Inference Stoch. Process. **15**(1), 81–104 (2012)
3. Dachian, S.: On limiting likelihood ratio processes of some change-point type statistical models. J. Stat. Plann. Inference **140**(9), 2682–2692 (2010)
4. Deshayes, J., Picard, D.: Lois asymptotiques des tests et estimateurs de rupture dans un modèle statistique classique. Ann. Inst. H. Poincaré Probab. Statist. **20**(4), 309–327 (1984)
5. Fujii, T.: On weak convergence of the likelihood ratio process in multi-phase regression models. Statist. Probab. Lett. **78**(14), 2066–2074 (2008)
6. Gihman, I.I., Skorohod, A.V.: The Theory of Stochastic Processes I. Springer, New York (1974)
7. Golubev, G.K.: Computation of the efficiency of the maximum-likelihood estimator when observing a discontinuous signal in white noise. Probl. Inform. Transm. **15**(3), 61–69 (1979)

8. Höpfner, R., Kutoyants, Yu.A.: Estimating discontinuous periodic signals in a time inhomogeneous diffusion (2009) Available via ArXiv. http://arxiv.org/abs/0903.5061
9. Ibragimov, I.A., Khasminskii, R.Z.: Estimation of a parameter of a discontinuous signal in a white Gaussian noise. Probl. Inform. Transm. **11**(3), 31–43, (1975)
10. Ibragimov, I.A., Khasminskii, R.Z.: Statistical estimation. Asymptotic Theory. Springer, New York (1981)
11. Küchler, U., Kutoyants, Yu.A.: Delay estimation for some stationary diffusion-type processes. Scand. J. Stat. **27**(3), 405–414 (2000)
12. Kutoyants, Yu.A.: Parameter Estimation for Stochastic Processes. Armenian Academy of Sciences, Yerevan, in Russian (1980). Translation of revised version, Heldermann-Verlag, Berlin (1984)
13. Kutoyants, Yu.A.: Identification of dynamical systems with small noise. Mathematics and its Applications, vol. **300**. Kluwer Academic Publishers, Dordrecht (1994)
14. Kutoyants, Yu.A.: Statistical Inference for Ergodic Diffusion Processes. Springer Series in Statistics, Springer, London (2004)
15. Rubin, H., Song, K.-S.: Exact computation of the asymptotic efficiency of maximum likelihood estimators of a discontinuous signal in a Gaussian white noise. Ann. Statist. **23**(3), 732–739 (1995)
16. Terent'yev, A.S.: Probability distribution of a time location of an absolute maximum at the output of a synchronized filter. Radioengineering Electron. **13**(4), 652–657 (1968)

Archetypal Symbolic Objects

5

M.R. D'Esposito, F. Palumbo, and G. Ragozini

Abstract
Symbolic Data Analysis has represented an important innovation in statistics since its first presentation by E. Diday in the late 1980s. Most of the interest has been for the statistical analysis of Symbolic Data that represent complex data structure where variables can assume more than just a single value. Thus, Symbolic Data allow to describe classes of statistical units as a whole. Furthermore, other entities can be defined in the realm of Symbolic data. These entities are the Symbolic objects, defined in terms of the relationships between two different knowledge levels. This article aims at introducing a new type of SO based on the archetypal analysis.

Keywords
Archetypal Analysis • Interval data • Symbolic Data Analysis • Two level paradigm

5.1 Introduction

At the end of the 1980s, Edwin Diday has introduced in statistics the new and revolutionary idea of Symbolic Data Analysis (SDA) [8]. SDA refers to a group of methods for managing and analyzing complex data sources.

M.R. D'Esposito (✉)
Dipartimento di Scienze Economiche e Statistiche, Università di Salerno, Via Ponte don Melillo, 84084 Fisciano, Italy
e-mail: mdesposito@unisa.it

F. Palumbo · G. Ragozini
Dipartimento di Scienze Politiche, Università degli Studi di Napoli Federico II, Via L. Rodinò 22, 80138 Napoli, Italy
e-mail: fpalumbo@unina.it; giragoz@unina.it

N. Torelli et al. (eds.), *Advances in Theoretical and Applied Statistics*,
Studies in Theoretical and Applied Statistics, DOI 10.1007/978-3-642-35588-2_5,

Symbolic data generalize the concept of variable to other types of complex variables that are called descriptions. In SDA, beyond the classical numerical or categorical single valued variables, three classes of complex variables have been identified [9]: (*i*) categorical multivalued, a set of categories; (*ii*) interval valued, a subset of $\Re$; (*iii*) *modal variables*, a set of categories with weights. Symbolic multi/interval-valued variables occur if there is uncertainty on individual values. They permit to take into account the uncertainty in the data coding when dealing with queries to charactere large and huge datasets or when variables have a natural intrinsic variation (e.g., the [min, max] blood pressure).

Practically speaking, symbolic data permit to describe more complex "statistical entities" such as *groups of units* (or clusters), *concepts* and *Symbolic Objects* (SO). A group represents a set of homogeneous statistical units; a concept may be the description of an animal species or of a product, and so on. To represent groups and concepts, it is necessary to use multivalued or interval-valued variables in order to take into account the whole set of possible values. Finally, Symbolic Objects are conjunctions of elementary "events." According to Touati et al. [18]: *Symbolic objects are defined by a logical conjunction of elementary "events" where values or variables may be single, multiple, or nonexistent and where objects are not necessarily defined on the same variables.*

Bock and Diday and then Diday and Noirhomme-Fraiture have theorized and defined different knowledge orders, the two level *paradigma* [1, 9]. First order knowledge corresponds to observable statistical units, which can be described both as symbolic data and as classical punctual data. Second order knowledge consists of concepts that are "intrinsically" symbolic data: groups of individuals are examples of the second order knowledge level. Third knowledge level refers to concepts that are naturally defined as complex concepts as they refer to more general ideas. A good example for introducing the third knowledge level is given by animal species. Species are naturally characterized by interval-valued descriptions because of their own natural variability. All races belonging to the same species represent the second order level, and the individuals belonging to the species are the first order knowledge level.

Let Ω be a set of n generic (first or second order) objects $\omega_1, \omega_2, \ldots, \omega_n$ that are defined by p descriptions, *Archetypal Objects* $\{\mathbf{a}_1, \ldots, \mathbf{a}_m\}$ are observed or unobserved objects in Ω that permit to "rebuild" any ω_i as a weighted sum of the archetypes. Archetypes are based on the definition given by Cutler and Breiman [3]: archetypes rely on the definition of "pure individual types," few points lying on the boundary of the data scatter and characterizing the archetypal pattern in the data. D'Esposito et al. [6, 7] and Corsaro and Marino [2] have generalized archetypes to interval-valued data that represent one of the three type of variable in Symbolic Data. Interval Archetypes $\{\mathbb{a}_1, \ldots, \mathbb{a}_m\}$ refer to archetypes where all descriptors are interval-valued variables.

Into the two level paradigm framework, this chapter aims at introducing interval archetypes in terms of higher order Symbolic Objects: the *Archetypal Symbolic Object*. According to the above definitions, each unit ω_i, $i = 1, \ldots, n$ is either associated with an interval archetype $\mathbb{a}_m$ or not. That is, γ_i is a mapping from

$\Omega \rightarrow [0, 1]$, if $\gamma_{i,m} > \gamma^*$, ω_i is associated with $\mathbb{a}_m$. The value γ^* is a threshold (with $\gamma^* \geq 0.5$ at least).

This chapter is organized as follows: In Sect. 5.2 we summarize the archetypal analysis. Within the two level paradigm framework we present the archetypal SOs as "basis concepts"; then we illustrate the relation between the interval archetypes and the interval-valued statistical units, and how it leads to the introduction of a new concept of Symbolic Objects (Sect. 5.3). Section 5.4 contains an illustrative example.

5.2 Archetypal Analysis in a Nutshell

Given a set of n statistical units described by p variables, archetypal analysis aims at synthesizing the data through a set of (not necessarily observed) points that are called archetypes, under the constraint that all points can be represented as a convex combination of the archetypes themselves, and that the archetypes are a convex combination of the data [3].

Let $\{\mathbf{x}_i, i = 1, \ldots, n\}$ be a set of multivariate data in $\Re^p$, $\mathbf{x}_i = (x_{i1}, \ldots, x_{ip})'$. Archetypal analysis looks for m p-vectors $\{\mathbf{a}_j, j = 1, \ldots, m\}$ that are convex combinations of the input dataset and such that each data point is a convex combination of the vectors $\mathbf{a}_j$. Formally, given the data matrix $\mathbf{X} = (\mathbf{x}_1, \ldots, \mathbf{x}_n)'$, $\mathbf{X} \in \mathbb{R}^{n \times p}$, the archetype matrix $\mathbf{A}(m) = (\mathbf{a}_1, \ldots, \mathbf{a}_m)'$, $\mathbf{A}(m) \in \mathbb{R}^{m \times p}$, and the convex combination coefficients $\boldsymbol{\beta}_j = (\beta_{j1}, \ldots, \beta_{jn})'$ and $\boldsymbol{\gamma}_i = (\gamma_{i1}, \ldots, \gamma_{im})'$, it is

$$\mathbf{a}_j' = \boldsymbol{\beta}_j'\mathbf{X} \qquad j = 1, \ldots, m, \quad \beta_{ji} \geq 0 \;\; \forall j, i \quad \boldsymbol{\beta}_j'\mathbf{1} = 1 \;\; \forall j \tag{5.1}$$

$$\mathbf{x}_i' = \boldsymbol{\gamma}_i'\mathbf{A}(m) \qquad i = 1, \ldots, n, \quad \gamma_{ij} \geq 0 \;\; \forall i, j \quad \boldsymbol{\gamma}_i'\mathbf{1} = 1 \;\; \forall i. \tag{5.2}$$

The above equations imply that the archetypes are the vertices of the data convex hull, and that the data points belong to the archetype convex hull. An exact solution of Eqs. (5.1) and (5.2) exists only if $m = V$, where V indicates the cardinality of the data convex hull, i.e., the archetypes coincide with its V vertices [17].

However, V is generally too large to properly synthesize the data. For this reason, looking for a smaller number of pure types, and wishing to preserve their closeness to the data, the archetypes can be defined as those m $\mathbf{a}_j$'s, with $m < V$, fulfilling as far as possible Eq. (5.2), and solving exactly Eq. (5.1).

If Eq. (5.2) is thus relaxed, data points can be only approximated through a convex combination of the archetypes, i.e., $\boldsymbol{\gamma}_i'\mathbf{A}(m) = \tilde{\mathbf{x}}_i'(m) \neq \mathbf{x}_i'$.

Define $\tilde{\mathbf{X}}(m) = (\tilde{\mathbf{x}}_1(m), \ldots, \tilde{\mathbf{x}}_n(m))'$, $\tilde{\mathbf{X}}(m) \in \mathbb{R}^{n \times p}$, $\boldsymbol{\Gamma}(m) = (\boldsymbol{\gamma}_1, \ldots, \boldsymbol{\gamma}_n)'$, $\boldsymbol{\Gamma}(m) \in \mathbb{R}^{n \times m}$, $\mathbf{B}(m) = (\boldsymbol{\beta}_1, \ldots, \boldsymbol{\beta}_m)$, $\mathbf{B}(m) \in \mathbb{R}^{n \times m}$, and

$$RSS(m) = \|\mathbf{X} - \tilde{\mathbf{X}}(m)\|_F = \|\mathbf{X} - \boldsymbol{\Gamma}(m)\mathbf{A}(m)\|_F = \left\|\mathbf{X} - \boldsymbol{\Gamma}(m)\mathbf{B}'(m)\mathbf{X}\right\|_F \tag{5.3}$$

where $\|\mathbf{Y}\|_F = \sqrt{Tr\,(\mathbf{Y}\mathbf{Y}')}$ is the Frobenius norm for a generic matrix $\mathbf{Y}$.

The m archetypes $\mathbf{a}_j$, for $m < V$, solve the minimization problem

$$\min_{\boldsymbol{\Gamma}(m),\mathbf{B}(m)} RSS(m) = \min_{\boldsymbol{\Gamma}(m),\mathbf{B}(m)} \left\|\mathbf{X} - \boldsymbol{\Gamma}(m)\mathbf{B}'(m)\mathbf{X}\right\|_F \tag{5.4}$$

holding all the other conditions on coefficients $\boldsymbol{\beta}'_j$ and $\boldsymbol{\gamma}'_i$.

A detailed discussion on archetypal analysis properties and usage is in [5, 17].

In the case of symbolic data described through interval-valued variables, archetypes are defined as follows. Consider an interval $\mathbb{x} = [\underline{x}, \overline{x}] = \{x : x \in \Re, \underline{x} \leq x \leq \overline{x}\}$, where $\underline{x}$ and $\overline{x}$ are the interval bound values [12], and the set of all intervals is usually written as $\Im\Re$.

Let $\mathbb{X} \in \Im\Re^{n\times p}$ be an interval matrix, $\mathbb{X} = (\mathbb{x}'_1, \ldots, \mathbb{x}'_n)$. Each observed data point is now a hyper-rectangle. In analogy with the classical case, the aim of archetypal analysis is to define m archetypal hyper-rectangles—that we will denote by $\mathbb{A}(m) \in \Im\Re^{m\times p}$, $\mathbb{A}(m) = (\mathbb{a}'_1, \ldots, \mathbb{a}'_m)$—which synthesize the locations and the shapes of all the other data. These archetypal hyper-rectangles are such that the other hyper-rectangles can be expressed as a convex combination of them, and they are a convex combination of all the others:

$$\mathbb{A}(m) = \mathbf{B}(m)\mathbb{X}, \tag{5.5}$$

$$\mathbb{X} = \boldsymbol{\Gamma}(m)\mathbb{A}(m). \tag{5.6}$$

Given the component-wise Hausdorff distance between intervals and the Frobenius norm of the distance matrix between two interval matrices [7], for each m, the m archetypal hyper-rectangles $\mathbb{A}(m)$ can be determined by minimizing the distance between the data interval matrix $\mathbb{X}$ and the matrix $\tilde{\mathbb{X}}(m) = \boldsymbol{\Gamma}(m)\mathbb{A}(m)$, $\tilde{\mathbb{X}}(m) \in \Im\Re^{n\times p}$, i.e., the data matrix reconstructed by m archetypal hyper-rectangle.

Thus, given m and the quantity:

$$\mathbb{RSS}(m) = \left\|d\left(\mathbb{X}, \tilde{\mathbb{X}}(m)\right)\right\|_F = \left\|\mathbb{X} - \left(\boldsymbol{\Gamma}(m)\mathbf{B}'(m)\mathbb{X}\right)\right\|_F, \tag{5.7}$$

the m archetypes solve the minimization problem:

$$\min_{\boldsymbol{\Gamma}(m),\mathbf{B}(m)} \mathbb{RSS}(m) = \min_{\boldsymbol{\Gamma}(m),\mathbf{B}(m)} \left\|\mathbb{X} - \left(\boldsymbol{\Gamma}(m)\mathbf{B}'(m)\mathbb{X}\right)\right\|_F \tag{5.8}$$

under the constraints on the convex combination coefficient matrices $\boldsymbol{\Gamma}(m)$ and $\mathbf{B}(m)$:

$$\boldsymbol{\Gamma}(m)\mathbf{1}_m = \mathbf{1}_n, \qquad \mathbf{B}\mathbf{1}_n = \mathbf{1}_m \tag{5.9}$$

where $\mathbf{1}_n$ and $\mathbf{1}_m$ are the all-ones vectors of dimension n and m, respectively (see [7]).

Up to now archetypal analysis has been applied in many fields. It has found application as a tool for image decomposition [14, 15], in marketing research to find archetypal consumers [16], for market segmentation and consumer fuzzy clustering [10, 13], and finally for product profiling [6]. In performance analysis, archetypes have been exploited to construct data driven benchmarks [17] and to analyze CPU performance [11].

5.3 Archetypes and SOs

One important and innovative aspect of archetypal analysis for interval data in the symbolic data analysis context concerns the possibility of viewing archetypes as SOs. In accordance to the Bock and Diday *two level paradigm* [1, page 43], this section aims at presenting the archetypal analysis as a new and innovative approach for defining Symbolic Objects. We will illustrate how the definition of archetypes implies the concept of Symbolic Objects, and how the archetypes for interval-valued data can be consistently defined as a *higher* knowledge level.

5.3.1 Interval Archetypes as SOs

Given the definition of interval archetype, an interval archetype corresponds to a junction of descriptions $\mathscr{A}_j$ $j = 1 \ldots, p$ and all the statistical units in the dataset are related to this description through a membership function $\mathscr{D}$ (we refer to [1, pages 5–7] for definitions). Whereas in the classical Boolean SO definition, the SOs are identified on the basis of a comparison function $\mathscr{R}$, in the archetypal framework statistical units are defined on the basis of distance function $\mathscr{D}$ from the archetypes. Hence the SO is defined by the triple $(\gamma_{ij}, \mathscr{D}, \mathscr{A}_j)$. The membership is evaluated on the basis of the similarity of the description of an observed generic point $\mathbf{x}_i$ with the description of the archetype, i.e., $\gamma_{ij} = [\mathbf{x}_i \mathscr{D} \mathscr{A}_j]$. If the two descriptions coincide, and hence the archetype corresponds to an observed statistical unit, $\gamma_{ij} = 1$. If a statistical unit is very different from the archetype, γ_{ij} tends to zero.

The reader can refer to Sect. 5.2 for a detailed description of the coefficients γ_{ij} in the case of single valued data and for the case of interval-valued data, i.e., $\gamma_{ij} = [\mathbb{x}_i \mathscr{D} \mathscr{A}_j]$. It is worth reminding that in the case of interval-valued data the function $\mathscr{D}$ is based on the Hausdorff distance.

As an example, in performance analysis archetypes make it possible to obtain descriptions of the ideal “best” and “worst” performers, and to compare all the other performers with such descriptions [4].

5.3.2 Interval Archetypes as Third Order SOs

In addition to the above discussion, interval archetypal analysis allows us to define third order SOs. Indeed, in this case statistical units are SOs and second order knowledge descriptions. One interval archetype is a second order “basis” SO; however, the whole set of archetypes represents a third order knowledge. Archetypal

SOs and all units can be then defined by the *conjunction* of the *weighted* sum of a set of interval-valued variables.

Formally, a generic interval archetype is a description defined as $\mathbb{a}'_j = \boldsymbol{\beta}'_j \mathbb{X} = \sum_i \beta_{ij} \mathbb{x}'_i$. Interval-valued data represents second order SOs, archetypes evaluated on such data are, hence, third order SOs.

What is important to remark is that, at the same time, any statistical unit is defined as a weighted sum of all the archetypes apart from a residual: $x_{ij} \approx \sum_m \gamma_{im} \mathbb{a}_{mj}$. Under these conditions, the generic statical units can be also described by the triple $(\gamma_{ij}, \mathscr{D}, Y_j)$ that represents a new SO definition. The quantity $\mathscr{D}$ is a distance function that corresponds to the distance function defined to identify the archetypes.

Next section will graphically show these relations through an example.

5.4 An Illustrative Example

This section presents the *Italian peppers* dataset as an example to empirically show the archetypal SO definition. Many results and computational details will be skipped for sake of space.

Data are reported in the Table 5.1 and refer to some characteristics describing eight different species of Italian peppers. They are natively defined as interval-valued variables and represent some of the chemio-physical characteristics of eight different species of Italian peppers. This is a good example of data in which we can distinguish two different sources of variability: variability among different species, variation admitted inside one specific breed. Variation associated with each species is represented by the range: difference between the maximum and the minimum value. Before the analysis, variables have been standardized: midpoints have been centered with respect their respective means, and both have been reduced with respect the S.D. of the midpoints.

Three interval archetypes have been determined to synthesize the eight pepper types, according to Corsaro and Marino's algorithm [2]. Using different random starts, solutions have been computed 100 times. Results shown here refer to the best one. It is worth pointing out that solutions appeared to be very stable.

The symbolic objects related to peppers are represented in the space spanned by the archetypal symbolic objects in Fig. 5.1. In such a space the coordinates of each symbolic objects are the gamma coefficients. In the Table 5.2 the descriptions of the archetypal symbolic objects are reported.

Figure 5.2 shows the unit number 5 (Pimiento) reconstructed as archetype symbolic object weighted sum. Of course, an exact reconstruction is not possible due to the residual component. A larger number of archetypes would reduce the deviation. The upper display shows the variables H_2O versus Protein, the lower part the variable Lipides versus Glucides.

Table 5.1 Italian peppers dataset

Id	H_2O	Protein	Lipid	Glucide
Corno di Bue	[90.45, 93.15]	[0.67, 0.95]	[0.23, 0.30]	[5.07, 7.76]
Cuban	[90.46, 91.55]	[0.97, 1.11]	[0.24, 0.33]	[6.42, 7.65]
Cuban Nano	[87.89, 91.40]	[0.89, 1.25]	[0.28, 0.35]	[6.80, 9.91]
Grosso di Nocera	[90.91, 92.55]	[0.52, 0.80]	[0.21, 0.27]	[5.98, 7.58]
Pimiento	[89.92, 93.43]	[0.61, 1.09]	[0.23, 0.24]	[5.23, 7.94]
Quadrato D'Asti	[91.31, 92.99]	[0.74, 0.90]	[0.20, 0.27]	[6.64, 7.10]
Sunnybrook	[89.65, 92.58]	[0.85, 1.50]	[0.20, 0.28]	[5.52, 8.52]
Yolo Wonder	[90.80, 94.26]	[0.73, 1.30]	[0.20, 0.25]	[4.39, 7.34]

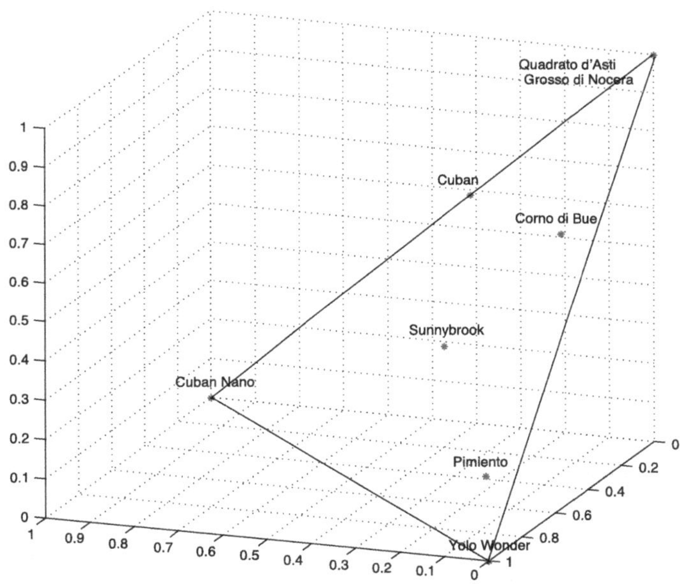

Fig. 5.1 Peppers represented in the space spanned by the archetypes. Coordinates are the γ coefficients

Table 5.2 Italian peppers: archetypal symbolic objects calculated on standardized data

Id	H_2O	Protein	Lipid	Glucide
Arc1	[−0.48, 1.46]	[−1.56, −0.08]	[−0.99, 0.73]	[1.46, 0.71]
Arc2	[−3.81, −0.05]	[−0.04, 2.40]	[0.71, 3.01]	[−0.11, 4.04]
Arc3	[−0.97, 2.94]	[−1.38, 2.48]	[−1.57, −0.20]	[−3.17, 0.89]

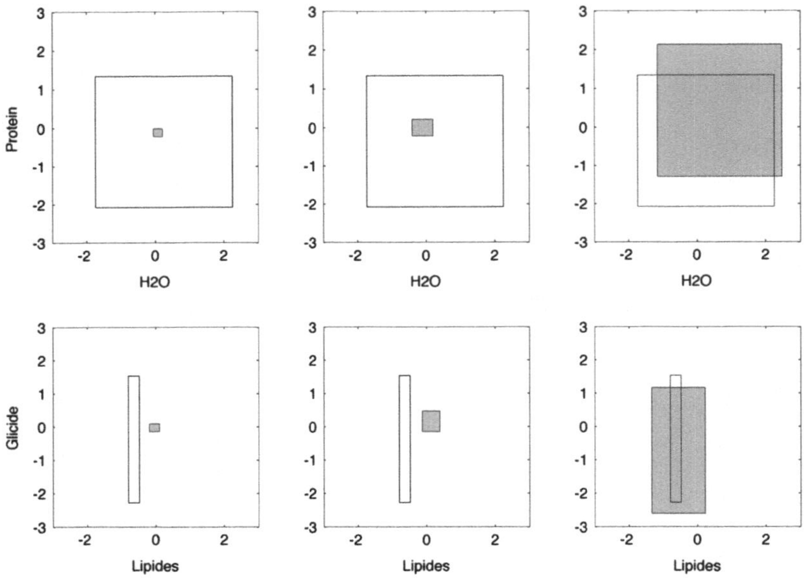

Fig. 5.2 The symbolic object 5 (Pimiento) reconstructed through the weighted sum of archetype symbolic objects

References

1. Bock, H.H., Diday, E.: Analysis of symbolic data. Exploratory Methods for Extracting Statistical Information from Complex Data. Springer, Heidelberg (2000)
2. Corsaro, S., Marino, M.: Archetypal analysis of interval data, Reliable Comput. **14**, 105–116 (2010)
3. Cutler, A., Breiman, L.: Archetypal analysis. Technometrics **36**, 338–347 (1994)
4. D'Esposito, M.R., Ragozini, G.: A New *R*-Ordering Procedure to Rank Multivariate Performances, Quaderni di Statistica **10**, 5–21 (2008)
5. D'Esposito, M.R., Ragozini, G, Vistocco.: Exploring data through archetypes. In: Locarek-Junge, H., Weihs, C. (eds.) Classification as Tool for Reasearch, Series in Studies in Classification, Data Analysis, and Knowledge Organization, pp. 287–289. Springer, Berlin Heidelberg (2010)
6. D'Esposito, M.R., Palumbo, F., Ragozini, G.: Archetypal analysis for interval data in marketing research. Ital. J. Appl. Stat. **18**, 343–358 (2006)
7. D'Esposito, M.R., Palumbo, F., Ragozini, G.: Interval archetypes: a new tool for interval data analysis. Stat. Anal. Data Min. **5**(4), 322–335 (2012)
8. Diday, E.: Introduction à l'approche symbolique en analyse des données. RAIRO **23**, 193–236 (1989)
9. Diday, E., Noirhomme-Fraiture, M.: Symbolic Data Analysis and the SODAS Software. Wiley, Chichester (2008)
10. Elder, A., Pinnel, J.: Archetypal analysis: an alternative approach to finding defining segments. In: 2003 Sawtooth Software Conference Proceedings, pp. 113–129, Sequim, WA (2003)

11. Heavlin, W.D.: Archetypal analysis of computer performance. In: 38th Symposium on the INTERFACE on Massive Data Sets and Stream, Pasedena, CA (2007)
12. Kearfott, R.B.: Rigorous Global Search: Continuous Problems. Kluwer Academic Publishers, Dordrecht (1996)
13. Li, S., Wang, P., Louviere, J., Carson, R.: Archetypal analysis: a new way to segment markets based on extreme individuals. In: A Celebration of Ehrenberg and Bass: Marketing Knowledge, Discoveries and Contribution, ANZMAC 2003 Conference Proceedings, pp. 1674–1679, Adelaide (2003)
14. Marinetti, S., Finesso, L., Marsilio, E.: Matrix factorisation methods: application to thermal NDT/E. NDT&E Int. **39**, 611–616 (2006)
15. Marinetti, S., Finesso, L., Marsilio, E.: Archetypes and principal components of an IR image sequence. Infrared Phys. Tech. **49**, 272–276 (2007)
16. Morris, L., Schmolze, R.: Consumer archetypes: a new approach to developing consumer understanding frameworks. J. Market. Res. **46**, 289–300 (2006)
17. Porzio, G.C., Ragozini, G. Vistocco, D.: On the use of archetypes as benchmarks. Appl. Stoch. Model Bus. Ind. **24**, 419–437 (2008)
18. Touati, M., Djedour, M., Diday, E.: Synthesis of objects. In: Gettler-Summa, M., Bottou, L., Goldfarb, B., Murtagh, F., Pardoux, C., Touati, M. (eds.) Statistical Learning and Data Science, Computer Science & Data Analysis series. CRC Press, Boca Raton (2011)

Lorenz Zonoids and Dependence Measures: A Proposal

6

Emanuela Raffinetti and Paolo Giudici

Abstract

Recently, the analysis of ordered and non-ordered categorical variables has assumed a relevant role, especially with regard to the evaluation of customer satisfaction, health and educational effectiveness. In such real contexts, the study of dependence relations among the involved variables represents an attractive research field. However, the categorical nature of variables does not always successfully allow the application of the existing standard dependence measures, since categorical data are not specified according to a metric scale. In fact, the aforementioned statistical methods are more appropriate in a purely quantitative setting, because based on the Euclidean distance. Our purpose aims at overcoming these restrictions by extending the dependence study in a quali–quantitative perspective. The idea is focused on employing specific statistical tools, such as the Lorenz curves and the so-called Lorenz zonoids. A novel Lorenz zonoids-based relative dependence measure is proposed as an alternative to the partial correlation coefficient to establish each categorical covariate contribution in a multiple linear regression model.

Keywords

Linear regression models • Lorenz zonoid • Relative Gini Index (RGI)

E. Raffinetti (✉)
Department of Economics, Business and Statistics, Università degli Studi di Milano, Via Conservatorio 7, Milano, Italy
e-mail: emanuela.raffinetti@unimi.it

P. Giudici
University of Pavia, Via Strada Nuova 65, Pavia, Italy
e-mail: giudici@unipv.it

N. Torelli et al. (eds.), *Advances in Theoretical and Applied Statistics*,
Studies in Theoretical and Applied Statistics, DOI 10.1007/978-3-642-35588-2_6,

6.1 Introduction

As well known, statistical dependence is a type of relation between any two features of units under study: these units may, for instance, be individuals, or objects, or various aspects of the environment.

In literature, several approaches addressed to the dependence and concordance study have been developed. In many cases, the dependence and concordance notions are strictly related especially in a quali–quantitative context, characterized by variables assuming mostly categorical nature.[1] In such a context, one of the main occurring problem concerns the application of standard dependence measures to capture information about the real existence of dependence relations among the involved variables. Let us suppose, for instance, to assume the employment of a multiple linear regression model built on a quantitative response variable and a set of categorical covariates. Typically, the existence of dependence relations among the response variable and the considered explanatory ones can be detected through the so-called multiple linear determination coefficient. This measure allows also to assess the data goodness of fit: anyway, the multiple linear determination coefficient is affected by some relevant restrictions since based on the Euclidean distance and then more appropriate in a quantitative context of analysis, as argued by Agresti (2002) (see, e.g., [1]). For this reason, an interesting research field is represented by the definition of novel and more reliable dependence measures, as proved in some recent contributions provided by Raffinetti and Giudici (2012) (see, e.g., [13]). More in detail, the dependence analysis has been further on extended by considering also the concordance notion, as initially illustrated in [10]. According to Raffinetti and Giudici's [13] recent suggestions, a useful tool in obtaining information about the goodness of fit of a multiple linear regression model is the so-called Multivariate Ranks-based Concordance Index. A related development is described in [5], where an alternative to residual analysis is proposed through a new Gini measure decomposition in terms of concordance and discordance. The recall to Gini measure and thus to the underlying Lorenz curve is motivated by the last decades research proposals in dependence analysis. In [11], for instance, the Lorenz curve tool was considered in order to define a new characterization of monotone dependence aimed at drawing a comparison between the regression function $E(Y|X)$ Lorenz curve and the Y Lorenz curve. The aforementioned notion is appropriate when asking the relation to be invariant under increasing transformation of X, but sensible to increasing transformation of Y. On the basis of such results, in [11] a partial ordering of monotone dependence on the class of nonnegative bivariate random vectors with given marginals was then defined. The proposed partial ordering need arises in the study of economic problems (e.g., taxation, see [10]) and in the study of several applications such as discriminant problems and statistical quality control.

[1]Hereafter we simply denote with "categorical" both ordered and non-ordered categorical variables.

In this chapter we refer to a partial ordering based on a specific statistical tool, named Lorenz zonoid. When considering multivariate data, the Lorenz zonoid represents the multidimensional extension of the Lorenz curve. The Lorenz zonoid has been introduced by Koshevoy (see, e.g., [6]) for empirical distributions and Mosler (see, e.g., [9]) for general probability distributions. More precisely, the Lorenz zonoid of a d-dimensional random vector corresponds to a convex set in $\mathbb{R}^{d+1}$, whose role is analyzing and comparing random vectors. Through the Lorenz zonoid representation one can establish an ordering of random vectors that reflects their variability: the investigation of such ordering is induced by the inclusion between Lorenz zonoids (see, e.g., [11] and [12]). This aspect provides a helpful support for our proposed developments in dependence relations analysis. In particular, our contributed approach is proved to provide a better measure of goodness of fit when the available data are characterized mainly by categorical nature.

Afterwards, the rest of this chapter is organized as follows. Section 6.2 illustrates the Lorenz zonoid and "*inclusion property*" definitions: in particular, the basic procedure needed in formalizing novel partial dependence measures will be applied with regard to a multiple linear regression model. Finally, Sect. 6.3 will be devoted to final comments.

6.2 Dependence Measures Based on Lorenz Zonoids

Let us begin by recalling the Lorenz curve definition of a random variable X, as reported in [8]. The Lorenz curve of a random variable X having expectation $\mu > 0$ is the graph of the function

$$t \mapsto \mu^{-1} \int_0^t F_X^{-1}(s)\mathrm{d}s, 0 \le t \le 1$$

where F_X^{-1} is the quantile function of X, $F_X^{-1} = \min\{x : F(x) \ge t\}$, $0 < t \le 1$.

The Lorenz zonoid of a general d-variate random vector is defined as follows (see, e.g., [8]). Consider the set $\mathscr{X}^d$ of random vectors in $\mathbb{R}^d$ that have finite expectation, the subset $\mathscr{X}^{d+} \subset \mathscr{X}^d$ of those vectors that have positive (in each component) expectation, and the subset $\mathscr{X}_+^{d+} \subset \mathscr{X}^{d+}$ of those that have, in addition, support in $\mathbb{R}_+^d$.

For $\mathbf{X} \in \mathscr{X}^{d+}$, we introduce the notation

$$\tilde{\mathbf{X}} = \left(\frac{X_1}{E(X_1)}, \dots, \frac{X_d}{E(X_d)}\right),$$

in order to point out the relative vector[2] that is the vector componentwise divided by its expectation.

[2]Relative data, that is data divided by their mean value is motivated by the classical definition of Lorenz curve. Since the Lorenz zonoid is exactly the Lorenz curve extension in the multidimensional context, one has to consider relative random vectors.

The *measurable function* definition is now needed (see, e.g., [2]).

Definition 1. A function $g : E \to \mathbb{R}$ is measurable if E is a measurable set and for each real number r, the set $\{x \in E : g(x) > r\}$ is measurable.

From Definition 1, it is clear that continuous and monotone functions are measurable.

The Lorenz zonoid of a random vector $\mathbf{X} \in \mathscr{X}^{d+}$ is a convex compact set in $\mathbb{R}^{d+1}$, defined as follows:

$$LZ(\mathbf{X}) = \left\{E[(g(\tilde{\mathbf{X}}), g(\tilde{\mathbf{X}})\tilde{\mathbf{X}}] : g : \mathbb{R}^d \to [0,1] \text{ measurable}\right\}.$$

In particular, if $\mathbf{X} \in \mathscr{X}_+^{d+}$, i.e., has support in $\mathbb{R}_+^d$, the Lorenz zonoid is contained in the hypercube of $\mathbb{R}^{d+1}$. Of our interest is the Lorenz zonoid in the univariate case, in fact when $d = 1$ the Lorenz zonoid corresponds to the area between the Lorenz curve and its dual,[3] and thus it is equivalent to the Gini measure, as proved by Koshevoy and Mosler (see [7]).

The Lorenz zonoid has many attractive properties which makes it useful for a broad range of applications: in order to introduce the following results in terms of Lorenz zonoids-based dependence measures, two relevant properties are needed. First, we show that linear dependence is related to inclusion of Lorenz zonoids. The definition of linear dependence between random vectors, can be found in many references (see, for instance, [3]).

Definition 2. A linear dependence preorder $\preceq_{ld}$ on $\mathscr{X}^{d+}$ is defined as follows:

$$\mathbf{Y} \preceq_{ld} \mathbf{X} \text{ if } LZ(\mathbf{X}) \subset LZ(\mathbf{Y}), \tag{6.1}$$

where $LZ(\mathbf{X})$ and $LZ(\mathbf{Y})$ are Lorenz zonoids of the random vectors $\mathbf{X}$ and $\mathbf{Y}$.

Then, the condition of linear dependence can be further on extended with regard to random variables whose linear relations can be investigated through a linear regression model. The aforementioned issue is illustrated by resorting to the so-called *inclusion property*. Let us suppose to consider a bivariate vector of random variables (Y, X) and to apply a simple linear regression model.

Proposition 1. *Denote respectively with* $L_Y(t)$ *and* $L'_Y(t)$ *the* Y *Lorenz curve and its dual, and with* $L_{E(Y|X)}(t)$ *and* $L'_{E(Y|X)}(t)$ *the* $E(Y|X)$ *Lorenz curve and its dual. One can prove that* $L_Y(t) \leq L_{E(Y|X)}(t) \leq L'_Y(t)$ *(see, e.g., [11] and [12]), where* $L'_Y(t) = \frac{1}{E(Y)} \int_{1-t}^{1} F_Y^{-1}(s)\mathrm{d}s$, $0 \leq t \leq 1$. *Furthermore,* $L'_{E(Y|X)}(t) \leq L'_Y(t)$.

[3]The dual Lorenz curve corresponds to the Lorenz curve built by ordering the underlying variable values in a decreasing sense.

Proof. Let $x_t = F_X^{-1}(t)$, where $F_X^{-1}(t) = \inf\{x : F_X(x) \geq t\}$ for $0 \leq t \leq 1$ (see [11]). In order to show that $L_Y(t) \leq L_{E(Y|X)}(t)$, note that

$$L_{E(Y|X)}(t) = \frac{1}{E(Y)} E(Y|X \leq x_t) F_X(x_t)$$

and

$$L_Y(t) = \frac{1}{E(Y)} E(Y|Y \leq y_t) F_Y(y_t).$$

Now, $(Y|X \leq x_t)$ is stochastically larger than $(Y|Y \leq y_t)$, that is

$$P(Y \leq y|X \leq x_t) \leq P(Y \leq y|Y \leq y_t) \text{ for all } y \in \mathbb{R},$$

and the result follows.

In the same manner, one can prove that the dual Lorenz curve of a general regression function always lies below the dual Lorenz curve of the response variable Y (see [12]).

Since the univariate Lorenz zonoid corresponds to the area between the Lorenz curves, Proposition 1 allows to conclude that $LZ(E(Y|X)) \subseteq LZ(Y)$, implying the same conclusion of Definition 1, even if in this case inclusion is full as consequence of Proposition 1. The existence of a linear dependence relation between Y and X translates into an inclusion between the response variable Lorenz zonoid and its linear estimated values Lorenz zonoid. Figure 6.1 shows this outcome in a depicted way. By establishing a linear relation between the response variable Y and the X covariate, through a simple linear regression function, one computes the response variable estimated values and then proceeds to the construction of the $E(Y|X)$ Lorenz zonoid. Since between the two random variables there exists linear dependence, the Y Lorenz zonoid contains the corresponding linear estimated values Lorenz zonoid, as one can deduce from Fig. 6.1. For these reasons, a direct implication is that the $E(Y|X)$ Lorenz zonoid represents a useful tool in order to define the total variability explained by the response variable linear estimated values.

In fact, in order to analyze the variability of random vectors in $\mathcal{X}^{d+}$, one can resort to the Lorenz zonoids and consider the order between random vectors that is induced by their inclusion. Let us suppose to consider multivariate data: an ordering able to stress the variability between two random vectors $\mathbf{X}$ and $\mathbf{Y}$ is the *dilation order*, which is related to the so-called *Lorenz zonoids order* (see, e.g., [7] and [12] for more details). Let us now introduce the Lorenz zonoid order definition (see, e.g., [8]).

Definition 3. For $\mathbf{X}$ and $\mathbf{Y} \in \mathcal{X}^{d+}$, the Lorenz zonoid order (Lorenz dominance) $\preceq_L$, is defined as:

$$\mathbf{Y} \preceq_L \mathbf{X} \text{ if } LZ(\mathbf{X}) \subset LZ(\mathbf{Y}).$$

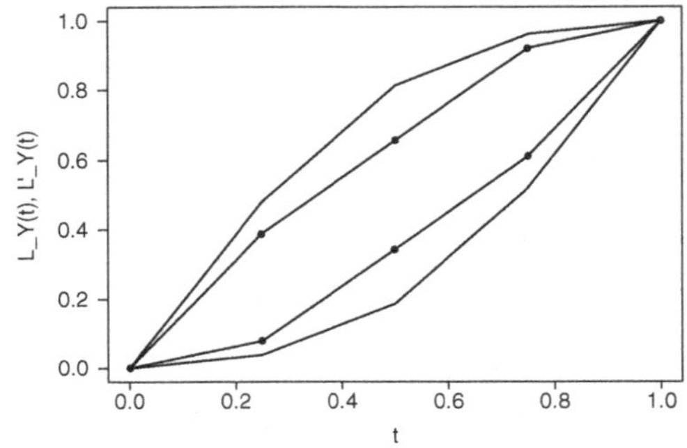

Fig. 6.1 Y Lorenz zonoid (area between continuous lines) and $E(Y|X)$ Lorenz zonoid (area between alternate circles and continuous lines)

Let us denote with $\preceq_{\text{dil}}$ the dilation order. One can show (see, e.g., [12]) a perfect equivalence between the dilation order and the Lorenz zonoid order, as proved by the following corollary:

Corollary 1. $\mathbf{X} \preceq_{\text{dil}} \mathbf{Y} \Rightarrow \mathbf{X} \preceq_L \mathbf{Y}$.

If $d = 1$, the reverse implication, $X \preceq_L Y \Rightarrow X \preceq_{\text{dil}} Y$ holds, meaning that through the Lorenz dominance one can provide an ordering among random variables according to their variability degree.

6.2.1 Single Partial Dependence Measures

Lorenz zonoids have been employed in statistical applications only in a bivariate case. In this context, a dependence measure for a general regression function $E(Y|X = x)$ was developed in an unpublished draft of Koshevoy and Muliere 2002 (see Koshevoy and Muliere, 2002, Dependence Orderings, Unpublished Draft, 2002).

Our current purpose consists in extending the dependence analysis when considering a linear regression function characterized by more than one independent variable. The proposal is to define single partial dependence measures whose role is addressed in expressing the Y Lorenz zonoid share "explained" by each single considered explanatory variable. Since the univariate Lorenz zonoid is equivalent to the Gini measure, it can be intended as a dispersion measure.

Let us define the response variable Y Lorenz zonoid and let us denote it with $LZ(Y)$. Assuming the application of a linear regression function to bivariate vector (Y, X_1), one can build the Lorenz zonoid of $\hat{Y}_{X_1} = E(Y|X_1)$, denoted with $LZ(\hat{Y}_{X_1})$. The next step is focused on operating with a trivariate random vector (Y, X_1, X_2). We can thus compute the Lorenz zonoid of the linear estimated Y variable values according to the second explanatory variable X_2, obtaining $LZ(\hat{Y}_{X_2})$. Furthermore, by exploiting the multiple linear regression tool, one can compute the linear estimated Y variable values $\hat{Y}_{X_1,X_2} = E(Y|X_1, X_2)$, considering both the covariates. The area that ranges between the $\hat{Y}_{X_1,X_2}$ Lorenz curve and its dual can be denoted with $LZ(\hat{Y}_{X_1,X_2})$. Forward stepwise inclusion of further covariates can be dealt in a similar manner without loss of generality. Here, Lorenz zonoids are particularly useful in order to define our proposed partial dependence measures. A partial measure representative of the Y Lorenz zonoid share "explained" by covariate X_1 alone, is derived through the following ratio:

$$G_{Y|X_1} = \frac{LZ(\hat{Y}_{X_1})}{LZ(Y)}, \tag{6.2}$$

and, in the same manner, a partial measure able to describe the share of the Y variable Lorenz zonoid "explained" by covariate X_2 alone is derived by the ratio

$$G_{Y|X_2} = \frac{LZ(\hat{Y}_{X_2})}{LZ(Y)}. \tag{6.3}$$

Since we are implementing a multiple linear regression model, in order to formalize partial dependence measures expressed in terms of Lorenz zonoids, our aim is based on determining each covariate possible partial contribution to the overall response variable explained variability. The aforementioned outcome is achieved by resorting to Definition 3 and Corollary 1.

6.2.2 Lorenz Zonoids-Based Approach: The Multivariate Case

Let us consider the general context characterized by k explanatory variables: our contribution is focused on defining the effect related to the introduction of a new $(k + 1)$-th explanatory variable into the linear regression model. The addition of a new explanatory variable can provide an enlargement of the $E(Y|\cdot)$ Lorenz zonoid. The Lorenz zonoid of the Y linear estimated values, denoted with $LZ(E(Y|X_1, \ldots, X_k))$, corresponds to the dilation measure of the Y response variable Lorenz zonoid $LZ(Y)$. In other words, the introduction of an additional covariate in a linear regression model translates into an increase of the "explained" Y variability: this result is confirmed by the Lorenz zonoids inclusion property.

In the well-known multiple linear regression model, properly, the contribution of a single variable to the regression plane is additive and, therefore, the inclusion of a new explanatory variable translates into an increase of the multiple determination coefficient (see, e.g., [4]). More precisely, suppose to build a linear regression model characterized by k explanatory variables. Let us introduce an additional $(k+1)-th$ explanatory variable: its contribution corresponds to an increase of the Y variable "explained" variability, defined as the difference between $Var(\hat{Y}_{X_1,\ldots,X_{k+1}})$ and $Var(\hat{Y}_{X_1,\ldots,X_k})$.[4] The squared partial correlation coefficient is expressed as

[4] $Var(\hat{Y}_{X_1,\ldots,X_k})$ denotes the Y variability "explained" by $X_1, \ldots, X_k$ whereas $Var(\hat{Y}_{X_1,\ldots,X_{k+1}})$ denotes the Y variability "explained" by $X_1, \ldots, X_{k+1}$.

$$r^2_{Y,X_{k+1}|X_1,\dots,X_k} = \frac{Var(\hat{Y}_{X_1,\dots,X_{k+1}}) - Var(\hat{Y}_{X_1,\dots,X_k})}{Var(Y) - Var(\hat{Y}_{X_1,\dots,X_k})}, \tag{6.4}$$

where $Var(Y) - Var(\hat{Y}_{X_1,\dots,X_k})$ identifies the Y variable variability not explained by covariates $X_1, \dots, X_k$.

Our purpose is building a partial dependence measure that "parallels" the partial correlation coefficient construction. In such a context, we aim at obtaining a ratio whose numerator is characterized by a term denoting one of the possible contribution generated by the $(k + 1)$-th explanatory variable and whose denominator is defined by a term describing the share of the Y Lorenz zonoid "not explained" by the $\hat{Y}_{X_k}$ Lorenz zonoid.

The possible additional contribution related to the $(k+1)$-th explanatory variable introduction can be measured by the difference between the Lorenz zonoid of $\hat{Y}_{X_1,\dots,X_{k+1}}$ and that of $\hat{Y}_{X_1,\dots,X_k}$, as follows:

$$LZ(\hat{Y}_{X_1,\dots,X_{k+1}}) - LZ(\hat{Y}_{X_1,\dots,X_k}). \tag{6.5}$$

A relative measure able to stress the possible additional contribution related to covariate X_{k+1} can be obtained in analogy with the partial correlation coefficient construction. Such a measure, called "*Relative Gini Index*", is expressed as:

$$RGI_{Y,X_{k+1}|X_1,\dots,X_k} = \frac{LZ(\hat{Y}_{X_1,\dots,X_{k+1}}) - LZ(\hat{Y}_{X_1,\dots,X_k})}{LZ(Y) - LZ(\hat{Y}_{X_1,\dots,X_k})}. \tag{6.6}$$

Let us prove this construction in the trivariate (Y, X_1, X_2) case.

Proof. When only two explanatory variables are considered, the multiple linear determination coefficient assumes the following expression:

$$R^2_{Y,X_1,X_2} = r^2_{Y,X_1} + r^2_{Y,X_2|X_1}(1 - R^2_{Y,X_1}), \tag{6.7}$$

where R^2_{Y,X_1} denotes the linear determination coefficient describing the X_1 covariate contribution in explaining the response variable variability.

In the considered linear regression model, the multiple linear determination coefficient R^2_{Y,X_1,X_2} defines the linear dependence among all the involved variables. In terms of Lorenz zonoids, the R^2_{Y,X_1,X_2} corresponds to $LZ(E(Y|X_1, X_2)) = LZ(\hat{Y}_{X_1,X_2})$. Thus, relation (6.7) becomes

$$LZ(E(Y|X_1, X_2)) = LZ(E(Y|X_1)) + LZ(E(Y, X_2|X_1))(LZ(Y) - LZ(E(Y|X_1))),$$

where $LZ(E(Y|X_1)) = LZ(\hat{Y}_{X_1})$ and $LZ(E(Y, X_2|X_1)) = RGI_{Y,X_2|X_1}$. By substituting to $RGI_{Y,X_2|X_1}$ its expression in (6.6), one gets

Table 6.1 Data

Y	350	202	404	263	451	304	275	385	244	102	74	346	53	395	430
X_1	1	2	3	2	1	3	5	1	4	2	2	3	4	3	3
X_2	4	4	1	5	2	4	4	4	3	3	2	4	2	1	3

$$
\begin{aligned}
LZ(E(Y|X_1, X_2)) &= LZ(\hat{Y}_{X_1}) + RGI_{Y.X_2|X_1}(LZ(Y) - LZ(\hat{Y}_{X_1}))\\
&= LZ(\hat{Y}_{X_1}) + \frac{LZ(\hat{Y}_{X_1.X_2}) - LZ(\hat{Y}_{X_1})}{LZ(Y) - LZ(\hat{Y}_{X_1})}(LZ(Y) - LZ(\hat{Y}_{X_1}))\\
&= LZ(\hat{Y}_{X_1}) + LZ(\hat{Y}_{X_1.X_2}) - LZ(\hat{Y}_{X_1}) = LZ(\hat{Y}_{X_1.X_2}). \qquad \square
\end{aligned}
$$

The obtained partial Lorenz dependence measure, *RGI*, defines one of the possible partial contribution to the Y Lorenz zonoid, related to the addition of a new explanatory variable into the model. We now discuss about the statistical interpretation of our proposed dependence measures. We will consider an example that combines multiple linear regression with Lorenz zonoids theory. Suppose to consider data in Table 6.1.

Since covariates X_1 and X_2 have categorical nature, we assign numeric labels to all their ordered or non-ordered assumed categories. By resorting both to multiple linear regression model and Lorenz zonoid tools, one obtains $LZ(Y) \cong 0.245$, $LZ(\hat{Y}_{X_1}) \cong 0.051$, $LZ(\hat{Y}_{X_2}) \cong 0.008$ and $LZ(\hat{Y}_{X_1,X_2}) \cong 0.054$. Through (6.2) and (6.3), the following single partial dependence measures can be derived: $G_{Y|X_1} \cong 0.209$ and $G_{Y|X_2} \cong 0.032$. The Lorenz zonoid of $E(Y|X_1) = G_{Y|X_1}$ represents the 20.9 % of the Y Lorenz zonoid and the Lorenz zonoid of $E(Y|X_2) = G_{Y|X_2}$ represents the 3.2 % of the Y Lorenz zonoid. The relative measure describing the possible additional contribution of covariate X_1 to the Y Lorenz zonoid is $RGI_{Y.X_1|X_2} \cong 0.194$, meaning that the introduction of explanatory variable X_1 allows to increase the dilation of the $E(Y|X_2)$ Lorenz zonoid in measure equivalent to 19.4 %. Finally, the relative measure describing the possible additional contribution of covariate X_2 to the Y Lorenz zonoid is $RGI_{Y.X_2|X_1} \cong 0.013$, meaning that the introduction of covariate X_2 into the model allows to increase the dilation of the $E(Y|X_1)$ Lorenz zonoid in measure equivalent to 1.3 %. Thus, we can conclude that an increase of the dilation measure implies a reduction of the unexplained response variable variability. The aforementioned example allows to highlight how our proposal represents an improvement of the classical model selection criteria, such as the AIC or BIC criterion, with regard to a quali–quantitative context. In fact, if we apply the AIC criterion, the stepwise forward procedure provides no covariates being relevant in explaining the response variable variability. On the contrary, our *RGI* measure denotes a substantial contribution of covariate X_1 to the response variable variability explanation. In fact, the X_1 covariate introduction allows to explain the 19.4 % of the total response variable variability.

6.3 Conclusions

In this research contribution we showed how Lorenz zonoids can be usefully employed to verify statistical dependence relationships among the involved variables.

The relevant feature of our proposal is provided by its adequacy in special cases of variables expressed through different measure scales. In fact, it assures a consistent standardization: this topic is very interesting in a model choice perspective, as illustrated in Sect. 6.2.2.

Our approach presents similarities with the R^2-based approach: indeed, both methods are built on a quantitative response variable. Furthermore, our contribution satisfies the property of invariance with respect to scale transformations of variables: for this reason, it can be intended as an alternative to the linear correlation measure in a quali–quantitative context.

According to all the previously discussed issues, we conclude that our novel dependence measure can assume also the role of measure of fit in cases of mainly categorical covariates. In fact, as provided, it allows to overcome the standard dependence measures restrictions when the relevant considered explanatory variables are not specified through a metric scale.

Acknowledgements A special acknowledge goes to referees for their helpful comments and suggestions.

References

1. Agresti, A.: Categorical Data Analysis. Wiley, Chichester (2002)
2. Billingsley, P.: Probability and Measure, 3rd edn. Wiley, Chichester (1995)
3. Dall'Aglio, M., Scarsini, M.: Zonoids, linear dependence, and size-biased distributions on the simplex. Adv. Appl. Probab. 35 (2003)
4. Giudici, P., Figini, S.: Applied Data Mining for Business and Industry, John Wiley & Sons, Second Edition (2009)
5. Giudici, P., Raffinetti, E.: On the Gini measure decomposition. Stat. Probab. Lett. **81**(1), 133–139 (2011)
6. Koshevoy, G.: Multivariate Lorenz majorization. Soc. Choice Welfare 12 (1995)
7. Koshevoy, G., Mosler, K.: The Lorenz zonoids of a multivariate distribution. J. Am. Stat. Assoc. **91**(434) (1996)
8. Koshevoy, G., Mosler, K.: Multivariate Lorenz dominance based on zonoids. AStA Adv. Stat. Anal. **91**(1), 57–76 (2007)
9. Mosler, K.: Majorization in economic disparity measures. Lin. Algebra Appl. **220** (1994)
10. Muliere, P.: Some remarks about the horizontal equity of a taxation (in Italian). Comunicazione **2**, (1986), Milano, ed. by Bocconi
11. Muliere, P., Petrone, S.: Generalized Lorenz curve and monotone dependence orderings. Metron **L**(3-4) (1992)
12. Raffinetti, E.: Multivariate Dependence Measures through Lorenz Curves and their Generalization. Ph.D. Thesis, Chapter 4, Università L, Bocconi (2011)
13. Raffinetti, E., Giudici, P.: Multivariate Ranks-based concordance indexes. In: Di Ciaccio, A., Coli, M., Angulo I., Jose M. (eds.) Advanced Statistical Methods for the Analysis of Large Data-Sets, "Studies in Theoretical and Applied Statistics" series. Springer, Berlin Heidelberg (2012)

Algebraic Generation of Orthogonal Fractional Factorial Designs

7

Roberto Fontana and Giovanni Pistone

Abstract

The joint use of counting functions, Hilbert basis, and Markov basis allows to define a procedure to generate all the fractional factorial designs that satisfy a given set of constraints in terms of orthogonality [Fontana, Pistone and Rogantin (JSPI, 2000), Pistone and Rogantin (JSPI, 2008)]. The general case of mixed level designs without restrictions on the number of levels of each factor (such as power of prime number) is studied. The generation problem is reduced to finding positive integer solutions of a linear system of equations [e.g., Carlini and Pistone (JSTP, 2007)]. This new methodology has been experimented on some significant classes of fractional factorial designs, including mixed level orthogonal arrays and sudoku designs [Fontana and Rogantin in Algebraic and Geometric Methods in Statistics, CUP (2009)]. For smaller cases the complete generating set of all the solutions can be computed. For larger cases we resort to the random generation of a sample solution.

Keywords

Algebraic statistics • Counting function • Design of experiments • Hilbert basis • Indicator polynomial • Markov basis

---

R. Fontana (✉)
DISMA Politecnico di Torino, Corso Duca degli Abruzzi 24, 10129 Torino, Italy
e-mail: roberto.fontana@polito.it

G. Pistone
de Castro Statistics Initiative, Collegio Carlo Alberto, 10024 Moncalieri, Italy
e-mail: giovanni.pistone@carloalberto.org

N. Torelli et al. (eds.), *Advances in Theoretical and Applied Statistics*,
Studies in Theoretical and Applied Statistics, DOI 10.1007/978-3-642-35588-2_7,

7.1 Introduction

The main result of this chapter is discussed in Sect. 7.3 where the problem of finding fractional factorial designs that satisfy a set of orthogonality conditions is translated into the problem of finding nonnegative integer solutions to a system of linear equations, so avoiding computations with complex numbers. Fractional factorial designs that satisfy a set of conditions in terms of orthogonality between factors have been described as the zero-set of a system of polynomial equations whose indeterminates are the complex coefficients of the counting polynomial functions, [7] and [4], see [5] for a short review. In Sect. 7.4, we use the software [10] to find the generators of classes of orthogonal arrays. Finally, in Sect. 7.5 we consider the problem of randomly sampling one fraction from a given class of orthogonality. Two simulation methods are considered, Simulated Annealing and Markov Chain Monte Carlo.

7.2 Full Factorial Design and Fractions of a Full Factorial Design

We recall notations and results from [7]:

- $\mathscr{D}_j$ is a *factor* with n_j levels coded with the n_j-th roots of the unity, $\mathscr{D}_j = \{\omega_0^{(n_j)}, \ldots, \omega_{n_j-1}^{(n_j)}\}$, $\omega_k^{(n_j)} = \exp\left(\sqrt{-1}\,\frac{2\pi}{n_j}\,k\right)$; $\mathscr{D} = \mathscr{D}_1 \times \cdots \mathscr{D}_j \cdots \times \mathscr{D}_m$ is the *full factorial design* with *complex coding* and $\#\mathscr{D}$ is its cardinality.
- X_j is the j-th component function, which maps a point to its j-th component, $X_j : \mathscr{D} \ni (\zeta_1, \ldots, \zeta_m) \longmapsto \zeta_j \in \mathscr{D}_j$; the function X_j is called *simple term* or, by abuse of terminology, *factor*. The *interaction term* is $X^\alpha = X_1^{\alpha_1} \cdots X_m^{\alpha_m}$, $\alpha \in L = \mathbb{Z}_{n_1} \times \cdots \times \mathbb{Z}_{n_m}$, i.e., the monomial function $X^\alpha : \mathscr{D} \ni (\zeta_1, \ldots, \zeta_m) \mapsto \zeta_1^{\alpha_1} \cdots \zeta_m^{\alpha_m}$.

We underline that L is both the full factorial design with integer coding and *the exponent set of all the simple factors and interaction terms* and α is both a treatment combination in the integer coding and a multi-exponent of an interaction term. These identifications make the complex coding especially simple.

A fraction $\mathscr{F}$ is a multiset $(\mathscr{F}_*, f_*)$ whose underlying set of elements $\mathscr{F}_*$ is contained in $\mathscr{D}$ and f_* is the multiplicity function $f_* : \mathscr{F}_* \rightarrow \mathbb{N}$ that for each element in $\mathscr{F}_*$ gives the number of times it belongs to the multiset $\mathscr{F}$.

Definition 1. If f is a $\mathbb{C}$-valued polynomial function defined on $\mathscr{F}$, briefly a response, then its *mean value on* $\mathscr{F}$ is $E_{\mathscr{F}}(f) = \frac{1}{\#\mathscr{F}} \sum_{\zeta \in \mathscr{F}} f(\zeta)$, where $\#\mathscr{F}$ is the total number of treatment combinations of the fraction. A response f is *centered* if $E_{\mathscr{F}}(f) = 0$. Two responses f and g are *orthogonal on* $\mathscr{F}$ if $E_{\mathscr{F}}(f\,\overline{g}) = 0$.

Remark 1. It should be noted that $\sum_{\zeta \in \mathscr{F}} f(\zeta)$ means $\sum_{\zeta \in \mathscr{F}_*} f_*(\zeta) f(\zeta)$.

With the complex coding the vector orthogonality of two interaction terms X^α and X^β, with respect to the Hermitian product $f \cdot g = E_{\mathscr{F}}(f\,\overline{g})$, corresponds to the combinatorial orthogonality as specified in Proposition 6. We consider the general case in which fractions can contain points that are replicated.

Definition 2. The *counting function* R of a fraction $\mathcal{F}$ is a complex polynomial defined on $\mathcal{D}$ so that for each $\zeta \in \mathcal{D}$, $R(\zeta)$ equals the number of appearances of ζ in the fraction. A $0-1$ valued counting function is called *indicator function* of a single replicate fraction $\mathcal{F}$. We denote by c_α the coefficients of the representation of R on $\mathcal{D}$ using the monomial basis $\{X^\alpha,\ \alpha \in L\}$: $R(\zeta) = \sum_{\alpha \in L} c_\alpha X^\alpha(\zeta),\ \zeta \in \mathcal{D},\ c_\alpha \in \mathbb{C}$.

Given a fraction $\mathcal{F} \equiv (\mathcal{F}_*, f_*)$, the relationship between f_* and its counting function R is

$$R(\zeta) = \begin{cases} f_*(\zeta) & \zeta \in \mathcal{F}_* \\ 0 & \text{elsewhere} \end{cases}$$

Proposition 1. *If $\mathcal{F}$ is a fraction of a full factorial design $\mathcal{D}$, $R = \sum_{\alpha \in L} c_\alpha X^\alpha$ is its counting function and $[\alpha - \beta]$ is the m-tuple made by the componentwise difference in the ring $\mathbb{Z}_{n_j}$, $\left([\alpha_1 - \beta_1]_{n_1}, \dots, [\alpha_j - \beta_j]_{n_j}, \dots, [\alpha_m - \beta_m]_{n_m}\right)$, then*

1. *The coefficients c_α are given by $c_\alpha = \frac{1}{\#\mathcal{D}} \sum_{\zeta \in \mathcal{F}} \overline{X^\alpha(\zeta)}$*
2. *The term X^α is centered on $\mathcal{F}$ if, and only if, $c_\alpha = c_{[-\alpha]} = 0$*
3. *The terms X^α and X^β are orthogonal on $\mathcal{F}$ if, and only if, $c_{[\alpha-\beta]} = 0$*

We now define projectivity and its relation with orthogonal arrays.

Definition 3. A fraction $\mathcal{F}$ *factorially projects* onto the I-factors, $I \subset \{1, \dots, m\}$, if the projection is a multiple full factorial design, i.e., a full factorial design where each point appears equally often. A fraction $\mathcal{F}$ is a *mixed orthogonal array* of strength t if it factorially projects onto any I-factors with $\#I = t$.

Proposition 2. *A fraction is an* orthogonal array of strength t *if, and only if, all the coefficients c_α of the counting function up to the order t are zero.*

7.3 Counting Functions and Strata

It follows from Proposition 1 and Proposition 2 that the problem of finding fractional factorial designs that satisfy a set of conditions in terms of orthogonality between factors can be written as a polynomial system in which the indeterminates are the complex coefficients c_α of the counting polynomial function.

Let us now introduce a different way to describe the full factorial design $\mathcal{D}$ and all its subsets. We consider the indicator functions 1_ζ of all the single points of $\mathcal{D}$. The counting function R of a fraction $\mathcal{F}$ can be written as $\sum_{\zeta \in \mathcal{D}} y_\zeta 1_\zeta$ with $y_\zeta \equiv R(\zeta) \in \{0, 1, \dots, n, \dots\}$. The particular case in which R is an indicator function corresponds to $y_\zeta \in \{0,1\}$. From Proposition 1 we obtain that the values of the counting function over $\mathcal{D}$, y_ζ, are related to the coefficients c_α by $c_\alpha = \frac{1}{\#\mathcal{D}} \sum_{\zeta \in \mathcal{D}} y_\zeta \overline{X^\alpha(\zeta)}$. As described in Sect. 7.2, we consider m factors, $\mathcal{D}_1, \ldots, \mathcal{D}_m$ where $\mathcal{D}_j \equiv \Omega_{n_j} = \{\omega_0^{(n_j)}, \ldots, \omega_{n_j-1}^{(n_j)}\}$, for $j = 1, \ldots, m$. From [7], we recall two basic properties which hold true for the full design $\mathcal{D}$.

Proposition 3. *Let X_j be the simple term with level set $\mathcal{D}_j = \Omega_{n_j} = \{\omega_0^{(n_j)}, \ldots, \omega_{n_j-1}^{(n_j)}\}$. Over $\mathcal{D}$, the term X_j^r takes all the values of Ω_{s_j} equally often, where $s_j = 1$ if $r = 0$ and $s_j = n_j / \gcd(r, n_j)$ if $r > 0$.*

Proposition 4. *Let $X^\alpha = X_1^{\alpha_1} \cdots X_m^{\alpha_m}$ be an interaction. $X_i^{\alpha_i}$ takes values in Ω_{s_i} where s_i is determined according to the previous Proposition 3. Over $\mathcal{D}$, the term X^α takes all the values of Ω_s equally often, where $s = \operatorname{lcm}(s_1, \ldots, s_m)$.*

Let us now define the strata that are associated with simple and interaction terms.

Definition 4. Given a term $X^\alpha, \alpha \in L = \mathbb{Z}_{n_1} \times \ldots \times \mathbb{Z}_{n_m}$, the full design $\mathcal{D}$ is partitioned into the strata $D_h^\alpha = \left\{\zeta \in \mathcal{D} : \overline{X^\alpha(\zeta)} = \omega_h^{(s)}\right\}$, where $\omega_h^{(s)} \in \Omega_s$ and s is determined according to the previous Propositions 3 and 4.

We use $n_{\alpha,h}$ to denote the number of points of the fraction $\mathcal{F}$ that are in the stratum D_h^α, $n_{\alpha,h} = \sum_{\zeta \in D_h^\alpha} y_\zeta$, $h = 0, \ldots, s-1$,. The following Proposition 5 (see [3] for proof) links the coefficients c_α with $n_{\alpha,h}$.

Proposition 5. *Let $\mathcal{F}$ be a fraction of $\mathcal{D}$ with counting fraction $R = \sum_{\alpha \in L} c_\alpha X^\alpha$. Each $c_\alpha, \alpha \in L$, depends on $n_{\alpha,h}, h = 0, \ldots, s-1$, as $c_\alpha = \frac{1}{\#\mathcal{D}} \sum_{h=0}^{s-1} n_{\alpha,h} \omega_h^{(s)}$, where s is determined by X^α (see Proposition 4).*

We now use a part of Proposition 3 of [7] to get conditions on $n_{\alpha,h}$ that makes X^α centered on the fraction $\mathcal{F}$.

Proposition 6. *Let X^α be a term with level set Ω_s on full design $\mathcal{D}$. Let $P(\zeta)$ be the complex polynomial associated with the sequence $(n_{\alpha,h})_{h=0,\ldots,s-1}$ so that $P(\zeta) = \sum_{h=0}^{s-1} n_{\alpha,h} \zeta^h$ and Φ_s the cyclotomic polynomial of the s-roots of the unity.*

1. *Let s be prime. The term X^α is centered on the fraction $\mathcal{F}$ if, and only if, its s levels appear equally often: $n_{\alpha,0} = n_{\alpha,1} = \ldots = n_{\alpha,s-1} = \lambda_\alpha$.*
2. *Let $s = p_1^{h_1} \ldots p_d^{h_d}$, p_i prime, $i = 1, \ldots, d$. The term X^α is centered on the fraction $\mathcal{F}$ if, and only if, the remainder $H(\zeta) = P(\zeta) \bmod \Phi_s(\zeta)$, whose coefficients are integer linear combinations of $n_{\alpha,h}, h = 0, \ldots, s-1$, is identically zero.*

We observe that, being D_h^α a partition of $\mathcal{D}$, if s is prime, we get $\lambda_\alpha = \frac{\#\mathcal{F}}{s}$.

If we remind that $n_{\alpha,h}$ are related to the values of the counting function R of a fraction $\mathscr{F}$ by $n_{\alpha,h} = \sum_{\zeta \in D_h^\alpha} y_\zeta$, this Proposition 6 allows to express the condition X^α *is centered on* $\mathscr{F}$ as integer linear combinations of the values $R(\zeta)$ of the counting function over the full design $\mathscr{D}$. In Sect. 7.4, we will show the use of this property to generate fractional factorial designs.

7.4 Generation of Fractions

We use strata to generate fractions that satisfy a given set of constrains on the coefficients of their counting functions. Formally, we give the following definition.

Definition 5. Given $\mathscr{C} \subseteq \mathbb{Z}_{n_1} \times \ldots \mathbb{Z}_{n_m}$, a counting function $R = \sum_\alpha c_\alpha X^\alpha$ associated with $\mathscr{F}$ is a $\mathscr{C}$-compatible counting function if $c_\alpha = 0, \quad \forall \alpha \in \mathscr{C}$.

We will denote by $OF(n_1 \ldots n_m, \mathscr{C})$ the set of all the fractions of $\mathscr{D}$ whose counting functions are $\mathscr{C}$-compatible. In the next sections, we will show our methodology on Orthogonal Arrays (other examples are in [3]). Let us consider $OA(n, s^m, t)$, i.e., orthogonal arrays with n rows and m columns where each column has s symbols, s prime, and with strength t. Using Proposition 2 we have that the coefficients of the corresponding counting functions must satisfy the conditions $c_\alpha = 0$ for all $\alpha \in \mathscr{C}$ where $\mathscr{C} \subseteq L = \{\alpha : 0 < \|\alpha\| \leq t\}$ and $\|\alpha\|$ is the number of non-null elements of α. It follows that $OF(s^m, \mathscr{C}) = \bigcup_n OA(n, s^m, t)$. Now using Proposition 6, we can express these conditions using strata. If we consider $\alpha \in \mathscr{C}$ we write the condition $c_\alpha = 0$ as $\sum_{\zeta \in D_h^\alpha} y_\zeta = \lambda, h = 0, \ldots, s-1$. To obtain all the conditions it is enough to vary $\alpha \in \mathscr{C}$. We therefore get the system of linear equations $AY = \lambda \underline{1}$ where A is the $(\#\mathscr{C} \times s^m)$ matrix whose rows contains the values, over $\mathscr{D}$, of the indicator function of the strata, $1_{D_h^\alpha}$, Y is the s^m column vector whose entries are the values of the counting function over $\mathscr{D}$, λ will be equal to $\frac{\#\mathscr{F}}{s}$ and $\underline{1}$ is the s^m column vector whose entries are all equal to 1. We can write an equivalent homogeneous system if we consider λ as a new variable. We obtain $\tilde{A}\tilde{Y} = 0$ where

$$\tilde{A} = \left[\begin{array}{c|c} A & \begin{array}{c} -1 \\ \ldots \\ -1 \end{array} \end{array} \right] = [A, -\underline{1}] \text{ and } \tilde{Y} = \begin{bmatrix} Y \\ \lambda \end{bmatrix} = (Y, \lambda)$$

It is now immediate to verify that the sum $Y = Y_1 + Y_2$ of two Orthogonal Arrays, $Y_1 \in OA(n_1, s^m, t)$ and $Y_2 \in OA(n_2, s^m, t)$, is an Orthogonal Array $OA(n_1 + n_2, s^m, t)$. The Hilbert Basis [9] is a minimal set of generators such that any $OA(n, s^m, t)$ becomes a linear combination of the generators with positive or null integer coefficients. This approach extends that of [1] where the conditions $c_\alpha = \frac{1}{\#\mathscr{D}} \sum_{\zeta \in \mathscr{F}} \overline{X^\alpha(\zeta)} = 0$ were used. The advantage of using strata is that we avoid computations with complex numbers $(\overline{X^\alpha(\zeta)})$. We explain this point in a couple of examples. For the computation we use 4ti2 [10].

- $OA(n, 2^5, 2)$ were investigated in [1]. We build the matrix $\tilde{A}$ that has 30 rows and 33 columns. We find the same 26,142 solutions as in the cited paper.
- For $OA(n, 3^3, 2)$ we build the matrix $\tilde{A}$ that has 54 rows and 28 columns. We find 66 solutions, 12 have 9 points, all different and 54 have 18 points, one replicated, i.e., support equal to 17.

Let us now consider the general case in which we do not put restrictions on the number of levels. We show our method for $OA(n, 4^2, 1)$. In this case the number of levels is a power of a prime, $4 = 2^2$. Using Proposition 2 we have that the coefficients of the corresponding counting functions must satisfy the conditions $c_\alpha = 0$ for all $\alpha \in \mathscr{C}$ where $\mathscr{C} \subseteq L = \{\alpha : \|\alpha\| = 1\}$. Let us consider $c_{1,0}$. From Proposition 3 we have that X_1 takes the values in Ω_s where $s = 4$. From Proposition 6, X_1 will be centered on $\mathscr{F}$ if, and only if, the remainder $H(\zeta) = P(\zeta) \bmod \Phi_4(\zeta)$ is identically zero. We have $\Phi_4(\zeta) = 1 + \zeta^2$ (see [6]) and so we can compute the remainder $H(\zeta) = n_{(1,0),0} - n_{(1,0),2} + (n_{(1,0),1} - n_{(1,0),3})\zeta$. The condition that $H(\zeta)$ must be identically zero translates into $n_{(1,0),0} - n_{(1,0),2} = 0$ and $n_{(1,0),1} - n_{(1,0),3} = 0$. Let us now consider $c_{2,0}$. From Proposition 3 we have that X_1^2 takes the values in Ω_s where $s = 2$. From Proposition 6, X_1^2 will be centered on $\mathscr{F}$ if, and only if, the remainder $H(\zeta) = P(\zeta) \bmod \Phi_2(\zeta)$ is identically zero. We have $\Phi_2(\zeta) = 1 + \zeta$ (see [6]) and so we can compute the remainder $H(\zeta) = n_{(2,0),0} - n_{(2,0),1}$.

If we repeat the same procedure for all the α such that $\|\alpha\| = 1$ and we recall that $n_{\alpha,h} = \sum_{\zeta \in D_h^\alpha} y_\zeta$, the orthogonal arrays $OA(n, 4^2, 1)$ become the positive integer solutions of the integer linear homogeneous system $AY = \mathbf{0}$, where

$$A = \begin{bmatrix} 1 & 0 & -1 & 0 & 1 & 0 & -1 & 0 & 1 & 0 & -1 & 0 & 1 & 0 & -1 & 0 \\ 0 & 1 & 0 & -1 & 0 & 1 & 0 & -1 & 0 & 1 & 0 & -1 & 0 & 1 & 0 & -1 \\ 1 & -1 & 1 & -1 & 1 & -1 & 1 & -1 & 1 & -1 & 1 & -1 & 1 & -1 & 1 & -1 \\ 1 & 0 & -1 & 0 & 1 & 0 & -1 & 0 & 1 & 0 & -1 & 0 & 1 & 0 & -1 & 0 \\ 0 & -1 & 0 & 1 & 0 & -1 & 0 & 1 & 0 & -1 & 0 & 1 & 0 & -1 & 0 & 1 \\ 1 & 1 & 1 & 1 & 0 & 0 & 0 & 0 & -1 & -1 & -1 & -1 & 0 & 0 & 0 & 0 \\ 0 & 0 & 0 & 0 & 1 & 1 & 1 & 1 & 0 & 0 & 0 & 0 & -1 & -1 & -1 & -1 \\ 1 & 1 & 1 & 1 & -1 & -1 & -1 & -1 & 1 & 1 & 1 & 1 & -1 & -1 & -1 & -1 \\ 1 & 1 & 1 & 1 & 0 & 0 & 0 & 0 & -1 & -1 & -1 & -1 & 0 & 0 & 0 & 0 \\ 0 & 0 & 0 & 0 & -1 & -1 & -1 & -1 & 0 & 0 & 0 & 0 & 1 & 1 & 1 & 1 \end{bmatrix}, \quad Y = \begin{bmatrix} y_{00} \\ y_{10} \\ y_{20} \\ y_{30} \\ y_{01} \\ y_{11} \\ y_{21} \\ y_{31} \\ y_{02} \\ y_{12} \\ y_{22} \\ y_{32} \\ y_{03} \\ y_{13} \\ y_{23} \\ y_{33} \end{bmatrix}$$

and $\mathbf{0}$ is the 10-rows column vector whose values are all equal to 0.

It should be noted that the matrix of the coefficients is not full rank, e.g., the first and the fourth rows are equal. This aspect is discussed in [3]. Anyhow the solution method used here does not require a reduction to a full rank matrix. Using 4ti2 we find 24 solutions that correspond to all the Latin Hypercupe Designs (LHD). Analogously for $OA(n, 6^2, 1)$ we find 620 solutions that correspond to all the LHD.

We conclude this section with three examples of mixed orthogonal arrays, $OA(n, 2 \cdot 6, 1)$, $OA(n, 2 \cdot 3 \cdot 4, 2)$ and $OA(n, 2 \cdot 3 \cdot 6, 1)$.

In the first case we have $\mathscr{C} \subseteq \mathbb{Z}_2 \times \mathbb{Z}_6 = \{(1,0), (0,1), (0,2), (0,3), (0,4), (0,5)\}$. As before, for each $\alpha \in \mathscr{C}$ we consider X^α. X^α takes the values in Ω_{s_α}, where s_α is determined using Proposition 3. From Proposition 6, X^α will be centered on $\mathscr{F}$ if, and only if, the remainder $H_\alpha(\zeta) = P_\alpha(\zeta) \bmod \Phi_{s_\alpha}(\zeta)$, where $P_\alpha(\zeta) = \sum_{h=0}^{s_\alpha - 1} n_{\alpha,h} X^\alpha(\zeta)$, is identically zero. We obtain that the orthogonal arrays $OA(n, 2 \cdot 6, 1)$ become the positive integer solutions of the integer linear homogenous system $AY = \mathbf{0}$ where

$$A = \begin{bmatrix} 0 & -1 & -1 & 0 & 1 & 1 & 0 & -1 & -1 & 0 & 1 & 1 \\ 1 & 1 & 0 & -1 & -1 & 0 & 1 & 1 & 0 & -1 & -1 & 0 \\ 0 & -1 & 1 & 0 & -1 & 1 & 0 & -1 & 1 & 0 & -1 & 1 \\ 1 & -1 & 0 & 1 & -1 & 0 & 1 & -1 & 0 & 1 & -1 & 0 \\ 1 & -1 & 1 & -1 & 1 & -1 & 1 & -1 & 1 & -1 & 1 & -1 \\ 0 & 1 & -1 & 0 & 1 & -1 & 0 & 1 & -1 & 0 & 1 & -1 \\ 1 & 0 & -1 & 1 & 0 & -1 & 1 & 0 & -1 & 1 & 0 & -1 \\ 0 & 1 & 1 & 0 & -1 & -1 & 0 & 1 & 1 & 0 & -1 & -1 \\ 1 & 0 & -1 & -1 & 0 & 1 & 1 & 0 & -1 & -1 & 0 & 1 \\ 1 & 1 & 1 & 1 & 1 & 1 & -1 & -1 & -1 & -1 & -1 & -1 \end{bmatrix}, \quad Y = \begin{bmatrix} y_{00} \\ y_{01} \\ y_{02} \\ y_{03} \\ y_{04} \\ y_{05} \\ y_{10} \\ y_{11} \\ y_{12} \\ y_{13} \\ y_{14} \\ y_{15} \end{bmatrix}$$

and $\mathbf{0}$ is the 10-rows column vector whose values are all equal to 0. Using 4ti2 we obtain 20 solutions that correspond to all the LHD.

Now we consider $OA(n, 2 \cdot 3 \cdot 4, 2)$. We have $\mathscr{C} \subseteq \mathbb{Z}_2 \times \mathbb{Z}_3 \times \mathbb{Z}_4 = \{\alpha : \alpha \neq (0,0,0) \text{ and } \|\alpha\| \leq 2\}$. By using the above procedure we obtain the matrix of the coefficients A that has 74 rows and 48 columns. Then using 4ti2 we find 1,860 solutions, all with solutions, all with 24 points, all different.

Finally we consider $OA(n, 2 \cdot 3 \cdot 6, 1)$. We find 117,360 solutions that, in Table 7.1, are classified with respect to cardinality of support, total number of points, and maximum number of replication.

7.4.1 Dual Characterization of $\mathscr{C}$-Compatible Counting Functions

We conclude this section with a new result. We observe that, given $\mathscr{C}$ and using the natural isomorphisms $\phi_i, i = 1, \ldots, m$ between $\mathbb{Z}_{n_i} \ni \alpha_i \mapsto \phi_i(\alpha_i) = \omega_{\alpha_i}^{(n_i)} \in \Omega_{n_i}$, we can define $\mathscr{C}^\star = \phi(\mathscr{C})$, where $\phi = \phi_1 \times \ldots \times \phi_n$.

Proposition 7. *Given $\mathscr{C} \subseteq \mathbb{Z}_{n_1} \times \ldots \mathbb{Z}_{n_m}$, a counting function $R = \sum_\alpha c_\alpha X^\alpha$ associated with $\mathscr{F}$ is a $\mathscr{C}$-compatible counting function if, and only if,*

$$\sum_{\alpha \in L} y_\alpha \overline{X^\alpha(\omega_\gamma)} = 0 \quad \omega_\gamma \in \mathscr{C}^\star$$

where $y_\alpha = R(\phi(\alpha))$ and $\omega_\gamma = (\omega_{\gamma_1}^{(n_1)}, \ldots, \omega_{\gamma_m}^{(n_m)})$.

Table 7.1 Support (supp), total, maximum replication (maxrep), for the Hilbert basis of the $OA(n, 2 \cdot 3 \cdot 6, 1)$

Supp.	Total	Maxrep	N
6	6	1	1,800
8	12	2	6,480
9	12	2	25,920
9	18	3	8,640
10	12	2	30,960
10	18	3	8,640
11	12	2	20,160
11	18	3	6,480
12	12	1	6,120
12	18	2	2,160

Proof. Let us consider $\gamma \in \mathscr{C} \subseteq \mathbb{Z}_{n_1} \times \ldots \mathbb{Z}_{n_m}$. We know from Proposition 1 that

$$c_\gamma = \frac{1}{\#\mathscr{D}} \sum_{\zeta \in \mathscr{F}} \overline{X^\gamma(\zeta)}$$

or, equivalently,

$$c_\gamma = \frac{1}{\#\mathscr{D}} \sum_{\zeta \in \mathscr{D}} R(\zeta)\overline{X^\gamma(\zeta)}$$

Using the isomorphism ϕ we get

$$c_\gamma = \frac{1}{\#\mathscr{D}} \sum_{\alpha \in L} R(\phi(\alpha))\overline{X^\gamma(\phi(\alpha))} = \frac{1}{\#\mathscr{D}} \sum_{\alpha \in L} y_\alpha \overline{X^\alpha(\phi(\gamma))} = \frac{1}{\#\mathscr{D}} \sum_{\alpha \in L} y_\alpha \overline{X^\alpha(\omega_\gamma)}$$

with $y_\alpha = R(\phi(\alpha))$ and $\omega_\gamma = (\omega_{\gamma_1}^{(n_1)}, \ldots, \omega_{\gamma_m}^{(n_m)})$. From the definition of $\mathscr{C}$-compatible counting function (Definition 5), we complete the proof.

This Proposition highlights an algebraic condition for a $\mathscr{C}$-compatible counting function. If we denote by Ideal $(\mathscr{C}^\star)$ the ideal of the polynomials that vanishes over $\mathscr{C}^\star$, we have that, if y_α represent the values of a $\mathscr{C}$-compatible counting function, the polynomial $\sum_{\alpha \in L} y_\alpha \overline{X^\alpha}$ must belong to Ideal $(\mathscr{C}^\star)$.

7.5 Sampling

Sometimes, given a set of conditions $\mathscr{C}$ we are interested in picking up a solution more than in finding all the generators. The basic idea is to generate somehow a starting solution and then to randomly walk in the set of all the solutions for a certain number of steps, taking the arrival point as a new but still $\mathscr{C}$-compatible counting function. We can combine the previous results on strata with Markov Chain Monte

Carlo Methods to sample one solution. We show the methods on some examples with indicator functions but it can be extended to counting functions.

Let us consider $OA(n, 3^3, 2)$ and let us suppose that we are searching for an orthogonal array with nine design points and no replications. It means that we are interested in an indicator function whose values $y_\zeta, \zeta \in \mathcal{D}$ satisfy the system of linear equation $AY = \underline{1}$ where A, Y and $\underline{1}$ have been defined as in see Sect. 7.4.

We now use standard simulated annealing to find one solution of our system [8]. We define the objective function to be maximized as the function that, for every indicator function defined over the design $\mathcal{D}$, counts the number of equations of the linear system $AY = \underline{1}$ that are satisfied. We have implemented this algorithm using SAS/IML. We found a solution in 2,702 iterations (a couple of seconds on a common laptop). We have also tried the algorithm on $OA(9, 3^4; 3)$ and 4×4 sudoku (we found one solution in 2,895 and 2,852 iterations respectively) and 9×9 sudoku where we did not find any solution in 100,000 iterations.

We observe that the algorithm can also be used to explore the set of solutions simply replacing *stop when an optimal solution is found* with *store the optimal solution and continue until the maximum number of iterations is reached*. Let us now use the previous results on strata to get a suitable set of *moves*. We will show this procedure in the case in which all the factors have the same number of levels s, s prime, but it can also be applied to the general case. In Sect. 7.4 we have shown that counting functions must satisfy the system of linear equations $AY = \lambda\underline{1}$, where A corresponds to the set of conditions $\mathcal{C}$ written in terms of strata.

It follows that if, given a $\mathcal{C}$-compatible solution Y, such that $AY = \lambda\underline{1}$, we search for an additive move X such that $A(Y + X)$ is still equal to $\lambda\underline{1}$, we have to solve the linear homogenous system $AX = 0$, with $X = (x_\zeta), \zeta \in \mathcal{D}$, $x_\zeta \in \mathbb{Z}$ and $y_\zeta + x_\zeta \geq 0$ for all $\zeta \in \mathcal{D}$. We observe that this set of conditions allows to determine new $\mathcal{C}$-compatible solutions *that give the same* λ. We know that $\lambda = \frac{\#\mathcal{F}}{s}$ so this homogenous system determines moves that *do not change the dimension of the solutions*.

Let us now consider the extended homogeneous system, where $\tilde{A}$ has already been defined in Sect. 7.4, $\tilde{A}\tilde{X} = 0$ with $\tilde{X} = (\tilde{x}_\zeta), \zeta \in \mathcal{D}$, $\tilde{x}_\zeta \in \mathbb{Z}$ and $\tilde{y}_\zeta + \tilde{x}_\zeta \geq 0$ for all $\zeta \in \mathcal{D}$. Given $\tilde{Y} = (Y, \lambda_Y)$, where Y is $\mathcal{C}$-compatible counting function and $\lambda_Y = \frac{\sum_\zeta y_\zeta}{s}$, the solutions of $\tilde{A}\tilde{X} = 0$ determine all the other $\tilde{Y} + \tilde{X} = (Y + X, \lambda_{Y+X})$ such that $\tilde{A}(\tilde{Y} + \tilde{X}) = 0$. $Y + X$ are $\mathcal{C}$-compatible counting functions whose sizes, $s\lambda_{Y+X}$, are, in general, *different from that of* Y. We use the theory of Markov basis (see for example [2] where it is also available a rich bibliography on this subject) to determine a set of generators of the moves. We use the following procedure in order to randomly select a $\mathcal{C}$-compatible counting function. We compute a Markov basis of $\ker(A)$ using 4ti2 [10]. Once we have determined the Markov basis of $\ker(A)$, we make a random walk on the *fiber* of Y, where Y, as usual, contains the values of the counting function of an initial design $\mathcal{F}$. The fiber is made by all the $\mathcal{C}$-compatible counting functions that have the same size of $\mathcal{F}$. The random walk is done randomly choosing one move among the feasible ones, i.e., among the moves for which we do not get negative values for the new counting function. In the next paragraphs we consider moves for the cases that we have already studied in Sect. 7.4.

We consider $OA(8, 2^5, 2)$. We use the matrix A, already built in Sect. 7.4 and give it as input to 4ti2 to obtain the Markov Basis, that we denote by $\mathcal{M}$. It contains 5,538 different moves. As an initial fraction $\mathcal{F}_0$, we consider the eight-run regular fraction whose indicator function is $R_0 = \frac{1}{4}(1+X_1X_2X_3)(1+X_1X_4X_5)$. We obtain the set $\mathcal{M}^f_{R_0}$ of the feasible moves from R_0, selecting from $\mathcal{M}$ the moves M such that $R_0(\zeta) + M(\zeta) \geq 0 \ \forall \zeta \in \mathcal{D}$ or $R_0(\zeta) - M(\zeta) \geq 0 \ \forall \zeta \in \mathcal{D}$. We find 12 moves. We randomly choose one move, M_{R_0}, out of the 12 available ones and move to $R_1 = R_0 + \epsilon_{M_{R_0}} M_{R_0}$ where $\epsilon_{M_{R_0}}$ is the proper plus or minus sign. We run 1.000 simulations repeating the same loop, generating R_i as $R_i = R_{i-1} + \epsilon_{M_{R_{i-1}}} M_{R_{i-1}}$. We obtain all the 60 different 8-run fractions, each one with 8 different points as in [1].

We now consider $OA(9, 3^3, 2)$. As before, we use 4ti2 to generate the Markov basis $\mathcal{M}$. It contains 81 different moves. As an initial fraction we consider the nine-run regular fraction $\mathcal{F}_0$ whose indicator function is $R_0 = \frac{1}{3}(1 + X_1X_2X_3 + X_1^2X_2^2X_3^2)$. Running 1.000 simulations we obtain all the 12 different 9-run fractions, each one with 9 different points as known in the literature and as found in Sect. 7.4.

7.6 Conclusions

We considered mixed level fractional factorial designs. Given the counting function R of a fraction $\mathcal{F}$, we translated the constraint $c_\alpha = 0$, where c_α is a generic coefficient of its polynomial representation $R = \sum_\alpha c_\alpha X^\alpha$, into a set of linear constraints with integer coefficients on the values y_ζ that R takes on all the points $\zeta \in \mathcal{D}$. We obtained the set of generators of the solutions of some problems using Hilbert bases. We underline that this kind of algebraic methods requires the use of algorithms of very high combinatorial complexity and so they do not scale well to high dimension problems. This prompts for less demanding methodologies like those for randomly sampling one solution.

For this reason we also studied moves between fractions. We characterized these moves as the solution of a homogeneous linear system. We defined a procedure to randomly walk among the solutions that is based on the Markov basis of this system. We showed the procedure on some examples. Computations have been made using 4ti2 [10]. Main advantages of the method are that we do not put restrictions on the number of levels of factors and it is not necessary to use software that deals with complex polynomials. Main limit is in the high computational effort that is required. In particular, only a small part of the Markov basis is used because of the requirement that counting functions can only take values greater than or equal to zero. The possibility to generate only the moves that are feasible could make the entire process more efficient and is object of current research. The use of the dual characterization of $\mathcal{C}$-compatible counting functions (Proposition 7) is also under study.

This work is an extended version of our contributed paper to the 45th Scientific Meeting of the Italian Statistical Society (University of Padua, 16–18 June 2010). The authors thank Prof. Giorgio Vittadini for the suggestions he gave on that version.

References

1. Carlini, E., Pistone, G.: Hibert bases for orthogonal arrays. J. Stat. Theory Pract. **1**(3–4), 299–309 (2007)
2. Drton, M., Sturmfels, B., Sullivant, S.: Lectures on Algebraic Statistics. Oberwolfach Seminars, vol. 39. Springer, Birkhäuser (2009)
3. Fontana, R., Pistone, G.: Algebraic strata for non symmetrical orthogonal fractional factorial designs and application. La matematica e le sue applicazioni 2010/1, Politecnico di Torino DIMAT (2010)
4. Fontana, R., Pistone, G., Rogantin, M.: Classification of two-level factorial fractions. J. Stat. Plann. Infer. **87**(1), 149–172 (2000)
5. Fontana, R., Rogantin, M.P.: Indicator function and sudoku designs. In: Gibilisco, P., Riccomagno, E., Rogantin, M.P., Wynn, H.P. (eds.) Algebraic and Geometric Methods in Statistics, pp. 203–224. Cambridge University Press, Cambridge (2010)
6. Lang, S.: Algebra. Addison Wesley, Reading, MA (1965)
7. Pistone, G., Rogantin, M.: Indicator function and complex coding for mixed fractional factorial designs. J. Stat. Plann. Infer. **138**(3), 787–802 (2008)
8. Ross, S.M.: Simulation, 3rd edn. Statistical Modeling and Decision Science. Academic Press, San Diego, CA (2002)
9. Schrijver, A.: Theory of linear and integer programming. Wiley-Interscience Series in Discrete Mathematics. Wiley, Chichester (1986). A Wiley-Interscience Publication
10. 4ti2 team: 4ti2—a software package for algebraic, geometric and combinatorial problems on linear spaces. Available at www.4ti2.de

Part II

Methods for Time Series, Spatial and Functional Data

A Functional Spatio-Temporal Model for Geometric Shape Analysis

8

Lara Fontanella, Luigi Ippoliti, and Pasquale Valentini

Abstract

In this chapter we consider a functional spatio-temporal model for shape objects represented by landmark data. The model describes a time-varying deformation of the ambient space in which the objects of interest lie. The use of basis functions, defined by principal warps in space and time, facilitates both the model specification and the fitting of the data in Procrustes tangent coordinates. The fitted model can be interpreted either just in terms of the finite set of landmarks at the given set of time points, or in terms of a deformation of the space which varies continuously in time. The method is illustrated on a facial expression dataset.

Keywords

Bending energy matrix • Procrustes tangent coordinates • Principal warps • Shape analysis • Spatio-temporal models • Splines

8.1 Introduction

Spatio-temporal modelling has largely been developed through applications in geostatistics, hydrology and meteorology [6]. More recent activities in the area include environment monitoring, tracking, functional MRI, health data and facial analysis [9]. Motivated by these applications, various modelling strategies have been developed, and they essentially depend on the underlying objective of the analysis and the scale and type of data.

In this chapter we are interested in shape analysis applications where data are represented by landmark coordinates, and the aim is to construct a spatio-temporal model which is able to describe the time-varying deformation of the shape space.

L. Fontanella · L. Ippoliti (✉) · P. Valentini
University G. d'Annunzio, Chieti-Pescara, DEc, Viale Pindaro 42, 65127 Pescara, Italy
e-mail: ippoliti@unich.it

N. Torelli et al. (eds.), *Advances in Theoretical and Applied Statistics*, Studies in Theoretical and Applied Statistics, DOI 10.1007/978-3-642-35588-2_8,

For simplicity, we assume that the observed data take the form of a regular array in space-time. That is, we have one-dimensional observations

$$y_{ij}, i = 1, \ldots, M,\ j = 1, \ldots, T$$

at sites $\mathbf{x}_i \in \mathcal{R}^d$ and time points $t_j \in \mathcal{R}$. Typically, we consider objects in $d = 2$ dimensions but the methodology extends to higher dimensions. The objective is to model the data as

$$y_{ij} = z(\mathbf{x}_i, t_j) + e_{ij}, \tag{8.1}$$

where $\{z(\mathbf{x}, t)\}$ is a deterministic space-time process and e_{ij} an error term. We usually assume the error terms e_{ij} are independent, identically distributed. Once a smooth process $\hat{z}(\mathbf{x}, t)$ has been fitted, it can be used for interpolation. A key-property of spatio-temporal data is that observations at nearby sites and times will tend to be similar to one another. This underlying smoothness characteristic of the space-time process $z(\mathbf{x}, t)$ can be captured by using a finite-dimensional space of deterministic drift functions, or using autocorrelation to make nearby values similar. In principle, both of these approaches can be applied in space and/or time and for a discussion see, for example, Ippoliti et al. [12] and Lopes et al. [13]. Since the specification of spatio-temporal covariance functions is not obvious for shape data, our space-time model results from a tensor product of drift in times and drift in space. This modelling approach is simple and, for example, compared with the approach described in Kume and Welling [11], shows some computational advantages. The outline of this chapter is as follows. In Sect. 8.2 we describe how to represent the data in a form suitable for fitting the space-time model. In Sect. 8.3 we describe the modelling strategy. Specifically, roughness penalties for directions of variation in space and for rate of variation in time are constructed and combined to give a joint space-time penalty. Also, a transformation of the data is given which facilitates the fitting of these smoothing models. Section 8.4 discusses a parametric version of the model and illustrates its application on a set of landmarks representing the dynamic of four different facial expressions. Finally, Sect. 8.5 concludes the chapter with a discussion.

8.2 The Data

Suppose landmark data are available on different individuals at a common set of time points, taking the form of a 4-way array $\{v_{lkmj}\}$ where:

> $l = 1, \ldots, N$ labels different individuals; $k = 1, \ldots, K$ labels different landmarks; $s = 1, \ldots, d$ labels different coordinates; $j = 1, \ldots, T$ labels different times $t_1, \ldots, t_T$.

It is convenient to represent these data as a collection $\{\mathbf{v}_{lj}\}$ of $K \times d$ matrices. Then, since the shape of an object determines its coordinates only up to similarity transformations, it is necessary to reduce the data to just the shape information. We do this using Procrustes tangent coordinates about a centred and scaled mean configuration $\boldsymbol{\mu}$. In general, a convenient choice for $\boldsymbol{\mu}$ is the Generalised Procrustes estimate [7] based on all NT configurations, but the exact choice does not matter [10].

Let $\{\mathbf{y}_{lj}\}$ denote the $K \times d$ (centred not Helmertized) matrix of Procrustes tangent coordinates of the configuration data $\{\mathbf{v}_{lj}\}$. Next, assuming all N individuals are i.i.d., we take a sample average of the Procrustes coordinates to get averaged data $\bar{\mathbf{y}}_j$. To fit our spatio-temporal model to the averaged data, it is convenient to rewrite them as a $M \times T$ matrix $\mathbf{Y}$, say, with $M = Kd$ and the jth column of $\mathbf{Y}$ defined by stacking the $d = 2$ columns of $\bar{\mathbf{y}}_j$ on top of one another.

8.3 The Drift–Drift Model

In this section we describe our spatio-temporal model which involves drift functions in space $\mathbf{x}_i \in \mathcal{R}^d$ and time $t \in \mathcal{R}$. There are three ingredients in this model:

(a) $\mathcal{F}$: a p-dimensional vector space of functions $\mathcal{R}^d \to \mathcal{R}$, specifying possible directions of shape variation. Let $\{f_\alpha(\mathbf{x}) : \alpha = 1, \ldots, p\}$ denote a basis.

(b) $\mathcal{G}$: a q-dimensional vector space of functions $\mathcal{R} \to \mathcal{R}$, specifying possible rates of variation in time. Let $\{g_\beta(t) : \beta = 1, \ldots, q\}$ denote a basis.

(c) r: a rank, $r > 0$ representing the complexity of the model.

Both $\mathcal{F}$ and $\mathcal{G}$ will usually include the constant function to accommodate an intercept term. From these ingredients, we consider a deterministic spatio-temporal process of the form

$$z(\mathbf{x}, t) = \sum_{\alpha=1}^{p} \sum_{\beta=1}^{q} a_{\alpha\beta} f_\alpha(\mathbf{x}) g_\beta(t),$$

where the $p \times q$ matrix of coefficients $\mathbf{A} = (a_{\alpha\beta})$ has rank r. Let the $(N \times p)$ matrix $\mathbf{F}$ denote the values of the spatial basis functions at sites (landmarks) $\mathbf{x}_i$, $i = 1, \ldots, N$. Similarly, let the $(T \times q)$ matrix $\mathbf{G}$ denote the values of the temporal basis functions at the times t_j, $j = 1, \ldots, T$. Thus, model (8.1) takes the form

$$\mathbf{Y} = \mathbf{FAG}' + \mathbf{E}, \tag{8.2}$$

where $\mathbf{Y} = (y_{ij})$ and $\mathbf{E} = (e_{ij})$. If the basis functions are chosen so that $\mathbf{F}$ and $\mathbf{G}$ have orthonormal columns, then $\mathbf{A}$ can be estimated using the dominant r components in a singular value decomposition of $\mathbf{F}'\mathbf{YG}$, though if an intercept term is separated out estimation can be a bit more involved (i.e. an alternating algorithm is needed to estimate the intercept and $\mathbf{A}$).

8.3.1 Choice of Drift Functions

There are various choices for the space of drift functions, such as polynomials, trigonometric functions, principal kriging functions or principal splines (see, for example, Ippoliti et al. [12] and Lopes et al. [13] and references therein).

Next, we describe how to construct and use the principal kriging functions in our model [2, 17]. For simplicity we largely focus on functions of time but a similar construction also holds in space. Essentially, there are two main components in the construction: a vector space of functions $\mathcal{G}_0$ of dimension $q_0 \geq 0$ called the "null space," which will form a subspace of $\mathcal{G}$, and a "potential" function $\tau(t)$, conditionally positive definite with respect to $\mathcal{G}_0$. That is, for all distinct times $t^{(h)}, h = 1, \ldots, h_0$, and all vectors of coefficients $\boldsymbol{\delta} = (\delta_h, h = 1, \ldots, h_0) \neq 0$,

$$\sum_{h_1, h_2 = 1}^{h_0} \delta_{h_1} \delta_{h_2} \tau(t^{(h_1)} - t^{(h_2)}) > 0 \qquad (8.3)$$

whenever $\sum_{h=1}^{h_0} \delta_h g(t^{(h)}) = 0$ for all $g \in \mathcal{G}_0$. Given the time points $t_j, j = 1, \ldots, T$, define the $(T \times T)$ matrix, $\mathbf{P}$, by $\mathbf{P} = \left(\tau(t_i - t_j)\right)$. Next, define a $(T \times q_0)$ "drift" matrix $\mathbf{U}$ by $u_{j\beta} = g_{0\beta}(t_j)$, where $\{g_{0\beta}(t_j), \beta = 1, \ldots, q_0\}$ is a basis of functions in $\mathcal{G}_0$. It is assumed that the data times are suitably spaced so that this matrix is of full rank q_0. These matrices can be combined into an $(T + q_0) \times (T + q_0)$ matrix $\mathbf{H} = \begin{bmatrix} \mathbf{P} & \mathbf{U} \\ \mathbf{U}' & \mathbf{0} \end{bmatrix}$, with partitioned inverse $\mathbf{H}^{-1} = \begin{bmatrix} \mathbf{B} & \mathbf{C} \\ \mathbf{C}' & \mathbf{L} \end{bmatrix}$, providing the two matrices $\mathbf{B}$ and $\mathbf{C}$.

Denote the eigenvectors of $\mathbf{B}$ by $\boldsymbol{\xi}_k, k = 1, \ldots, K$, with the corresponding eigenvalues written in nondecreasing order. The first q_0 eigenvalues of $\mathbf{B}$ are 0 and the associated eigenvectors are given by the q_0 columns of $\mathbf{U}$. Finally, let $\boldsymbol{\tau}_0(t)$ denote the vector function of t with the jth element $\tau_0(t)_j = \tau(t - t_j)$, $j = 1, \ldots, T$, and let $\mathbf{u}_0(t)$ denote the vector function of t with βth component $g_{0\beta}(t)$, $\beta = 1, \ldots, q_0$. The kth principal kriging function is defined by

$$g_k^{\mathrm{PKF}}(t) = \boldsymbol{\xi}_k' \left[\mathbf{B}\boldsymbol{\tau}_0(t) + \mathbf{C}\mathbf{u}_0(t)\right].$$

It can be shown that $g_k^{\mathrm{PKF}}(t)$ is an interpolating function with $g_k^{\mathrm{PKF}}(t_j) = (\boldsymbol{\xi}_k)_j$. The vector space $\mathcal{G}$ is defined to be the span of $g_\beta^{\mathrm{PKF}}(t)$, $1 \leq \beta \leq q$, where q is a specified dimension, $q_0 \leq q \leq T$. Different possible choices for $\tau(t)$ are available; in general, any valid covariance function for which any null space of functions $\mathcal{G}_0$ will suffice is a possible choice (see, for example, Mardia et al. [16]).

In this chapter, we consider $\tau(t) = |t|^3$, which is conditionally positive definite whenever $\mathcal{G}_0$ contains linear functions, $span(1, t)$. In this case, it turn outs that the kth principal kriging function is an interpolating cubic spline which minimises the penalty function

$$\Phi(g) = \int_{-\infty}^{\infty} \left(\frac{\mathrm{d}^2 g(t)}{\mathrm{d}t^2} \right) \mathrm{d}t,$$

subject to the constraints $g_k(t_j) = (\boldsymbol{\xi}_k)_j$, $j = 1, \ldots, T$. This motivates the alternative name "principal spline" for a principal kriging function.

Principal splines are also easily dealt with for two-dimensional data at landmarks $\mathbf{x}_i$, $i = 1, \ldots, N$. In this case, it is common [1] to use the thin-plate spline penalty

$$\Phi(f) = \int \left\{ \left(\frac{\partial^2 f}{\partial x[1]^2} \right)^2 + \left(\frac{\partial^2 f}{\partial x[1] \partial x[2]} \right)^2 + \left(\frac{\partial^2 f}{\partial x[2]^2} \right)^2 \right\} \mathrm{d}\mathbf{x},$$

where the integral is over $\mathcal{R}^2$ and $\mathbf{x} = (x[1], x[2])$ denotes the components of $\mathbf{x}$. The above construction of principal splines can be carried out with little change in this setting as well. Replace the potential function by $\tau(\mathbf{x}) = |\mathbf{x}|^2 \log |\mathbf{x}|$, $\mathbf{x} \in \mathcal{R}^2$, and the null space by $\mathcal{F}_0 = span(1, x[1], x[2])$. Principal splines in $\mathcal{R}^2$ were introduced by Bookstein [1] in the context of deformations in shape analysis, and he used the term "principal warps". The matrix $\mathbf{B}$, instead, was called the "bending energy matrix".

8.3.2 Model Details

In this section we describe how to obtain the matrices $\mathbf{F}$ and $\mathbf{G}$ in Eq. (8.2). Let $t_1, \ldots, t_T$ denote the common set of times at which the data are observed. The associated $(T \times T)$ symmetric positive semidefinite bending energy matrix $\mathbf{B}$ has rank $T - 2$. For our modelling purposes, it is useful to combine all the eigenvectors (principle warp vectors) of $\mathbf{B}$ (excluding the constant vector) into the $T \times (T - 1)$ column orthonormal matrix $\mathbf{G}$. Denote the corresponding eigenvalues of $\mathbf{B}$ by $\beta_j, j = 1, \ldots, T - 1$, with $\beta_1 = 0$, arranged in nondecreasing order. The columns of $\mathbf{G}$ roughly correspond to the effects of orthogonal polynomials at the data times. A similar construction can be carried out in space for the $\mathbf{F}$ matrix. In particular, let $\{\mathbf{x}_1, \ldots, \mathbf{x}_K\}$ denote a fixed collection of landmarks in $\mathcal{R}^2$. In our application we shall use the K rows of $\boldsymbol{\mu}$. Then, a $(K \times K)$ bending energy matrix can be constructed with analogous properties to the above setting, though this time of rank $K - 3$. Then, since a deformation can be constructed from a pair of functions $\mathcal{R}^2 \to \mathcal{R}$, we combine together two copies of the eigenvectors. After removing 4 degrees of freedom for the constraints in Procrustes tangent space, we are left with the $2K \times (2K - 4)$ column orthonormal matrix $\mathbf{F}$. The first two columns of $\mathbf{F}$ represent linear functions orthogonal to the similarity transformations—i.e. the Bookstein's uniform components; see Bookstein [3] and Mardia [15, Sect. 7]. The remaining columns of $\mathbf{F}$ come in pairs: $\begin{bmatrix} \boldsymbol{\xi} & \mathbf{0} \\ \mathbf{0} & \boldsymbol{\xi} \end{bmatrix}$, where $\boldsymbol{\xi}$ is the $K \times 1$ eigenvector of the bending energy matrix for the thin-plate spline. Also, let $\alpha_{2j-1} = \alpha_{2j}$, $j = 1, \ldots, K - 2$ denote the corresponding eigenvalues in nondecreasing order, each listed twice. Note $\alpha_1 = \alpha_2 = 0$.

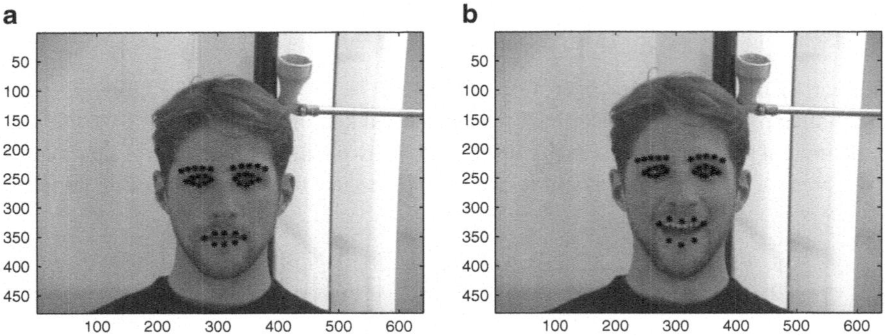

Fig. 8.1 Frames at point times t_1 and t_7. The frames represent: (**a**) neutral and (**b**) surprise expressions respectively

Thus, the $(2K-4)\times(T-1)$ matrix of coefficients, $\mathbf{A} = \{a_{ij}\}$, in equation (8.2) is obtained as $\mathbf{A} = \mathbf{F}'\mathbf{Y}\mathbf{G}$. As shown in Sect. 8.4, this matrix is of key importance for specifying different models. Finally, if the eigenvectors in $\mathbf{F}$ and $\mathbf{G}$ are interpolated to yield a pair of space-time functions $(\Psi_1(x[1], t), \Psi_2(x[2], t))$, then the penalty function

$$\Phi(\Psi_1, \Psi_2) = \sum_{i=1}^{2} \int \left\{ \left(\frac{\partial^4 \Psi_i}{\partial x[1]^2 \partial t^2} \right)^2 + 2 \left(\frac{\partial^4 \Psi_i}{\partial x[1] \partial x[2] \partial t^2} \right)^2 + \left(\frac{\partial^4 \Psi_i}{\partial x[2]^2 \partial t^2} \right)^2 \right\} \mathrm{d}\mathbf{x}\, \mathrm{d}t,$$

where the integral is over $\mathbf{x} \in \mathcal{R}^2$, $t \in \mathcal{R}$, reduces to $\sum_{i=1}^{2K-4} \sum_{j=1}^{T-1} a_{ij}^2 \alpha_i \beta_j$, which depends just on the coefficients and the eigenvalues.

8.4 Modelling Facial Expressions

In this section we consider the FG-NET Database with Facial Expressions and Emotions from the Technical University of Munich. This is an image database containing face images showing a number of subjects performing the six different basic emotions defined by Ekman and Friesen [8]. Specifically, the database contains material gathered from 18 different individuals, each performing all six desired actions three times. Additionally three sequences doing no expressions at all are recorded. All together this gives an amount of 399 sequences. Depending on the kind of emotion, a single recorded sequence can take up to several seconds. From each recorded sequence we have extracted five frames summarising the dynamic of the expression and on each of them we have manually placed a set of 34 landmarks. For our modelling purposes we have considered here four different expressions, namely: disgust, happiness, sadness and surprise. There is thus complete information on $N = 18$ subjects at $T = 5$ times on $K = 34$ landmarks. As an example Fig. 8.1 shows, for a specific subject, two frames representing the first (t_1—neutral) and

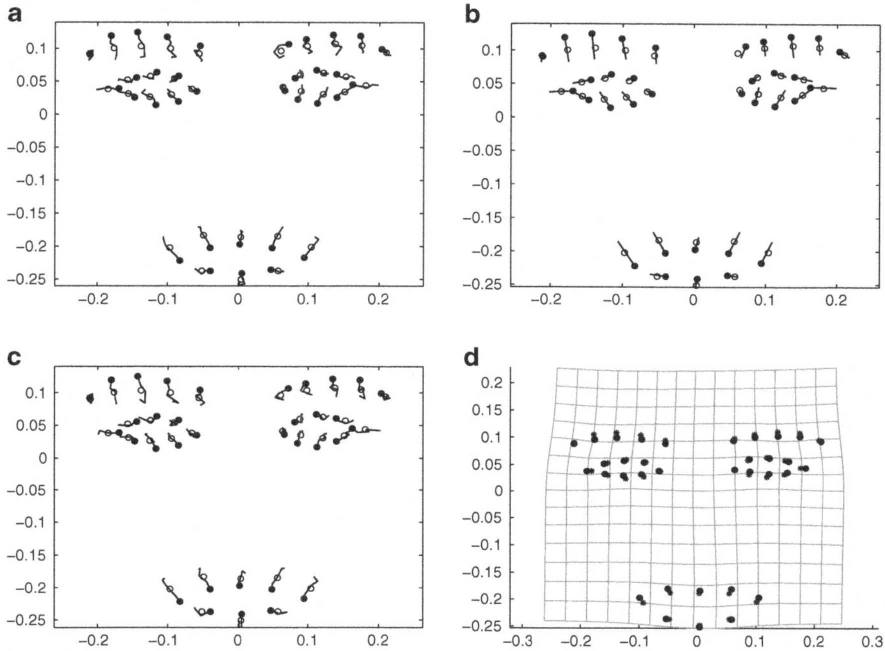

Fig. 8.2 Fitted models for the FGNET data (Disgust): (**a**) *full*, (**b**) *full-lin*, (**c**) *spec*, with the growth patterns blown up by a factor of 3 for clarity. Part (**d**) shows the grid deformation, without an expansion factor, between the initial and final times for the *special* model

last (t_7—surprise) frames. For the purposes of this chapter we shall ignore any differences between the subjects. Furthermore, we also ignore changes in size and limit attention only to changes in the shape of the object.

For this dataset, let $\mathbf{A} = \mathbf{F}'\mathbf{Y}\mathbf{G}$ denote the 64×4 doubly rotated version of the Procrustes tangent matrix for the mean data. The rows of $\mathbf{A}$ label the eigenvectors in space and the columns label the eigenvectors in time. The mean shape $\boldsymbol{\mu}$ is obtained as the generalized Procrustes estimate based on the $T = 5$ configurations of the neutral expression such that the model provides insights on its deformation.

Some important models and their application to the expression data are described below, and a selection of fitted models is plotted in Figs. 8.2–8.5 where each "o" represents a landmark of the Procrustes mean shape $\boldsymbol{\mu}$, and each closed circle represents the position of a landmark at the initial time.

1 *Full Rank*. These models can be specified in terms of the nonzero entries in the estimated $\mathbf{A}$. For example, the notation $[1 : J, 1 : L]$ means that the block of entries specified by the first J rows and first L columns is allowed to be nonzero.

FULL = [1:64, 1:4]. In this case there is no data reduction (other than the averaging over individuals in the first place). For this model, the residual sum of squares (RSS) is zero (Figs. 8.2–8.5a).

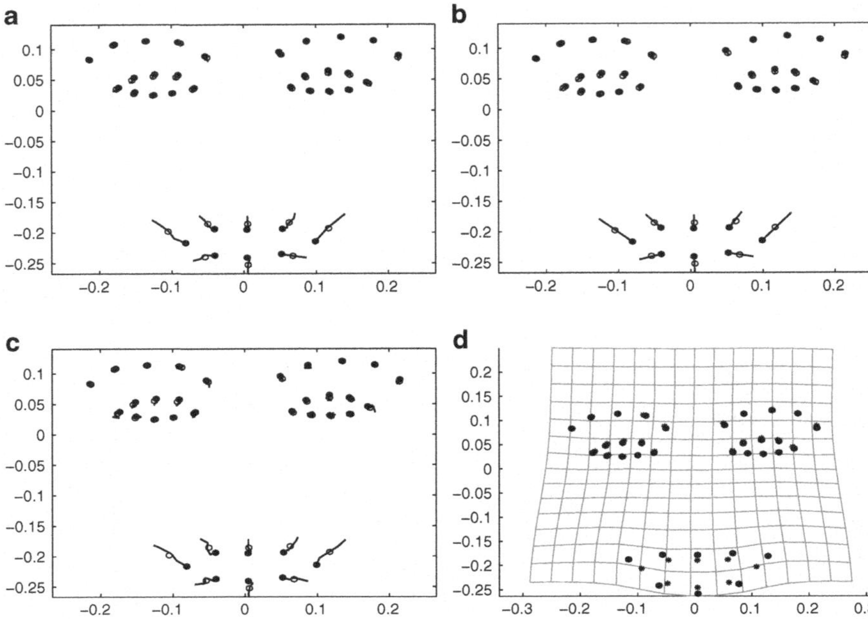

Fig. 8.3 Fitted models for the FGNET data (Happiness): (**a**) *full*, (**b**) *full-lin*, (**c**) *spec*, with the growth patterns blown up by a factor of 5 for clarity. Part (**d**) shows the grid deformation, without an expansion factor, between the initial and final times for the *special* model

LIN–LIN = [1:2,1]. This is the simplest possible model, linear in space (a two-dimensional subspace) and linear in time (one-dimensional). Thus it consists only of a constant variation in a single direction arising from a linear transformation in two dimensions. This model captures some of the main features of the data, but fails to capture the curvature of the paths, and does not in general fit well at several landmarks.

FULL-LIN = [1:64, 1]. In this case there is full flexibility in the spatial variation but the variation in time is still linear. In Figs. 8.2–8.5b we see that the main pattern of variation is captured, but the curvature of the paths is not captured at all.

SPEC = [1:64, 1] + [1:2, 2]. This model is called *special* because the nonzero parameter values do not form a rectangular block. It captures the curvature in the paths through a term which is second order in time and linear in space. Figures 8.2–8.5c show that this model yields quite a good fit to the data.

LIN-FULL = [1:2, 1:4]. This model allows variation in the two linear directions in space, but at arbitrary rates in time. The paths are curved but do not match the data very well.

NULL = [1:64, 1] + [1:2, 2:4] FULLLIN + LINFULL. This is the most general model that has 0 roughness penalty and fits the data surprisingly well. Visually the fit is similar to the special model.

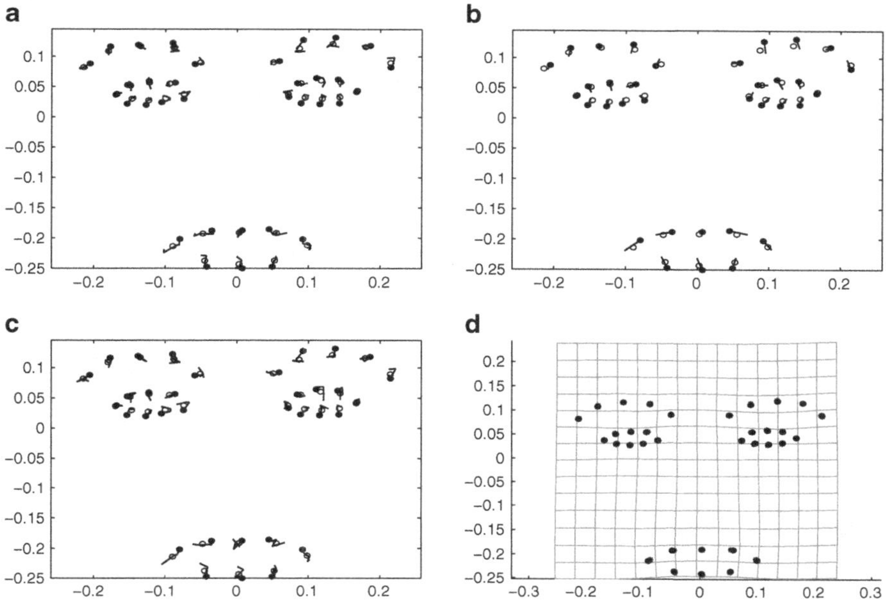

Fig. 8.4 Fitted models for the FGNET data (Sadness): (**a**) *full*, (**b**) *full-lin*, (**c**) *spec*, with the growth patterns blown up by a factor of 6 for clarity. Part (**d**) shows the grid deformation, without an expansion factor, between the initial and final times for the *special* model

2 *Reduced Rank*. Start with a rectangular block of coefficients in A, take a singular value decomposition and retain the dominant components. The simplest such model which fits the data well can be written as FULL.2 = [1:64, 1:4, 2]; i.e. take the dominant 2 components of the singular value decomposition of the whole matrix **A**. This fitted model is very closely related to the exploratory analysis based on two principal components in Le and Kume [14].

Figure 8.2 shows results for some of the fitted models for the disgust expression. We notice that this expression is mainly characterized by mouth and eyebrow movements. Specifically, we observe that the eyebrows are lowered and drawn together while the lower eyelids cheeks and the upper lip are raised.

Figure 8.3 shows results for some of the fitted models for the happiness expression. Here, we observe that the deformation is mainly characterized by the movements of the mouth since the eyes move only slightly. Specifically, we observe a slight narrowing of the eyelids and a raising of the lip corners describing an upward curving of mouth and expansion on vertical and horizontal direction.

Figure 8.4 shows results for some of the fitted models for the sadness expression. In this case, as expected, we notice that the features for this expression are represented by narrowed eyes, eyebrows brought together and a down-turned mouth.

Finally, Fig. 8.5 shows results for some of the fitted models for the surprise expression. In this case, we observe that this expression appears with a vertical

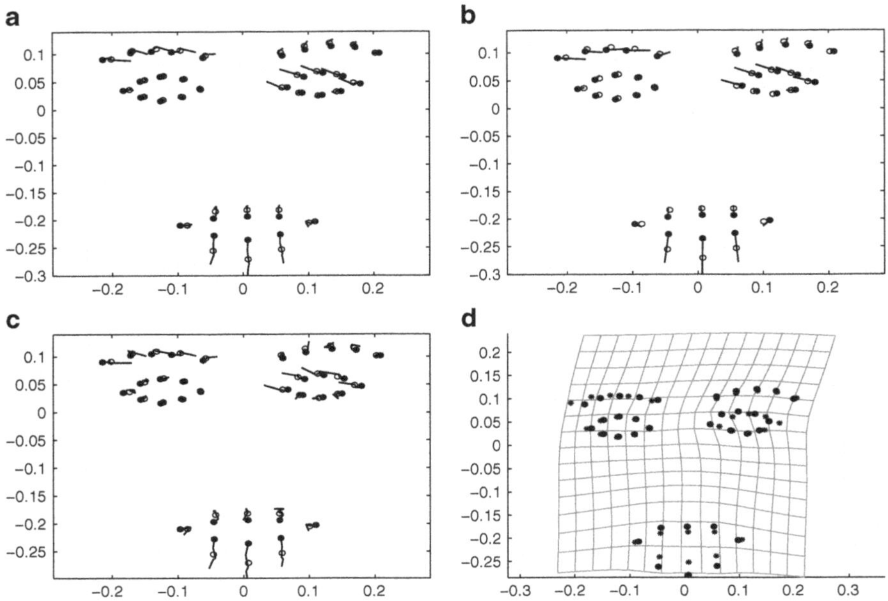

Fig. 8.5 Fitted models for the FGNET data (Surprise): (**a**) *full*, (**b**) *full-lin*, (**c**) *spec*, with the growth patterns blown up by a factor of 3 for clarity. Part (**d**) shows the grid deformation, without an expansion factor, between the initial and final times for the *special* model

expansion of the mouth, widened eyes and slightly raised eyelids and eyebrows. However, an asymmetry is surprisingly observed between the movement of the upper eyelids for the two eyes.

8.5 Discussion

Accurate and efficient analysis of the expressions can be facilitated by proper identification based on their dominant dynamic characteristics. In general, it is required to process the acquired landmark configurations to get the relevant information "buried" in it. As a solution of the outlined problem, we have proposed to model the temporal dynamics of each landmark by a functional spatio-temporal model. The proposed model is simple to estimate and is computationally appealing respect to other models (see, for example, Kume and Welling [11]). However, results obtained via the Procrustes tangent coordinates are based on the induced distribution on the tangent space of the Procrustes mean. In the general covariance case and in relatively dispersed shape data this approximation may not be reasonable and the approach described in Kume and Welling [11] may be preferred, although the modelling of the temporal correlation is not obvious here.

A glance at the **A** matrix shows that the largest values fall in the first column and the first two entries in the second column. This observation suggests that the "special" parametric model will provide a good-fitting parsimonious model, as verified by a small RSS value. Further, the reduced rank model "full.2" also captures these effects and yields a similar fit. Overall the most useful fitted model seems to be the "special" one. The variation can be decomposed into two components: a linear (in time) variation in a general spatial direction, together with a quadratic component (in time) which is restricted to a linear transformation in space. An illustration of the deformation involved between the starting and final times under this model is plotted (with no expansion factor) in Figs. 8.2–8.5d.

By estimating the model for each subject, we can also provide a compact, parametrized description of shape for any instance of a face. Hence, in principle, the model parameters in **A** could be used as features of an accurate classifier for "shape" (expression) allocation. In this context, useful inferential tools which facilitate a comparison of the different expressions are also provided by Brombin and Salmaso [4] and Brombin et al. [5].

References

1. Bookstein, F.L.: Principle Warps: thin plate splines and the decomposition of deformations. IEEE Trans. Pattern Anal. Mach. Intell. **16**, 460–468 (1989)
2. Bookstein, F.L.: Morphometric Tools for Landmark Data. Cambridge University Press, Cambridge (1991)
3. Bookstein, F.L.: A standard formula for the uniform shape component in landmark data. In: NATO Proceedings, 1995
4. Brombin, C., Salmaso, L.: Multiaspect permutation tests in shape analysis with small sample size. Comput. Stat. Data Anal. **53**, 3921–3931 (2009)
5. Brombin, C., Salmaso, L., Ferronato, G., Galzignato, P.F.: Multi-aspect procedures for paired data with application to biometric morphing. Comm. Stat. Simulat. Comput. **40**, 1–12 (2011)
6. Cressie, N., Wikle, C.K.: Statistics for Spatio-Temporal Data. Wiley, Chichester (2011)
7. Dryden, I.L., Mardia, K.V.: Statistical Shape Analysis. Wiley, Chichester (1998)
8. Ekman, P., Friesen, W.V: Constants across cultures in the face and emotions. J. Pers. Soc. Psychol. **17**, 124–129 (1971)
9. Friston, K.L., Penny, W.D., Ashburner, J.T.,Kiebel, S.J., Nichols, T.E.: Statistical Parametric Mapping: The Analysis of Functional Brain Images. Elsevier, Amsterdam (2007)
10. Kent, J.T., Mardia, K.V., Morris, R.J., Aykroyd, R.G.: Functional models of growth for landmark data. In: Mardia, K.V., Kent, J.T., Aykroyd, R.G. (eds) Proceedings in Functional and Spatial Data Analysis. Leeds University Press, Leeds (2001). ISBN 0-85316-220-4
11. Kume, A., Welling, M.: Maximum-likelihood estimation for the offset normal shape distributions using EM. J. Comput. Graph. Stat. **19**, 702–723 (2010)
12. Ippoliti, L., Valentini, P., Fontanella, L. and Gamerman, D.: Space-time modelling of coupled environmental variables. In: *Statistics for Spatio-Temporal Modelling, Proceedings of the 4th International Workshop on Spatio-Temporal Modelling*, Alghero, 24–26 September, Editrice democratica sarda, Sassari (2008). ISBN: 88-6025-098-6
13. Lopes, H.F., Salazar, E., Gamerman, D.: Spatial dynamic factor analysis. Bayesian Anal. **3**, 759–792 (2008)
14. Le, H., Kume, A.: Detection of shape changes in biological features. J. Microsc. **200**, 140–147 (2000)

15. Mardia, K.V.: Shape advances and future perspectives. In: Mardia, K.V., Gill, C.A. (eds.) Proceedings in Current Issues in Statistical Shape Analysis, pp. 57–75. Leeds University Press, Leeds (1995)
16. Mardia, K.V., Kent, J.T., Goodall, C.R., Little, J.A.: Kriging and spline with derivative information. Biometrika **83**, 217–285 (1996)
17. Mardia, K.V., Goodall, C., Redfern, E.J., Alonso, F.J.: The kriged kalman filter (with discussion). Test **7**, 217–252 (1998)

Vector Threshold Moving Average Models: Model Specification and Invertibility

9

Marcella Niglio and Cosimo Damiano Vitale

Abstract

In this chapter we propose a class of nonlinear time series models in which the underlying process shows a threshold structure where each regime follows a vector moving average model. We call this class of processes Threshold Vector Moving Average. The stochastic structure is presented even proposing alternative model specifications. The invertibility of the model is discussed detail and, in this context, empirical examples are proposed to show some features that distinguish the stochastic structure under analysis from other linear and nonlinear time series models widely investigated in the literature.

Keywords

Invertibility • Moving average • Vector threshold process

9.1 Introduction

In time series analysis the study of large datasets has motivated the introduction of multivariate models that allow to investigate the interdependence among variables which is lost in the univariate domain. Among the others [8, 9] give a wide presentation of the linear multivariate parametric models that generalize the so-called Autoregressive Moving Average (ARMA) univariate structure largely discussed in [2].

In this multivariate context, our interest will be given to a class of nonlinear models whose stochastic structure has not been examined in detail until now. We call these models Threshold Vector Moving Average (TVMA) that can be seen

M. Niglio (✉) · C.D. Vitale
Di.S.E.S., University of Salerno
Via Ponte Don Melillo 84084 Fisciano (SA), Italy
e-mail: mniglio@unisa.it; cvitale@unisa.it

N. Torelli et al. (eds.), *Advances in Theoretical and Applied Statistics*,
Studies in Theoretical and Applied Statistics, DOI 10.1007/978-3-642-35588-2_9,

as multivariate representations of the TMA model proposed in [10] and largely investigated in [4, 6] and [5].

More precisely, let X_t a stochastic process, it is said to follow a TMA $(\ell; q_1, \ldots, q_\ell)$ model if:

$$X_t = \sum_{i=1}^{\ell} \left(\mu^{(i)} + \sum_{j=0}^{q_i} \theta_j^{(i)} u_{t-j}^{(i)} \right) I_{i,t-d}, \tag{9.1}$$

where $\theta_0^{(i)} = 1$, $I_{i,t-d}$ is the indicator function:

$$I_{i,t-d} \equiv \mathbb{I}(X_{t-d} \in \mathscr{R}_i) = \begin{cases} 0 & \text{if} \quad X_{t-d} \notin \mathscr{R}_i \\ 1 & \text{if} \quad X_{t-d} \in \mathscr{R}_i, \end{cases}$$

with $\mathscr{R}_i = [r_{i-1}, r_i)$, for $i = 1, 2, \ldots, \ell$, no-overlapping subsets of the real line $\mathscr{R}$ such that $\bigcup_{i=1}^{\ell} \mathscr{R}_i = \mathscr{R}$ and $-\infty = r_0 < r_1 < \ldots < r_{\ell-1} < r_\ell = \infty$, d is the threshold delay, $\mu^{(i)}$ and $\theta_j^{(i)}$ are constants and $\{u_t^{(i)}\}$ is a sequence of independent and identically distributed (i.i.d.) random variables with zero mean and finite variance.

Interesting properties related to this model have been investigated in [3, 6, 7] and [5]. In this chapter the attention will be given to its multivariate analog. In particular, after the presentation of the model, in Sect. 9.2 we investigate the structure of the moments up to order two, whereas in Sect. 9.3 we focus the attention on its invertibility that is an inescapable requirement under which the forecasts can be generated. Some examples are proposed in Sect. 9.4 and the concluding remarks are given at the end.

9.2 The Threshold Vector Moving Average Model

The TVMA model can be seen as extension, in nonlinear domain, of the VMA(q) model whose main features have been presented in [8]. In particular, a stochastic process $\boldsymbol{y}_t$ is said to follow a VMA(q) model if:

$$\boldsymbol{y}_t = \boldsymbol{u}_t + \boldsymbol{M}_1 \boldsymbol{u}_{t-1} + \boldsymbol{M}_2 \boldsymbol{u}_{t-2} + \ldots + \boldsymbol{M}_q \boldsymbol{u}_{t-q} \tag{9.2}$$

where $\boldsymbol{y}_t = (y_{1t}, y_{2t}, \ldots, y_{Kt})'$ is a $(K \times 1)$ vector of variables observed at time t, $\boldsymbol{u}_t = (u_{1t}, u_{2t}, \ldots, u_{Kt})'$ is a $(K \times 1)$ zero mean white noise with nonsingular covariance matrix Σ_u and $\boldsymbol{M}_j$ is a $(K \times K)$ matrix of coefficients, for $j = 1, 2, \ldots q$.

Starting from model (9.2) and following the so-called *threshold principle* presented in [10], the TVMA model can be seen as a local linear structure characterized by the introduction of regimes whose switching among them is regulated by a threshold variable.

In more detail, let $\boldsymbol{y}_t$ a $(K \times 1)$ random vector, it is said to follow a TVMA$(\ell; q_1, \ldots, q_\ell)$ model if it has form:

$$\boldsymbol{y}_t = \sum_{i=1}^{\ell} \left(\boldsymbol{M}_1^{(i)} \boldsymbol{u}_{t-1} + \boldsymbol{M}_2^{(i)} \boldsymbol{u}_{t-2} + \ldots + \boldsymbol{M}_{q_i}^{(i)} \boldsymbol{u}_{t-q_i} \right) \mathbb{I}(z_{t-d} \in \mathcal{R}_i) + \boldsymbol{u}_t \quad (9.3)$$

where $\boldsymbol{M}_j^{(i)}$, for $j = 1, 2, \ldots, q_i$ and $i = 1, 2, \ldots, \ell$, are the $(K \times K)$ matrices of coefficients in regime i, $\{\boldsymbol{u}_t\}$ is a sequence of i.i.d. random variables and z_{t-d} is the univariate threshold variable.

To explore some features of model (9.3) it is easy to note that the joint distribution of $\boldsymbol{y}_t$ is uniquely determined by $\boldsymbol{u}_t$.

This model can be further generalized introducing regimes with different errors:

$$\boldsymbol{y}_t = \sum_{i=1}^{\ell} \left(\boldsymbol{u}_t^{(i)} + \boldsymbol{M}_1^{(i)} \boldsymbol{u}_{t-1}^{(i)} + \boldsymbol{M}_2^{(i)} \boldsymbol{u}_{t-2}^{(i)} + \ldots + \boldsymbol{M}_{q_i}^{(i)} \boldsymbol{u}_{t-q_i}^{(i)} \right) \mathbb{I}(z_{t-d} \in \mathcal{R}_i) \quad (9.4)$$

where $\{\boldsymbol{u}_t^{(i)}\}$ is a sequence of i.i.d. random variables with density $f_i(\boldsymbol{u}_t^{(i)}) : \mathbb{R}^K$, for $i = 1, 2, \ldots, \ell$ and $f_i(\cdot)$ independent from $f_j(\cdot)$, for $i \neq j$. Model (9.4) can be brought back to the parametrization of model (9.3) noting that if $f_i(\cdot) = f_j(\cdot)$, $\boldsymbol{u}_{t-s}$ can be given as $\boldsymbol{u}_{t-s} = \sum_{i=1}^{\ell} \boldsymbol{u}_{t-s}^{(i)} \mathbb{I}(z_{t-d} \in \mathcal{R}_i)$, for $s = 0, 1, \ldots, q_i$.

In order to simplify the notation of model (9.3), but without constrain its general representation, in the following we assume that all regimes have the same order q (it can be easily met including null matrices in model (9.3)).

Reminding the assumptions on $\boldsymbol{u}_t$, the first and the second moment of model (9.3), with $\ell = 2$, are:

$$E[\boldsymbol{y}_t] = \boldsymbol{0}, \quad (9.5)$$

$$\begin{aligned} \Gamma(h) = E[(\boldsymbol{y}_t \boldsymbol{y}_{t-h}')] = \sum_{j=1}^{q} \Big[& \boldsymbol{M}_j^{(1)} \Sigma_u \left(\boldsymbol{M}_{j-h}^{(1)} \right)' p_h + \boldsymbol{M}_j^{(1)} \Sigma_u \left(\boldsymbol{M}_{j-h}^{(2)} \right)' (p - p_h) \\ & + \boldsymbol{M}_j^{(2)} \Sigma_u \left(\boldsymbol{M}_{j-h}^{(1)} \right)' (p - p_h) + \boldsymbol{M}_j^{(2)} \Sigma_u \left(\boldsymbol{M}_{j-h}^{(2)} \right)' (1 - 2p + p_h) \Big], \end{aligned} \quad (9.6)$$

where, $\boldsymbol{M}_0^{(i)} = \boldsymbol{I}_K$, $\boldsymbol{M}_{j-h}^{(i)} = \boldsymbol{0}$, for $j < h$, $\Gamma(h) = \boldsymbol{0}$ for $h > q$, and as widely explored in [1]:

$$\begin{aligned} p = E[\mathbb{I}(z_{t-d} \in \mathcal{R}_1)], \quad & p_h = E[\mathbb{I}(z_{t-d} \in \mathcal{R}_1)\mathbb{I}(z_{t-d-h} \in \mathcal{R}_1)] \\ p - p_h = & \; E[\mathbb{I}(z_{t-d} \in \mathcal{R}_1)(1 - \mathbb{I}(z_{t-d-h} \in \mathcal{R}_1))] \\ 1 - 2p + p_h = & \; E[(1 - \mathbb{I}(z_{t-d} \in \mathcal{R}_1))(1 - \mathbb{I}(z_{t-d-h} \in \mathcal{R}_1))], \end{aligned} \quad (9.7)$$

with $E[\mathbb{I}(z_{t-d} \in \mathscr{R}_1)] = P(z_{t-d} \leq r_1)$ that, in other words, corresponds to the probability that z_{t-d} belongs to the interval $(-\infty, r_1)$.

Further note that the autocovariance (9.6) can be even extended to the more general TVMA($\ell; q$) model only using a more heavy notation.

Another interesting feature of model (9.3) is related to its Markovian representation. It is a Kq-dimensional TVMA(ℓ; 1) process, where all moving average regimes are of order one:

$$\boldsymbol{Y}_t = \boldsymbol{U}_t + \sum_{i=1}^{\ell} \boldsymbol{B}^{(i)} \boldsymbol{U}_{t-1} \mathbb{I}(z_{t-d} \in \mathscr{R}_i), \tag{9.8}$$

with

$$\boldsymbol{Y}_t = \begin{bmatrix} \boldsymbol{y}_t \\ \boldsymbol{u}_{t-1} \\ \vdots \\ \boldsymbol{u}_{t-q+1} \end{bmatrix}_{(Kq\times 1)} \qquad \boldsymbol{U}_t = \begin{bmatrix} \boldsymbol{u}_t \\ \boldsymbol{u}_{t-1} \\ \vdots \\ \boldsymbol{u}_{t-q+1} \end{bmatrix}_{(Kq\times 1)} \qquad \boldsymbol{B}^{(i)} = \begin{bmatrix} \boldsymbol{M}_1^{(i)} & \boldsymbol{M}_2^{(i)} & \dots & \boldsymbol{M}_q^{(i)} \\ \boldsymbol{0} & \boldsymbol{0} & \dots & \boldsymbol{0} \\ \vdots & \vdots & \ddots & \vdots \\ \boldsymbol{0} & \boldsymbol{0} & \dots & \boldsymbol{0} \end{bmatrix}_{(Kq\times Kq)}$$

Note that this last representation of the TVMA($\ell; q$) model allows to remark that model (9.3) can be seen as a vector moving average process with time-dependent coefficients that are conditionally dependent to t through $\mathbb{I}(z_{t-d} \in \mathscr{R}_i)$. It becomes more evident if we consider a TVMA($2; q$) model whose form becomes:

$$\boldsymbol{Y}_t = \boldsymbol{U}_t + \boldsymbol{B}_{t-d} \boldsymbol{U}_{t-1}, \tag{9.9}$$

with

$$\boldsymbol{B}_{t-d} = \begin{bmatrix} \boldsymbol{M}_1^{(1)} & \boldsymbol{M}_2^{(1)} & \dots & \boldsymbol{M}_q^{(1)} \\ \boldsymbol{0} & \boldsymbol{0} & \dots & \boldsymbol{0} \\ \vdots & \vdots & \ddots & \vdots \\ \boldsymbol{0} & \boldsymbol{0} & \dots & \boldsymbol{0} \end{bmatrix} I_{t-d} + \begin{bmatrix} \boldsymbol{M}_1^{(2)} & \boldsymbol{M}_2^{(2)} & \dots & \boldsymbol{M}_q^{(2)} \\ \boldsymbol{0} & \boldsymbol{0} & \dots & \boldsymbol{0} \\ \vdots & \vdots & \ddots & \vdots \\ \boldsymbol{0} & \boldsymbol{0} & \dots & \boldsymbol{0} \end{bmatrix} (1 - I_{t-d})$$

and $I_{t-d} = \mathbb{I}(z_{t-d} \in \mathscr{R}_1)$.

After n iterations, the TVMA($2; q$) model (9.9) has form:

$$\boldsymbol{U}_t = \boldsymbol{Y}_t + \sum_{j=1}^{n} \prod_{i=0}^{j-1} (-1)^j \boldsymbol{B}_{t-d-i} \boldsymbol{Y}_{t-j} + (-1)^n \prod_{i=0}^{n} \boldsymbol{B}_{t-d-i} \boldsymbol{U}_{t-n-1}, \tag{9.10}$$

whose notation can be simplified introducing the $(K \times Kq)$ matrix $\boldsymbol{J} = (\boldsymbol{I}_K, \boldsymbol{0}, \dots, \boldsymbol{0})$, with $\boldsymbol{I}_K$ an identity matrix and $\boldsymbol{0}$ a $(K \times K)$ null matrix, such that $\boldsymbol{y}_t = \boldsymbol{J}\boldsymbol{Y}_t$, $\boldsymbol{u}_t = \boldsymbol{J}\boldsymbol{U}_t$ and model (9.10) becomes:

$$\boldsymbol{u}_t = \boldsymbol{y}_t + \boldsymbol{J}\sum_{j=1}^{n}\prod_{i=0}^{j-1}(-1)^j \boldsymbol{B}_{t-d-i}\boldsymbol{Y}_{t-j} + (-1)^n \boldsymbol{J}\prod_{i=0}^{n}\boldsymbol{B}_{t-d-i}\boldsymbol{U}_{t-n-1}. \qquad (9.11)$$

9.3 The TVMA Invertibility

The form presented in (9.11) for the TVMA(2; q) model allows to state the conditions under which the process $\boldsymbol{y}_t$ is invertible.

Theorem 1. *Let* $\boldsymbol{y}_t$ *~TVMA(2;* q*) with independent white noise errors* $\{\boldsymbol{u}_t\}$ *and let* z_{t-d} *a strictly stationary and ergodic stochastic process, then* $\boldsymbol{y}_t$ *is invertible if:*

$$|\lambda(\boldsymbol{B}^{(1)})|^p |\lambda(\boldsymbol{B}^{(2)})|^{(1-p)} < 1$$

with $|\lambda(\boldsymbol{B}^{(i)})|$ *the dominant eigenvalue of* $\boldsymbol{B}^{(i)}$*, for* $i = 1, 2$ *and* $p = E[\mathbb{I}(z_{t-d} \in \mathscr{R}_1)]$.

Proof. Before to show the main steps of the proof, it is interesting to remark that, given the model representation (9.11), $\boldsymbol{y}_t$ is invertible if $\prod_{i=0}^{n}|\boldsymbol{B}_{t-d-i}|$ converges, in some sense, to a null matrix.

In more detail, given the TVMA(2; q) process:

$$\boldsymbol{u}_t - \boldsymbol{y}_t - \boldsymbol{J}\sum_{j=1}^{n}\prod_{i=0}^{j-1}(-1)^j \boldsymbol{B}_{t-d-i}\boldsymbol{Y}_{t-j} = (-1)^n \boldsymbol{J}\prod_{i=0}^{n}\boldsymbol{B}_{t-d-i}\boldsymbol{U}_{t-n-1},$$

$\boldsymbol{y}_t$ is invertible if

$$\prod_{i=0}^{n}|\boldsymbol{B}_{t-d-i}| \xrightarrow{p} \boldsymbol{0}, \qquad (9.12)$$

as $n \to \infty$.

The proof takes advantage of the definition of dominant sequence of matrices: let $\boldsymbol{A}_n$ a sequence of matrix and let $\lambda_i(\boldsymbol{A}_n)$ the i-th eigenvalue of $\boldsymbol{A}_n$, then we say that the sequence $\boldsymbol{B}_n$ is dominant with respect to $\boldsymbol{A}_n$ ($\boldsymbol{A}_n \prec \boldsymbol{B}_n$), if the dominant eigenvalue of $\boldsymbol{B}_n$, denoted $|\lambda(\boldsymbol{B}_n)|$, is greater or equal to the dominant eigenvalue of $\boldsymbol{A}_n$, $|\lambda(\boldsymbol{A}_n)|$, for $n = 1, 2, \ldots$.

In other words:

$$\boldsymbol{A}_n \prec \boldsymbol{B}_n, \qquad \text{if} \quad |\lambda(\boldsymbol{A}_n)| \leq |\lambda(\boldsymbol{B}_n)| \qquad (9.13)$$

From (9.13) it follows that:

1. If $|\lambda(\boldsymbol{B}_n)| \to 0$, as $n \to \infty$, then $\boldsymbol{B}_n \to^p \boldsymbol{0}$ and $\boldsymbol{A}_n \to^p \boldsymbol{0}$, with $\boldsymbol{0}$ a null matrix.
2. Given the sequence $\boldsymbol{A}_n$, then $\boldsymbol{A}_n \prec |\lambda(\boldsymbol{A}_n)|\boldsymbol{I}$, where $\boldsymbol{I}$ is the identity matrix.

Using these last arguments and starting from the definition of $\boldsymbol{B}_{t-d-i}$:

$$\begin{aligned}
\prod_{i=0}^{n} |\boldsymbol{B}_{t-d-i}| &= \prod_{i=0}^{n} |I_{t-d-i}\boldsymbol{B}^{(1)} + (1 - I_{t-d-i})\boldsymbol{B}^{(2)}| \\
&\prec \prod_{i=0}^{n} \left| I_{t-d-i}|\lambda(\boldsymbol{B}^{(1)})|\boldsymbol{I} + (1 - I_{t-d-i})|\lambda(\boldsymbol{B}^{(2)})|\boldsymbol{I} \right| \\
&= \exp\left\{ \sum_{i=0}^{n} \ln \left| I_{t-d-i}\lambda(\boldsymbol{B}^{(1)})\boldsymbol{I} + (1 - I_{t-d-i})\lambda(\boldsymbol{B}^{(2)})\boldsymbol{I} \right| \right\} \\
&= \exp\left\{ \ln \left|\lambda(\boldsymbol{B}^{(1)})\boldsymbol{I}\right| \sum_{i=0}^{n} I_{t-d-i} + \ln \left|\lambda(\boldsymbol{B}^{(2)})\boldsymbol{I}\right| \sum_{i=0}^{n}(1 - I_{t-d-i}) \right\}.
\end{aligned}$$

Using the assumptions on the process and noting that $\frac{1}{n+1}\sum_{i=0}^{n} I_{t-d-i} = p + O_p(n^{-1/2})$, then:

$$\begin{aligned}
\prod_{i=0}^{n} |\boldsymbol{B}_{t-d-i}| &\prec \exp\left\{ \ln \left|\lambda(\boldsymbol{B}^{(1)})\boldsymbol{I}\right| (n+1)(p + O_p(n^{-1/2})) \right. \\
&\quad \left. + \ln \left|\lambda(\boldsymbol{B}^{(2)})\boldsymbol{I}\right| (n+1)(1 - p + O_p(n^{-1/2})) \right\} \\
&= \left|\lambda(\boldsymbol{B}^{(1)})\boldsymbol{I}\right|^{(n+1)(p+O_p(n^{-1/2}))} \left|\lambda(\boldsymbol{B}^{(2)})\boldsymbol{I}\right|^{(n+1)(1-p+O_p(n^{-1/2}))} \\
&= \left(\left|\lambda(\boldsymbol{B}^{(1)})\right|^{p} \left|\lambda(\boldsymbol{B}^{(2)})\right|^{1-p} \right)^{(n+1)} \left|\lambda(\boldsymbol{B}^{(1)})\lambda(\boldsymbol{B}^{(2)})\right|^{(n+1)O_p(n^{-1/2})} \boldsymbol{I},
\end{aligned}$$

where the convergence is reached if

$$|\lambda(\boldsymbol{B}^{(1)})|^{p}|\lambda(\boldsymbol{B}^{(2)})|^{(1-p)} < 1. \qquad \square$$

The results of Theorem 1 can be extended to the more general TVMA($\ell; q$) model:

Theorem 2. *Let $\boldsymbol{y}_t \sim TVMA(\ell; q)$, under the assumptions of Theorem 1, $\boldsymbol{y}_t$ is invertible if:*

$$\prod_{i=1}^{\ell} |\lambda(\boldsymbol{B}^{(i)})|^{p_i} < 1 \tag{9.14}$$

where $|\lambda(\boldsymbol{B}^{(i)})|$ is the dominant eigenvalue of $\boldsymbol{B}^{(i)}$, $p_i = E[I_{i,t-d}]$ with

$$I_{i,t-d} = \begin{cases} 1 & \textit{if} \quad z_{t-d} \in \mathscr{R}_i \\ 0 & \textit{otherwise}, \end{cases} \qquad \text{for} \quad i = 1, 2, \ldots, \ell.$$

Proof. The proof follows the same steps proposed in Theorem 1. Starting from model (9.8), a TVMA($\ell; q$) model can be always represented through the form (9.9) and the corresponding form (9.11), where the matrix B_{t-d-i} now becomes:

$$\boldsymbol{B}_{t-d-i} = \sum_{j=1}^{\ell} \boldsymbol{B}^{(i)} I_{j,t-d}, \qquad \text{for} \quad i = 1, 2, \ldots, n.$$

It allows to state that a TVMA($\ell; q$) model is invertible if:

$$\prod_{i=0}^{n} \left| \sum_{j=1}^{\ell} \boldsymbol{B}^{(i)} I_{j,t-d} \right| \xrightarrow{p} \boldsymbol{0}, \qquad \text{as} \quad n \to \infty, \tag{9.15}$$

where the conditions under which this last convergence holds are obtained from the arguments presented in Theorem 1. □

The results of the two previous theorems are of interest for different reasons: (i) they play an important role when one-step and/or multi-steps ahead predictors are generated for TVMA($\ell; q$) models; (ii) they allow to distinguish between *local* and *global* invertibility. More precisely we say that a process is locally invertible if all VMA regimes are invertible. From the results of the previous theorems it follows that a TVMA process can be *globally* invertible even in presence of *locally* non invertible regimes if the inequality (9.14) holds. Further, the invertibility allows to differently characterize model (9.2). In fact, we can see the TVMA model as a Vector Autoregressive structure of infinite order with conditionally time dependent parameters so generalizing what has been done in [8] for the Vector Moving Average models.

In more detail, let $\boldsymbol{y}_t \sim TVMA(\ell; q)$, under the invertibility conditions of Theorem 2 it can be given as:

$$\boldsymbol{u}_t = \boldsymbol{y}_t + \sum_{j=1}^{\infty} \boldsymbol{\Pi}_{j,t} \boldsymbol{y}_{t-j}, \tag{9.16}$$

where

$$\boldsymbol{\Pi}_{j,t} = -\sum_{i=1}^{q} \boldsymbol{\Pi}_{j-i,t} \left(\sum_{k=1}^{\ell} \boldsymbol{M}_i^{(k)} I_{k,t-d-(j-i)} \right),$$

with $\boldsymbol{\Pi}_{j-i,t} = \boldsymbol{0}$ if $j < i$ and $\boldsymbol{\Pi}_{j-i,t} = \boldsymbol{I}_K$ if $i = j$.

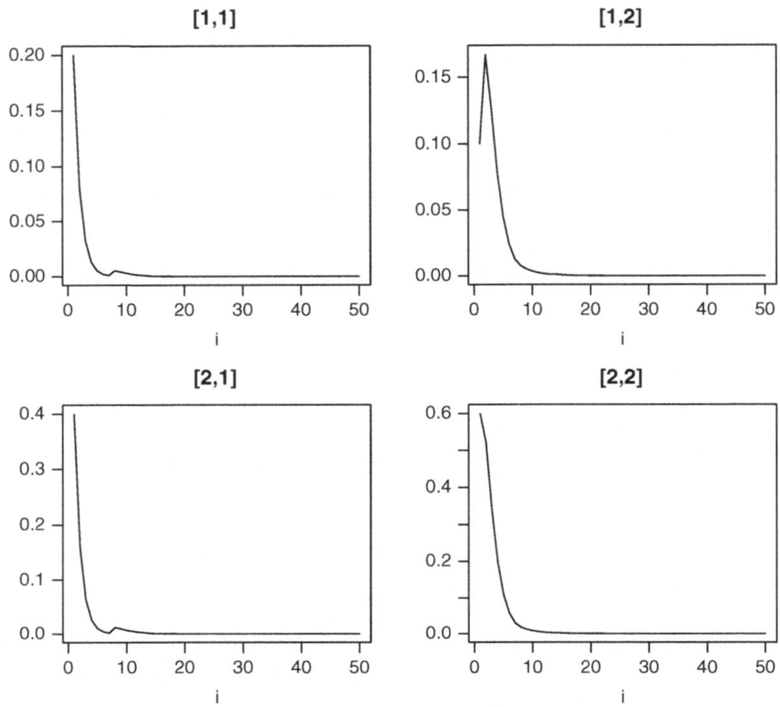

Fig. 9.1 Empirical evidence of the convergence to a null matrix of the (2×2) matrix $\boldsymbol{B}_{t-d-i}$, for $i = 1, 2, \ldots, 50$, in $\prod_{i=0}^{n} \boldsymbol{B}_{t-d-i}$, in the presence of model (9.17). Each frame shows the convergence of the four elements of $\boldsymbol{B}_{t-d-i}$ to zero

9.4 Examples

To better clarify the distinction between *local* and *global* invertibility, we now propose three examples where we show that, under the conditions of Theorem 1, the TVMA$(2; q)$ model becomes:

$$\boldsymbol{u}_t \approx \boldsymbol{y}_t + \boldsymbol{J} \sum_{j=1}^{n} \prod_{i=0}^{j-1} (-1)^j \boldsymbol{B}_{t-d-i} \boldsymbol{Y}_{t-j},$$

for n adequately large.

Example 1. ***TVMA(2;1) model with both regimes locally invertible***
Consider a TVMA(2;1) model:

$$\boldsymbol{y}_t = \boldsymbol{M}_1^{(1)} \boldsymbol{u}_{t-1} I_{t-d} + \boldsymbol{M}_1^{(2)} \boldsymbol{u}_{t-1} (1 - I_{t-d}) + \boldsymbol{u}_t, \tag{9.17}$$

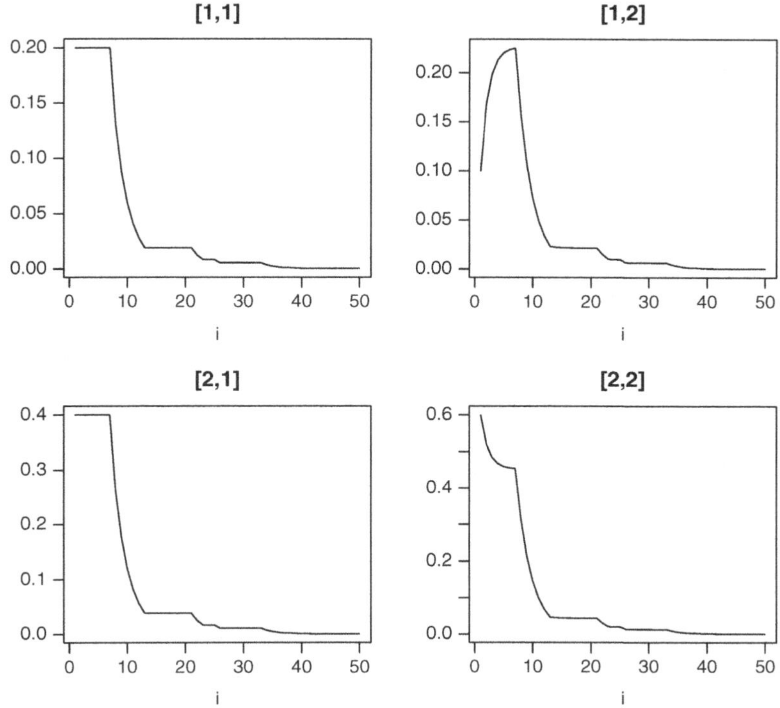

Fig. 9.2 Empirical evidence of the convergence to a null matrix of the (2×2) matrix $\boldsymbol{B}_{t-d-i}$, for $i = 1, 2, \ldots, 50$, in $\prod_{i=0}^{n} \boldsymbol{B}_{t-d-i}$ when the time series is simulated from the model in Example 2. Each frame shows the convergence of the four elements of $\boldsymbol{B}_{t-d-i}$ to zero

where $\boldsymbol{y}_t = (y_{1t}, y_{2t})$, $\boldsymbol{u}_t$ has mean zero and covariance matrix $E[\boldsymbol{u}_t \boldsymbol{u}_t'] = \boldsymbol{I}$, z_{t-1} follows a stationary AR(1) process with $\phi = 0.65$, the threshold delay and the threshold value are $d = 1$ and $r = 0$ respectively and the matrices of coefficients are:

$$\boldsymbol{M}_1^{(1)} = \begin{bmatrix} 0.4 & 0.6 \\ 0 & 0.47 \end{bmatrix} \quad \text{and} \quad M_1^{(2)} = \begin{bmatrix} 0.2 & 0.1 \\ 0.4 & 0.6 \end{bmatrix}.$$

Under these conditions it is easy to note that both regimes are *locally* invertible, with $|\lambda(\boldsymbol{B}^{(1)})| = 0.47$ and $|\lambda(\boldsymbol{B}^{(2)})| = 0.68$ where, as expected from Theorem 1, the local invertibility is sufficient for the global one and further in model (9.11) the product $\prod_{i=0}^{n} \boldsymbol{B}_{t-d-i} \rightarrow^{p} \boldsymbol{0}$, as n grows.

To better clarify this last remark we have simulated from model (9.17) a time series of length 1,000 (with burn-in 1,000) and in Fig. 9.1 we have presented the four elements of the (2×2) matrix $\boldsymbol{B}_{t-d-i}$, for $i = 1, \ldots, 50$. It can be noted that in presence of two *locally* invertible regimes the convergence to a null matrix of the previous product is very fast.

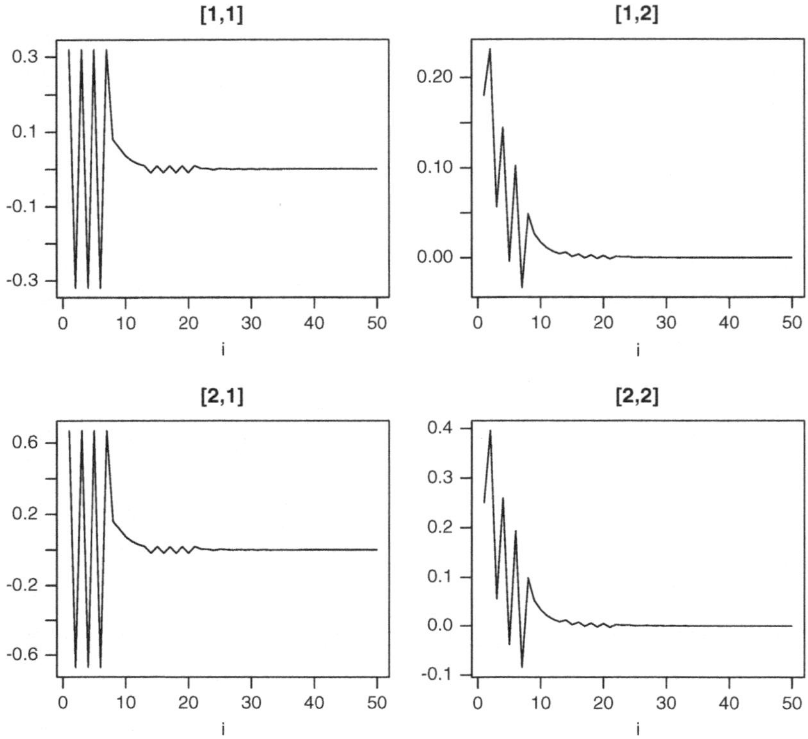

Fig. 9.3 Empirical evidence of the convergence to a null matrix of the (2×2) matrix $\boldsymbol{B}_{t-d-i}$, for $i = 1, 2, \ldots, 50$, in $\prod_{i=0}^{n} \boldsymbol{B}_{t-d-i}$ when the time series is simulated from the model in Example 3. Each frame shows the convergence of the four elements of $\boldsymbol{B}_{t-d-i}$ to zero

Example 2. ***TVMA(2;1) model globally (but not locally) invertible***

Starting from model (9.17) we now assume that the two matrices of coefficients are:

$$\boldsymbol{M}_1^{(1)} = \begin{bmatrix} 1 & 0.6 \\ 0 & 0.47 \end{bmatrix} \quad \text{and} \quad \boldsymbol{M}_1^{(2)} = \begin{bmatrix} 0.2 & 0.1 \\ 0.4 & 0.6 \end{bmatrix}$$

with $|\lambda(\boldsymbol{B}^{(1)})| = 1$ and $|\lambda(\boldsymbol{B}^{(2)})| = 0.68$, where the first regime is *locally* non-invertible.

From Theorem 1 it can be noted that this condition does not affect the *global* invertibility of the model. It can be empirically evaluated if we simulate a time series of length 1,000 (with burn-in 1,000) from the current model. In Fig. 9.2 it can be noted that the four elements of the matrix obtained from the product $\prod_{i=0}^{n} \boldsymbol{B}_{t-d-i}$ become null for a moderate value of n ($n \approx 40$).

The main consequence of this result is that in the presence of TVMA models, the underlying process can be *globally* invertible even in presence of *locally* non-invertible regimes.

Example 3. ***TVMA(2;2) model globally (but not locally) invertible***

The results of the previous example are considered even in the presence of a more complicated model. Let $\boldsymbol{y}_t \sim TVMA(2;2)$ model:

$$\boldsymbol{y}_t = (\boldsymbol{M}_1^{(1)}\boldsymbol{u}_{t-1} + \boldsymbol{M}_2^{(1)}\boldsymbol{u}_{t-2})I_{t-d} + (\boldsymbol{M}_1^{(2)}\boldsymbol{u}_{t-1} + \boldsymbol{M}_2^{(2)}\boldsymbol{u}_{t-2})(1 - I_{t-d}) + \boldsymbol{u}_t,$$

where $\boldsymbol{u}_t$, z_{t-1} have the same structure described in the first example whereas

$$\boldsymbol{M}_1^{(1)} = \begin{bmatrix} -1 & 0.33 \\ 0 & 0.7 \end{bmatrix}, \quad \boldsymbol{M}_2^{(1)} = \begin{bmatrix} 0.45 & -1 \\ 0.12 & 0.2 \end{bmatrix}, \quad \boldsymbol{M}_1^{(2)} = \begin{bmatrix} 0.32 & 0.18 \\ 0.67 & 0.25 \end{bmatrix}, \quad \boldsymbol{M}_2^{(2)} = \begin{bmatrix} 0.25 & 0.12 \\ 0.31 & 0.06 \end{bmatrix}.$$

Under these conditions $|\lambda(\boldsymbol{B}_1)| = 1$ and $|\lambda(\boldsymbol{B}_2)| = 0.63$ and the process $\boldsymbol{y}_t$, which is *locally* non-invertible, is characterized by *global* invertibility. Even in this case, if we simulate a time series of length 1,000 from the current model, the convergence to a null matrix of the product $\prod_{i=0}^{n} \boldsymbol{B}_{t-d-i}$ is empirically presented in Fig. 9.3 where it can be noted that the (2×2) elements of $\boldsymbol{B}_{t-d-i}$ decrease to zero for $n \approx 25$.

9.5 Conclusions

We have proposed a multivariate nonlinear time series model, called Threshold Vector Moving Average, that at the same time extends the threshold moving average model (introduced in [10]) in the multivariate domain and generalizes, in nonlinear context, the vector moving average model [8]. The covariance structure has been presented and sufficient conditions under which a TVMA($\ell;q$) model is invertible have been given and illustrated through empirical examples. These last results have emphasized that an interesting feature of the TVMA model is that it can be invertible even in the presence of one or more regimes with characteristic polynomials having roots on the unit circle.

The use of the proposed model can be advantageous in different contexts: for example in hydrology the TVMA model can be able to catch the behavior of some variables related to the rivers flow with respect to the rainfall level (the threshold variable) in a given area; in economics it can be able to catch the prices fluctuation when an exogenous threshold variable assumes values within well-defined bounds (that could be defined, among the others, by central banks, national governments, etc.). This empirical feature of the model will be object of future research.

References

1. Amendola, A., Niglio, M., Vitale, C.D.: The moments of SETARMA model. Stat. Probab. Lett. **76**, 625–633 (2006)
2. Box, G.E.P., Jenkins, G.M.: Time Series Analysis, Forecasting and Control. Holden-Day, San Francisco (1976)
3. Chan, K.S., Tong, H.: A note on the inveribility of nonlinear ARMA models. J. Stat. Plann. Infer. **140**, 3709–3714 (2010)
4. De Gooijer, J.G.: On threshold moving-average models. J. Time Anal. **19**, 1–18 (1998)
5. Li, D.: A note on moving-average models with feedback. J. Time Anal. **33**, 873–879 (2012)
6. Li, D., Ling, S., Tong, H.: On moving average models with feedback. Bernoulli **18**, 735–745 (2012)
7. Ling, S., Tong H., Li, D.: Ergodicity and invertibility of threshold moving-average models. Bernoulli **13**, 161–168 (2007)
8. Lütkepohl, H.: Introduction to Multiple Time Series Analysis. Springer, Berlin (1991)
9. Reinsel, G.C.: Elements of Multivariate Time Series Analysis. Springer, New York (1997)
10. Tong, H.: Non-linear time series. A Dynamical System Approach. Clarendon Press, Oxford (1990)

A Regionalization Method for Spatial Functional Data Based on Variogram Models: An Application on Environmental Data

10

Elvira Romano, Antonio Balzanella, and Rosanna Verde

Abstract

This chapter proposes a Dynamic Clustering Algorithm (DCA) as a new regionalization method for spatial functional data. The method looks for the best partition optimizing a criterion of spatial association among functional data. Furthermore it is such that a summary of the variability structure of each cluster is discovered. The performance of the proposal is checked through an application on real data.

Keywords

Clustering • Functional data • Spatial data • Variogram model

10.1 Introduction and Problematic

Nowadays many phenomena are monitored by sensor networks, let's think for instance, to security, transport, environment (pollutant and radioactivity monitoring), etc. Usually these sensed data are spatially correlated. With this in mind, recently, in spatial data analysis, there has been an increasing interest to the development of methods that attempt to exploit such correlation and among them, regionalization methods.

Regionalization methods, also known as spatially constrained clustering, aim to aggregate basic spatial units into larger units (regions) in order to preserve confidentiality, to minimize population differences, to reduce the effects of outliers or inaccuracies in the data, or simply, to facilitate the visualization and interpretation of information in maps [10].

E. Romano (✉) · A. Balzanella · R. Verde
Dipartimento di Studi Europei e Mediterranei, Facoltá di Studi Politici, Seconda Universitá degli Studi di Napoli, Via del Setificio 15, 81100 Caserta, Italy
e-mail: elvira.romano@unina2.it; balzanella2@alice.it; rosanna.verde@unina2.it

N. Torelli et al. (eds.), *Advances in Theoretical and Applied Statistics*,
Studies in Theoretical and Applied Statistics, DOI 10.1007/978-3-642-35588-2_10,

Essentially regionalization is a special form of classification where the spatial units are grouped together according to a particular criteria and a set of contiguity or adjacency constraints [2]. The wide range of fields in which these methods can be used makes it difficult to provide a single definition.

The main characteristics which are common to all methods are:

- They aggregate geographical areas into a predefined number of regions optimizing an aggregation criterion.
- They assume a prior knowledge of the aggregation process, especially of relevant variables for the aggregation (if are several), the number of clusters, the spatial contiguity constraint and the existence of an aggregation criterion.

These methods can be distinguished into two groups: methods in which the results of conventional clustering algorithms are revised in terms of spatial contiguity and methods which force spatial contiguity by maximizing regional compactness.

To our knowledge, in the framework of spatial functional data analysis, only two clustering methods exist that consider the spatial information about the curves.

The first one [7] is a hierarchical clustering algorithm for geographically referenced functional data developed to give a solution to the problem of classifying spatially correlated curves. The method allows to find groups of curves which are spatially homogeneous by weighting the dissimilarities between curves by the trace-variogram [6].

The second one [9] is a clustering algorithm based on the minimization of the spatial variability among the curves in each cluster and on discovering a prototype that is a spatial functional linear model in a point of the space. The main characteristic of the approach is that the prototype, obtained by an ordinary kriging prediction, has a representative spatial location in the geographic cluster. The main motivation of the method proposed in this chapter is that whenever an analyzed region is spatially wide, global measure of correlation are not locally representative. To deal with this issue, we propose to partition the whole spatial region into clusters which are internally homogeneous in terms of their spatial association.

A measure of spatial association able to emphasize the average spatial dependence over the studied area is the variogram.

We propose to use a Dynamic Clustering Algorithm to find the best partition of the spatial functional data. Especially the algorithm optimizes, for each cluster, the fitting between the empirical and the theoretical variogram function for functional data which is chosen as the prototype of the cluster.

According to the variability structure of the regions a suitable variogram model (linear, exponential, spherical, Gaussian, Mathern) is chosen.

The estimated variogram models are assumed as the prototypes of the clusters of the partition. The allocation of the curves to clusters is performed according to the minimum of the mean squared errors with respect to the estimated variogram model of the several clusters.

The chapter is organized as follows. Section 10.2 includes details on spatial functional data, Sect. 10.3 introduces the proposed method, and finally Sect. 10.4 ends the chapter with an application of the proposed methodology on a real dataset.

10.2 Spatial Variability Measures for Geostatistical Functional Data

Geostatistical functional data are a particular kind of spatially dependent functional data. They may be defined as the data for which the measurements on each observation that is a curve are part of a single underlying continuous spatial functional process defined as

$$\{\chi_s : s \in D \subseteq \mathbb{R}^d\} \tag{10.1}$$

where s is a generic data location in a fixed d-dimensional Euclidean space $D \subseteq \mathbb{R}^d$ with positive volume, where n points $(s_1, \ldots, s_i, \ldots, s_n)$ in D are chosen to observe the random functions $(\chi_{s_1}(t), \ldots, \chi_{s_i}(t), \ldots, \chi_{s_n}(t))$. In particular each function is defined on $T = [a, b] \subseteq \mathbb{R}$ and assumed to belong to an Hilbert space

$$L_2(T) = \{f : T \to \mathbb{R}, \text{ such that } \int_T f(t)^2 \mathrm{d}t < \infty\}.$$

with the inner product $\langle f, g\rangle = \int_T f(t)g(t)\mathrm{d}t$.

For a fixed site s_i it is assumed that the observed data follow the model:

$$\chi_{s_i}(t) = \mu_{s_i}(t) + \epsilon_{s_i}(t), \quad i = 1, \ldots, n \tag{10.2}$$

where $\epsilon_{s_i}(t)$ are zero-mean residuals and $\mu_{s_i}(\cdot)$ is the mean function which summarizes the main structure of χ_{s_i}. For each t, the random process is assumed to be second order stationary and isotropic: that is, the mean and variance functions are constant and the covariance depends only on the distance between sampling sites.

Formally we have:

$\mathbb{E}(\chi_s(t)) = m(t)\ \forall\, t \in T,\ s \in D,$
$\mathbb{V}(\chi_s(t)) = \sigma^2(t),\ \forall\, t \in T,\ s \in D,$ and
$\mathbb{C}ov(\chi_{s_i}(t), \chi_{s_j}(t)) = \mathbb{C}(h, t)$ where $h = \|s_i - s_j\|\ \ \forall\, s_i, s_j \in D$

Moreover, since we are assuming that the mean function is constant over D, the semivariogram function $\gamma(h, t) = \gamma_{s_i s_j}(t) = \frac{1}{2}\mathbb{V}(\chi_{s_i}(t) - \chi_{s_j}(t))$ where $h = \|s_i - s_j\|\ \forall\, s_i, s_j \in D$ can be expressed by

$$\gamma(h, t) = \gamma_{s_i s_j}(t) = \frac{1}{2}\mathbb{V}(\chi_{s_i}(t) - \chi_{s_j}(t)) = \frac{1}{2}\mathbb{E}\left[\chi_{s_i}(t) - \chi_{s_j}(t)\right]^2 \tag{10.3}$$

By considering the integral on T of this expression, using Fubini's theorem and following [5], a measure of spatial variability can be considered

$$\gamma(h) = \frac{1}{2}\mathbb{E}\left[\int_T (\chi_{s_i}(t) - \chi_{s_j}(t))^2 \mathrm{d}t\right], \text{ for } s_i, s_j \in D \text{ with } h = \|s_i - s_j\|$$

which is the so-called trace-variogram. This can be estimated as

$$\hat{\gamma}(h) = \frac{1}{2|N(h)|} \sum_{i,j \in N(h)} \int_T (\chi_{s_i}(t) - \chi_{s_j}(t))^2 dt, \tag{10.4}$$

where $N(h) = \{(s_i, s_j) : \|s_i - s_j\| = h\}$, and $|N(h)|$ is the number of distinct elements in $N(h)$. For irregularly spaced data there are generally not enough observations separated by exactly h. Then $N(h)$ is modified to $\{(s_i, s_j) : \|s_i - s_j\| \in (h - \varepsilon, h + \varepsilon)\}$, with $\varepsilon > 0$ being a small value. Note that the estimation of the empirical variogram for functional data using (10.4) involves the computation of integrals that can be simplified by considering that the functions are expanded in terms of some basis functions. The empirical variograms cannot be computed at every lag distance h, and due to variation in the estimation, it is not ensured that it is a valid variogram. In applied geostatistics, the empirical variograms are thus approximated [by ordinary least squares (OLS) or weighted least squares (WLS)] by model functions, ensuring validity [3]. Some widely used models include: Spherical, Gaussian, exponential or Mathern [2]. The variogram, as defined before, is used to describe the spatial variability among functional data across an entire spatial domain. However, this spatial variability may be strongly influenced by an unusual or changing behavior within this wide area.

Thus, in order to describe these spatial variability substructures, we introduce the concept of the spatial functional variability components with regards to a specific location by defining a centered variogram for functional data. Coherently with the above definition, given a curve $\chi_{s_i}(t)$, at a specific spatial location s_i, we define the centered variogram as

$$\gamma^{s_i}(h) = \frac{1}{2}\mathbb{E}\left[\int_T (\chi_{s_i}(t) - \chi_{s_j}(t))^2\right] \tag{10.5}$$

for each $s_j \neq s_i \in D$. Differently from the simple variogram function for functional data, the centered variogram is defined as the variability function of a set of curves respect to a fixed location s_i. Similarly to the variogram function, the centered variogram of the curve $\chi_{s_i}(t)$, as a function of the lag h, can be estimated through the method of moments:

$$\hat{\gamma}^{s_i}(h) = \frac{1}{2\,|N^{s_i}(h)|} \sum_{i,j \in N^{s_i}(h)} \int_T \left(\chi_{s_i}(t) - \chi_{s_j}(t)\right)^2 dt \tag{10.6}$$

where $N^{s_i}(h) \subset N(h) = \left\{(s_i, s_j) : \left\|s_i - s_j\right\| = h\right\}$, and it is such that $N(h) = \cup N^{s_i}(h)$ and $|N(h)| = \sum_i |N^{s_i}(h)|$.

Through straightforward algebraic operations, it is possible to show that the variogram function is a weighted average of centered variograms.

10.3 A Variogram-Based Dynamic Clustering Approach

A Dynamic Clustering Algorithm (DCA) [1, 4] is an unsupervised learning algorithm, which partitions a set of objects into internally dense and sparsely connected clusters. The main characteristic of the DCA is that it finds, simultaneously, the partition of data into a fixed number of clusters and a set of representative syntheses, named prototypes, obtained through the optimization of a fitting criterion. Formally, let E be a set of n objects. The Dynamic Clustering Algorithm finds a partition $P^* = (C_1, \ldots, C_k, \ldots, C_K)$ of E in K nonempty clusters and a set of representative prototypes $L^* = (G_1, \ldots, G_k, \ldots, G_K)$ for each C_k cluster of P so that both P^* and L^* optimize the following criterion:

$$\Delta(P^*, L^*) = \text{Min}\,\{\Delta(P, L) \;/\; P \in P_K, L \in \Lambda_K\} \tag{10.7}$$

with P_K the set of all the K-cluster partitions of E and Λ_K the representation space of the prototypes. $\Delta(P, L)$ is a function, which measures how well the prototype G_k represents the characteristics of the objects in a cluster, and it can usually be interpreted as a measure of goodness of fit between G_k and C_k. The definition of the algorithm is performed according to two main tasks:

- *Representation function* allowing to associate a set of prototypes $L = (G_1, \ldots, G_k, \ldots, G_K)$ of the representation space Λ_K, to each partition $P \in P_K$ of the data in K classes C_k $(k = 1, \ldots, K)$.
- *Allocation function* allowing to assign to each $G_k \in L$, a set of elements C_k.

The first choice concerns the representation for the classes $C_1, \ldots, C_K \in P$.

Let $\{\chi_{s_1}(t), \ldots, \chi_{s_n}(t)\}$ (with $t \in T$ and $s \in D$) be the sample of spatially located functional data. The proposed method aims at partitioning them into clusters in order to minimize, in each cluster, the spatial variability.

Following this aim, the method optimizes a best fit criterion between the centered variogram function $\gamma_k^{s_i}(h)$ and a theoretical variogram function $\gamma_k^*(h)$ for each cluster as follows:

$$\Delta(P, L) = \sum_{k=1}^{K} \sum_{\chi_{s_i}(t) \in C_k} \sum_{h} (\gamma_k^{s_i}(h) - \gamma_k^*(h))^2 \tag{10.8}$$

where $\gamma_k^{s_i}$ describes the spatial dependence between a curve $\chi_{s_i}(t)$ at the site s_i and all the other curves $\chi_{s_j}(t)$ in the cluster C_k, at each spatial lag h. This criterion allows to evaluate the membership of the curves $\chi_{s_i}(t)$ to the variability structures of the clusters.

As already mentioned, starting from a random initialization, the algorithm alternates *representation* and *allocation* steps until it reaches the convergence to a stationary value of the criterion $\Delta(P, L)$.

In the *representation* step, the theoretical variogram $\gamma_k^*(h)$ of the set of curves $\chi_{s_i}(t) \in C_k$, for each cluster C_k is estimated. This involves the computation of the empirical variogram and its model fitting by the Ordinary Least Square method.

In the *allocation* step, the function $\gamma_k^{s_i}$ is computed for each curve $\chi_{s_i}(t)$. Then a curve $\chi_{s_i}(t)$ is allocated to a cluster C_k by evaluating its matching with the spatial variability structure of the clusters according to the following rule:

$$\sum_{h<h^*\in[m_k;M_k]} (\gamma_k^{s_i}(h) - \gamma_k^*(h))^2 \rho_k < \sum_{h<h^*\in[m_k;M_k]} (\gamma_{k'}^{s_i}(h) - \gamma_{k'}^*(h))^2 \rho_{k'} \quad \forall k \neq k' \tag{10.9}$$

where:

- $\rho_k = \frac{|N^{s_i}{}_k|}{|N_k|}$ and $\rho_{k'} = \frac{|N_{k'}^{s_i}|}{|N_{k'}|}$ are the weights computed respectively, considering, for a fixed s_i, the number of location pairs $N_k^{s_i}$, $N_{k'}^{s_i}$ that are separated by a distance h in a cluster k, and k'.
- $m_k = \min_k h_k^*$, $M_k = \max_k h_k^*$ where h_k^* is the spatial distance at which the variogram γ_k^* for each cluster k reaches its sill.

According to the above allocation criterion, only one level h^* is chosen such that for $h > h^*$, there is no spatial correlation. This rule facilitates the spatial aggregation process leading to a tendency to form regions of spatially correlated curves. Especially, h^* is set in the range $[m_k, M_k]$.

The consistency between the representation of the clusters and the allocation criterion guarantees the convergence of the criterion to a stationary minimum value [1]. In the context of the proposed method, this is verified when:

$$\gamma_k^*(h) = \text{argmin} \sum_{\chi_{s_i}(t)\in C_k} (\gamma_k^{s_i}(h) - \gamma_k^*(h))^2 \tag{10.10}$$

Thus, since the allocation of each curve $\chi_{s_i}(t)$ to a cluster C_k is based on computing the squared Euclidean distance between $\gamma_k^{s_i}(h)$ and $\gamma_k^*(h)$; since the variogram $\gamma_k^*(h)$ is the average of the functions $\gamma_k^{s_i}(h)$, then $\gamma_k^*(h)$ minimizes the spatial variability of each cluster.

10.4 Application

The studied dataset provides climate monitoring series from a wide set of meteorological stations throughout Europe and Mediterranean by the European Climate Assessment & Dataset (ECA&D) available at http://eca.knmi.nl/.

We focus on maximum temperature data recorded from 1 January 2000 to 28 February 2010. Even though the whole dataset contains 1, 110 series, we performed our tests only on 730 stations available on the whole time period. Each time series records a new observations each day, whenever single measurements are not available, these are obtained by infilling from nearby stations.

Due to the extension of the spatial region involved in the monitoring activity, several spatial variability structures are present in the data.

Algorithm 1 A variogram-based Dynamic Clustering Algorithm

Initialization:
Start from a random partition $P = (C_1, \ldots, C_k, \ldots, C_K)$
Representation step:
for all clusters C_k **do**
Compute the prototype $\gamma_k^*(h)$ which optimizes the best fitting criterion:

$$\min \sum_{\chi_{s_i}(t) \in C_k} (\gamma_k^{s_i}(h) - \gamma_k^*(h))^2$$

end for
Allocation step:
for all $\chi_{s_i}(t)$ with $i = 1, \ldots, n$ **do**
find the cluster index k, for $h^* \in [m_k; M_k]$:
$\chi_{s_i}(t) \rightarrow C_k$
$if \ \sum_{h<h^*} (\gamma^{s_i}{}_k(h) - \gamma_k^*(h))^2 \rho_k < \sum_{h<h^*} (\gamma_{k'}^{s_i}(h) - \gamma_{k'}^*(h))^2 \rho_{k'} \ \forall k \neq k'$
end for

Thus, we evaluate how the proposed strategy is able to discover such variability structures and their associated spatial regions. With this aim, we analyze, for each cluster, the fitting between the empirical trace-variogram and the representative theoretical trace-variogram according to the optimized criterion in (10.8).

In order to provide the results, the following choices have to be performed:

- The basis functions to use for series representation.
- The parameters to use in the computation of the empirical variogram.
- The number of clusters K.
- The trace variogram model function $\gamma^*{}_k$.

The first step is to construct the set of functions expanded in terms of B-Spline Basis functions. We need to choose an appropriate order of expansion Z, taking into account that a large Z causes overfitting and a Z too small we may miss important aspects of the function that we are estimating [8]. We consider a procedure based on a classical nonparametric cross-validation analysis. Especially we evaluate cubic splines for each series and get a collection of smooth curves that is able to take into account the variability of the data.

Whenever the data are detected on irregularly spaced locations, the computation of the empirical trace-variogram may be performed considering only one pair of locations that is h apart. Averages based on only one or two points are poor estimates with large uncertainties. It is usual to increase the accuracy of the point estimates by defining tolerance regions and group the sample pairs into these regions prior to averaging. Each region has to be small enough so that we retain enough spatial resolution to define the structure of the semivariogram, but also large enough so that we base each point estimate on a relative large number of paired differences.

On the studied dataset we impose that the empirical trace-variogram has to be computed using 10 regions. Such value has allowed to get a reasonable number of curves in the averaging process. The choice of the number of clusters is based on tests which evaluate the optimized criterion for several values of K. As shown in

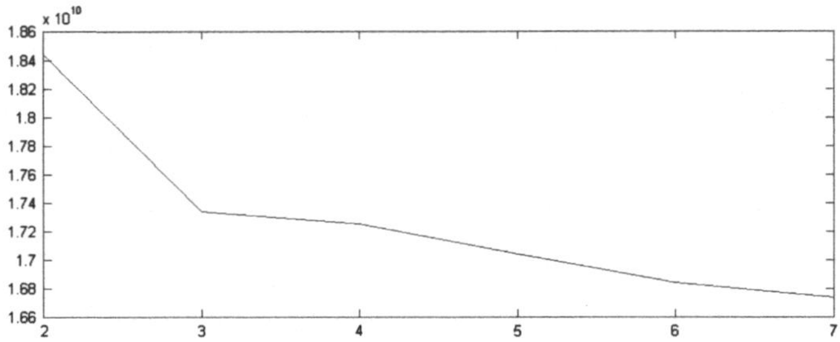

Fig. 10.1 Value of the optimized criterion for $K = 2, \ldots, 7$

Table 10.1 Criterion evaluation for several Trace-variogram model functions

Trace-variogram model & $\Delta(P, L)$	
Exponential	$1.73e^{+10}$
Gaussian	$1.75e^{+10}$
Spherical	$1.76e^{+10}$

Fig. 10.1, the optimal number of clusters is $K = 3$ since it provides the highest decrease of $\Delta(P, L)$.

A further preliminary decision concerns the trace-variogram model function to be fitted in the clustering procedure. Since we have no a priori information on what is the best fitting trace-variogram model function, we tested the value of the criterion for the Exponential model, the Spherical model and the Gaussian model. The results, which are available in Table 10.1, highlight that the Exponential model is the most suitable in our context.

Starting from the chosen input parameters, the algorithm run on the dataset detects the spatial regions available in Fig. 10.2. The value of the optimized criterion is $\Delta(P, L) = 1.73e^{+10}$, the number of iterations until convergence has been 7.

It is possible to note that the three discovered variability structures split the studied area into three spatial regions which include:

- The Northern Europe
- he most of the continental Europe and the Great Britain
- The Northern Africa, the Spain, the penisola balcanica

The discovered trace-variogram model functions which are representative of each clusters are shown in Fig. 10.3.

Looking at the plots, it is possible to highlight that the variability in the first cluster rises at an higher rate, when it is compared to the two other clusters. This is indicative of an high spatial dependence. At the opposite, the third cluster shows a reduced spatial dependence due to the low rate of growth of the spatial variability. It is still possible to note that the third cluster includes locations at an higher spatial distance when compared with the two other clusters.

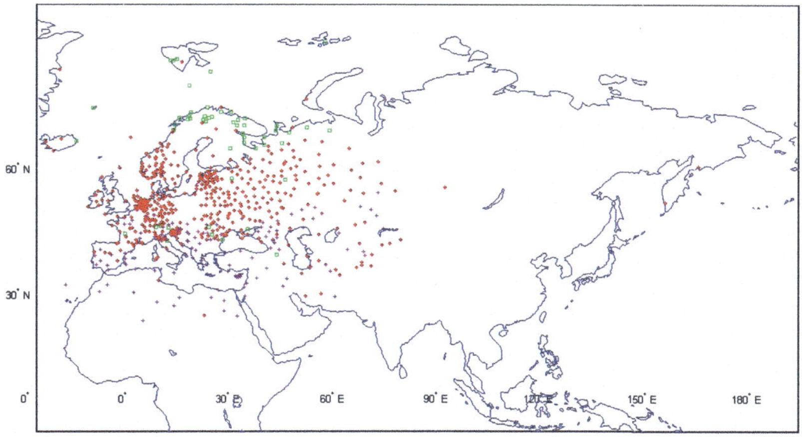

Fig. 10.2 Discovered clusters plotted on the map

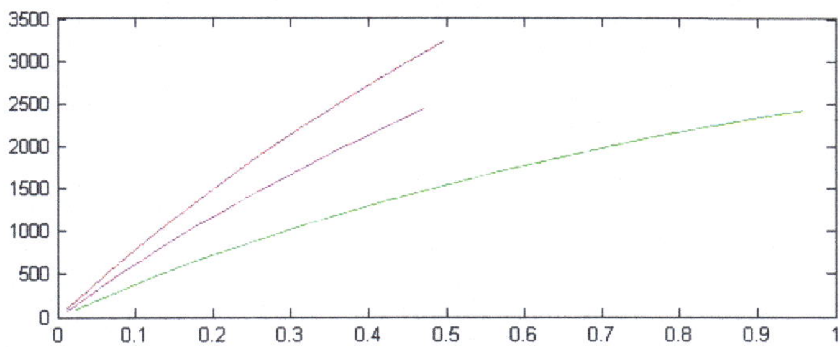

Fig. 10.3 Theoretical trace-variogram models for each cluster

10.5 Conclusion

In this chapter we have introduced a new regionalization method for spatially dependent functional data. The method can be applied to a wide range of domains. It is able to discover both the spatial partition of the data and the spatial variability structures representative of each cluster. The main novelty is the use of the trace-variogram model function as prototype of each cluster along the iterations of the clustering process. We have analyzed an environmental dataset to assess the performance of the method. We have focused on OLS techniques as classical estimation method; however, further attention will be given to alternative methods parameter and nonparametric, like weighted ordinary least squares, generalized least squares and maximum likelihood estimation [2].

References

1. Celeux, G., Diday, E., Govaert, G., Lechevallier, Y., Ralambondrainy, H.: Classification automatique des donnees: Bordas, Paris (1989)
2. Cressie, N.: Statistics for Spatial Data. Wiley Interscience (1993)
3. Chiles, J.P., Delfiner, P.: Geostatistics: Modelling Spatial Uncertainty. Wiley Series in Probability and Statistics (1999). ISBN: 0-471-08315-1
4. Diday, E.: La methode des Nuees dynamiques. Revue de Statistique Appliquee **19**(2), 19–34 (1971)
5. Delicado, P., Giraldo, R., Comas, C., Mateu, J.: Statistics for spatial functional data: some recent contributions. Environmetrics **21**, 224–239 (2010). doi:10.1002/env.1003
6. Giraldo, R., Delicado, P., Mateu, J.: Geostatistics for functional data: an ordinary kriging approach. Technical Report http://hdl.handle.net/2117/1099, Universitat Polit'ecnica de Catalunya. Submitted to Environmental and Statistics (2007)
7. Giraldo, R., Delicado, P., Comas, C., Mateu, J.: Hierarchical clustering of spatially correlated functional data. Technical Report, http://www.ciencias.unal.edu.co/estadistica/reporte02.pdf (2009)
8. Ramsay, J.E., Silverman, B.W.: Functional Data Analysis, 2nd edn. Springer, New York (2005)
9. Romano, E., Balzanella, A., Verde, R.: Clustering Spatio-functional data: a model based approach. Studies in Classification, Data Analysis, and Knowledge Organization. Springer, Berlin-Heidelberg (2010)
10. Wise, S.M., Haining, R.P., MA, J.: Regionalization tools for the exploratory spatial analysis of health data. In: Fischer, M., Getis, A. (eds.) Recent Developments in Spatial Analysis: Spatial Statistics, Behavioural Modelling and Neuro-Computing, pp. 83-100. Springer, Berlin (1997)

Spectral Decomposition of the AR Metric 11

Maria Iannario and Domenico Piccolo

Abstract

This work investigates a spectral decomposition of the *AR* metric proposed as a measure of structural dissimilarity among *ARIMA* processes. Specifically, the metric will be related to the variance of a stationary process so that its behaviour in the frequency domain will help to detect how unobserved components generated by the parameters of both phenomena concur in specifying the obtained distance. Foundations for the metric are briefly reminded and the main consequences of the proposed decomposition are discussed with special reference to some specific stochastic processes in order to improve the interpretative content of the *AR* metric.

Keywords

AR metric • *ARIMA* processes • Spectral decomposition

11.1 Introduction

In recent years, we register an increasing interest towards dissimilarity measures among dynamic phenomena for clustering and discrimination purposes. Motivation stems from the current availability of data collected at very low frequencies and by the recent computer efficiency in dealing with large amount of time series.

This interest is often involved with *data mining* procedures which operates on massive data archives originated from various fields of study: Economics and Finance, Signal processing, Environmental Sciences, Medicine, Demography, Hydrology, Meteorology, Physics, Astronomy, Geology and so on. In this respect,

---

M. Iannario (✉) · D. Piccolo
Department of Statistical Sciences, University of Naples Federico II, Via L. Rodin 22, I-80138 Naples, Italy
e-mail: maria.iannario@unina.it; domenico.piccolo@unina.it

N. Torelli et al. (eds.), *Advances in Theoretical and Applied Statistics*, Studies in Theoretical and Applied Statistics, DOI 10.1007/978-3-642-35588-2_11,

extensive reviews of the current research on clustering and discrimination of time series can be found in [11–13, 15, 17, 25, 26].

In this chapter, moving from a formal result about the variance of a stationary process, a spectral decomposition of the *AR* metric between linear processes is considered. Specifically, the metric will be related to the variance of a convenient stationary process involving the parameters of both processes to be compared; so its behaviour in the frequency domain helps to detect how unobserved components of this new process concur in specifying the obtained distance.

From a logical point of view, this approach is similar to results of [2] who obtained a frequential decomposition of the linear determinism index [3]. However, it should be remarked that, in our case, a correct decomposition of a real number over a set of discrete/continuous components strictly implies a parametric model for both processes to be compared.

The chapter is organized as follows: in the next section, we summarize the foundations for the metric. In Sect. 11.3 we establish the connection between the *AR* metric and the variance of a suitable stationary process, and derive a spectral decomposition of such measure. Section 11.4 is concerned with the study and comments of the decomposition of some processes often encountered in the applications. Some concluding remarks end the chapter.

11.2 Foundations for the *AR* Metric

In order to set a convenient notation, for a given stochastic process X_t, we consider *ARIMA* models of a transformed process $Z_t = g(X_t) - f_t$, which is obtained after a possibly non-linear transformation of X_t and then by removing any deterministic component f_t: trading days, calendar effects, outliers and mathematical functions of time, including constants. Such transformations are completely general and should improve both linearity and Gaussianity of Z_t.

Hereafter, Z_t is a zero-mean invertible $ARIMA(p, d, q)$ process,[1] that is:

$$\phi(B)\,\nabla^d\, Z_t \;=\; \theta(B)\,a_t, \qquad\qquad a_t \sim WN(0, \sigma_a^2)\,,$$

where a_t is a sequence of zero-mean uncorrelated and homoscedastic random variables (defined White Noise) and the backshift operator B applies to any function h_t of t in such a way that $B^k\, h_t = h_{t-k},\ \forall k = 0, \pm 1, \ldots$.

The *AR* and *MA* polynomials $\phi(B) = 1 - \phi_1 B - \cdots - \phi_p B^p$ and $\theta(B) = 1 - \theta_1\, B - \cdots - \theta_q\, B^q$, respectively, have *no common factors* and, moreover, all the roots of $\phi(B)\theta(B) = 0$ lie outside the unit circle. Finally, we denote by $\mathscr{L}$ the class of all linear *causal and invertible* stochastic processes $Z_t \sim$

[1] In order to simplify notation, in this chapter we are not formally considering multiplicative seasonal *ARIMA* processes; however, all the subsequent results straightforwardly follow. In addition, we are strictly using classical notation as in [5].

ARIMA(p, d, q). Notice that $\mathcal{L}$ includes (long-memory) fractional difference processes $Z_t \sim ARFIMA(p, d + d^*, q)$, as long as $d^* \in (-0.5, 0.5)$ [4].

For convenient initial values and a constant $\sigma_a^2 < +\infty$, any process $Z_t \in \mathcal{L}$ admits a well-defined *AR* operator $\pi(B) = 1 - \pi_1 B - \pi_2 B^2 - \ldots$ characterized by the square-convergent sequence $\{\pi_1, \pi_2, \ldots\}$, which specifies the *forecast function* $F_t = \mathbb{E}(Z_t \mid Z_{t-1}, Z_{t-2}, \ldots)$. Indeed, for all $Z_t \in \mathcal{L}$, the following orthogonal representation holds:

$$F_t = \pi_1 Z_{t-1} + \pi_2 Z_{t-2} + \ldots; \qquad Z_t = F_t + a_t, \quad F_t \perp a_t .$$

In this context, a fundamental theorem [6, 130–133] assesses that: "for *any* stationary process with a continuous spectrum $f(\omega)$ there exists a finite order $AR(p)$ whose spectrum $f_{AR}(\omega)$ is as close as possible in absolute value to $f(\omega)$ uniformly on $[-\pi, \pi]$". This theorem has been extended also to *MA* and, for numerical efficiency, to *ARMA* processes. It confirms that the *AR* operator is the simplest and effective approximation for any stationary process or for any process that may be transformed to a stationary one. Indeed, the theorem applies to both linear and non-linear stationary processes.

Taking these results into account, Piccolo [22, 24] introduced a measure of *structural dissimilarity* among two processes $X_t \in \mathcal{L}$ and $Y_t \in \mathcal{L}$ by defining the *AR metric* as the Euclidean distance among the coefficients $\boldsymbol{\pi}_x = (\pi_{1,x}, \pi_{2,x}, \ldots,)'$ and $\boldsymbol{\pi}_y = (\pi_{1,y}, \pi_{2,y}, \ldots,)'$, obtained by the *AR* expansions of X_t and Y_t, respectively.

With obvious notation, we get:

$$d(X, Y) = \sqrt{\left[(\boldsymbol{\pi}_x - \boldsymbol{\pi}_y)'(\boldsymbol{\pi}_x - \boldsymbol{\pi}_y)\right]} = \sqrt{\sum_{j=1}^{\infty} (\pi_{j,x} - \pi_{j,y})^2} . \tag{11.1}$$

Then, Piccolo [23] and Corduas [8, 9] derived the asymptotic distributions of the corresponding estimator for *AR* and *ARIMA* processes, respectively. Other results on these topics are reported by [16, 25] and [12], who provide efficient algorithms for computation.

The main interpretation of the *AR* metric (which is also verified for *ARFIMA* processes: [10]) is the following: *the distance between two ARIMA processes is zero if*, given the same set of initial values, *the corresponding models produce the same forecasts*. Notice that, as a measure of structural dissimilarity, the *AR* metric cannot take *WN* variances into account since they are just scale factors and do not affect the dynamic of the processes to be compared. This feature is shared by cepstral distances [14] when computed between cointegrated processes.

A useful example of the *AR* metric is obtained by comparing $X_t \sim ARMA(1, 1)$ and $Y_t \sim ARMA(1, 1)$, both belonging to $\mathcal{L}$. Thus, with obvious notation, we get:

$$d^2(X, Y) = \frac{(\phi_x - \theta_x)^2}{1 - \theta_x^2} + \frac{(\phi_y - \theta_y)^2}{1 - \theta_y^2} - 2\,\frac{(\phi_x - \theta_x)(\phi_y - \theta_y)}{1 - \theta_x \theta_y} .$$

If in this formula we let some *AR* parameters equal to 0 and/or 1, we find distances for several *ARIMA* subclasses including the *ARIMA*$(0,1,1)$ processes implied by the *exponential smoothing* procedures.

We observe that the introduction of $d(X,Y)$ transforms $\mathscr{L}$ in a *metric space* and this supports several statistical consequences. Specifically, we are able to introduce multivariate techniques (cluster analysis, discrimination, multidimensional scaling and so on). Thus, the diameter of a class of processes seems relevant for sensible applications of the metric if one have to choose representative time series or anomalous structures, for instance. Moreover, innovative proposals aimed at extending the use of *AR* metric to more complex econometric models have been extensively discussed by [18–21, 27].

11.3 Frequential Analysis of the *AR* Metric

The starting point of our derivation is a theorem obtained by Corduas [7] who proposed to compute the *AR* metric by means of the variance of a related *ARMA* process, by using efficient algorithms for obtaining the autocovariance function [1, 28]. We slightly modify her results for saving the symmetry of the metric and for a more immediate fitting to our objectives.

We denote by

$$\delta_j = \mid \pi_{j,x} - \pi_{j,y} \mid, \qquad j = 1, 2, \ldots .$$

the absolute difference of the *AR* expansion of processes $X_t \in \mathscr{L}$ and $Y_t \in \mathscr{L}$. Then, we introduce the stationary process:

$$W_t = \eta_t + \alpha_1\, \eta_{t-1} + \alpha_2\, \eta_{t-2} + \cdots = \alpha(B)\, \eta_t\,, \tag{11.2}$$

where $\eta_t \sim WN(0, \delta_1^2)$ and the coefficients α_j are obtained as[2]:

$$\alpha_j = \frac{\delta_{j+1}}{\delta_1}, \qquad j = 1, 2, \ldots$$

Then, it is immediate to show that

$$Var(W_t) = \sum_{j=1}^{\infty} \mid \pi_{j,x} - \pi_{j,y} \mid^2 = d^2(X, Y)\,. \tag{11.3}$$

The process W_t is obtained by filtering the *WN* process η_t by means of a linear operator $\alpha(B)$ where the differences of the *AR* expansions of both processes X_t

[2] In fact, to be correct we should define as δ_1 *the first non-zero absolute difference* between the *AR* expansions of X_t and Y_t processes. This consideration is important when comparing pure *AR* seasonal processes, for instance; however, we are omitting this point for simplicity of notation.

and Y_t are taken into account. It is evident that the time dynamics of W_t reflects characteristics of both processes to be compared. As a matter of fact, two processes X_t and Y_t with similar behaviour produce small α_j coefficients while processes with quite dissimilar structures give large coefficients α_j. Then, although W_t is a fictitious process, its statistical analysis helps in investigating the structural dissimilarity among X_t and Y_t as synthesized by the *AR* metric.

The process W_t may be non-invertible but it is *always stationary*. Thus, it admits a symmetric spectral function defined by:

$$g(\omega) = \frac{\delta_1^2}{\pi} \mid \alpha(e^{-\iota\omega}) \mid^2, \qquad 0 < \omega < \pi .$$

Then, the spectral decomposition of the *AR* metric immediately follows:

$$d^2(X, Y) = \int_0^{\pi} g(\omega)\, d\omega . \tag{11.4}$$

This result allows to ascertain whether some components at low, high, periodic, seasonal frequencies contribute to the determination of the value of $d^2(X, Y)$.

In addition, since $g(\omega)/d^2(X, Y)$ shares all the properties of a well-defined density function over $(0, \pi)$, we may also consider the spectral distribution function:

$$G(\omega) = \int_0^{\omega} \frac{g(z)}{d^2(X, Y)}\, dz . \tag{11.5}$$

In this way, for any fixed angular frequency $\omega_0 \in (0, \pi)$, the function $G(\omega_0)$ measures the relative proportion of distance explained by angular frequencies less or equal to ω_0. Similarly, for any interval $(\omega_1, \omega_2) \subset (0, \pi)$, the quantity $G(\omega_2) - G(\omega_1)$ is the proportion of the measured *AR* metric accounted by the frequencies of the W_t process belonging to (ω_1, ω_2).

Finally, for an adequate interpretation of such results, one should consider that a high accumulation of variance of W_t around some range of angular frequencies means that such components are important for explaining the computed distance among processes, and thus for increasing dissimilarities among them.

11.4 Discussion of Some Specific Decompositions

We study some specific comparisons by examining the implied spectral decomposition when the *AR* metric is computed between the most common processes.

- *AR processes*. If $X_t \sim AR(p_x)$ and $Y_t \sim AR(p_y)$, then W_t is a pure *MA* process of order $q_w = \max(p_x, p_y) - 1$ characterized by a finite set of coefficients:

$$\alpha_j = \frac{\left|\phi_{j+1,x} - \phi_{j+1,y}\right|}{\left|\phi_{1,x} - \phi_{1,y}\right|}, \qquad j = 1, 2, \ldots, q_w .$$

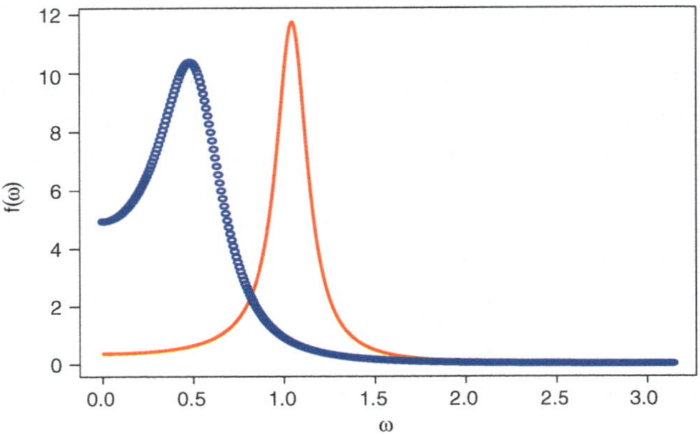

Fig. 11.1 Spectra of two $AR(2)$ processes with stochastic periods 12 (*circle*) and 6 (*line*)

As a consequence, we get a constant spectrum $g(\omega)$ when $\max(p_x, p_y) = 1$; this result is also true for the comparison of pure seasonal $AR(1)$ processes because of the *isometry* of the *AR* metric [24, 25].

A different situation happens in environmental studies, where it may be of interest to compare phenomena characterized by different cycles. These time series may be well fitted by $AR(2)$ processes with different stochastic periods $P_x \geq 2$ and $P_y \geq 2$, and amplitudes $\rho_x > 1$ and $\rho_y > 1$, respectively. Thus, the *AR* coefficients for X_t and Y_t are:

$$\phi_{1,x} = 2\rho_x^{-1} \cos\left(\frac{2\pi}{P_x}\right); \quad \phi_{2,x} = -\rho_x^{-2}; \qquad \phi_{1,y} = 2\rho_y^{-1} \cos\left(\frac{2\pi}{P_y}\right); \quad \phi_{2,y} = -\rho_y^{-2};$$

respectively. In this case, the W_t process is specified as:

$$W_t = \eta_t + \alpha_1 \, \eta_{t-1}, \qquad \eta_t \sim WN(0, \mid \phi_{1,x} - \phi_{1,y} \mid^2),$$

where $\alpha_1 = \mid \phi_{2,x} - \phi_{2,y} \mid / \mid \phi_{1,x} - \phi_{1,y} \mid$.

Then, the spectral decomposition (4) of $d^2(X, Y)$ derives from the spectrum:

$$g(\omega) = \frac{\delta_1^2}{\pi} \left[1 + \alpha_1^2 + 2\alpha_1 \cos(\omega)\right], \qquad 0 < \omega < \pi .$$

It is also possible to give a close formulation of spectral distribution function (5):

$$G(\omega) = \frac{\delta_1^2}{\pi \, d^2(X, Y)} \left[(1 + \alpha_1^2)\, \omega + 2\alpha_1 \sin(\omega)\right], \qquad 0 < \omega < \pi .$$

As an instance, Fig. 11.1 displays the spectra of two $AR(2)$ processes such that: $\phi_{1,x} = 1.3964$; $\phi_{2,x} = -0.65$; and $\phi_{1,y} = 0.9$; $\phi_{2,y} = -0.81$, which correspond to stochastic periods $P_x = 12$ and $P_y = 6$, respectively. Computation of the *AR* metric gives $d^2(X, Y) = 0.27201$.

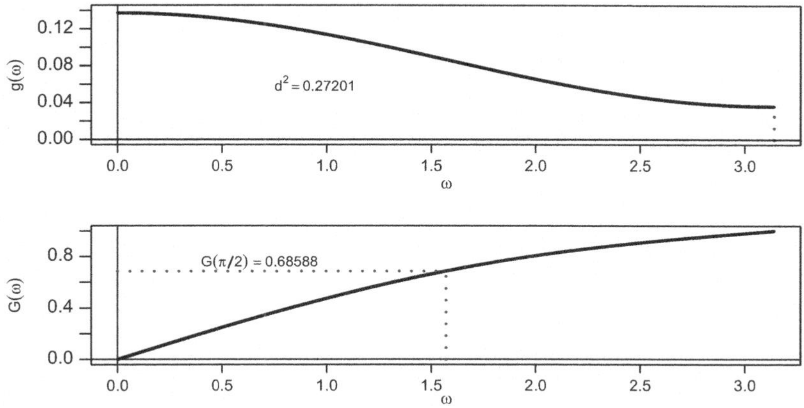

Fig. 11.2 Spectral decomposition and distribution function of the *AR* metric between $AR(2)$ processes

The W_t process is completely specified by:

$$W_t = \eta_t + 0.32232\,\eta_{t-1}, \qquad \eta_t \sim WN(0, 0.4964^2)$$

Thus, Fig. 11.2 (upper panel) shows the spectral decomposition of the distance $d^2(X, Y)$ between these $AR(2)$ processes and the corresponding spectral distribution function of the induced process W_t (lower panel).

It is evident that the main contributions to the distance among the two processes are given by low frequencies (that is high periods) of the composite process W_t. As a matter of fact, what really makes the differences between the given $AR(2)$ processes is concentrated over the angular frequencies $\omega \in (0, \pi/2)$ since they accounts for 69% of the total variance of W_t: in fact, $G(\pi/2) = 0.686$.

Further general considerations may be derived from these results: for instance, if two processes are characterized by stochastic periods such that $P_x = 2\,P_y/(P_y-2)$, then $g(\omega)$ is uniformly constant, since this implies $\phi_{2,x} = \phi_{2,y}$. As a consequence, two $AR(2)$ processes with stochastic periods are more and more distant inasmuch as these periods are different from 4 time units (which corresponds to $\omega = \pi/2$). This result is consistent with the circumstance that the sign of ϕ_1 coefficient in $AR(2)$ processes changes when the stochastic period moves around $\omega = \pi/2$, and of course this change increases d^2.

- *ARMA(1,1) processes.* The *AR* expansions of two $X_t \sim ARMA(1,1)$ and $Y_t \sim ARMA(1,1)$ processes, both belonging to $\mathscr{L}$, with a standard notation, are given by:

$$\pi_{j,x} = (\phi_x - \theta_x)\,\theta_x^{j-1}; \quad \pi_{j,y} = (\phi_y - \theta_y)\,\theta_y^{j-1}; \qquad j = 1, 2, \ldots$$

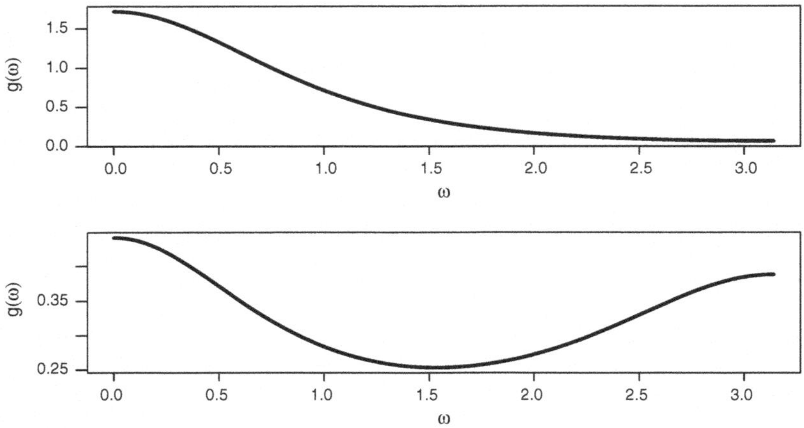

Fig. 11.3 Spectral decompositions of the *AR* metric for two $ARMA(1,1)$ processes

Then, $\delta_1 = \mid (\phi_x - \theta_x) - (\phi_y - \theta_y) \mid$ and the sequence of coefficients:

$$\alpha_j = \delta_1^{-1} \left| (\phi_x - \theta_x)\, \theta_x^{j-1} - (\phi_y - \theta_y)\, \theta_y^{j-1} \right|; \qquad j = 1, 2, \ldots,$$

converges to 0 when $\delta_1 \neq 0$. We can approximate the $\alpha(B)$ operator by omitting α_j coefficients for $j > M$, where the integer M is such that $\max(\theta_x^M, \theta_y^M) < \epsilon$, for a small $\epsilon > 0$. In this way, we consider as a convenient approximation for $g(\omega)$ the spectral decomposition $g_M(\omega)$ based on a truncated $\alpha_M(B)$ operator.

The upper panel of Fig. 11.3 shows the spectral decomposition obtained when we compare two $ARMA(1,1)$ processes which are quite similar, as those characterized by parameters: $\phi_x = 0.9; \theta_x = 0.3; \phi_y = 0.7; \theta_y = 0.2$, for instance. On the contrary, in the lower panel of Fig. 11.3 we display the decomposition for two very dissimilar $ARMA(1,1)$ processes, as those characterized by parameters: $\phi_x = -0.9; \theta_x = 0.3; \phi_y = 0.6; \theta_y = -0.4$, for instance.

The function $g(\omega)$ is mainly concentrated on the low frequencies in the first instance whereas it is shared almost evenly between high and low frequencies in the second case. As far as dissimilarity is measured by the *AR* metric, the distance in the first case is mainly accounted by low frequencies (and thus by the amount of difference about parameters which are concordant in signs); instead, in the second case, the distance increases with both low and high angular frequencies (since now the parameters are discordant in signs).

11.5 Concluding Remarks

This chapter provides a frequential analysis of the *AR* metric in order to evaluate the relevance and persistence of possible unobserved components in determining the observed distance. Among the possible fields of applications we quote the comparison of monthly indexes of industrial production (for allocating similarities among series to short or long movements in the economic cycle), the study of EEG time series data (for discriminating among safe and risky patterns) and so on.

Apparently, these issues are all achievable by other spectral measures proposed by many authors (see: [26], for some comments). However, spectral metrics are not always well defined for all processes belonging to $\mathscr{L}$ as it happens when processes with difference operators of different orders are to be compared. Thus, it seems worthwhile to explore the spectral features of a measure as the *AR* metric which is always well defined on $\mathscr{L}$.

As the first contribution on the topic, this work is mainly devoted to obtain useful results based on a full knowledge of the probability structure of the processes to be compared; further researches should translate such achievements in an inferential framework when models are estimated on real time series datasets. However, the availability of consistent and efficient maximum likelihood estimators should make such further extensions immediate.

In this respect, some empirical work on real dataset seems encouraging; however, further research is necessary for using such results in an effective way.

Acknowledgements The research has been supported by Department of Statistical Sciences, University of Naples Federico II.

References

1. Anderson, B., Moore, J.: Optimal Filtering. Prentice-Hall, Englewood Cliffs (1979)
2. Barbieri, M.M.: Decomposizione frequenziale dell'indice di determinismo lineare. Atti della XXXIV Riunione Scientifica della SIS, vol. 2, pp. 59–66. Nuova Immagine Editrice, Siena (1988)
3. Battaglia, F.: Inverse autocovariances and a measure of linear determinism for a stationary process. J. Time Anal. **4**, 79–87 (1983)
4. Beran, J.: Statistics for Long Memory Processes. Chapman & Hall, New York (1994)
5. Box, G.E.P., Jenkins, G.M.: Time Series Analysis: Forecasting and Analysis, revised edition, 1976. Holden-Day, San Francisco (1970)
6. Brockwell, P.J., Davies, R.A.: Time Series: Theory and Methods, 2nd edn. Springer, New York (1991)
7. Corduas, M.: Misure di distanza tra serie storiche e modelli parametrici. Quaderni dell'Istituto Economico Finanziario, n.3. Universit di Napoli Federico II, Naples (1992)
8. Corduas, M.: Uno studio sulla distribuzione asintotica della metrica autoregressiva. Statistica **LVI**, 321–332 (1996)
9. Corduas, M.: La metrica autoregressiva tra modelli ARIMA: una procedura in linguaggio GAUSS. Quaderni di Statistica **2**, 1–37 (2000)
10. Corduas, M.: Classifying daily streamflow time series by AR metric. In: 4th International Workshop on Space-Temporal Modelling, EDES, Sassari, 171–176 (2008)
11. Corduas, M.: Clustering streamflow time series for regional classification. J. Hydrol. **407**, 73–80 (2011)
12. Corduas, M., Piccolo, D.: Time series clustering and classification by the autoregressive metric. Comput. Stat. Data Anal. **52**, 1860–1872 (2007)
13. Di Iorio, F., Triacca, U.: Testing for non-causality by using Autoregressive metric. Dipartimento di Scienze Statistiche, Working paper (2010)

14. Gray, A., Markel, J.: Distance measures for speech processing. IEEE Trans. Acoust. Speech Signal Process. **ASSP-24**, 380–391 (1976)
15. Liao, T.: Clustering time series data: a survey. Pattern Recogn. **38**, 1857–1874 (2005)
16. Maharaj, E.A.: A significance test for classifying ARMA models. J. Stat. Comput. Simulat. **54**, 305–311 (1996)
17. Maharaj, E.A.: Clusters of time series. J. Classif. **17**, 297–314 (2000)
18. Otranto, E.: Classifying the markets volatility with ARMA distance measures. Quaderni di Statistica **6**, 1–19 (2004)
19. Otranto, E.: Clustering heteroskedastic time series by model-based procedures. Comput. Stat. Data Anal. **52**, 4685–4698 (2007)
20. Otranto, E.: Identifying financial time series with similar dynamic conditional correlation. Comput. Stat. Data Anal. **55**, 1–15 (2010)
21. Otranto, E., Triacca, U.: Testing for equal predictability of stationary ARMA processes. J. Appl. Stat. **34**, 1091–1108 (2007)
22. Piccolo, D.: Una topologia per la classe dei processi ARIMA. Statistica **XLIV**, 47–59 (1984)
23. Piccolo, D.: On a measure of dissimilarity between ARIMA models. In: Proceedings of the ASA Meetings, Business and Economic Section, pp. 231–236, Washington, DC (1989)
24. Piccolo, D.: A distance measure for classifying ARIMA models. J. Time Anal. **11**, 153–164 (1990)
25. Piccolo, D.: Statistical issues on the AR metric in time series analysis. In: Proceedings of the SIS 2007 Conference on "Risk and Prediction", pp. 221–232, CLEUP, Padova (2007)
26. Piccolo, D.: The autoregressive metric for comparing time series models. Statistica **LXX**, 459–480 (2010)
27. Triacca, U.: Feedback, causality and distance between ARMA models. Math. Comput. Simulat. **64**, 679–685 (2004)
28. Tunnicliffe Wilson, G.: Some efficient procedure for high order ARMA models. J. Stat. Comput. Simulat. **8**, 301–309 (1979)

12 Nonlinear Nonstationary Model Building by Genetic Algorithms

Francesco Battaglia and Mattheos K. Protopapas

Abstract

Many time series exhibits both nonlinearity and nonstationarity. Though both features have been often taken into account separately, few attempts have been proposed for modeling them simultaneously. We consider threshold models and present a general model allowing for several different regimes both in time and in levels, where regime transitions may happen according to self-exciting, or smoothly varying, or piecewise linear threshold modeling. Since fitting such a model involves the choice of a large number of structural parameters, we propose a procedure based on genetic algorithms, evaluating models by means of a generalized identification criterion. The proposed model building strategy is applied to a financial index.

Keywords

Evolutionary computation • Threshold model • Time series

12.1 Introduction

Traditional linear models such as ARMA models cannot fit successfully nonlinear or nonstationary time series. An interesting and simple alternative might be multi-regime threshold models. A time series that follows such a model is generated by several alternative linear autoregressive equations (the regimes). A generating process switches from one regime to another according to the value of an indicator, which may be related to time or to another time series (called the driving variable). In the models used in this chapter the driving variable is a delayed value of the time series itself, as well as time. When there are different regimes in time we have a

F. Battaglia (✉) · M.K. Protopapas
Dipartimento di Scienze Statistiche, Sapienza Universita di Roma, Italy
e-mail: Francesco.Battaglia@uniroma1.it; Mattheos.Protopapas@gmail.com

N. Torelli et al. (eds.), *Advances in Theoretical and Applied Statistics*,
Studies in Theoretical and Applied Statistics, DOI 10.1007/978-3-642-35588-2_12,

nonstationary but linear model [also called structural change, for a description see 1, 14]. When a delayed observation determines the regime the model is nonlinear but stationary, under suitable choices of the parameter values. These are usually called threshold models, see Tong [19].

It has been pointed out that a complex dynamic behavior could be reproduced by a nonstationary as well as a nonlinear model, and often a confusion may arise between the two kinds of models [see 7, 13]. This suggests that a parsimonious model might be obtained allowing simultaneous nonlinearity and nonstationarity. In this chapter we try to fit models that have different regimes, both in time and according to a delayed observation of the time series as the driving variable. Lundberg et al. [15] proposed a model based on an AR(1) structure with the parameter changing both according to time and to a driving variable alternating between two regimes only, with change determined by the logistic function [in the same way as smooth transition autoregressive models of 18]. Battaglia and Protopapas [5] introduced a change of the autoregressive parameters in a piecewise linear fashion [piecewise linear threshold multi-regime model, 3], as well as a smooth transition, and proposed a genetic algorithm to build a two regime model as well. Limitation to only two regimes is a serious constraint, especially with long series, where several regime changes in time are likely; but the implementation of multiple-regime models is much more difficult and makes the use of meta-heuristic methods still more convenient. We present here a genetic algorithm which allows up to four regimes (both in time and in levels), this involves nontrivial complication and requires a completely new coding strategy. Such a limitation seems appropriate for most practical applications, but could be removed at the cost of larger encoding complexity and computation time.

We stress our view, similar to that of [17], that models are only partial tools for describing some features of the data, and our aim is to discover not the absolutely correct generating process, but a reasonably parsimonious model, satisfactorily fitting the time series; we shall explain how genetic algorithms may be helpful for that search.

12.2 The Model and Estimates

The original autoregressive threshold model proposed by Tong [19] has at each t an autoregressive structure, where parameters change according to the value of another series (the driving variable). If the driving variable is given by the delayed series itself, we have a self exciting threshold model (SETAR). Let r_L denote the number of regimes, and $R_k = (l_{k-1}, l_k], k = 1, \dots, r_L$ a partition of the real line, the SETAR model is described by:

$$X_t = \phi_1^{(k)} X_{t-1} + \ldots + \phi_p^{(k)} X_{t-p} + \epsilon_t \quad \text{if } X_{t-d} \in R_k \ , \ k = 1, \dots, r_L \qquad (12.1)$$

where d is called the delay. Teräsvirta [18] generalized the SETAR model to the smooth transition autoregressive model (STAR) model, in which the parameters change in a continuous, smooth way, since the transition from one regime to the next is driven by a continuous function (usually a logistic). The STAR equation may be written:

$$X_t = \sum_{k=1}^{r_L} \sum_{j=1}^{p} \phi_j^{(k)} G_{k-1}(X_{t-d}) X_{t-j} + \epsilon_t \tag{12.2}$$

where $G_0(x) = 1$ and $G_k(x) = [1 + \exp(-\gamma_L(x - l_k))]^{-1}$. The autoregressive coefficients are essentially constant in each regime, with a continuous smooth change between regimes, whose speed is controlled by the constant $\gamma_L (> 0)$. The SETAR model may be interpreted as a special case of the STAR model, when γ tends to infinity.

Baragona et al. [3] proposed the piecewise linear threshold model (PLTAR), where the autoregressive coefficients change linearly and continuously with the driving variable X_{t-d}, but with different slope in each regime. PLTAR is described by:

$$X_t = \sum_{j=1}^{p} [\phi_j^{(0)} + \phi_j^{(1)} X_{t-d} + \sum_{k=2}^{r_L} \phi_j^{(k)} \max(0, X_{t-d} - l_{k-1})] X_{t-j} + \epsilon_t \,. \tag{12.3}$$

Here the autoregressive coefficient behaves like a linear spline across regimes. The PLTAR may be written in a similar fashion to the STAR letting $S_k(x) = \max(0, x - l_{k-1})$, $S_1(x) = x$, $S_0(x) = 1$ and

$$X_t = \sum_{k=1}^{r_L+1} \sum_{j=1}^{p} \phi_j^{(k)} S_{k-1}(X_{t-d}) X_{t-j} + \epsilon_t \,. \tag{12.4}$$

Note however that there is an additional parameter here (the linear term in X_{t-d}) for each lag. Therefore the sum over k ranges from 1 to the number of regimes plus 1. If the model has nonzero or varying means, intercept terms $\phi_0^{(k)}$ may be added. To model nonstationarity, we allow each of the coefficients $\phi_j^{(k)}$ to depend also on time, in a STAR or PLTAR way. Let r_T denote the number of regimes in time, and t_j denote the thresholds so that the regimes are defined by the partition $R'_k = (t_{k-1}, t_k]$, where $1 = t_0 < t_1 < \ldots < t_{r_T} = N$ (where N is the series length).

We introduce STAR dependence on time using

$$\phi_j^{(k)} = \sum_{i=1}^{r_T} \beta_j(i,k) G'_{i-1}(t) \ , \ j = 1, \ldots, p \tag{12.5}$$

with $G'_0(t) = 1$, $G'_i(t) = [1 + \exp(-\gamma_T(t - t_i))]^{-1}$, $i = 1, \ldots, r_T$.

Alternatively, a time nonstationarity following a PLTAR structure may be defined as

$$\phi_j^{(k)} = \sum_{i=1}^{r_T+1} \beta_j(i,k) S'_{i-1}(t) \ , \ j = 1, \ldots, p \tag{12.6}$$

where $S'_i(t) = \max(0, t - t_{i-1})$, $S'_1(t) = t$, $S'_0(t) = 1$.

After the combination of the different types of models that are driven by levels of the driving variable, and models driven by time, nine different kinds of models result: the type of model in time is either stationary, STAR, or PLTAR, while the type of the model in levels is either linear, STAR, or PLTAR. Since the final model equation is linear in the elementary parameters $\beta_j(i,k)$, the β's may be estimated by least squares.

The identification of a multi-regime model requires the selection of the solution that possesses the best properties, out of a large and discrete space of elements. Such kind of problems have been recently addressed by means of meta-heuristic methods [for a review see 4]. We prefer evolutionary computation methods, since they are population-based, and evolve several solutions in parallel; and among them, the genetic algorithm appears most suitable because the space of solutions is discrete, while others, e.g., differential evolution [16], particle swarm [12], or ITO algorithm [8], are employed for continuous spaces.

12.3 Genetic Algorithms

Genetic algorithms are heuristics that aim at finding near-optimal solutions in optimization problems. They are called that way because they are inspired by natural evolution [11]. Potential solutions of the problem are encoded as chromosomes (strings of binary digits). The algorithm works on a set of chromosomes called "population"; it evaluates these candidate solutions using a "fitness function" and evolves them using transformation operators: crossover—where two chromosomes of the population called "parent chromosomes" are recombined to form two "children chromosomes"—and mutation—where parts of a single chromosome are altered in a random way—in steps called generations, to find a near optimal solution to the problem at hand. For an introduction to genetic algorithms, the reader is referred to Holland [11] or Goldberg [9].

Chromosome Encoding. In the multi-regime threshold models analyzed here, we have the following decision variables: the type of model in time and in levels, the order, and in the case of STAR and PLTAR models the number of regimes in time and in levels, the delay of the driving variable and the thresholds in time and in levels. Finally, there are also the γ coefficients in time and/or in levels, in the case of a STAR model.

All the relevant decision variables are encoded in the chromosome in distinct bit-strings (genes). For determining the number of thresholds, the type of the model, and most other variables, we used standard binary encoding methods. Two further genes encode the γ parameters, allowing values between a minimum and a maximum. Since the γ parameter controls the speed of change from 0 to 1 in the logistic function, we can select a maximum value that makes the STAR model essentially indistinguishable from a SETAR, i.e., such that, for a sufficiently small value ϵ the logistic function has value ϵ immediately before the threshold, and value $(1-\epsilon)$ immediately after. On the other side, in order to select the minimum gamma values, we assume that the change from ϵ to $1-\epsilon$ in the logistic function requires an interval not longer than $1/q$ of the full observation interval. Details may be found in [5]. We chose $\epsilon = 0.01$ and $q = 10$.

The most critical point is coding the thresholds, since many intuitive methods lead to several illegal chromosomes and/or to large redundancy [see e.g. 2]. We use for each threshold a gene consisting of 12 bits, which is first transformed to a real value $u_i, 0 \leq u_i \leq 1$. Then the range of possible values for the threshold is computed and the gene u_i is mapped to that range. Since we assume a minimum of m observations in each regime, the allowable value for each threshold depends upon the number of regimes and the previous thresholds. For example, for 2 regimes in time, the first m observations should fall in the first regime, and the last m into the second; thus a value of zero in u_1 corresponds to a threshold given by the mth observation of the time series and a value of one corresponds to observation $N - m$ (N is the total number of observations). If there are three regimes, at least the last m values should fall in the third regime, and at least m preceding observations should fall into the second. Then, $u_1 = 0$ determines the first threshold equal to m as before, while a value $u_1 = 1$ corresponds to the fact that the first regime is as large as possible: in that case the first threshold corresponds to observation $N - 2m$, the second regime consists of the observations $N - 2m + 1, \ldots, N - m$, and the third of the observations $N - m + 1, \ldots, N$: essentially, we map the interval (0, 1) to the interval $(m, N - 2m)$ of admissible values for the first threshold. Finally, if there are four regimes, we must account for a minimum of m observations in each regime, therefore the admissible values for the first thresholds are between m and $N - 3m$. A similar technique is applied to the coding of the second and third threshold. The whole process may be described in the following equations.

- Two regimes. If the gene is denoted by u_1, the threshold is $t_1 = m + (N - 2m)u_1$
- Three regimes. Genes u_1 and u_2. Thresholds: $t_1 = m + (N - 3m)u_1$, and $t_2 = t_1 + m + (N - 2m - t_1)u_2$.
- Four regimes. The genes are denoted by u_1, u_2, u_3. The thresholds are obtained from: $t_1 = m + (N - 4m)u_1$; $t_2 = t_1 + m + (N - 3m - t_1)u_2$, and $t_3 = t_2 + m + (N - 2m - t_2)u_3$.

The values t_i determine the integer values of the thresholds in time (a number between m and $N - m$). For thresholds in levels, an extra step is required because the thresholds are the actual observations themselves. If the decoded gene value is τ, the corresponding threshold is the observed value Y_τ, where $\{Y_t\}$ denotes the observations arranged in increasing order (this definition has essentially the same effect as if choosing any real number in-between two consecutive—in terms of magnitude—observations, as the threshold).

In our implementation we set the minimum number of observations per regime equal to $m = N/5$. The choice of m is important since it balances between the flexibility of the model and the variability in parameter estimation.

Fitness, Crossover and Mutation. In order to compute the fitness, for each given chromosome the model is estimated by least squares and the residual variance estimate $\hat{\sigma}^2$ is computed. We use an identification criterion (generalized AIC) to determine the fitness. This is equivalent to computing a penalized gaussian likelihood: if model M has p parameters, the quantities $IC(M) = N \log \hat{\sigma}_M^2 + c(p)$ are evaluated and the model with minimum value is selected. Choosing a linear penalization function $c(p) = cp$ corresponds to the generalized AIC criterion [6]: the original Akaike's criterion is obtained for $c = 2$, while other criteria correspond to other choices of c; the Schwartz criterion for example corresponds to $c = \log N$. The value $c = 3$ is selected here, as in Battaglia and Protopapas [5], because in this case the behavior of the criterion is equivalent to a test of linearity–stationarity against two-regimes alternatives proposed by Lundberg et al. [15].

The fitness function must be related to the identification criterion through a monotone decreasing transformation, and is positive. We use a simple negative exponential transformation: $\text{fitness}(M) = \exp\{-IC(M)/N\} = \hat{\sigma}_M^{-2} \exp\{-cp/N\}$. We put p equal to the total number of estimated parameters, i.e., the β's in (12.5) and (12.6), plus the number of thresholds and gammas: thus, increasing the number of regimes implies multiplication of the value of p, which makes our procedure relatively conservative against nonlinearity and nonstationarity.

The rule used for parent selection is "roulette wheel selection," i.e., the probability of a chromosome to be selected as a parent is proportional to its fitness.

The crossover operator used is random point crossover [9]; potential cutting points are the boundary points between genes. Consequently the children chromosomes inherit each gene from only one parent. After crossover, a bit-wise independent mutation operator is used to—possibly—alter binary digits of the children chromosomes; the probability of mutating a bit (from 0 to 1, or vice-versa) is selected to be 0.025. We adopted a population size equal to 50.

Finally, a form of elitism is employed: after the next generation's population is formed, the best chromosome of the previous generation is implanted as it is into the population, thus replacing a random chromosome of the new generation's population.

12.4 Application: Dow Jones Index

We consider the daily closure values of the Dow Jones Industrial Average Index ($\hat{}DJL$) for years 2005–2009. Figure 12.1a displays the data, and Fig. 12.1b the returns. We only use data from January 2005 to September 2009 for the model (1,195 observations), leaving the last 3 months for out-of-sample forecasts.

The series shows an apparent instability both in mean and in variance, and is not well fitted by linear models. The best ARIMA model according to the Schwarz criterion is an ARI(2,1) but its residual variance, 18,895, is nearly equal to that of the first difference series (19,402). This corroborates the position of several analysts that the daily variations are essentially independent, and the best forecast of tomorrow is today's value (the *random walk* hypothesis).

We applied our genetic algorithm-based procedure using alternatively the original values or their first differences as driving variable, and repeated the applications

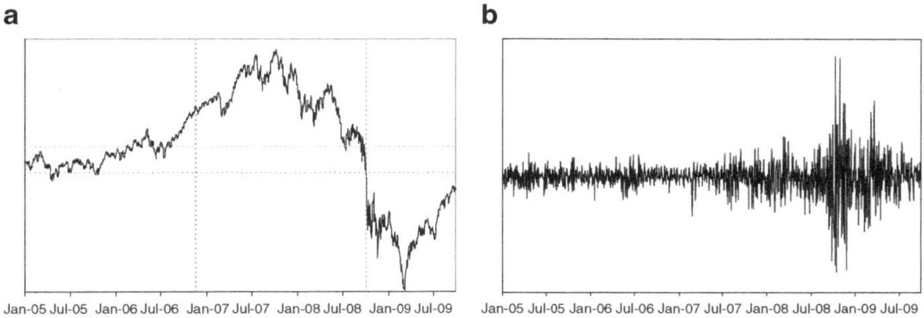

Fig. 12.1 Dow Jones Industrial Average Index, daily 2005–2009. (**a**) Values, (**b**) returns

Table 12.1 Number of observations in each regime

level	<10,252	10,252–11,090	>11,090
$t < 478$	19	300	151
$479 < t < 948$	2	15	453
$949 < t < 1195$	247	0	0

on the logs, transforming back to the original scale using the correction factor of Guerrero [10]. The best fitting results were obtained for original values, and will be exposed here.

The resulting optimal model is nonstationary and nonlinear, of the smooth transition type in both domains, with 3 regimes in levels (and thresholds 10,252, 11,090) and also 3 regimes in time (thresholds 478 and 948), order 2, delay 4, and residual variance 16111. The regimes are delimited by dotted lines in Fig. 12.1. We also recorded the best found linear nonstationary and the best found nonlinear stationary models, but their residual variance (18,668 and 18,428 respectively) are similar to that of the ARI(2,1) model, indicating that both features should be taken into account. However, looking at Fig. 12.1 and at Table 12.1, where the number of observations falling into each combination of regimes in time and levels are reported, an interesting fact arises: the regimes in time and levels interact, so that essentially in the second and third time regime the data are contained in only one level regime. Thus, the model suggests that for the first time period the series follows essentially a nonlinear two-regime model with threshold 11,090, while in the second and third period the fitting model is essentially a linear autoregressive model. The estimated model equations are (after a slight simplification) as follows:

- For the first time regime (January 2005 to November 22, 2006)

$$\nabla X_t = \{-0.03\nabla X_{t-1} - 70.5\}\{1 - G(X_{t-4} - 11090)\} + \{0.05\nabla X_{t-1} - 223.5\}G(X_{t-4} - 11090) + \varepsilon_t$$

- For the second time period (November 24, 2006 to October 7, 2008)

$$X_t = 0.86X_{t-1} + 0.125X_{t-2} + 181.5 + \varepsilon_t$$

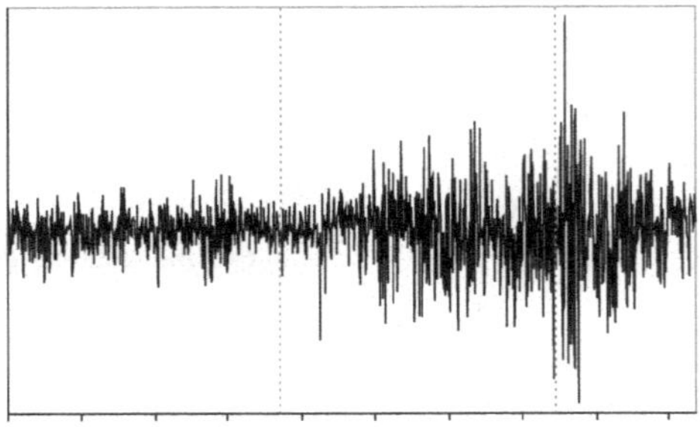

Fig. 12.2 Residuals of the STAR–STAR model

Table 12.2 Mean absolute errors of the lead-1 to lead-5 forecasts obtained with the ARI(2,1) model, the STAR–STAR fitted model and the random walk; observations from October, 1st to December 30h, 2009

model	lead-1	lead-2	lead-3	lead-4	lead-5
STAR–STAR	67.3	94.1	108.8	116.9	124.2
Random Walk	70.4	101.6	120.6	132.7	143.1
ARI(2,1)	75.6	108.4	128.3	147.4	165.8

- For the third time period (October 8, 2008 to September 30, 2009)

$$X_t = 0.815X_{t-1} + 0.162X_{t-2} + 240.8 + \varepsilon_t$$

where $G(x) = [1 + \exp\{-0.7x\}]^{-1}$. The models for the second and third regime are stationary autoregressive, though one root is very near to unity. The residuals are plotted in Fig. 12.2.

Note that though the search space allows for very complicated models, the optimal choice is relatively simple, suggesting three different time periods, and describing the first one with a two-regime STAR model, and the next two with ordinary autoregressive structures.

On considering the forecasts obtained with this model (in Table 12.2 we display the mean absolute errors of lead-1 to lead-5 forecasts for the out-of-sample observations from October 1, 2009 to December 30, 2009) we see that there is a substantial improvement over both the random walk and the ARI(2,1) model fitted to the entire series, and the advantage increases for larger lead times.

It may be observed that one of the most interesting features of the financial indexes is heteroskedasticity, and also the present data, as may be seen from the returns plot in Fig. 12.1b, exhibit an unstable volatility. We suggest that our procedure may also contribute to the analysis of heteroskedasticity: if the residual series of the three time regimes of the fitted model are considered separately, and an Arch-LM test performed on each of them, only for the third series the test is significant; moreover, also the Ljung-Box test does not reject the null hypothesis for the first and the second time regime. Therefore we are led to conclude that a relevant heteroskedasticity is present only in the more recent observations (beginning from

October 2008). It may be checked that this last residual subseries is satisfactorily fitted by a GARCH(1,1), or even an ARCH(2) model.

Some other interesting features concerning volatility may be examined considering the series of the square residuals of the STAR–STAR model. First of all, for checking the asymmetric volatility hypothesis, implying that the conditional variance follows two different models according to the sign of the residual, we applied the proposed procedure to the squared residuals, using the actual residuals as driving variable. The best obtained model is a STAR stationary model, with 3 regimes and thresholds -77 and -12. It suggests that the square residuals are best fitted by different parameters for positive and negative residuals (and a further different behavior for very small negative residuals), but the variance explained by this model is only about 8 %, therefore the volatility asymmetry is not so evident.

In addition, it may be of interest to check if the segmentation in time and level regimes, found for the data, is relevant also for volatility changes. To see this, we have applied our procedure to the series of the squared residuals, using the original series as driving variable. The selected model has a STAR structure in time and in levels, and explains nearly one half of the variance. It has three regimes in time and thresholds 475, 948, and three regimes in levels with thresholds 9,875 and 11,479. This segmentation is very similar to that obtained for the original Dow Jones series, suggesting that the identified regimes are relevant both for the index values and for its volatility.

Acknowledgements This work was supported by European Commission through Marie Curie Research and Training Network COMISEF Computational Methods in Statistics, Econometrics and Finance, and by Italian Ministry of Education through a national research grant PRIN2007.

References

1. Bai, J., Perron, P.: Estimating and testing linear models with multiple structural changes. Econometrica **66**, 47–78 (1998)
2. Baragona, R., Battaglia, F. (2006) Genetic algorithms for building double threshold generalized autoregressive conditional heteroscedastic models of time series. In: Rizzi, A., Vichi, M. (eds.) COMPSTAT 2006 Proceedings in Computational Statistics, pp. 441–452. Physica-Verlag, Heidelberg (2006)
3. Baragona, R., Battaglia, F., Cucina, D.: Fitting piecewise linear threshold autoregressive models by means of genetic algorithms. Comput. Stat. Data Anal. **47**, 277–295 (2004)
4. Baragona, R., Battaglia, F., Poli, I.: Evolutionary Statistical Procedures. Springer, Berlin (2011)
5. Battaglia, F., Protopapas, M.K.: Time-varying multi-regime models fitting by genetic algorithms. J. Time Anal. **32**, 237–252 (2011)
6. Bhansali, R.J., Downham, D.Y.: Some properties of the order of an autoregressive model selected by a generalization of Akaike's EPF criterion. Biometrika **64**, 547–551 (1977)
7. Carrasco, M.: Misspecified structural change, thresholds and Markov-switching models. J. Econometrics **109**, 239–273 (2002)
8. Dong, W., Hu, Y.: Time series modeling based on ITO algorithm. In: ICNC 2007 Third International Conference on Natural Computation, vol. 5, pp. 671–678. IEEE Computer Society (2007)

9. Goldberg, D.E.: Genetic Algorithms in Search, Optimization and Machine Learning. Addison-Wesley, Reading (1989)
10. Guerrero, V.M.: Time series analysis supported by power transformations. J. Forecasting **12**, 37–48 (1993)
11. Holland, J.H.: Adaptation in Natural and Artificial Systems. University of Michigan Press, Ann Arbor (1975)
12. Kennedy, J., Eberhart, R.: Swarm Intelligence. Morgan Kaufman, San Mateo, CA (2001)
13. Koop, G., Potter, S.: Are apparent finding of nonlinearity due to structural instability in economic time series? Econometrics J. **4**, 37–55 (2001)
14. Lin, C., Teräsvirta, T.: Testing the constancy of regression parameters against continuous structural change. J. Econometrics **62**, 211–228 (1994)
15. Lundberg, S., Teräsvirta, T., van Dijk, D.: Time-varying smooth transition autoregressive models. J. Bus. Econ. Stat. **21**, 104–121 (2003)
16. Price, K.V., Storn, R., Lampinen, J.: Differential Evolution, a Practical Approach to Global Optimization. Springer, Berlin (2005)
17. Rissanen, J.: Information and Complexity in Statistical Models. Springer, Berlin (2007)
18. Teräsvirta, T.: Specification, estimation and evaluation of smooth transition autoregressive models. J. Am. Stat. Assoc. **89**, 208–218 (1994)
19. Tong, H.: Non Linear Time Series: A Dynamical System Approach. Oxford University Press, Oxford (1990)

Using the Autodependogram in Model Diagnostic Checking

13

Luca Bagnato and Antonio Punzo

Abstract
In this chapter the autodependogram is contextualized in model diagnostic checking for nonlinear models by studying the lag-dependencies of the residuals. Simulations are considered to evaluate its effectiveness in this context. An application to the Swiss Market Index is also provided.

Keywords
χ^2-test • Autodependogram • Model diagnostic checking • Nonlinear time series • Serial independence

13.1 Introduction

As for serial correlation (better known as *autocorrelation*) in the linear case, the analysis of serial dependence (that, for analogy, could be defined as *autodependence*) is fundamental in the nonlinear approach. The term "serial" emphasizes that the dependence/correlation structure is analyzed as a function of the time lags. The study of the serial dependence is particularly useful in *model diagnostic checking* (also known as *model validation*) when the residuals from a nonlinear model are considered.

L. Bagnato (✉)
Dipartimento di Discipline Matematiche, Finanza Matematica e Econometria
Università Cattolica del Sacro Cuore, Italy
e-mail: luca.bagnato@unicatt.it

A. Punzo
Dipartimento di Economia e Impresa
Università di Catania, Italy
e-mail: antonio.punzo@unict.it

N. Torelli et al. (eds.), *Advances in Theoretical and Applied Statistics*,
Studies in Theoretical and Applied Statistics, DOI 10.1007/978-3-642-35588-2_13,

This chapter addresses a particular aspect of model validation, that is, the evaluation of the autodependencies (in the following alternatively said lag-dependencies) of the estimated residuals. With this aim, the *autodependogram* introduced in [4] is considered. This diagram is based on the lag-independence test proposed in [2] which is, substantially, a serial version of the well-known Pearson χ^2-test of independence. In analogy with the *autocorrelogram*, the autodependogram is obtained representing the lags on the x-axis and the values of the χ^2-statistic, corresponding to each lag, on the y-axis. Similar to the autocorrelogram, the autodependogram has a critical line emphasizing the lags at which there is evidence against the null hypothesis of independence.

Although the autocorrelogram and the autodependogram appear to be very similar in aspect, they are substantially different. First, the autocorrelogram is sensitive only to linear lag-dependencies while the autodependogram can detect linear/nonlinear lag-dependencies (for this reason, it is defined as *omnibus* in [4]). Second, the bars of the autocorrelogram measure the "strength" of the linear lag-dependencies, while those of the autodependogram quantify the "evidence" about the presence of dependence at the corresponding lags. Although the strength of the dependence and its evidence are strictly related, we think that there is a deep conceptual distinction about them: the strength of the dependence is a descriptive aspect, while the evidence about the dependence is a purely inferential concept. This fact makes the autodependogram a perfect tool in identification and validation of nonlinear models.

To summarize, Sect. 13.2 presents the problem of graphically checking autodependencies of the residuals from a statistical model. Here the contextualization is in the nonlinear frame and the autodependogram is briefly illustrated. The method proposed in [4] to select the number of classes of the contingency tables used to construct the bars of the autodependogram is also recalled. In Sect. 13.3, some convenient experiments on popular models in the time series literature are performed to compare the performances of the autocorrelogram and the autodependogram in terms of size and power of each of their bars. In Sect. 13.4 a financial dataset is considered to appreciate the advantages in using the autodependogram for model validation. Conclusions are finally given in Sect. 13.5.

13.2 The Autodependogram Applied to Residuals

Let $\{X_t\}_{t\in\mathbb{N}}$ represent a strictly stationary and ergodic stochastic process. Let $(X_t, \ldots, X_{t-n+1})$ be a single realization, of length n, of the process at the time t. In order to describe the process, suppose to use the nonlinear (parametric/nonparametric) model

$$X_t = f(\boldsymbol{\xi}_t) + \sigma(\boldsymbol{\xi}_t)\,\varepsilon_t, \tag{13.1}$$

where $\boldsymbol{\xi}_t = (X_{t-1}, \ldots, X_{t-d}, \varepsilon_{t-1}, \ldots, \varepsilon_{t-v})'$, $f(\cdot)$ and $\sigma(\cdot)$ are known/unknown functions of d past X_t's and v past ε_t's, with $d < n$ and $v < n$, and where $\{\varepsilon_t\}$

Table 13.1 Lagged residuals ε_{t-r}, $r = 1, \ldots, l$, $l < n - p$, and contingency table related to the shaded residuals

a Lagged residuals

ε_t	ε_{t-1}	$\cdots$	ε_{t-r}	$\cdots$	ε_{t-l}
$\widehat{\varepsilon}_{t-n+p+1}$					
$\widehat{\varepsilon}_{t-n+p+2}$	$\widehat{\varepsilon}_{t-n+p+1}$				
$\vdots$	$\vdots$	$\ddots$			
$\widehat{\varepsilon}_{t-n+p+r+1}$	$\widehat{\varepsilon}_{t-n+p+r}$	$\cdots$	$\widehat{\varepsilon}_{t-n+p+1}$		
$\vdots$	$\vdots$		$\vdots$	$\ddots$	
$\widehat{\varepsilon}_{t-n+p+l+1}$	$\widehat{\varepsilon}_{t-n+p+l}$	$\cdots$	$\widehat{\varepsilon}_{t-n+p+l+1-r}$	$\cdots$	$\widehat{\varepsilon}_{t-n+p+1}$
$\vdots$	$\vdots$		$\vdots$		$\vdots$
$\widehat{\varepsilon}_{t-1}$	$\widehat{\varepsilon}_{t-2}$	$\cdots$	$\widehat{\varepsilon}_{t-r-1}$	$\cdots$	$\widehat{\varepsilon}_{t-l-1}$
$\widehat{\varepsilon}_{t}$	$\widehat{\varepsilon}_{t-1}$	$\cdots$	$\widehat{\varepsilon}_{t-r}$	$\cdots$	$\widehat{\varepsilon}_{t-l}$

b Contingency table

ε_t \ ε_{t-r}	$D_1^{(r)}$	$\cdots$	$D_j^{(r)}$	$\cdots$	$D_k^{(r)}$	
$C_1^{(r)}$	$n_{11}^{(r)}$	$\cdots$	$n_{1j}^{(r)}$	$\cdots$	$n_{1k}^{(r)}$	
$\vdots$	$\vdots$		$\vdots$		$\vdots$	
$C_i^{(r)}$	$n_{i1}^{(r)}$	$\cdots$	$n_{ij}^{(r)}$	$\cdots$	$n_{ik}^{(r)}$	
$\vdots$	$\vdots$		$\vdots$		$\vdots$	
$C_k^{(r)}$	$n_{k1}^{(r)}$	$\cdots$	$n_{kj}^{(r)}$	$\cdots$	$n_{kk}^{(r)}$	
						$n^{(r)}$

is a serial independent process. Model (13.1) can be viewed, in the nonparametric context, as a generalization of a NARCH model adopted in [6] where $f(\cdot)$ and $\sigma(\cdot)$ depend only on the lagged variables $(X_{t-1}, \ldots, X_{t-d})$. Naturally a simpler model can be used; here, formulation (13.1) is considered to highlight the great flexibility, in nonlinear model validation, provided by the autodependogram.

Now, let $p = \max(d, v)$. Once estimated model (13.1), we can obtain the residual vector $\left(\widehat{\varepsilon}_t, \ldots, \widehat{\varepsilon}_{t-n+p+1}\right)$, of length $n - p$, in which the elements are obtained from the simple relations

$$\widehat{\varepsilon}_t = \frac{X_t - \hat{f}\left(\boldsymbol{\xi}_t\right)}{\hat{\sigma}\left(\boldsymbol{\xi}_t\right)}, \ldots, \widehat{\varepsilon}_{t-n+p+1} = \frac{X_{t-n+p+1} - \hat{f}\left(\boldsymbol{\xi}_{t-n+p+1}\right)}{\hat{\sigma}\left(\boldsymbol{\xi}_{t-n+p+1}\right)}.$$

Naturally, $\left(\widehat{\varepsilon}_t, \ldots, \widehat{\varepsilon}_{t-n+p+1}\right)$ can be seen as a realization of $\{\varepsilon_t\}$ in (13.1).

Suppose to be interested in studying the lag-dependencies of $\{\varepsilon_t\}$ in model (13.1) using the observed counterpart $\left(\widehat{\varepsilon}_t, \ldots, \widehat{\varepsilon}_{t-n+p+1}\right)$. With this aim, we can adopt the autodependogram, presented in [4], in the new context of model validation. Thus, let l be an integer, with $0 < l < n - p$, and consider Table 13.1a where, for each lag r, $r = 1, \ldots, l$, the estimated residuals up to time $t - r$ are contained in the respective column. The autodependogram allows to study the generic dependence of lag r by using the well-known and general χ^2 statistic of independence. With this aim, the first step consists in creating the contingency table referred to the variables ε_t and ε_{t-r}, as shown in Table 13.1b, considering the $n^{(r)} = n - p - r$ elements highlighted in Table 13.1a. Before proceeding, the following remark should be made: if one focuses the attention on a fixed value of r, then evaluating the dependence between ε_t and ε_{t-r} is equivalent to evaluate it between ε_{t-1} and ε_{t-r-1}, or between ε_{t-2} and ε_{t-r-2}, and so on. This issue can be easily comprehended considering adjacent columns in Table 13.1a. Thus, in order to have the largest number of observations for each comparison, it is always convenient to fix ε_t that, from a practical point of view, it is equivalent to fix the first column in Table 13.1a.

In Table 13.1b, k is the number of (marginal) classes supposed, according to [2–4], to be the same for each variable and for each lag r, and $\left\{C_i^{(r)}\right\}_{i=1}^k$ and $\left\{D_j^{(r)}\right\}_{j=1}^k$ are two fixed sets of adjacent intervals for each r. Naturally, the procedure can be generalized by setting k as a function of r. Thus, the simple χ^2-test can be used to evaluate the dependence of lag r at level α. In particular, the test statistic

$$\widehat{\delta}_r = \sum_{i=1}^{k}\sum_{j=1}^{k}\frac{\left[n_{ij}^{(r)} - \hat{n}_{ij}^{(r)}\right]^2}{\hat{n}_{ij}^{(r)}} \tag{13.2}$$

is referred in [4] to AutoDependence Function (ADF), where $n_{ij}^{(r)}$ are the observed frequencies and $\hat{n}_{ij}^{(r)}$ are the theoretical frequencies under the (null) hypothesis of independence of lag r, with $i, j = 1, \ldots, k$. The asymptotic well-known results for a χ^2-statistic of independence, as shown via simulation in [4], are preserved in the serial context. Thus, we can state that $\widehat{\delta}_r$ is asymptotically distributed as a χ^2 with $(k-1)^2$ degrees of freedom. Note that all considerations and results described so far can also be extended to a vector of estimated residuals $\left(\widehat{\varepsilon}_t, \ldots, \widehat{\varepsilon}_{t-n+p+1}\right)$ with missing values.

To completely specify $\widehat{\delta}_r$ it is necessary to define the partitions $\left\{C_i^{(r)}\right\}_{i=1}^k$ and $\left\{D_j^{(r)}\right\}_{j=1}^k$. As in [4], we will adopt the so-called *equifrequency intervals*, which assign equal frequencies to each interval $C_i^{(r)}$ and to each interval $D_j^{(r)}$. It is straightforward to note that this partition rule is not deterministic but stochastic. Moreover, equifrequency intervals can be seen as the sample counterpart, via the empirical distribution function, of the equiprobability intervals defined on the true, but unknown, distribution function. Using the equifrequency intervals, only the value of k must be selected. Following [4], k is chosen such that

$$k = \min\left\{k_s, k_p\right\}, \qquad \text{with} \quad k_s = \left\lfloor \left(\frac{n^{(l)}}{5}\right)^{\frac{1}{2}} \right\rfloor \quad \text{and} \quad k_p = \left\lfloor 2^{\frac{11}{10}} \left(\frac{n^{(l)}-1}{|z_{1-\alpha}|}\right)^{\frac{1}{5}} \right\rfloor . \tag{13.3}$$

The operator $\lfloor\cdot\rfloor$ denotes the floor function while $z_{1-\alpha}$ stands for the $(1-\alpha)$-quantile of the standard normal distribution. The value of k_s is chosen in order to assure a sufficient adherence between the actual and nominal sizes of the test and it is a stronger version of the rule in [5] requiring at least five expected frequencies in each cell of the contingency table. The value of k_p is an adaptation, to the contingency tables, of the well-known formula of [7] introduced in order to maximize the power of the χ^2-test of goodness of fit. Simulations in [4] have confirmed the validity of the rule (13.3).

Finally note that, since k is fixed and equal for each lag, the critical value $\chi^2_{[(k-1)^2;1-\alpha]}$ can be used for each of the l tests of lag-independence. Then, a horizontal line at height $\chi^2_{[(k-1)^2;1-\alpha]}$ can be added to the autodependogram. In conformity with the autocorrelogram, the usefulness of such a line is clear since it provides a graphical representation of the acceptance/rejection of the hypothesis of lag-independence for each lag.

13.3 Simulation Study

In this section, Monte Carlo experiments are considered to evaluate the performance of the autodependogram, in terms of size and power of each of its bars, in the context of model validation. A comparison with the autocorrelogram is also considered. To perform these experiments, we use the `R` computing environment [8]. The adopted `R`-code is available from the authors upon request. The ACF was computed using the `acf` command in the `TSA` package. It considers the asymptotic confidence bands $\mp z_{1-\alpha/2}/\sqrt{n}$ which are usually taken into account when the ACF is used to test for linear independence. In what follows, the nominal size will be fixed at 0.05 and a maximum lag $l = 30$ will be considered.

The experimental scheme can be summarized as follows.

Step 1: Raw data are randomly generated from the following common models:

$$\text{AR}(1) \rightarrow X_t = 0.3X_{t-1} + \varepsilon_t \tag{13.4}$$

$$\text{ARCH}(1) \rightarrow X_t = \sqrt{h_t}\varepsilon_t, \quad h_t = 1 + 0.4X_{t-1}^2 \tag{13.5}$$

$$\text{GARCH}(1,1) \rightarrow X_t = \sqrt{h_t}\varepsilon_t, \quad h_t = 0.01 + 0.8h_{t-1} + 0.15X_{t-1}^2 \tag{13.6}$$

$$\text{Quadratic MA}(1) \rightarrow X_t = 0.5\varepsilon_{t-1}^2 + \varepsilon_t \tag{13.7}$$

The white noise $\{\varepsilon_t\}$ is generated from a standard normal.
For each model, 10,000 samples are generated and rejection rates, of both ACF and ADF, are recorded for each lag r, $r = 1, \ldots, l$ (here, the rth rejection rate corresponds to the proportion of times the rth bar exceeds the confidence bands). Each sample has length 900 even if only the final $n = 800$ observations are used. The value of k in each repetition is selected according to (13.3). To facilitate the size evaluation, in all displayed plots, a dotted horizontal line is placed at height 0.05.

Step 2: For each repetition, an AR(1) is fitted on raw data. Here rejection rates arise from ACF and ADF computed on the resulting residuals.

Step 3: For each repetition, an ARMA(1,1)-GARCH(1,1) is considered as a parametric example of model (13.1). With the term ARMA(1,1)-GARCH(1,1) we mean that a GARCH(1,1) is fitted on the residuals arising from an ARMA(1,1) estimated on the raw data. Rejection rates arise, in this case, from ACF and ADF computed on the resulting residuals from the GARCH(1,1).

Note that ARMA and GARCH models are estimated in `R`, with the maximum likelihood approach, by respectively using the `arima` command in the `stats` package, and the `garch` command in the `tseries` package.

Figure 13.1 displays rejection rates for model (13.4), on the left, and model (13.5), on the right. Here, each line of plots corresponds to one of the above steps. With reference to raw data for model (13.4), Fig. 13.1a shows as the autocorrelogram performs better (its bars are in fact higher for the first lags) than the autodependogram. As mentioned in [4], this is due to the generality of the dependence structure investigated by the ADF. In particular, because of this generality, the autodependogram is less powerful than the autocorrelogram when the structure of dependence characterizing the observed time series is the linear one, like for model (13.4). However, when in step 2 the AR(1) is estimated on raw data, it captures this linear dependence and, consequently, there is not more dependence structure in the resulting residuals; this can be observed in Fig. 13.1c. The same reasoning can be applied for the results in Fig. 13.1e. The ARMA$(1,1)$-GARCH$(1,1)$ is indeed a generalization of the AR(1) used in the second step; consequently, it will capture at least the same dependence structure captured by the AR(1).

Regarding model (13.5), Fig. 13.1b shows that the performance on the raw data of the two diagrams is inverted with respect to the previous one. More specifically, although the same general pattern can be observed, it is also possible to note that for the active lags, actually characterized only by nonlinear dependence, the ACF-size is greater than 0.05 (value which the ACF should assume). This is due to the fact that when the underlying process is uncorrelated but dependent, the asymptotic distribution of the ACF is not that which is taken into account to define the confidence bands. By considering the second step, the fitted AR(1) can only capture the linear dependence of the series; unfortunately, this is not the underlying dependence structure induced by the ARCH(1). The ADF recognizes this aspect since its power, on the first lag, is approximately maintained by moving from Fig. 13.1b to Fig. 13.1d. Finally, in the third step, the ADF of Fig. 13.1f reflects serial independence since the fitting of the ARMA$(1,1)$-GARCH$(1,1)$ allows to remove the true underlying dependence structure.

The same general considerations stated for model (13.5) hold also for model (13.6) since the fitted ARMA$(1,1)$-GARCH$(1,1)$ is more than able to capture the serial dependence structure induced by the GARCH(1,1). Finally, in Fig. 13.2f we can observe as the ADF preserves its power in correspondence to the first lag since the ARMA$(1,1)$-GARCH$(1,1)$ cannot capture the nonlinear dependence induced by model (13.7).

13.4 A Real Application to Financial Data

In this section we consider an application to a financial time series. The aim is to appreciate how the use of the autodependogram, applied on the residuals from a nonlinear (parametric) model, can detect significant lag-dependencies.

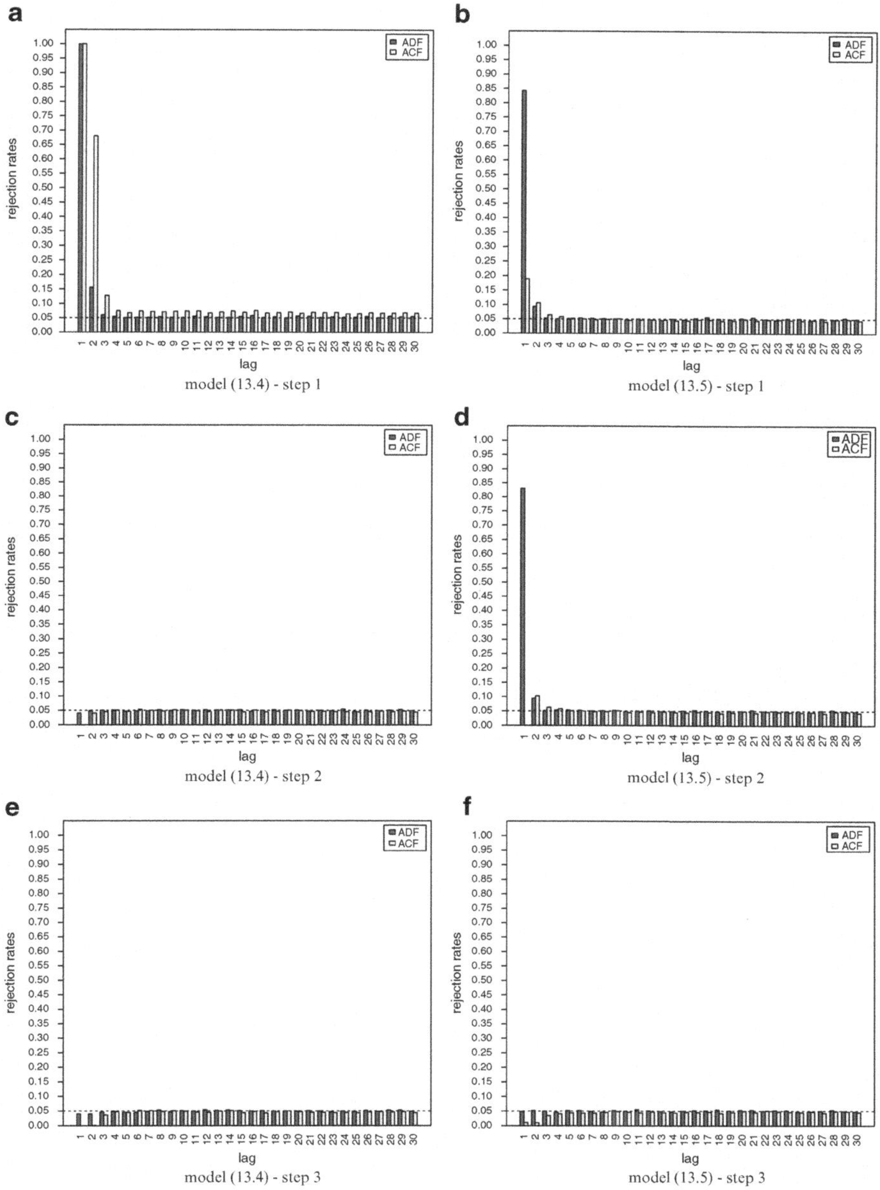

Fig. 13.1 Rejection rates of ACF and ADF for model (13.4), on the left, and model (13.5), on the right, in the various steps of the simulation procedure. Nominal size 0.05, $l = 30$, number of replications 10,000 and sample size $n = 800$

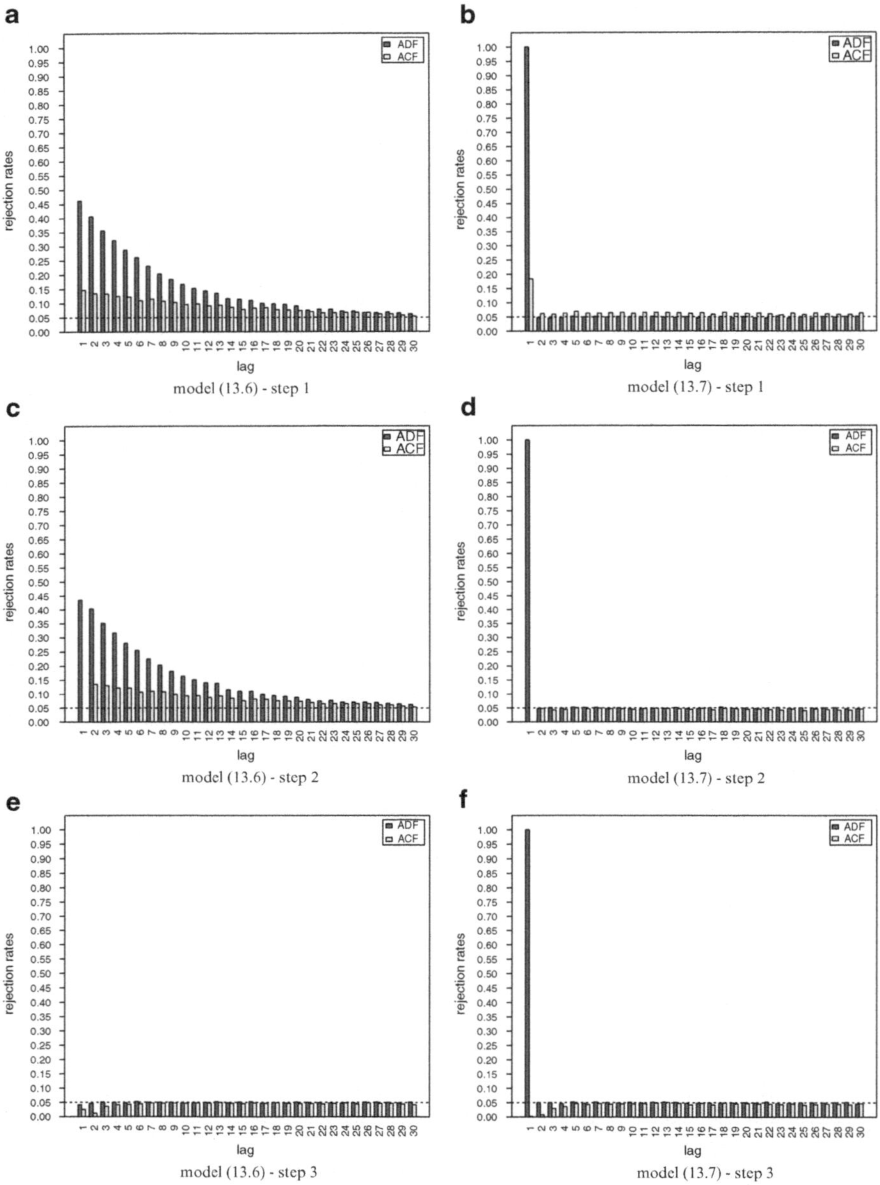

Fig. 13.2 Rejection rates of ACF and ADF for model (13.6), *on the left*, and model (13.7), *on the right*, in the various steps of the simulation procedure. Nominal size 0.05, $l = 30$, number of replications 10,000 and sample size $n = 800$

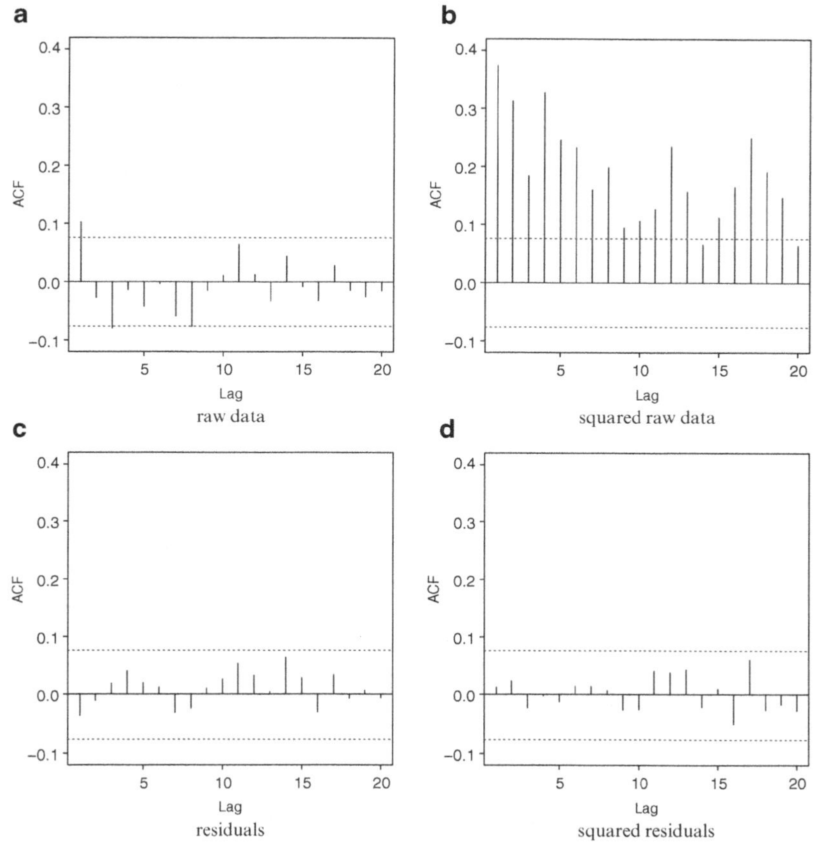

Fig. 13.3 Autocorrelograms related to the rates of return of the Swiss Market Index (SMI) for the period spanning from 7 August 2009 to 6 March 2012 ($n = 663$). By row, *from top to bottom*, the plots are referred to raw data and residuals from the ARMA(3,3)-GARCH(3,3). By column, *from left to right*, are reported the ACF on the series and ACF on the squared residuals ($l = 20$)

In particular, we consider daily returns of the Swiss Market Index (SMI), spanning the period from 7 August 2009 to 6 March 2012 ($n = 663$ observations downloadable from http://it.finance.yahoo.com/borse-mercati). Figure 13.3 shows a 2×2 matrix of plots related to the autocorrelogram ($l = 20$). In the first row, it is applied on raw data and on their squares; in Fig. 13.3b a clear presence of nonlinear dependence can be observed. So, we filter the time series with a sufficient flexible nonlinear (parametric) ARMA(3,3)-GARCH(3,3) in order to remove serial dependence. The last row of Fig. 13.3 shows, respectively, the ACF on the residuals and on the squared residuals from the ARMA(3,3)-GARCH(3,3). A simple graphical inspection and the Ljung-Box test computed by considering the first 10 lags (p-value equal to 0.932, on the residuals, and equal to 0.995 on

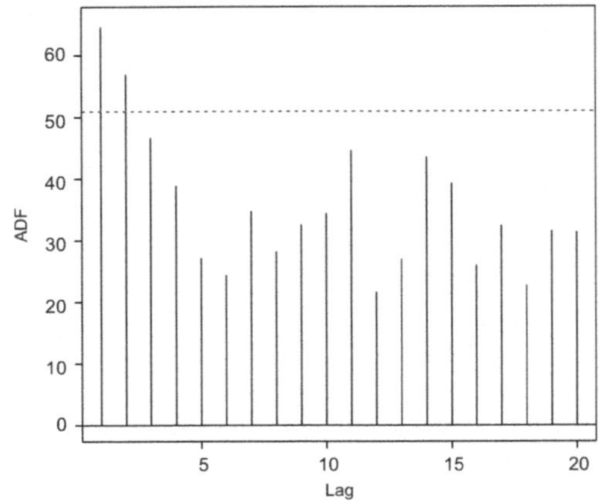

Fig. 13.4 Autodependogram on the residuals from the ARMA(3,3)-GARCH(3,3) fitted on rates of return of the Swiss Market Index (SMI) for the period spanning from 7 August 2009 to 6 March 2012 ($n = 663$)

the squared ones) confirm that both the residuals and the squared residuals are uncorrelated. On the contrary, the autodependogram on the residuals (see Fig. 13.4) highlights the presence of ADF-values statistically significant for the first two lags. This denotes a more complex type of nonlinear dependence due, for example, to a different kind of heteroscedasticity in the return series. Then, a more flexible model with nonparametric conditional mean and variance could be fitted on the observed data [1, 9]. By adding the information gained from the autodependogram in Fig. 13.4, such a type of dependence appears to be characterized by decaying memory properties.

13.5 Concluding Remarks

In this chapter the autodependogram proposed in [4] is applied in the context of nonlinear model diagnostic checking. Some experimental results on famous models in time series literature, and a real application to financial data, have shown the usefulness of this graphical device.

References

1. Bagnato, L.: Nonparametric ARCH with additive mean and multiplicative volatility: A new estimation procedure. Statistica & Applicazioni **VII**(1), 63–76 (2009)
2. Bagnato, L., Punzo, A.: On the use of χ^2-test to check serial independence. Statistica & Applicazioni **VIII**(1), 57–74 (2010)
3. Bagnato, L., Punzo, A.: Checking serial independence of residuals from a nonlinear model. In: Gaul, W., Geyer-shulz, A., Schmidt-thieme, L., Kunze, J. (eds.) Challenges at the Interface of Data Analysis, Computer Science, and Optimization, Studies in Classification, Data Analysis and Knowledge Organization, pp. 203–211. Springer, Berlin-Heidelberg (2012)
4. Bagnato, L., Punzo, A., Nicolis, O.: The autodependogram: a graphical device to investigate serial dependences. J. Time Anal. **33**(2), 233–254 (2012)

5. Cochran, W.G.: Some methods for strengthening the common χ^2 tests. Biometrics **10**(4), 417–451 (1954)
6. Jianqing, F., Qiwei, Y.: Nonlinear Time Series: Nonparametric and Parametric Methods. Springer, New York (2003)
7. Mann, H.B., Wald, A.: On the choice of the number of intervals in the application of the chi-square test. Ann. Math. Stat. **13**(3), 306–317 (1942)
8. R Development Core Team: R: A Language and Environment for Statistical Computing. R Foundation for Statistical Computing, Vienna (2011). ISBN 3-900051-07-0
9. Yang, L., Härdle, W., Nielsen, J.P.: Nonparametric autoregression with multiplicative volatility and additive mean. J. Time Anal. **20**(5), 579–604 (1999)

Part III

Statistical Modelling and Data Analysis

Refined Estimation of a Light Tail: An Application to Environmental Data

14

M. Ivette Gomes, Lígia Henriques-Rodrigues, and Frederico Caeiro

Abstract

In this chapter, we consider a recent class of *generalized negative moment* estimators of a negative extreme value index, the primary parameter in *statistics of extremes*. Apart from the usual integer parameter k, related to the number of top order statistics involved in the estimation, these estimators depend on an extra real parameter θ, which makes them highly flexible and possibly second-order unbiased for a large variety of models. In this chapter, we are interested not only on the adaptive choice of the *tuning* parameters k and θ, but also on an application of these semi-parametric estimators to the analysis of sets of environmental and simulated data.

Keywords

Bias reduction • Extreme value index • Moment estimator • Semi-parametric estimation

14.1 Introduction and Outline of the Paper

Whenever interested in extreme large events, the most common models in *statistics of extremes* are semi-parametric or even non-parametric in nature, with the

M.I. Gomes
CEAUL and FCUL, Campo Grande, 1749-016, Lisboa, Portugal
e-mail: ivette.gomes@fc.ul.pt

L. Henriques-Rodrigues (✉)
CEAUL and IPT, Tomar, Portugal
e-mail: Ligia.Rodrigues@aim.estt.ipt.pt

F. Caeiro
CMA and FCT, Universidade Nova de Lisboa, Portugal
e-mail: fac@fct.unl.pt

N. Torelli et al. (eds.), *Advances in Theoretical and Applied Statistics*, Studies in Theoretical and Applied Statistics, DOI 10.1007/978-3-642-35588-2_14,

imposition of a few "regularity conditions" in the right-tail, $\overline{F}(x) := 1 - F(x)$, as $x \to +\infty$, of an unknown model F underlying the available data. The primordial parameter is the *extreme value index* (EVI). For large values, the EVI is the shape parameter γ in the distribution function (d.f.)

$$\mathrm{EV}_\gamma(x) = \begin{cases} \exp\left(-(1+\gamma x)^{-1/\gamma}\right), \ 1+\gamma x > 0 & \text{if } \gamma \neq 0 \\ \exp(-\exp(-x)), \ x \in \mathbb{R} & \text{if } \gamma = 0, \end{cases} \tag{14.1}$$

the (unified) *extreme value* (EV) distribution.

The parameter γ needs to be estimated in a precise way, because such an estimation is one of the basis for the estimation of other parameters of extreme and large events, like the *right endpoint*, whenever finite, of the model F underlying the data (for further details on the subject, see Chaps. 1 and 4 of [11]).

In this chapter we are interested in an application of *statistics of extremes* to a set of environmental data. Due to the specificity of the data, we pay special attention to the estimation of the EVI γ, in (14.1), based on a quite recent estimator of a negative EVI, the *generalized negative moment* estimator [1]. In Sect. 14.2, we present some preliminary results in *extreme value theory* (EVT). In Sect. 14.3, we introduce the EVI-estimators under consideration, giving special emphasis to the *generalized negative moment* estimator. In Sect. 14.4, we provide an *Algorithm* for the adaptive choice of the *tuning* parameters under play in the semi-parametric estimation of the EVI through such an estimator. Finally, in Sect. 14.5, we apply the *Algorithm*, in Sect. 14.4, to the analysis of a set of environmental data, the daily average wind speeds in knots (one nautical mile per hour), collected in Dublin airport, in the period 1961–1978, as well as to a set of simulated data. Due to the seasonality of *Wind* data, we restrict ourselves to the Spring and Summer months.

14.2 Preliminary Results in EVT

One of the main results in EVT is related to the possible limiting laws, as $n \to \infty$, of the sequence $X_{n:n} := \max(X_1, X_2, \ldots, X_n)$, of maximum values of either independent, identically distributed random variables (r.v.'s) or possibly weakly dependent and stationary from a model F. In order to obtain a possible non-degenerate behaviour for $X_{n:n}$, we need to normalize it. Similar to the *central limit theorem* for sums or means, we know that if the maximum $X_{n:n}$, linearly normalized, converges to a non-degenerate r.v., then there exist real constants $\{a_n\}_{n\geq 1}$ $(a_n > 0)$ and $\{b_n\}_{n\geq 1}$, the so-called *attraction coefficients* of F, such that

$$\lim_{n\to\infty} \mathbb{P}\big((X_{n:n} - b_n)/a_n \leq x\big) = \mathrm{EV}_\gamma(x), \tag{14.2}$$

for some $\gamma \in \mathbb{R}$, with $\mathrm{EV}_\gamma(x)$ given in (14.1) [6, 9]. We then say that F is in the *domain of attraction* (for maxima) of EV_γ and we use the notation $F \in \mathscr{D}_{\mathscr{M}}(\mathrm{EV}_\gamma)$.

The EVI γ, in (14.1), measures essentially the weight of the right-tail $\overline{F} = 1-F$. If $\gamma < 0$, the right tail is *light*, i.e. F has a finite right endpoint ($x^F := \sup\{x : F(x) < 1\} < +\infty$). If $\gamma > 0$, the right-tail is *heavy*, of a *negative polynomial* type, i.e. F has an infinite right endpoint. If $\gamma = 0$, the right tail is of an *exponential* type and the right endpoint can be either finite or infinite.

Let us define $U(t) := F^{\leftarrow}(1 - 1/t)$ $(t > 1)$, with $F^{\leftarrow}(x) := \inf\{y : F(y) \geq x\}$ denoting thus the generalized inverse function of F. It is possible to prove (see [10, 11]) that, for all $x > 0$, we have the validity of the first-order condition,

$$F \in \mathscr{D}_{\mathcal{M}}(\mathrm{EV}_\gamma) \iff \lim_{t\to\infty} \frac{U(tx) - U(t)}{a(t)} = \begin{cases} \frac{x^\gamma - 1}{\gamma} & \text{if } \gamma \neq 0 \\ \ln x & \text{if } \gamma = 0. \end{cases}$$

We have $a(t) \equiv a_t \equiv a_{[t]}$, $[t]$ = integer part of t, with a_n the scale attraction coefficient in (14.2). Moreover, the location attraction coefficient b_n, also in (14.2), comes similarly from the relation $b_t = U(t)$.

In order to derive the asymptotic behaviour of semi-parametric EVI-estimators, we need to assume, for all $x > 0$, the validity of a second-order condition, like

$$\lim_{t\to\infty} \frac{\frac{U(tx)-U(t)}{a(t)} - \frac{x^\gamma - 1}{\gamma}}{A(t)} = H_{\gamma,\rho}(x) := \frac{1}{\rho}\left(\frac{x^{\gamma+\rho} - 1}{\gamma+\rho} - \frac{x^\gamma - 1}{\gamma}\right),$$

where $\rho \leq 0$ is a second order parameter controlling the speed of convergence in the first-order condition, and $|A(t)| \in RV_\rho$, with RV_α standing for the class of regularly varying functions at infinity with an index of regular variation α, i.e. positive measurable functions g such that $\lim_{t\to\infty} g(tx)/g(t) = x^\alpha$, for all $x > 0$.

14.3 Semi-parametric Estimation of Some Parameters of Rare Events

On the basis of the available random sample, $(X_1, \ldots, X_n)$, and the sample $(X_{1:n} \leq \cdots \leq X_{n:n})$ of associated ascending order statistics (o.s.'s), let us see how to estimate the EVI γ, the primordial parameter in *statistics of extremes*. For any integer $j \geq 1$, let us denote

$$M_{k,n}^{(j)} := \frac{1}{k}\sum_{i=1}^{k} \{\ln X_{n-i+1:n} - \ln X_{n-k:n}\}^j . \tag{14.3}$$

For the estimation of γ, we shall first refer the *moment* (M) estimator [3], valid for all $\gamma \in \mathbb{R}$. The M-estimator has the functional form,

$$\hat{\gamma}_{k,n}^{\mathrm{M}} := M_{k,n}^{(1)} + \tfrac{1}{2}\left\{1 - \left(M_{k,n}^{(2)}/[M_{k,n}^{(1)}]^2 - 1\right)^{-1}\right\}, \tag{14.4}$$

with $M_{k,n}^{(j)}$, $j = 1, 2$ defined in (14.3). Note that the statistic $M_{k,n}^{(1)}$, in (14.3), which plays also a main role in the *moment* estimator, in (14.4), is the well-known *Hill* estimator [12], often denoted $\hat{\gamma}^{\rm H}_{k,n} \equiv M^{(1)}_{k,n}$, the average of the log-excesses and valid, i.e. consistent, only for $\gamma \geq 0$. We can thus write the *moment* estimator in (14.4), as

$$\hat{\gamma}^{\rm M}_{k,n} = \hat{\gamma}^{\rm H}_{k,n} + \hat{\gamma}^{\rm NM}_{k,n}, \qquad \hat{\gamma}^{\rm NM}_{k,n} := \tfrac{1}{2}\Big\{1 - \Big(M^{(2)}_{k,n}/[M^{(1)}_{k,n}]^2 - 1\Big)^{-1}\Big\}, \tag{14.5}$$

with NM standing for *negative moment* estimator. Indeed, whereas the H-estimator, $\hat{\gamma}^{\rm H}_{k,n}$, is consistent for the estimation of $\gamma_+ := \max(0,\gamma)$ in the whole $\mathscr{D}_{\mathscr{M}}(\mathrm{EV}_\gamma)$, the NM-estimator, in (14.5), is consistent for the estimation of $\gamma_- := \min(0,\gamma)$ in the whole $\mathscr{D}_{\mathscr{M}}(\mathrm{EV}_\gamma)_{\gamma<0}$, provided that $k = k_n$ is an intermediate sequence, i.e. a sequence of integers such that $k = k_n \to \infty$ and $k_n = o(n)$, as $n \to \infty$. Under these same conditions, the M-estimator, in (14.4), is consistent for the estimation of any real γ, which can be written as $\gamma = \gamma_+ + \gamma_-$.

Most of the classical EVI-estimators have usually a high variance for small values of k and a high bias when k is large. This problem affects both the *moment* and the *Hill* estimators, and leads to a difficult choice of the "optimal" k, in the sense of the value k that minimizes the mean squared error (MSE). On the basis of a comment in [5], where it is noticed that when $\gamma > 0$, $\hat{\gamma}^{\rm H}_{k,n}$ has a smaller asymptotic variance than $\hat{\gamma}^{\rm M}_{k,n}$, and when $\gamma < 0$, $\hat{\gamma}^{\rm NM}_{k,n}$ and $\hat{\gamma}^{\rm M}_{k,n}$ have the same asymptotic variance, the authors in [1] were led to the introduction of a semi-parametric class of consistent estimators for $\gamma < 0$, which generalizes both the M-estimator, in (14.4), and the NM-estimator, in (14.5). Such a class, denoted NM(θ), is given by

$$\hat{\gamma}^{\mathrm{NM}(\theta)}_{k,n} := \hat{\gamma}^{\rm NM}_{k,n} + \theta\, M^{(1)}_{k,n}, \quad \theta \in \mathbb{R}. \tag{14.6}$$

The dependence on the tuning parameter $\theta \in \mathbb{R}$ makes such a class highly flexible, even able to become second-order reduced-bias for a great variety of models in $\mathscr{D}_{\mathscr{M}}(\mathrm{EV}_\gamma)_{\gamma<0}$. Indeed, with the appropriate choice of θ, $\hat{\gamma}^{\mathrm{NM}(\theta)}_{k,n}$, in (14.6), enables us to have access to an estimator of a negative EVI with a smaller asymptotic bias and the same asymptotic variance as the M-estimator. Note that we get the *negative moment* estimator in (14.5) for $\theta = 0$ and the *moment estimator*, in (14.4), for $\theta = 1$.

Figure 14.1 illustrates, for a few values of θ, $\theta = 0, 0.5, 0.9$ and 1, the behaviour of $\hat{\gamma}^{\mathrm{NM}(\theta)}_{k,n}$, in (14.6), versus k, for a sample of size $n = 6{,}574$ (the size of the whole sample of real data analysed in Sect. 14.5) from the *generalized Pareto* (GP) distribution with $\gamma = -0.25$. The GP d.f. is strongly related to the EV d.f. in (14.1), through the equation $\mathrm{GP}_\gamma(x) = 1 + \ln \mathrm{EV}_\gamma(x)$, $1 + \gamma x > 0$, $x > 0$.

For a large variety of models and under mild second-order conditions, like the one mentioned at the end of Sect. 14.2, we can guarantee the asymptotic normality of all the above-mentioned estimators and build approximate confidence intervals (CI's) for γ, as well as for all other parameters of extreme events. Indeed, for a negative EVI, adequate conditions on k and θ (see [1], for details), and with $\mathscr{N}(\mu,\sigma^2)$ denoting a normal r.v. with mean value μ and variance σ^2, we get

Fig. 14.1 Sample paths of $\hat{\gamma}_{k,n}^{\mathrm{NM}(\theta)}$, denoted NM($\theta$), against k, for one sample of size $n = 6{,}574$ from the GP$_\gamma$ model, with $\gamma = -0.25$

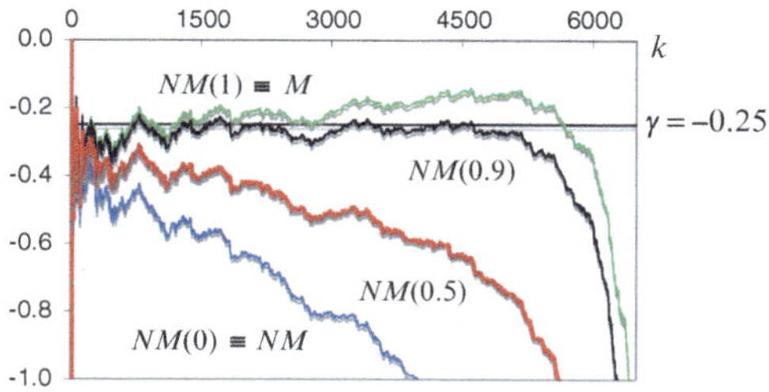

$$\sqrt{k}(\hat{\gamma}_{k,n}^{\mathrm{NM}(\theta)} - \gamma) \underset{n\to\infty}{\overset{d}{\longrightarrow}} \mathcal{N}\left(0,\ \sigma_{\mathrm{NM}}^2 = \frac{(1-\gamma)^2(1-2\gamma)(1-\gamma+6\gamma^2)}{(1-3\gamma)(1-4\gamma)}\right). \quad (14.7)$$

14.4 Adaptive Choice of the *Tuning* Parameters Under Play

The objective of this section is to provide an *Algorithm* for the choice of k and θ for the EVI-estimation through the *generalized negative moment* estimator in (14.6). Due to the fact that the main idea underlying this estimator is related to a possible reduction of bias, and γ is unknown, it is sensible to use any auxiliary statistic, strongly and directly related to the estimator under consideration, but going to the known value zero. This can be done in a way similar to the one considered in [7] for the choice of k in the *Hill* estimator, through the bootstrap methodology. Indeed, in the present setup, the most obvious auxiliary statistic is

$$T_{k,n}(\theta) := \gamma_{[k/2],n}^{\mathrm{NM}(\theta)} - \gamma_{k,n}^{\mathrm{NM}(\theta)} = \left(\hat{\gamma}_{[k/2],n}^{\mathrm{NM}} - \hat{\gamma}_{k,n}^{\mathrm{NM}}\right) + \theta\left(M_{[k/2],n}^{(1)} - M_{k,n}^{(1)}\right) \quad (14.8)$$

$$=: r_k + \theta s_k, \qquad k = 2, \ldots, n-1. \quad (14.9)$$

The stability of $T_{k,n}(\theta)$ around zero for moderate values of k, say $k \in [k_1, k_2]$, with $k_1 := [n^{0.05}] + 1$ and $k_2 := [n^{0.95}]$, enable us to choose

$$\hat{\theta} \equiv \hat{\theta}(k_1, k_2) := \arg\min_{\theta} \sum_{k=k_1}^{k_2} (r_k + \theta s_k)^2 = -\sum_{k=k_1}^{k_2} r_k s_k \Big/ \sum_{k_1}^{k_2} s_k^2, \quad (14.10)$$

where r_k and s_k have been defined in (14.9).

The choice of k for the EVI-estimation is next done on the basis of the bootstrap methodology, in a way similar to the one in [2,4,7] and more recently [8], and it is written algorithmically in the following:

1. Compute $\hat{\gamma}_{k,n}^{\mathrm{NM}(\hat{\theta})}$, $k = 1, 2, \ldots, n-1$, with $\hat{\gamma}_{k,n}^{\mathrm{NM}(\theta)}$ and $\hat{\theta}$ defined in (14.6) and (14.10), respectively.
2. Next, consider sub-sample sizes $n_1 = o(n)$ and $n_2 = [n_1^2/n] + 1$.
3. For l from 1 until $B = 250$, generate independently B bootstrap samples $(x_1^*, \ldots, x_{n_2}^*)$ and $(x_1^*, \ldots, x_{n_2}^*, x_{n_2+1}^*, \ldots, x_{n_1}^*)$, of sizes n_2 and n_1, respectively,

from the empirical d.f., $F_n^*(x) = \frac{1}{n}\sum_{i=1}^{n} I_{\{X_i \le x\}}$, associated with the observed sample $(x_1, \ldots, x_n)$.

4. Denoting $T_{k,n}^*(\hat{\theta})$ the bootstrap counterpart of $T_{k,n}(\hat{\theta})$, with $T_{k,n}(\theta)$ defined in (14.8), obtain $(t_{k,n_1,l}^*,\ 1 < k < n_1)$, $(t_{k,n_2,l}^*,\ 1 < k < n_2)$, $1 \le l \le B$, the observed values of the statistics T_{k,n_i}^*, $i = 1,2$. For $k = 1,2,\ldots,n_i - 1$, compute $\mathrm{MSE}^*(n_i,k) = \sum_{l=1}^{B} \left(t_{k,n_i,l}^*\right)^2/B,\ i = 1,2.$
5. Obtain
$$\hat{k}_{0|T}^*(n_i) := \arg\min_{1<k<n_i} \mathrm{MSE}^*(n_i,k), \quad i = 1,2. \tag{14.11}$$
6. For the estimation of the second-order parameter ρ, in (14.2), consider the bootstrap estimator of [2], given by
$$\hat{\rho}^* := \ln \hat{k}_{0|T}^*(n_1)/\big(2\ln(\hat{k}_{0|T}^*(n_1)/n_1)\big). \tag{14.12}$$
7. Compute the threshold estimate
$$\begin{aligned}\hat{k}_0^* &\equiv \hat{k}_0^*(n;n_1)\\ &:= \min\left(n-1, \left[(1-2^{\hat{\rho}^*})^{2/(1-2\hat{\rho}^*)}\left(\hat{k}_{0|T}^*(n_1)\right)^2/\hat{k}_{0|T}^*(n_1^2/n)\right]+1\right), \end{aligned} \tag{14.13}$$
with $\hat{k}_{0|T}^*(n_i)$ and $\hat{\rho}^*$ given in (14.11) and (14.12), respectively (see equation (29) and Sect. 4. of [7] for the theoretical details).
8. Obtain $\hat{\gamma}^* \equiv \hat{\gamma}^*(n;n_1) := \hat{\gamma}_{\hat{k}_0(n;n_1),n}^{\mathrm{NM}(\hat{\theta})}$, with $\hat{\gamma}_{k,n}^{\mathrm{NM}(\theta)}$, $\hat{\theta}$ and $\hat{k}_0(n;n_1)$ given in (14.6), (14.10) and (14.13), respectively.

Remarks:

1. If there are negative elements in the sample, the value of n should be replaced by $n^+ = \sum_{i=1}^{n} I_{\{X_i>0\}}$, the number of positive values in the sample.
2. The use of the sample of size n_2, $(x_1^*, \ldots, x_{n_2}^*)$, and of the extended sample of size n_1, $(x_1^*, \ldots, x_{n_2}^*, x_{n_2+1}^*, \ldots, x_{n_1}^*)$, lead us to increase the precision of the result with a smaller B, the number of bootstrap samples generated in Step 3. This is quite similar to the use of the simulation technique of *"Common Random Numbers"* in comparison problems, when we want to decrease the variance of a final answer to $z = y_1 - y_2$, inducing a positive dependence between y_1 and y_2.
3. The Monte Carlo procedure in Steps 3. up to 7. or 3. up to 8. of the *Algorithm* in this section can be replicated if we want to associate easily a standard error with the estimated parameters. The value of B can also be adequately chosen.
4. The main practical question which can be raised under this setup is related to the dependence of the method on the choice of the sub-sample size n_1, in Step 2. of the *Algorithm*. A possible choice seems then to be for instance $n_1 = [n^{0.90}]$ or $n_1 = [n^{0.95}]$. The first one will be the one used later on in Sect. 14.5. If the sample size n is large, values of n_1 larger than $[n^{0.95}]$ do not seem convenient, as we shall see later on in Sect. 14.5. Nevertheless, and particularly if we get rid

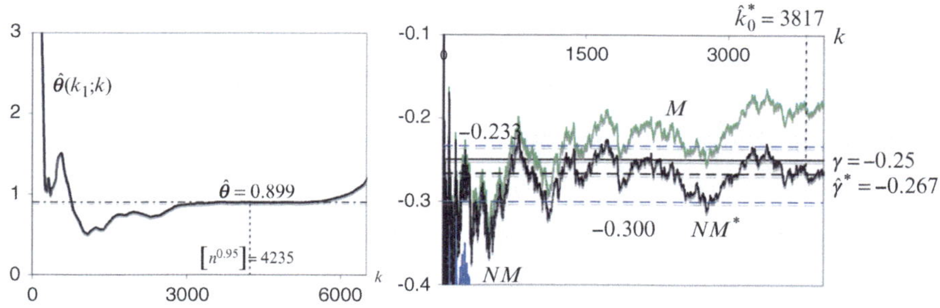

Fig. 14.2 Estimates $\hat{\theta}(k_1;k)$ (*left*), and estimates of the EVI provided by the estimators under consideration (*right*), for the simulated GP(−0.25) sample

of discrepant estimates of either $k^*_{0|T}(n_1)$ or $k^*_{0|T}(n_2)$, easy to detect and discard, the sensitivity of the estimates on n_1 becomes reasonably weak.

5. However, in practical applications, and contrarily to what has happened for positive γ (see [7, 8]), one should always do a sensitivity analysis for n_1.

14.5 Applications to Simulated and Environmental Data Sets

We shall first provide an application to a simulated data sample, a case where we know the value of γ. We next provide an application not only to a subset related to the periods from March 21 until June 20 of the *Wind* data set mentioned in Sect. 14.1, the so-called *Spring-Wind* data, but also to the *Summer-Wind* data set.

Simulated GP Data

The analysis of a randomly simulated sample from a GP model with $\gamma = -0.25$, with a size $n = 6{,}574$, the size of the *Wind* data set, led us to the choice $\hat{\theta} = 0.899$, as shown in Fig. 14.2, *left*.

It is sensible to note that $\hat{\theta}(k_1;k)$ exhibit a stable behaviour for a wide region of k-values. For this data set, we have got $\hat{\theta} = 0.9$ for $k \in [2793, 5884]$. The application of the *Algorithm*, in Sect. 14.4, for $n_1 = [n^{0.9}]$, led then to $\hat{k}_0^* = 3{,}817$ and the adaptive EVI-estimate $\hat{\gamma}^* = -0.267$, quite close to the true value $\gamma = -0.25$, and pictured in Fig. 14.2, *right*. In this case there is not a big discrepancy between the M and the NM*-estimators. On the basis of (14.7), an approximate 95 % confidence interval (CI) for γ is given by $\big(\hat{\gamma}^* - 1.96\,\hat{\sigma}_{\mathrm{NM}}/\sqrt{k},\ \hat{\gamma}^* + 1.96\,\hat{\sigma}_{\mathrm{NM}}/\sqrt{k}\big)$. In this case, we have got $\hat{\sigma}_{\mathrm{NM}} = 1.058$ and the approximate 95 % CI, (−0.300, −0.233), also pictured in Fig. 14.2, *right*.

Wind Data

The wind data set under analysis is the daily average wind speeds (in knots), collected in Dublin airport, in the period 1961–1978, with a size $n = 6{,}574$. To avoid seasonality we have split the original data on four data sets, according to the

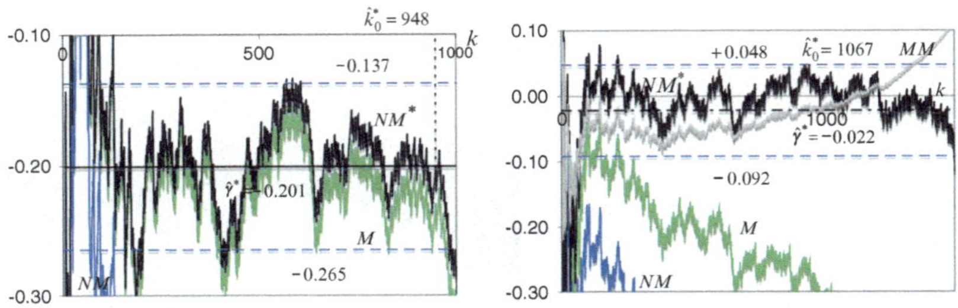

Fig. 14.3 Estimates of the EVI provided by the estimators under consideration, for the wind data (Spring period, at the *left*, and Summer period, at the *right*)

Table 14.1 Estimates associated with a GP_γ model, with $\gamma = -0.25$

p	$n_1 = [n^p]$	$n_2 = n_1^2/n$	$\hat{k}^*_{0\|T}(n_1)$	$\hat{k}^*_{0\|T}(n_2)$	$\hat{k}_0^*$	$\hat{\gamma}^*$
0.8	1,133	195	662	116	3,770	−0.22
0.9	2,729	1,132	1,576	650	3,817	−0.24
0.95	4,235	2,728	2,427	1,562	3,768	−0.22
0.99	6,020	5,512	3,442	3,162	3,744	−0.20
0.999	6,516	6,458	2,169	2,160	2,128	−0.22

season of the year, and we consider in this chapter the *Spring-Wind* data set, with a size $n = 1{,}656$, as well as the *Summer-Wind* data set, with a size $n = 1{,}692$.

- For the *Spring-Wind data*, we were led to $\hat{\theta} = 1.411$. The application of the *Algorithm*, in Sect. 14.4, for $n_1 = [n^{0.9}]$, led then to $\hat{k}_0^* = 948$ and the adaptive EVI-estimate $\hat{\gamma}^* = -0.201$, the value pictured in Fig. 14.3, *left*. In this case, the M and the NM^*-estimators almost overlap. We have got $\hat{\sigma}_{\text{NM}} = 1.005$ and the approximate 95 % CI, $(-0.265, -0.137)$, also pictured in Fig. 14.3, *left*, and clearly indicating a light right-tail.
- For the *Summer-Wind data*, we were led to the choice $\hat{\theta} = 1.655$. The application of the *Algorithm*, in Sect. 14.4, for $n_1 = [n^{0.9}]$, led then to $\hat{k}_0^* = 1{,}067$ and the adaptive EVI-estimate $\hat{\gamma}^* = -0.022$, the value pictured in Fig. 14.3, *right*. In this case, we have got $\hat{\sigma}_{\text{NM}} = 0.982$ and the approximate 95 % CI, $(-0.092, +0.048)$, also pictured in Fig. 14.3, *right*, indicating now the possibility of an exponential-type tail, i.e. of an EVI $\gamma = 0$.

Sensitivity of the Algorithm to the Choice of n_1

We shall now briefly address the sensitivity of the *Algorithm* in Sect. 14.4 to the choice of the sub-sample size n_1. In Table 14.1, as an illustration, we present a few estimates of $\hat{k}_0^*$ and $\hat{\gamma}^*$, for the *GP* data set.

Bootstrap estimates of the optimal sample fractions (OSFs) and of the EVI, γ, as functions of n_1, for n_1 running from $[n^{0.8}]$ until $[n^{0.999}]$, are pictured in Figs. 14.4–14.6, for the *GP*, the *Spring-Wind* and the *Summer-Wind* data, respectively. In those

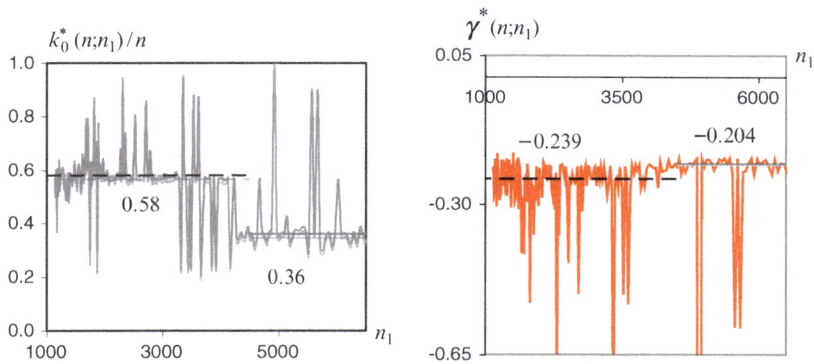

Fig. 14.4 Bootstrap estimates of the optimal sample fraction (*left*) and of the extreme value index γ (*right*), as a function of the sub-sample size n_1, for the *GP* data

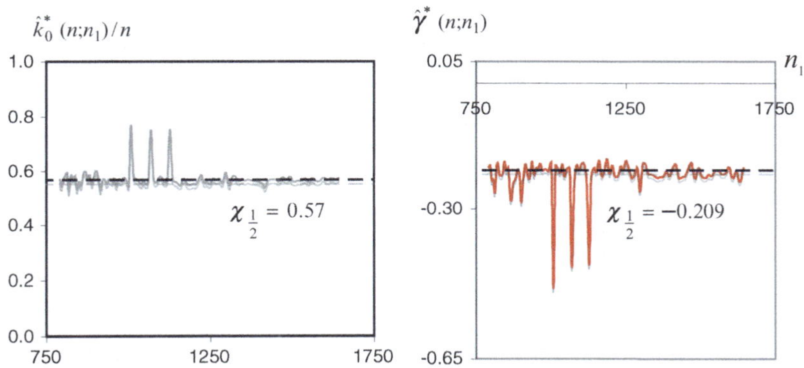

Fig. 14.5 Bootstrap estimates of the optimal sample fraction (*left*) and of the extreme value index γ (*right*), as a function of the sub-sample size n_1, for the *Spring-Wind* data

figures we also picture horizontal line(s), with the overall median of all values or the medians of different stability regions.

- In the case of an underlying GP model, with $\gamma = -0.25$, it seems sensible to advise the choice of sub-sample sizes $n_1 \leq [n^{0.95}]$. We then get a median equal to -0.239 (slightly over-estimating the true γ). For larger values of n_1 there appears a clear over-estimation of the EVI γ. A few sub-sample sizes n_1 lead to some extremely small estimates of γ, the estimates -0.84, -0.87 and -3.90, neatly discrepant from all the other estimates.
- For the *Spring-Wind* data, we can now go up to sub-sample sizes $n_1 = [n^{0.999}]$. The median of the bootstrap EVI-estimates associated with all these sub-sample sizes is equal to -0.209. There are however a few discrepant values, related to a weak estimation of $\hat{k}^*_{0|T}(n_2)$.
- For the *Summer-Wind* data, there is a high volatility in the estimation of the OSFs. A few sub-sample sizes n_1 (around 6.5 % of all sub-sample sizes considered)

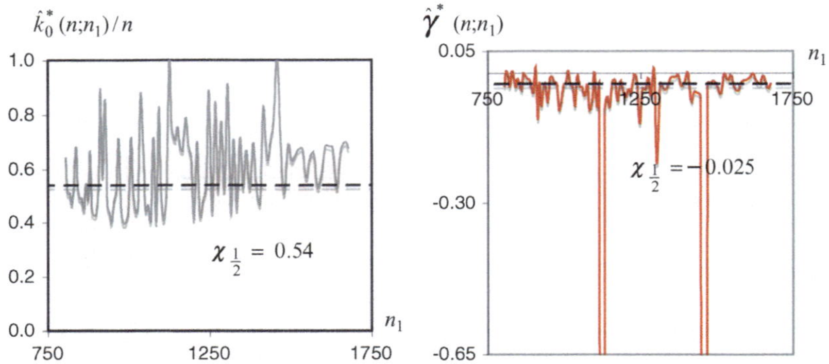

Fig. 14.6 Bootstrap estimates of the optimal sample fraction (*left*) and of the extreme value index γ (*right*), as a function of the sub-sample size n_1, for the *Summer-Wind* data

lead even to small positive values of $\hat{\gamma}^\star$, and there were two extremely small estimates of γ, the estimates -1.35 and -7.69, neatly discrepant from all the other estimates. The volatility of the bootstrap estimates of γ is also high. The main reason for the volatility of the estimates on n_1 is the bad bootstrap estimation of OSFs, but as mentioned above, we can easily get rid of such a volatility.

These results claim for a simulation study of the *Algorithm* in Sect. 14.4, a topic out of the scope of this chapter.

Acknowledgements Research partially supported by EXTREMA, PTDC/MAT/101736/2008, and National Funds through **FCT** — Fundação para a Ciência e a Tecnologia, projects PEst-OE/MAT/UI0006/2011 (CEAUL) and PEst-OE/MAT/UI0297/2011 (CMA/UNL).

References

1. Caeiro, F., Gomes, M.I.: An asymptotically unbiased moment estimator of a negative extreme value index. Discuss. Mathe. Probabil. Stat. **30**(1), 5–19 (2010)
2. Danielsson, J., Haan, L. de, Peng, L., Vries, C.G. de: Using a bootstrap method to choose the sample fraction in tail index estimation. J. Multivariate Anal. **76**, 226–248 (2001)
3. Dekkers, A.L.M., Einmahl, J.H.J., Haan, L. de: A moment estimator for the index of an extreme-value distribution. Ann. Stat. **17**(4), 1833–1855 (1989)
4. Draisma, G., Haan, L. de, Peng, L., Themido Pereira, T.: A bootstrap-based method to achieve optimality in estimating the extreme value index. Extremes **2**(4), 367–404 (1999)
5. Fraga Alves, M.I.: Weiss-Hill estimator. Test **10**, 203–224 (2001)
6. Gnedenko, B.V.: Sur la distribution limite du terme maximum d'une série aléatoire. Ann. Math. **44**, 423–453 (1943)
7. Gomes, M.I., Oliveira, O.: The bootstrap methodology in statistics of extremes — choice of the optimal sample fraction. Extremes **4**(4), 331–358 (2001)
8. Gomes, M.I., Figueiredo, F., Neves, M.M.: Adaptive estimation of heavy right tails: the bootstrap methodology in action. Extremes, **15**(4), 463–489 (2012)

9. Haan, L. de: On Regular Variation and its Application to the Weak Convergence of Sample Extremes. Mathematical Centre Tract 32, Amsterdam (1970)
10. Haan, L. de: Slow variation and characterization of domains of attraction. In: Tiago de Oliveira (ed.) Statistical Extremes and Applications, pp. 31–48. D. Reidel, Dordrecht (1984)
11. Haan, L. de, Ferreira, A.: Extreme Value Theory: An Introduction. Springer, New York (2006)
12. Hill, B.M.: A simple general approach to inference about the tail of a distribution. Ann. Stat. **3**(5), 1163–1174 (1975)

Model-Based Classification of Clustered Binary Data with Non-ignorable Missing Values

15

Francesco Lagona

Abstract
A hierarchical logistic regression model with nested, discrete random effects is proposed for the unsupervised classification of clustered binary data with non-ignorable missing values. An E-M algorithm is proposed that essentially reduces to the iterative estimation of a set of weighted logistic regressions from two augmented datasets, alternated with weights updating. The proposed approach is exploited on a sample of Chinese older adults, to cluster subjects according to their cognitive impairment and ability to cope with a Mini-Mental State Examination questionnaire.

Keywords
E-M algorithm • Latent classes • Multilevel data • Non-ignorable missing values • Random effects

15.1 Introduction

Clustered (or repeated) binary data arise when a battery of binary variables (or items) is observed on a sample of subjects, together with a number of subject- or item-specific covariates. These data are typically examined by generalized linear mixed models [12], specified on the basis of continuous (often normally distributed) random effects, which account for the dependence between the outcomes that have been observed on the same subject. In classification studies of repeated normal data, mixtures of linear mixed models are widely exploited to identify clusters of subjects according to a number of subpopulations with different correlation structures of the random effects [1]. Because mixtures of linear mixed models are hierarchical

---

F. Lagona (✉)
DIPES – University Roma Tre, via G. Chiabrera 199, 00145 Rome, Italy
e-mail: lagona@uniroma3.it

N. Torelli et al. (eds.), *Advances in Theoretical and Applied Statistics*, Studies in Theoretical and Applied Statistics, DOI 10.1007/978-3-642-35588-2_15,

models where the correlation structure of the clustered responses at the item level is allowed to vary between clusters of subjects, the classification outcomes depend on the distributional assumptions on the item-specific random effects, which are often difficult to motivate in real case studies. Misspecification of the random effect distribution can be avoided by considering mixtures of semiparametric generalized linear mixed models, also known as multilevel latent class models [14]. In the case of complete data information, maximum likelihood estimation of multilevel models can be obtained by extending standard E-M algorithms for finite mixture models to allow for the nested structure of the latent classes [14]. When observations are incomplete, i.e., some of the clustered outcomes on a subject are missing, the missing mechanism should be taken into account, in order to obtain efficient MLEs [13]. If the data are ignorably missing, i.e., the probability of not observing a value does not depend on the unobserved value, then efficient MLEs can be found by maximizing the marginal likelihood, obtained by integrating the likelihood function of the complete data with respect to the missing values. If, otherwise, the data are non-ignorably missing, then a missing value is informative of the unobserved value and a more complex likelihood function must be maximized, obtained by specifying the joint distribution of the complete data and the missing pattern for each subject. Such joint models can be classified into either "pattern mixture models" or "selection models." A pattern mixture model factors the joint distribution into the conditional distribution of the response given the missingness pattern, and the marginal distribution of missing indicators. A selection model factors the joint distribution into the marginal distribution of the response and the conditional distribution of missing indicators given the response. Because both specifications have advantages and disadvantages, the choice of a specific approach relies on the purpose of the analysis [10]. In classification studies the interest lies in the identification of clusters according to the marginal distribution of the response, and the selection model has the advantage of directly parameterizing the marginal distribution of the response. Moreover, when the dimension of the response is high, such as in the case considered in this chapter, pattern mixture models are weakly identifiable if some of the missing patterns are either observed on a small part of the sample or not observed. These issues motivate the choice of a selection approach. A number of different selection models have been proposed in the literature, depending on the assumptions on the process that generates the missing values. Shared random effect models are parsimonious selection models where the missingness is dependent on an unobserved random effect underlying the observed and unobserved response variables. This constraint can be extended by allowing for different but correlated random effects, because the latent variables affecting the missingness could be different from those affecting the response. On the other hand, they are correlated due to the common latent risk factors associated with both the missing mechanism and the response model. Under these correlated random-effects models, ignorable missingness is obtained when the two random effects are independent. Correlated random effect models for the analysis of clustered normal responses have been recently introduced in the literature, under the assumption of normal bivariate random effects [9]. In this chapter we extend this strand of the literature by considering mixtures of correlated random effects models for clustered binary data, using a nonparametric distribution of the bivariate random effects and hence avoiding the assumption of normal random effects. A suitable E-M algorithm is exploited for maximum-likelihood estimation, by adapting the algorithm developed by Hunt and Jorgensen [6] to handle ignorable missing values. The model is exploited on a sample of Chinese older adults, obtained from the Chinese Longitudinal Health and Longevity Survey (CLHLS; [15]), to cluster subjects according to their cognitive impairment and ability to cope with a Mini-Mental State Examination questionnaire.

15.2 A Two-Levels Hierarchical Logistic Regression

Clustered binary data arise when a battery of J binary variables (or items) is observed in a sample of n subjects and a binary vector $\mathbf{y}_i = (y_{i1}, \ldots, y_{ij}, \ldots, y_{iJ})$ is obtained for each subject (or cluster) $i = 1, \ldots, n$. In classification studies, these data are often examined in conjunction with a number of covariates, which are available at the item level and/or at the subject level. As a result, each subject is associated with the J rows $\mathbf{x}_{ij}$ of a $J \times K$ design matrix $\mathbf{X}_i$.

To allow for missing outcomes on each subject i, we introduce a binary vector $\mathbf{r}_i = (r_{i1}, \ldots, r_{ij}, \ldots, r_{iJ})$, where $r_{ij} = 1$ if y_{ij} is missing and 0 otherwise. When the data are non-ignorably missing, missingness of a value is informative of the unobserved value and, as a result, missing patterns $\mathbf{r}_i$ and complete data vectors $\mathbf{y}_i$ should be jointly modeled. The joint distribution of the profile $(\mathbf{y}_i, \mathbf{m}_i)$ is specified through a hierarchical model that is based on the introduction of a univariate random effect at the subjects level, to account for subjects heterogeneity, and a set of question-specific, correlated random effects at the item level, which accommodate non-ignorable missing values and, simultaneously, allow for latent heterogeneity between different items.

Discrete random effects can be conveniently introduced through latent multinomial random variables with classes that are associated with the support points of the random effect. Formally, at the subject's level, we introduce a multinomial random vector of binary coordinates $\mathbf{z} = (z_1, \ldots, z_s, \ldots, z_S)$, where the event $z_s = 1$ indicates subject's memberships to the latent class (cluster) s, with distribution

$$p(\mathbf{z}) = \prod_{s=1}^{S} \pi_s^{z_s}, \quad \sum_{s=1}^{S} \pi_s = 1.$$

At the item level, and conditionally on $\mathbf{z}$, we introduce J (item-specific) pairs of correlated random vectors, say $(\mathbf{u}_{sj}, \mathbf{v}_{sj})$, where $\mathbf{u}_{sj} = (u_{sj1}, \ldots, u_{sjh}, \ldots, u_{sjH})$ is a multinomial random variable with one trial and H classes, and $\mathbf{v}_{sj} = (v_{sj1}, \ldots, v_{sjk}, \ldots, v_{sjK})$ is a multinomial random variable with one trial and K classes. The conditional distribution of the jth pair $(\mathbf{u}_{sj}, \mathbf{v}_{sj})$, given membership to class s ($z_s = 1$) is given by

$$p_j(\mathbf{u}_{sj}, \mathbf{v}_{sj}) = \prod_{h=1}^{H}\prod_{k=1}^{K} w_{sjhk}^{u_{sjh}v_{sjk}}, \quad \sum_{h=1}^{H}\sum_{k=1}^{K} w_{sjhk} = 1.$$

Conditionally on the covariates, profile $(\mathbf{y}_i, \mathbf{m}_i)$ of subject i is modeled by

$$p(\mathbf{y}_i, \mathbf{m}_i | \mathbf{x}_{ij}) = \sum_{s=1}^{S} \pi_s \prod_{j=1}^{J} \sum_{h=1}^{H}\sum_{k=1}^{K} w_{sjhk}\, p(y_{ij}|\mathbf{x}_{ij}; \boldsymbol{\beta}_j) p(m_{ij}|\mathbf{x}_{ij}; \boldsymbol{\gamma}_j), \quad i = 1, \dots, n \tag{15.1}$$

where

$$\text{logit } p(y_{ij}|\mathbf{x}_{ij}, z_s = 1, u_{sjh} = 1; \boldsymbol{\beta}) = \beta_{0sjh} + \mathbf{x}_{ij}^{\mathsf{T}}\boldsymbol{\beta}_j \quad j = 1, \dots, J$$

$$\text{logit } p(m_{ij}|\mathbf{x}_{ij}, z_s = 1, v_{sjk} = 1; \boldsymbol{\gamma}) = \gamma_{0sjk} + \mathbf{x}_{ij}^{\mathsf{T}}\boldsymbol{\gamma}_j \quad j = 1, \dots, J,$$

whereas β_{0sjh} and γ_{0sjk} are random intercepts, while $\boldsymbol{\beta}_j$ and $\boldsymbol{\gamma}_j$ are parametric vectors that separately capture the effect of the covariates on question-specific binary outcomes and missing patterns, respectively. According to model (15.1), subjects are clustered into S clusters. Further, binary outcomes and missing patterns that relate to different questions are assumed conditionally independent, given subject's membership to cluster s. Within each subject-level cluster s, item-specific binary outcomes and missing patterns are clustered according to $H \times K$ classes, associated with the random intercepts $(\beta_{0sjh}, \gamma_{0sjk})$, allowing for flexible modeling when subjects are more homogeneous with respect to the outcomes than to the missing patterns, or viceversa, capturing the variation in question-specific outcomes and unanswered items that is not explained by the available covariates. Finally, the correlation between β_{0sjh} and γ_{0sjk} allows for non-ignorable missing values, given subject's membership to class s.

15.3 Estimation

For each subject i, let $M(i)$ and $O(i)$ be the subsets of the missing and the observed items, respectively, $O(i) \cup M(i) = \{1, 2, \dots, J\}$. Maximum likelihood estimates of the parameters $\theta = (\pi, \mathbf{w}, \boldsymbol{\beta}, \boldsymbol{\gamma})$ in model (15.1) can be obtained by maximization of the marginal log-likelihood function

$$\begin{aligned} l(\theta|\mathbf{y}, \mathbf{x}) &= \sum_{i=1}^{n} \log \sum_{\mathbf{y}_{M(i)}} p(\mathbf{y}_i, \mathbf{m}_i | \mathbf{x}_{ij}) \\ &= \sum_{i=1}^{n} \log \sum_{s=1}^{S} \pi_s \prod_{j=1}^{J} \sum_{h=1}^{H}\sum_{k=1}^{K} w_{sjhk} \left(p(y_{ij}|\mathbf{x}_{ij}; \boldsymbol{\beta})\right)^{1-m_{ij}} p(m_{ij}|\mathbf{x}_{ij}, \boldsymbol{\gamma}), \end{aligned} \tag{15.2}$$

obtained after integrating the log-likelihood with respect to the missing values. Because direct maximization of (15.2) is computationally intractable, we use an E-M algorithm, based on the iterative maximization of the expected complete log-likelihood function. Accordingly, let $\mathbf{z}_i = (z_{i1}, \ldots, z_{is}, \ldots, z_{iS})$ indicate the unobserved class membership of subject i and, for each item j in cluster i, let $\mathbf{u}_{isj} = (u_{isj1}, \ldots, u_{isjh}, \ldots, u_{isjH})$ and $\mathbf{v}_{isj} = (v_{isj1}, \ldots, v_{isjk}, \ldots, v_{isjK})$ respectively indicate the latent class membership of item responses and missing patterns, within the subjects class. The complete likelihood function given data and class membership is given by

$$L_c(\boldsymbol{\theta}) = \prod_{i=1}^{n} \prod_{s=1}^{S} \left(\pi_s \prod_{j=1}^{J} \prod_{h=1}^{H} \prod_{k=1}^{K} \left(w_{sjhk} \left(p(y_{ij}|\mathbf{x}_{ij}; \boldsymbol{\beta}) \right)^{1-m_{ij}} p(m_{ij}|\mathbf{x}_{ij}, \boldsymbol{\gamma}) \right)^{u_{isjh} v_{isjk}} \right)^{z_{is}}.$$

Given a value $\boldsymbol{\theta}^{(g-1)}$ of the parameters, the expected complete log-likelihood, computed with respect to the conditional distribution of the random effects given the outcomes and the missing patterns, takes the form of an updating function $Q(\boldsymbol{\theta}|\boldsymbol{\theta}^{(g-1)})$, which depends on the parameter estimate obtained at iteration $g-1$, and which provides an update of this estimate when it is maximized. More precisely:

$$\begin{aligned} Q(\boldsymbol{\theta}|\boldsymbol{\theta}^{(g-1)}) &= \mathsf{E}(\log L_c(\boldsymbol{\theta})) \\ &= \sum_{i=1}^{n} \sum_{s=1}^{S} \pi_{is}^{(g-1)} \log \pi_s \\ &+ \sum_{i=1}^{n} \sum_{s=1}^{S} \pi_{is}^{(g-1)} \sum_{j=1}^{J} \sum_{h=1}^{H} \sum_{k=1}^{K} w_{isjhk}^{(g-1)} \log w_{sjhk} \\ &+ \sum_{i=1}^{n} \sum_{s=1}^{S} \pi_{is}^{(g-1)} \sum_{j=1}^{J} \sum_{h=1}^{H} \sum_{k=1}^{K} w_{isjhk}^{(g-1)} (1 - m_{ij}) \log p(y_{ij}|\mathbf{x}_{ij}, \boldsymbol{\beta}) \\ &+ \sum_{i=1}^{n} \sum_{s=1}^{S} \pi_{is}^{(g-1)} \sum_{j=1}^{J} \sum_{h=1}^{H} \sum_{k=1}^{K} w_{isjhk}^{(g-1)} \log p(m_{ij}|\mathbf{x}_{ij}, \boldsymbol{\gamma}) \end{aligned}$$

where

$$\begin{aligned} \pi_{is}^{(g-1)} &= p(Z_s = 1|\mathbf{y}_i, \mathbf{m}_i, \mathbf{x}, \boldsymbol{\theta}^{(g-1)}) \\ &= \frac{p(\mathbf{y}_i, \mathbf{m}_i, \mathbf{x}, \boldsymbol{\theta}^{(g-1)}|Z_s = 1)\pi_s^{(g-1)}}{\sum_{s=1}^{S} p(\mathbf{y}_i, \mathbf{m}_i, \mathbf{x}, \boldsymbol{\theta}^{(g-1)}|Z_s = 1)\pi_s^{(g-1)}} \end{aligned}$$

and

$$w_{isjhk}^{(g-1)} = p(u_{sjh} = 1, v_{sjk} = 1 | z_s = 1, y_{ij}, m_{ij}, \mathbf{x}_{ij}, \boldsymbol{\theta}^{(g-1)})$$
$$= \frac{p(y_{ij}, m_{ij}, u_{sjh} = 1, v_{sjk} = 1 | z_s = 1) w_{sjhk}}{\sum_{h=1}^{H} \sum_{k=1}^{K} p(y_{ij}, m_{ij}, u_{sjh} = 1, v_{sjk} = 1 | z_s = 1) w_{sjhk}}.$$

The E-M algorithm reduces to the iterative maximization of $Q(\boldsymbol{\theta} | \boldsymbol{\theta}^{(g-1)})$ (Maximization step) to obtain a new parameter vector $\boldsymbol{\theta}^{(g)}$, alternated with an update of the weights (Expectation Step) $\pi_{is}^{(g)}$ and $w_{isjhk}^{(g)}$, evaluated at $\boldsymbol{\theta}^{(g)}$, according to the equations stated above. We observe that the four additive components of $Q(\boldsymbol{\theta} | \boldsymbol{\theta}^{(g-1)})$ can be separately maximized, because they are functions of independent parameters. In particular, the first component is maximized by

$$\hat{\pi}_s^{(g)} = \frac{\sum_{i=1}^{n} \sum_{s=1}^{S} \pi_{is}^{(g-1)}}{n},$$

while the second component is maximized by

$$\hat{w}_{sjhk}^{(g)} = \frac{\sum_{i=1}^{n} \pi_{is}^{(g-1)} w_{sijhk}^{(g-1)}}{\sum_{i=1}^{n} \pi_{is}^{(g-1)}}.$$

The remaining components of the expected log-likelihood function can be maximized by fitting a weighted logistic model on an augmented dataset that includes $H \times S$ pseudo-subjects, each given a weight $\pi_{is} \sum_k w_{isjhk}$, and, respectively, a weighted logistic model on an augmented dataset that includes $K \times S$ pseudo-subjects, each given a weight $\pi_{is} \sum_h w_{isjhk}$. The algorithm is iterated up to the convergence of the estimates, whose limit is a local maximum point of the log-likelihood.

When the number $H \times K \times S$ of latent classes is small, standard errors of the estimates can be obtained by taking the square root of the diagonal of the observed information matrix, as obtained from the complete log-likelihood by standard methods of the analysis of finite mixture models [11]. If the number of classes is large, stable standard errors of the parameters of interest can be computed [3] on the basis of the predictive distribution of the missing values, as obtained from the estimates of the last iteration of the E-M algorithm, to compute naive point estimates and variance estimates of the parameters. Finally, the variance of the EM estimator is obtained as a weighted sum of the mean of the imputation variances and the empirical variance of the imputation point estimates, with weights 1 and N, where N is the number of imputations used.

15.4 Application

15.4.1 Data

The data that motivated this chapter are drawn from the study No. 3891 of the Inter-University Consortium for Political and Social Research [15]. The study was carried out on 7,352 Chinese subjects, aged between 80 and 106, whose cognitive impairment was examined through the Chinese Mini-Mental State Examination (MMSE) questionnaire. Questions in the MMSE questionnaire are typically compound and include a number of single items to be separately asked to the subject. Scores on each item are binary (e.g., 1 for a correct answer and 0 otherwise). With respect to the popular 30-items MMSE [2], the 23-items Chinese MMSE adopts some appropriate adjustments to make the questions more understandable and answerable among ordinary oldest old Chinese, the majority of whom are illiterate [15]. Overall, respondents were asked a 5-items orientation-related question (naming the current time, animal year, season, festival, and county), a 12-items language-related question (6 items on word recalling, 3 items on word repetition, and 3 items on sentence comprehension), a 5-items calculation question (respondents are asked to subtract 3 from 20, then 3 from the previous result, and so on), and a single-item drawing question (drawing a figure that is shown to the respondent). Only 57 % of the questionnaires were complete. Partial questionnaires with a number of missing items between 1 and 10 missing items were about the 30 % of the sample. The rest of the sample (13 %) left more than 10 items unanswered. Of these, 7 % are subjects who did not answer any questions of the questionnaire.

An outcome often reported in the literature is the observation that missing scores on tests of cognitive impairment occur more frequently among cognitively impaired patients [5, 17]. Missing values cannot therefore be ignored in the analysis, as they are informative of the level of cognitive impairment in a subject.

For each subject, relevant covariates were obtained at the time of the MMSE interview: age (in months), gender, type of residence (rural or urban), whether the subject is sedentary or active, and limits in activities of daily living (ADL; six activities including bathing, dressing, eating, indoor transferring, toileting and continence), categorized into three levels: no, one, two or more limits. In this sample, the median age at the MMSE interview is 92 years, while the lower and the upper quartiles are respectively 91 and 100 years, 59 % of the subjects are males, 64 % are rural residents, 45 % have a sedentary lifestyle, 14 % have one limit in ADL and finally 20 % have two or more ADL limits.

15.4.2 Results

Model (15.1) has been estimated from the CLHLS data by assuming different combinations of H, K, and S. For each model, we computed the minimum of the BIC criterion, $\mathrm{BIC}(H, M, S) = 2\ \log L_{H,M,S}(\boldsymbol{\theta}) + q \log n$, where n is the number of subjects in the sample and q is the number of unknown parameters. According to the above criterion, a model with $H \times K \times S = 3 \times 2 \times 2$ latent classes seems a good compromise between goodness of fit and parsimony. Although preliminary simulation studies have shown a reasonable performance of BIC in choosing the dimension of model (15.1), the general behavior of BIC (and other alternative criteria such as AIC) in the case of clustered binary data with non-ignorable missing values is still an area of open research [7].

Under the chosen model, subjects are clustered in two latent groups, with $\hat{\pi}_1 = 0.77$ and $\hat{\pi}_2 = 0.23$. Figure 15.1 displays the bivariate, question-specific distribution of the correlated random effects $(\beta_{0sjh}, \gamma_{0sjk})$ within group $s = 1$ (left column) and group $s = 2$ (right column), as obtained after the last iteration of the E-M algorithm. Due to the relative homogeneity of the MMSE items, the support points of the random effects are similar, although probability masses on these support points are extremely heterogeneous, indicating that the missing mechanism is question-specific and heterogeneous within classes of subjects with similar levels of cognitive functioning, even after controlling for relevant covariates. As a result, the presence (absence) of unanswered questions is not necessarily an indication of poor (good) cognitive functioning, because unpredicted differences in the ability to cope with the questionnaire might reflect the absence of important covariates, such as education level, which influence the probability of answering specific questions. In a small number of cases, we observe distributions with a significant negative correlation, where most of the MMSE questionnaires are clustered according to classes that include larger (smaller) numbers of unanswered items and wrong answers than those predicted by the available covariates. This indicate that only in a small number of cases missing values are strongly non-ignorable, given class membership. In the remaining cases, the low level of correlation seems to indicate that for certain questions, data are essentially missing at random within groups of subjects.

Table 15.1 displays the estimated effects of the covariates on cognitive functioning and missing value occurrence in five multi-items questions (standard errors within brackets). There are noticeable differences between question-specific estimates, although signs are concordant. Overall, older adults are more cognitively impaired than younger adults, while males are less cognitively impaired than females, confirming the results on gender differentials found by Zhang [16] on the same data used in this chapter. In keeping with the outcomes reported by Gu and Qui [4], urban residents are less cognitively impaired than the rural residents, whereas physical disabilities and a sedentary life style negatively influence a subject's cognitive functioning, even after correcting for age at the interview. Factors that negatively influence cognitive functioning increase the probability of leaving an item unanswered, in keeping with the results reported by Lagona and Zhang [8]. Interestingly, males answer the MMSE questionnaire items more often than females, although males answer correctly less often than females.

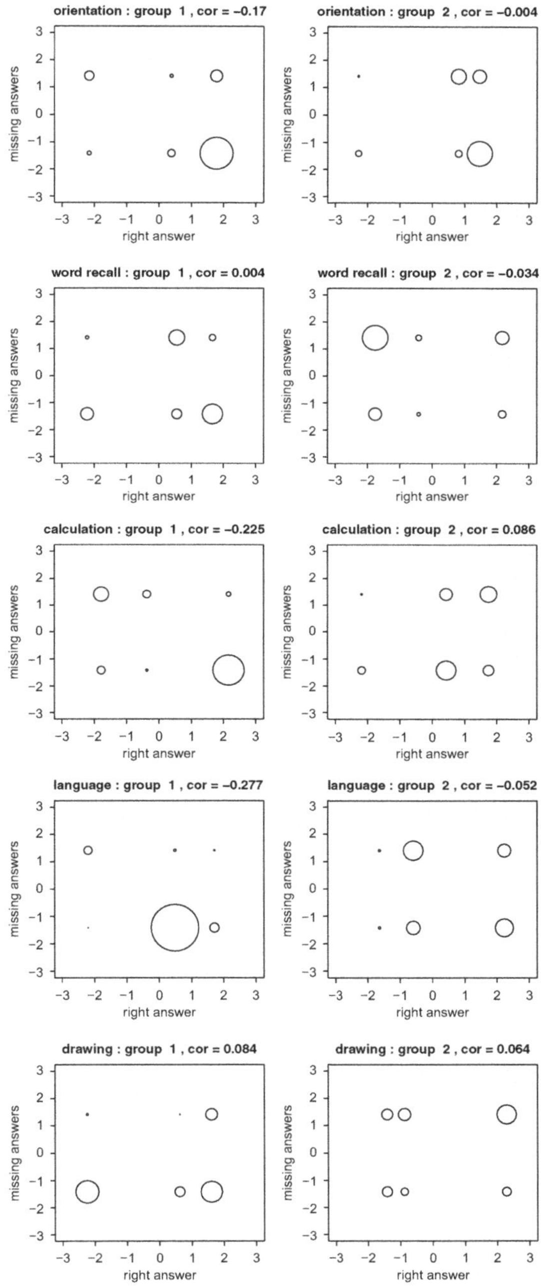

Fig. 15.1 Question-specific nonparametric distribution of the random intercepts $(\beta_{0sjh}, \gamma_{0sjk})$ within two groups of subjects; *circles* are proportional to the estimated latent-class probability w_{sjhk}

Table 15.1 Covariate effects on cognitive functioning and missing mechanism

Covariate		Cognitive functioning Orientation	 Word recall	 Calculation	 Language	 Drawing
Age (months)		−0.0058	−0.0046	−0.0060	−0.0050	−0.0024
		(0.0003)	(0.0002)	(0.0003)	(0.0003)	(0.0004)
Gender	Male	−0.6514	−0.2029	−1.6447	−0.3704	−0.7768
(ref: female)		(0.0530)	(0.0299)	(0.0453)	(0.0518)	(0.0632)
Residence	Rural	−0.0818	−0.0788	−0.7160	−0.1956	−0.3922
(ref: urban)		(0.0506)	(0.0304)	(0.0437)	(0.0516)	(0.0650)
Lifestyle	Sedentary	−0.7673	−0.4315	−1.3120	0.7541	−0.7219
(ref: active)		(0.0570)	(0.0320)	(0.0457)	(0.0572)	(0.0678)
ADL limits						
(ref: no limits)	One	−0.3441	−0.3389	−0.2942	−0.4318	−0.1353
		(0.0690)	(0.0422)	(0.0595)	(0.0694)	(0.0987)
	Two or more	−0.9107	−0.7682	−0.9239	−1.0167	−0.4260
		(0.0573)	(0.0392)	(0.0554)	(0.0576)	(0.1105)
		Missing mechanism				
Covariate		Orientation	Word recall	Calculation	Language	Drawing
Age (months)		0.0044	0.0024	0.0033	0.0036	0.0031
		(0.0004)	(0.0004)	(0.0009)	(0.0004)	(0.0004)
Gender	Male	0.4135	0.0967	0.7956	0.1619	0.5457
(ref: female)		(0.0685)	(0.0636)	(0.1401)	(0.0569)	(0.0736)
Residence	Rural	0.2137	0.1263	0.0375	0.1792	0.2546
(ref: urban)		(0.0664)	(0.0635)	(0.1442)	(0.0562)	(0.0749)
Lifestyle	Sedentary	0.8662	0.4797	0.8527	1.2930	1.2405
(ref: active)		(0.0785)	(0.0705)	(0.1507)	(0.0610)	(0.0780)
ADL limits						
(ref: no limits)	One	0.3524	0.1978	0.5863	0.4958	0.1966
		(0.0892)	(0.0898)	(0.2076)	(0.0774)	(0.1038)
	Two or more	0.4531	0.1502	0.3069	0.9128	0.7994
		(0.0696)	(0.0673)	(0.1485)	(0.0581)	(0.0941)

References

1. Celeux, G., Martin, O., Lavergne, C.: Mixture of linear mixed models for clustering gene expression profiles from repeated microarray experiments. Stat. Model. **5**, 243–267 (2005)
2. Folstein, M.F., Folstein, S.E., McHugh, P.R.: Mini-mental state: a practical method for grading the cognitive state of patients for the clinician. J. Psychiat. Res. **12**, 189–198 (1975)
3. Goetghebeur, E., Ryan, L.: Semiparametric regression analysis of interval-censored data. Biometrics **56**, 1139–1144 (2000)
4. Gu, D., Qiu, L.: Cognitive functioning and its determinants among the oldest-old in China. J. Nanjing College Populat. Manag. **2**, 3–9 (2003)
5. Herzog, A.R., Wallace, R.B.: Measures of cognitive functioning in the AHEAD Study. J. Gerontol. Psycol. Sci. Soc. Sci. **B52**, 37–48 (1997)

6. Hunt, L., Jorgensen, M.: Mixture model clustering for mixed data with missing information. Comput. Stat. Data Anal. **41**, 429–440 (2003)
7. Ibrahim, J.G., Zhu, H., Tang, N.: Model selection criteria for missing-data problems using the EM algorithm. J. Am. Stat. Assoc. **103**, 1648–1658 (2008)
8. Lagona, F., Zhang, Z.: A missing composite covariate in survival analysis: a case study of the Chinese longitudinal health and longevity survey. Stat. Med. **29**, 248–261 (2010)
9. Lin, H., Liu, D., Zhou, X.-H.: Correlated random-effects model for normal longitudinal data with nonignorable missingness. Stat. Med. **29**, 236–247 (2010)
10. Little, R.J.A.: Modeling the drop-out mechanism in repeated-measures studies. J. Am. Stat. Assoc. **90**, 1112–1121 (1995)
11. Louis, T.A.: Finding the observed information matrix when using the EM algorithm. J. Roy. Stat. Soc. B, **44**, 226–233 (1982)
12. McCullagh, C.E.: Generalized Linear Mixed Models. Institute of Mathematical Statistics, Beachwood (Ohio) (2003)
13. Rubin, D.B.: Inference and missing data. Biometrika **63**, 81–92 (1976)
14. Vermunt, J.K.: Latent class and finite mixture models for multilevel datasets. Stat. Methods Med. Res. **17**, 33–51 (2008)
15. Zeng, Y., Vaupel, J.W.: Functional capacity and self-evaluation of health and life of oldest old in China. J. Soc. Issues **58**, 733–748 (2002)
16. Zhang, Z.: Gender differentials in cognitive impairment and decline in the oldest old in China. J. Gerontol. Soc. Sci. **B61**, S107–S115 (2006)
17. Zimmer, Z., Martin, L.G., Chang, M.-C.: Changes in functional limitation and survival among older Taiwanese, 1993, 1996, and 1999. Popul. Stud. **56**, 265–276 (2002)

A Model for Correlated Paired Comparison Data

16

Manuela Cattelan and Cristiano Varin

Abstract

Paired comparison data arise when objects are compared in couples. This type of data occurs in many applications. Traditional models developed for the analysis of paired comparison data assume independence among all observations, but this seems unrealistic because comparisons with a common object are naturally correlated. A model that introduces correlation between comparisons with at least a common object is discussed. The likelihood function of the proposed model involves the approximation of a high dimensional integral. To overcome numerical difficulties a pairwise likelihood approach is adopted. The methodology is illustrated through the analysis of the 2006/2007 Italian men's volleyball tournament and the 2008/2009 season of the Italian water polo league.

Keywords

Paired comparison data • Pairwise likelihood • Random effects • Thurstone–Mosteller model

16.1 Paired Comparison Data

Paired comparison data arise from the comparison of objects in couples. This type of data occurs in many applications. In general, it is much easier for people to compare objects in couples than to rank a list of items. For this reason, paired comparison data often arise when the judgement of people is involved. Examples include the

---

M. Cattelan (✉)
Department of Statistical Sciences, University of Padua, via C. Battisti 241, 35121 Padova, Italy
e-mail: manuela.cattelan@unipd.it

C. Varin
D.A.I.S., University Ca' Foscari, San Giobbe, Cannaregio 873, 30121 Venice, Italy
e-mail: sammy@unive.it

N. Torelli et al. (eds.), *Advances in Theoretical and Applied Statistics*,
Studies in Theoretical and Applied Statistics, DOI 10.1007/978-3-642-35588-2_16,

comparison of sounds in pairs to analyze which features affect their unpleasantness [7], the investigation of preferences of students for different universities [6], and the judgement of the painfulness of different injections to patients under regular haemodialysis [10].

Paired comparison data can be observed also when there is no judgement made by people, but a sort of comparison takes place between two items from which a winner and a loser can be identified. This happens in sport contests where two players or teams compete and a winner is determined at the end of the match. Another instance is animal behavior experiments that are performed by zoologists to investigate which features influence the results of fights between animals [16].

The traditional models developed for the analysis of paired comparison data are the Bradley–Terry model [2] and the Thurstone–Mosteller model [13, 17]. In both models, the probability that an object is preferred over another is a function of the difference of the true "worth" of the objects. The main difference between the two models lies in the link function: logit for the Bradley–Terry model and probit for the Thurstone–Mosteller model. Extensions of these models have been developed to take into account specific features of paired comparison data as the existence of an order effect that advantages the object presented first, or situations in which there are three possible outcomes of the comparisons, i.e., preference for one of the two objects or impossibility to express a preference.

Commonly, models for paired comparison data are fitted by maximum likelihood under the assumption of independence among all paired comparisons. This assumption is rarely fulfilled in real applications. An example, later discussed, is sports tournaments where results of two matches involving a common player are naturally correlated. This is an important limitation also in other contexts, such as animal behavior experiments or judgements performed by the same person. It is reasonable to believe that comparisons performed by the same person involving a common object are dependent. In the following section, we illustrate a model that allows for correlation between paired comparisons.

16.2 Mixed Effects Models for Paired Comparison Data

Let Y_{ij}, $j > i = 1, \ldots, n$, be a binary random variable taking value 1 if object i is preferred to object j, and 0 otherwise. In traditional models for paired comparison data, the following generalized linear model is assumed. The density of Y_{ij} is distributed as a Bernoulli random variable whose mean is related to the worth of objects through

$$g\left\{\Pr\left(Y_{ij} = 1\right)\right\} = \lambda_i - \lambda_j,$$

where g is a suitable link function and λ_i is the worth parameter for object $i = 1, \ldots, n$. The link functions commonly employed are the probit link, describing the Thurstone–Mosteller model [13,17], and the logit link, employed in the Bradley–Terry model [2], but it is possible to adopt any symmetric link as, for example, the cauchit link. In case a probit link is specified, then the model becomes

$\Phi^{-1}\left\{\Pr\left(Y_{ij} = 1\right)\right\} = \lambda_i - \lambda_j$, where Φ denotes the cumulative distribution function of a standard normal random variable. The worth parameter may depend on explanatory variables through the relation

$$\lambda_i = x_i^{\mathrm{T}}\beta,$$

where x_i is a p-dimensional vector of explanatory variables related to object i and β is a vector of p regression parameters. Note that the linear predictor does not include an intercept because this is not identifiable in paired comparison models.

Correlation between observations with a common object can be introduced by including an object-specific zero mean random effect u_i

$$\lambda_i = x_i^{\mathrm{T}}\beta + u_i.$$

Accordingly, the conditional mean of an observation given the object-specific random effects is expressed as

$$g\left\{\Pr\left(Y_{ij} = 1|u_i, u_j\right)\right\} = (x_i - x_j)^{\mathrm{T}}\beta + u_i - u_j.$$

The binary observation Y_{ij} is equivalently represented as a censored continuous random variable $Y_{ij} = I\left\{Z_{ij} > 0\right\}$, where $I\{A\}$ denotes the indicator function of the set A and

$$Z_{ij} = (x_i - x_j)^{\mathrm{T}}\beta + u_i - u_j + \epsilon_{ij},$$

where ϵ_{ij} are independent zero mean continuous random variables.

To proceed with likelihood inference, we assume that the random effects u_i are independent, identically distributed normal random variables with zero mean and variance σ^2, the latent errors ϵ_{ij} are independent, identically distributed standard normal variables and they are uncorrelated with the random effects. In other words, the proposed model is a mixed effects version of the Thurstone–Mosteller model. Accordingly, the correlation between a pair of censored random variables Z_{ij} and Z_{kl} is

$$\operatorname{corr}\left(Z_{ij}, Z_{kl}\right) = \begin{cases} \sigma^2/(1+2\sigma^2) & \text{if } i = k \text{ or } j = l, \\ 0 & \text{if } i \neq j \neq k \neq l, \\ -\sigma^2/(1+2\sigma^2) & \text{if } i = l \text{ or } j = k, \end{cases} \tag{16.1}$$

thus, the model allows for dependence between pairs of observations sharing an object.

The inclusion of the random effects is useful not only to model dependence in paired comparison data, but it also allows to account for the imperfect representation of the worth λ_i by the linear predictor $x_i^{\mathrm{T}}\beta$.

Unfortunately, the mixed effects Thurstone–Mosteller model has an intractable likelihood function, which results from integrating out all the random effects

$$\mathscr{L}(\theta;y)=\int_{\mathbb{R}^n}\left\{\prod_{i=1}^{n-1}\prod_{j=i+1}^{n}P(Y_{ij}=y_{ij}|u_i,u_j;\theta)\right\}\left\{\prod_{i=1}^{n}\frac{1}{\sigma}\phi\left(\frac{u_i}{\sigma}\right)\right\}\,\mathrm{d}u_1\cdots\,\mathrm{d}u_n,\tag{16.2}$$

where $\theta=(\beta^{\mathrm{T}},\sigma^2)^{\mathrm{T}}$ is the parameter vector and $\phi(\cdot)$ denotes the density function of a standard normal variable. Thus, the full likelihood consists in a complicated integral of dimension equal to the number of objects being compared. Except for small n, a direct approximation of the likelihood can yield numerical difficulties, or even be impractical. We propose to adopt pairwise likelihood inference to reduce the computational complexity while retaining part of the likelihood properties.

16.3 Pairwise Likelihood Inference

A composite likelihood is a class of pseudo-likelihoods constructed by compounding marginal or conditional probabilities for subsets of events [9, 18]. In our specific case, it is convenient to consider a particular example of composite likelihood known as pairwise likelihood. This consists of the product of bivariate marginal probabilities associated with each pair of observations

$$\mathscr{L}_{\text{pair}}(\theta;y)=\prod_{\{i,j,k,l\in\mathscr{D}\}}\Pr(Y_{ij}=y_{ij},Y_{kl}=y_{kl};\theta),$$

where $\mathscr{D}$ denotes the set of indexes i,j,k,l identifying two different observations, that is with $i<j$, $k<l$, excluding the case in which both $i=k$ and $j=l$, and with $k\geqslant i$ in order to include all couples of observations only once. Under the model assumptions each of the above bivariate marginal probabilities is a two-dimensional normal integral. Indeed, the joint distribution of the pair of censored random variables (Z_{ij},Z_{kl}) is bivariate normal with zero mean, variance $1+2\sigma^2$ and correlation as in (16.1). Let $\mu_{ij}=(x_i-x_j)^{\mathrm{T}}\beta/\sqrt{1+2\sigma^2}$ and $\nu=\sigma^2/1+2\sigma^2$, then the probability that object i loses against both j and k is

$$\Pr(Y_{ij}=0,Y_{ik}=0;\theta)=\Pr(Z_{ij}<0,Z_{ik}<0;\theta)=\Phi_2\left(-\mu_{ij},-\mu_{ik};\nu\right),$$

where $\Phi_2(\cdot,\cdot;\rho)$ denotes the cumulative distribution function of a bivariate normal random variable with standardized marginals and correlation ρ. The probabilities of the other possible outcomes (win–loss, loss–win, and win–win) are

$$\Pr(Y_{ij}=1,Y_{ik}=0;\theta)=\Phi\left(-\mu_{ik}\right)-\Phi_2\left(-\mu_{ij},-\mu_{ik};\nu\right),$$
$$\Pr(Y_{ij}=0,Y_{ik}=1;\theta)=\Phi\left(-\mu_{ij}\right)-\Phi_2\left(-\mu_{ij},-\mu_{ik};\nu\right),$$
$$\Pr(Y_{ij}=1,Y_{ik}=1;\theta)=1-\Phi\left(-\mu_{ij}\right)-\Phi\left(-\mu_{ik}\right)+\Phi_2\left(-\mu_{ij},-\mu_{ik};\nu\right).$$

Hence, pairwise likelihood considerably reduces the computational effort as it involves a set of bivariate normal integrals in place of the high-dimensional integral of the full likelihood. Bivariate normal integrals are computed with very high numerical accuracy using routines in the R [15] package `mvtnorm` [8].

The logarithm of the pairwise likelihood is denoted by $\ell_{\text{pair}}(\theta; y) = \log \mathscr{L}_{\text{pair}}(\theta; y)$ and its maximum, $\hat{\theta}$, is the maximum pairwise likelihood estimator. Under mild regularity conditions, the maximum pairwise likelihood estimator is consistent and asymptotically distributed as a normal random variable with mean θ and covariance matrix $\mathrm{G}(\theta) = \mathrm{H}(\theta)^{-1}\mathrm{J}(\theta)\mathrm{H}(\theta)^{-1}$, where $\mathrm{J}(\theta) = \text{var}\left\{\nabla \ell_{\text{pair}}(\theta; Y)\right\}$ and $\mathrm{H}(\theta) = \mathrm{E}\left\{-\nabla^2 \ell_{\text{pair}}(\theta; Y)\right\}$, see [4, 12].

Hypothesis testing and interval estimation can be based on the pairwise likelihood analogue of the likelihood ratio statistic. Suppose that δ is a q-dimensional subvector of the whole parameter vector $\theta = (\delta^{\mathrm{T}}, \lambda^{\mathrm{T}})^{\mathrm{T}}$ and that it is of interest to test hypothesis $\mathrm{H}_0 : \delta = \delta_0$. This hypothesis can be assessed through the pairwise likelihood ratio statistic defined as

$$\mathscr{W}_{\text{pair}}(Y) = 2\left[\ell_{\text{pair}}(\hat{\theta}; Y) - \ell_{\text{pair}}(\delta_0, \hat{\lambda}(\delta_0); Y)\right],$$

where $\hat{\lambda}(\delta_0)$ denotes the maximum pairwise likelihood estimator in the subspace where $\delta = \delta_0$. The pairwise log-likelihood ratio statistic has asymptotic distribution given by the weighted sum $\sum_{i=1}^{q} \xi_i \chi^2_{i(1)}$, where $\chi^2_{i(1)}$ are independent chi-square random variables with 1 degree of freedom and the ξ_i are the eigenvalues of $(\mathrm{H}^{\delta\delta})^{-1}\mathrm{G}_{\delta\delta}$, where $\mathrm{H}^{\delta\delta}$ denotes the block of the inverse of $\mathrm{H}(\theta)$ pertaining to δ and $\mathrm{G}_{\delta\delta}$ is the block of the matrix $\mathrm{G}(\theta)$ pertaining to δ [12, 18]. The evaluation of this nonstandard limit distribution is difficult. Hence, proposals for adjustments of the statistic in order to recover an approximate chi-square distribution [3, 14] or the use of parametric bootstrap techniques have been suggested.

16.3.1 Simulations

The performance of the pairwise likelihood estimator is evaluated through a simulation study. Data are simulated from a single round robin tournament in which each of n objects is compared once with all the other objects. The worth parameter of the objects is assumed to be

$$\lambda_i = \beta_1 x_{1i} + \beta_2 x_{2i} + u_i,$$

where covariates x_{1i} are independently simulated from a normal distribution with mean 0 and standard deviation 0.1 and covariates x_{2i} are independently simulated from a Bernoulli distribution with probability of success 0.6. Table 16.1 reports empirical means and standard deviations of 500 simulated parameter estimates in data sets with $n = 30$ for various values of the random effects standard deviation $\sigma \in \{0.2, 0.4, 0.6, 0.8, 1.0\}$.

Table 16.1 Empirical means and standard deviations of 500 estimates of the parameters of the mixed effects Thurstone–Mosteller model for increasing values of $\sigma \in \{0.2, 0.4, 0.6, 0.8, 1\}$

		σ				
		0.2	0.4	0.6	0.8	1.0
β_1	Mean	−1.975	−2.098	−2.021	−2.070	−2.018
	Std. dev.	0.781	1.028	1.659	2.190	2.597
β_2	Mean	1.008	1.006	1.021	1.013	1.023
	Std. dev.	0.130	0.201	0.237	0.338	0.415
σ	Mean	0.185	0.376	0.592	0.787	0.985
	Std. dev.	0.061	0.087	0.111	0.132	0.186

True values for β_1 and β_2 are −2 and 1, respectively

The results of the simulations seem satisfactory. Biases of the regression parameters are relatively small. The estimate of σ is slightly downward biased, as expected in variance components models. In fact, also the full likelihood is known to produce downward biased estimates of this parameter. Finally, as expected, inflating σ implies higher variability in the estimates of the regression coefficients.

16.4 Applications

In this section, we illustrate pairwise likelihood inference in the mixed effects Thurstone–Mosteller model with application to two sports tournaments.

16.4.1 Volleyball

Sports data are a natural field of application of models for paired comparison data. The first application considered regards the results of the 2006/2007 Italian men's volleyball A1 league. The league is composed of 14 teams that compete in a double round-robin tournament, that is, each team competes twice against all the other teams in the league, for a total of 182 competitions. The matches cannot end in a tie, so there are only two possible outcomes for each contest. At the end of the regular season, the best eight teams access to the playoffs to compete for the title of Italian Champion. The information available about the volleyball teams are the number of accesses to the playoffs in the previous 8 years and the mean age of the players. The home effect is a further covariate which accounts for the advantage deriving from playing in a home field. In fact, it is commonly recognized that a team playing at home enjoys the benefits of the acquaintance with the playing field and a larger number of supporters. None of the matches played during the season took place in a neutral field, so it seems important to account for this effect. The interest lies in determining whether these covariates affect the result of the matches.

The first two columns of Table 16.2 display the estimates of the traditional Thurstone–Mosteller model termed independence model and corresponding to the

Table 16.2 Estimates (`est.`) and standard errors (`s.e.`) of independence (first two columns) and mixed effects (last two columns) models for the volleyball data

	Independence		Mixed effects	
	est.	s.e.	est.	s.e.
Playoffs	0.101	0.032	0.116	0.061
Home effect	0.446	0.098	0.513	0.103
Mean age	0.136	0.059	0.157	0.114
σ	–	–	0.401	0.121
Pairwise log-likelihood	−20073.86		−20062.33	

Last row reports the value of the maximized pairwise log-likelihood

restriction $\sigma^2 = 0$, and the last two columns present the estimates of the proposed mixed effects model. The estimated home effect and its standard error confirm that, in this tournament, teams playing at home have an important advantage. The significance of the parameter relating to the number of accesses to the playoffs in the previous 8 years reveals that teams which were strong in the past tend to remain strong also in the present season. The independence model states that the mean age of the team has a positive influence on its ability. This covariate has a narrow range, it lies between 25.25 and 29.31 years old, but it seems that teams with older players, who are probably more experienced, have higher probability of winning. The fitted mixed effects model leads to a different conclusion. Indeed, the mean age effect is not significant anymore, while the other covariates remain significant. Finally, the estimate of the random effect standard deviation σ is 0.401 with standard error 0.121.

The main interest is on testing whether the variance component is null, or in other terms if correlation between matches with a common player is relevant or not. The test of the hypothesis $\mathrm{H}_0 : \sigma^2 = 0$ against $\mathrm{H}_1 : \sigma^2 > 0$ is complicated because the parameter value under the null hypothesis lies on the boundary of the parameter space and thus standard asymptotic results do not apply. Even though, when just one variance component is tested, it is possible to halve the p-values of a chi-square distribution with one degree of freedom [11], this approximation is very poor [5] and bootstrap methods are more robust. Moreover, if pairwise likelihood is employed, it is necessary to compute a scaling parameter that depends on the matrixes $H(\theta)$ and $G(\theta)$, so it is even more convenient to resort to parametric bootstrap as in [1]. First, compute the observed value of the pairwise likelihood ratio statistic

$$\mathscr{W}_{\text{pair}}(y) = 2\left[\ell_{\text{pair}}\{\hat{\theta}(y); y\} - \ell_{\text{pair}}\{\hat{\theta}_0(y); y\}\right],$$

where $\hat{\theta}_0(y)$ is the maximum pairwise likelihood estimator under the null hypothesis. Then, setting $\theta = \hat{\theta}_0(y)$, generate M data sets $y^{(1)}, \ldots, y^{(M)}$. For each simulated data set, compute the maximum pairwise likelihood estimator $\hat{\theta}(y^{(m)})$, the maximum pairwise likelihood estimator under the null hypothesis $\hat{\theta}_0(y^{(m)})$ and the relative pairwise log-likelihood ratio statistic $\mathscr{W}_{\text{pair}}(y^{(m)})$. The p-value of the test is then estimated by quantity

$$p = \frac{\sum_{m=1}^{M} I\left\{\mathscr{W}_{\text{pair}}(y^{(m)}) \geq \mathscr{W}_{\text{pair}}(y)\right\} + 1}{M + 1}.$$

In the volleyball data, this parametric bootstrap test based on $1,000$ simulations yields a p-value smaller than 0.01, thus not supporting the null hypothesis $\text{H}_0 : \sigma^2 = 0$.

16.4.2 Water Polo

The second data set considered here consists of the results of the water polo matches played by teams in the male A1 league during the 2008/2009 regular season. The water polo tournament has a double round robin structure, so in each half of the season every team competes once against all the other teams in the league. The A1 league includes 12 teams playing altogether 132 matches. At the end of the regular season, the best eight teams access to the playoffs to compete for the title of Italian Champion. The available covariate is the number of accesses to the playoffs in the previous 6 years. The analysis is focused on determining whether there is a sort of "tradition effect" in water polo, that is whether teams strong in past seasons tend to be strong also in the present one.

Water polo matches can also end in ties, hence the model presented so far needs a further extension in order to account for the three possible outcomes of the matches. This extension can be accomplished through the introduction of a threshold parameter τ. Thus, the probability that i loses against j is equal to the probability that the corresponding latent random variable Z_{ij} is smaller than $-\tau$. The probability of a tie between i and j is equal to the probability that the corresponding latent variable Z_{ij} is between $-\tau$ and τ. Finally, the probability that i wins against j is equal to the probability that Z_{ij} is larger than τ. Let $t = \tau/\sqrt{1 + 2\sigma^2}$, then the probability that i loses both the matches against j and k is

$$P(Z_{ij} < -\tau, Z_{ik} < -\tau) = \Phi_2\left(-t - \mu_{ij}, -t - \mu_{ik}; \nu\right),$$

while the probability that i and j draws and i loses against k is equal to

$$\begin{aligned} P(-\tau < Z_{ij} < \tau, Z_{ik} < -\tau) &= P(Z_{ij} < \tau, Z_{ik} < -\tau) - P(Z_{ij} < -\tau, Z_{ik} < -\tau) \\ &= \Phi_2\left(t - \mu_{ij}, -t - \mu_{ik}; \nu\right) - \Phi_2\left(-t - \mu_{ij}, -t - \mu_{ik}; \nu\right). \end{aligned}$$

The probabilities of the other possible outcomes are similarly computed.

Again, besides the accesses to the playoffs in the previous 6 years, the effect of playing at home is taken into account. Table 16.3 shows the results of the estimates of an independence model, corresponding to the restriction $\sigma^2 = 0$, (first two columns) and the mixed effects model (last two columns).

Both models confirm that the team playing at home has actually an advantage over the away team. The estimate of the accesses to the playoffs in the previous 6 years is also significant in both models, denoting that teams which were strong in

Table 16.3 Estimates (`est.`) and standard errors (`s.e.`) of independence (first two columns) and mixed effects (last two columns) models for the water polo data

	Independence		Mixed effects	
	est.	s.e.	est.	s.e.
Playoffs	0.238	0.034	0.315	0.090
Home effect	0.223	0.116	0.294	0.129
Threshold	0.218	0.051	0.290	0.066
σ	–	–	0.616	0.134
Pairwise log-likelihood	−12972.09		−12948.95	

Last row reports the value of the maximized pairwise log-likelihood

the recent past tend to be strong also in the present season. Finally, the estimated random effect standard deviation is 0.616 with standard error 0.134. The testing of $H_0 : \sigma^2 = 0$ presents the same problems specified for the volleyball data, hence parametric bootstrap is used. The bootstrap test for validating the hypothesis $H_0 : \sigma^2 = 0$ yields a p-value smaller than 0.001 based on 1,000 simulations. Therefore, also in water polo the hypothesis of null variance of the random effects may not be accepted.

16.5 Discussion

In this chapter, traditional models for paired comparison data are extended in order to introduce correlation among observations with common objects. In many instances, as for example in sports data, it is evident that a model which allows for correlation is more realistic. In the volleyball and water polo data analyzed here, the presence of correlation between matches with common teams is borne out by the significance of the variance component. Modeling the dependence in the paired comparison model is natural given the structure of the data. As illustrated in this chapter, accounting for dependence may change the significance of a parameter as found in the volleyball application.

The mixed effects Thurstone–Mosteller model can be usefully applied also in the other areas mentioned in Sect. 16.1. For example, in biological studies scientists are interested in determining whether some specific covariates affect the outcomes of contests between animals [16]. In this instance it seems important to account for dependence between fights involving the same animal in order to ascertain at which extent covariates are associated with the outcomes of fights. Is it important to consider the inclusion of dependence also when comparisons are performed by people. For example, in [6] a data set about pairwise evaluations of universities situated in different European countries is analyzed and university-specific and student-specific covariates are taken into account. In this case it seems important to include in the model dependence between comparisons performed by the same student involving a common university in order to determine which covariates really influence the decisions.

The model for sports data can be further extended allowing for a temporal evolution of abilities of teams. For example, it is possible to include time-varying covariates which yield different abilities of teams in different matches. An alternative currently under study is the specification of a temporal evolution of the random effects which induce a temporal variation of abilities.

Acknowledgements The authors would like to thank David Firth for valuable discussion and suggestions.

References

1. Bellio, R., Varin, C.: A pairwise likelihood approach to generalized linear models with crossed random effects. Stat. Model. **5**, 217–227 (2005)
2. Bradley, R.A., Terry, M.E.: The rank analysis of incomplete block designs. I. The method of paired comparisons. Biometrika **39**, 324–345 (1952)
3. Chandler, R.E., Bate, S.: Inference for clustered data using the independence log-likelihood. Biometrika **94**, 167–183 (2007)
4. Cox, D.R., Reid, N.: A note on pseudolikelihood constructed from marginal densities. Biometrika **91**, 729–737 (2004)
5. Crainiceanu, C.M., Ruppert, D., Vogelsang, T.J.: Some properties of likelihood ratio tests in linear mixed models. Working Paper, Cornell University (2003)
6. Dittrich, R., Hatzinger, R., Katzenbeisser, W.: Modelling the effect of subject-specific covariates in paired comparison studies with an application to university rankings. Appl. Stat. **47**, 511–525 (1998)
7. Ellermeier, W., Mader, M., Daniel, P.: Scaling the unpleasantness of sounds according to the BTL model: ratio-scale representation and psychoacoustical analysis. Acta Acust. United Ac. **90**, 101–107 (2004)
8. Genz, A., Bretz, F., Miwa, T., Mi, X., Leisch, F., Scheipl, F., Hothorn, T.: `mvtnorm`: Multivariate normal and t distributions. R package version 0.9–92 (2010)
9. Lindsay, B.G.: Composite likelihood methods. Contemp. Math. **80**, 221–239 (1988)
10. Matthews, J.N.S., Morris, K.P.: An application of Bradley–Terry-type models to the measurement of pain. Appl. Stat. **44**, 243–255 (1995)
11. McCulloch, C.E., Searle, S.R.: Generalized, Linear, and Mixed Models. Wiley, New York (2001)
12. Molenberghs, G., Verbeke, G.: Models for Discrete Longitudinal Data. Springer, New York (2005)
13. Mosteller, F.: Remarks on the method of paired comparisons. I. The least squares solution assuming equal standard deviations and equal correlations. Psychometrika **16**, 203–206 (1951)
14. Pace, L., Salvan, A., Sartori, N.: Adjusting composite likelihood ratio statistics. Stat. Sin. **21**, 129–148 (2011)
15. R Development core team: R: A language and environment for statistical computing. R foundation for statistical computing, Vienna, Austria. ISBN 3-900051-07-0 (2011)
16. Stuart-Fox, D.M., Firth, D., Moussalli, A., Whiting, M.J.: Multiple signals in chameleon contests: designing and analysing animal contests as a tournament. Anim. Behav. **71**, 1263–1271 (2006)
17. Thurstone, L.L.: A law of comparative judgement. Psychol. Rev. **34**, 273–286 (1927)
18. Varin, C., Reid, N., Firth, D.: An overview of composite likelihood methods. Stat. Sin. **21**, 5–42 (2011)

Closed Skew Normal Stochastic Frontier Models for Panel Data

17

Roberto Colombi

Abstract

We introduce a stochastic frontier model for longitudinal data where a subject random effect coexists with a time independent random inefficiency component and with a time dependent random inefficiency component. The role of the closed skew normal distribution in this kind of modeling is stressed.

Keywords

Longitudinal data analysis • Statistical modeling • Variance components models

17.1 Introduction

A *stochastic frontier* defines a random upper bound of a response variable (*output*) and it is equal to a function of independent variables (*inputs*) plus an idiosyncratic random error. In a *stochastic frontier model* the observable response is given by the stochastic frontier minus a nonnegative random variable which is called *random inefficiency component*. Stochastic frontiers have been mainly used as statistical models for production functions [12] but we believe that they are random effects models of a wider applicability. In this chapter a four random component stochastic frontier model for panel data is introduced. Subject unobservable heterogeneity is modeled by a random component and the presence of both a time invariant and a time dependent random inefficiency component is allowed. In this way, subject heterogeneity is not wrongly modeled as inefficiency [11], and it is possible to disentangle a time persistent component from the total inefficiency. The main

R. Colombi (✉)
Dipartimento di Ingegneria dell'Informazione e Metodi Matematici, viale Marconi 5 24044 Dalmine
e-mail: roberto.colombi@unibg.it

N. Torelli et al. (eds.), *Advances in Theoretical and Applied Statistics*, Studies in Theoretical and Applied Statistics, DOI 10.1007/978-3-642-35588-2_17,

contribution of this work is that time independent inefficiency can coexist with time dependent inefficiency and with a subject random effect and that it is not necessary to resort to simulated maximum likelihood methods as suggested by Greene [11]. Models with either time varying inefficiency or time persistent inefficiency are very diffused in the econometric literature [3, 12, 13] but as far as we know, no attempt was made before to define a multiple random component stochastic frontier in order to consider subject heterogeneity and different aspect of technical inefficiency. Previous Stochastic Frontier models are likely to be misspecified because either confound random subject effects with time invariant inefficiency or do not separate time invariant inefficiency from time dependent inefficiency. In this chapter we show that these problems are avoided by a more general Stochastic Frontier model which encompasses many used Stochastic Frontier models that can be tested against the proposed more general model.

The model here presented is related but not identical to the stochastic frontiers introduced by Dominguez-Molina et al. [7] and to the linear mixed models with skew normal errors of Lin and Lee [14] and Arellano-Valle et al. [2].

17.2 A Stochastic Frontier Model for Panel Data

We consider the stochastic frontier model:

$$y_{it} = \beta_0 + \boldsymbol{x}'_{it}\boldsymbol{\beta} + b_i - u_{i0} - \delta_t u_{i1} + e_{it} \tag{17.1}$$

where the index i, $i = 1, 2, \ldots, n$, denotes n units or subjects and t, $t = 1, 2, \ldots, T$, identifies the T time points at which every subject is observed. The variable y_{it} is the logarithm of the output of the i-th unit at time t, $\boldsymbol{x}'_{it}$ is a row vector of p regressors and $\boldsymbol{\beta}$ is a column vector of unknown parameters. The random variable e_{it} is the *idiosyncratic random component* and b_i is the *random subject effect*. Finally u_{i0} is the *time invariant stochastic inefficiency* and $\delta_t u_{i1}$ is the *time dependent stochastic inefficiency*. As in Lee and Schmidt [13], the nonnegative parameters δ_t ($\delta_1 = 1$ for identifiability reasons) represent the impact of the random inefficiency u_{i1} at time $t, t = 1, 2, \ldots, T$. We note that $y^{\mathrm{s}}_{it} = \beta_0 + \boldsymbol{x}'_{it}\boldsymbol{\beta} + b_i + e_{it}$ is the subject specific stochastic frontier and that $y^{\mathrm{p}}_{it} = \beta_0 + \boldsymbol{x}'_{it}\boldsymbol{\beta} + e_{it}$ is the population stochastic frontier.

The specification of the time varying inefficiency component $\delta_t u_{i1}$ is often criticized [12] on the ground that the parameters δ_t are the same for every unit. We do not agree with this point of view because the specific unit effect is taken into account by the random variable u_{i1} and the parameters δ_t are intended to represent time factors which are independent from the subject efficiency. An alternative approach to modeling the time varying inefficiency can be found in Colombi et al. [6] and Colombi [5] and is briefly introduced in Sect. 5. As discussed in the just quoted papers the greater generality unfortunately implies a significative computational burden in the maximization of the log-likelihood.

Many interesting models can be obtained from (17.1) by omitting the subject random component or a random inefficiency component. Every model can be identified by a three letters label. The first letter pertains to the presence (T=True) or absence (F=False) of the random subject effect, the second letter do the same for the time dependent stochastic inefficiency and in the same way the third letter relates to the time independent stochastic inefficiency. For example TTT is model (17.1), FFT is the Pitt and Lee model I [17], TFT is a generalization of the Pitt and Lee model I obtained by introducing a subject random component, and FTF is the Lee and Schmidt [13] stochastic frontier without time independent inefficiency and subject specific component. The models TTT, FTT, TTF, and TFT do not seem to have been previously examined.

We observe that testing one of the previous models against the general model TTT is a nonstandard problem because under the null hypothesis one or two parameters are on the boundary of the parametric space. In fact, under reasonable assumptions the asymptotic distribution of the log-likelihood ratio test statistic is a mixture of chi-squared distributions known as *chi-bar-squared distribution* [19]. For the sake of simplicity we consider only the case of balanced panel data with a fixed number of observations per unit but the results of the following sections are easily extended to unbalanced panels.

We assume that: (A1) for $i = 1, 2, \ldots, n$, the random variables u_{i0}, b_i, u_{i1} and e_{it}, $t = 1, 2, \ldots, T$, are independent in probability; (A2) the random vectors $(u_{i0}, u_{i1}, b_i, e_{i1}, e_{i2}, \ldots, e_{iT})$, $i = 1, 2, \ldots, n$ are independent in probability; (A3) for every i, the random inefficiency component u_{i0} is a normal random variable with null expected value, variance σ^2_{0u} and left truncated at zero; (A4) for every i, the random inefficiency component u_{i1} is a normal random variable with null expected value, variance σ^2_{1u} and left truncated at zero; (A5) for every i, the subject random effect b_i is a normal random variable with null expected value and variance σ^2_b ; (A6) for every i and t, e_{it} is a normal random variable with null expected value and variance σ^2_e; (A7) the vectors $\boldsymbol{x}_{it}$ are vectors of known constants.

The following matricial representation of model (17.1) is used in the next sections. Let $\mathbf{1}_T$ be a vector of T ones, $\mathbf{0}_T$ be a vector of T zeros and $\boldsymbol{I}_T$ be the identity matrix of dimension T. Moreover $\boldsymbol{y}_i$ is the vector of the T observations on the i-th unit, $\boldsymbol{X}_i$ is the $T \times p$ matrix with rows $\boldsymbol{x}'_{it}$, $\boldsymbol{u}_i$ is the vector with components u_{i0}, u_{i1} and $\boldsymbol{e}_i$ is the vector of the T idiosyncratic random components of the i-th unit. From (17.1) it follows that: $\boldsymbol{y}_i = \mathbf{1}_T(\beta_0 + b_i) + \boldsymbol{X}_i\boldsymbol{\beta} + \boldsymbol{A}\boldsymbol{u}_i + \boldsymbol{e}_i$ where the matrix $\boldsymbol{A}$ is so defined: $\boldsymbol{A} = -[\mathbf{1}_T\ \boldsymbol{\delta}]$ and $\boldsymbol{\delta}$ is the vector of the parameters $\delta_t, t = 1, 2, \ldots, T$. In the next section we will derive the joint density of the random component $\mathbf{1}_T b_i + \boldsymbol{A}\boldsymbol{u}_i + \boldsymbol{e}_i$.

17.3 Log Likelihood of the Closed Skew Normal Stochastic Frontier

In this section some important consequences of assumptions (A1)–(A7) are examined in order to derive the log-likelihood of model (17.1). With $\phi_q(\boldsymbol{x}, \boldsymbol{\mu}, \boldsymbol{\Omega})$ we denote the density function of a q-dimensional normal random variable with expected value $\boldsymbol{\mu}$ and variance $\boldsymbol{\Omega}$ and $\bar{\Phi}_q(\boldsymbol{\mu}, \boldsymbol{\Omega})$ is the probability that a q-variate normal random variable of expected value $\boldsymbol{\mu}$ and variance matrix $\boldsymbol{\Omega}$ belongs to the positive orthant. For an easy reference we report the definition of the closed skew normal distribution [1, 8].

Definition 1. A random vector $\boldsymbol{x}$ has a (p, q) closed skew normal distribution with parameters $\boldsymbol{\mu}$, $\boldsymbol{\Gamma}$, $\boldsymbol{D}$, $\boldsymbol{\nu}$, $\boldsymbol{\Delta}$ if its probability density function is:

$$f(\boldsymbol{x}, \boldsymbol{\mu}, \boldsymbol{\Gamma}, \boldsymbol{D}, \boldsymbol{\nu}, \boldsymbol{\Delta}, p, q) = \frac{\phi_p(\boldsymbol{x}, \boldsymbol{\mu}, \boldsymbol{\Gamma})\bar{\Phi}_q(\boldsymbol{D}(\boldsymbol{y} - \boldsymbol{\mu}) - \boldsymbol{\nu}, \boldsymbol{\Delta})}{\bar{\Phi}_q(-\boldsymbol{\nu}, \boldsymbol{\Delta} + \boldsymbol{D}\boldsymbol{\Gamma}\boldsymbol{D}')}. \tag{17.2}$$

The moment generating function of the previous random variable is:

$$E(\exp\{\boldsymbol{t}'\boldsymbol{x}\}) = \frac{\bar{\Phi}_q(\boldsymbol{D}\boldsymbol{\Gamma}\boldsymbol{t} - \boldsymbol{\nu}, \boldsymbol{\Delta} + \boldsymbol{D}\boldsymbol{\Gamma}\boldsymbol{D}')}{\bar{\Phi}_q(-\boldsymbol{\nu}, \boldsymbol{\Delta} + \boldsymbol{D}\boldsymbol{\Gamma}\boldsymbol{D}')} \exp\{\boldsymbol{t}'\boldsymbol{\mu} + \frac{1}{2}\boldsymbol{t}'\boldsymbol{\Gamma}\boldsymbol{t}\}. \tag{17.3}$$

When $\boldsymbol{\nu} = \boldsymbol{0}$ the previous density simplifies to

$$f(\boldsymbol{x}, \boldsymbol{\mu}, \boldsymbol{\Gamma}, \boldsymbol{D}, \boldsymbol{\nu}, \boldsymbol{\Delta}, p, q) = \frac{\phi_p(\boldsymbol{x}, \boldsymbol{\mu}, \boldsymbol{\Gamma})\bar{\Phi}_q(\boldsymbol{D}(\boldsymbol{y} - \boldsymbol{\mu}), \boldsymbol{\Delta})}{2^{-q}}.$$

In the next Propositions we will need the following matrices:

$$\boldsymbol{V} = \begin{bmatrix} \sigma_{1u}^2 & 0 \\ 0 & \sigma_{2u}^2 \end{bmatrix}, \quad \boldsymbol{\Sigma} = \sigma_e^2 \boldsymbol{I}_T + \sigma_b^2 \boldsymbol{1}_T \boldsymbol{1}_T'$$

$$\boldsymbol{\Lambda} = \boldsymbol{V} - \boldsymbol{V}\boldsymbol{A}'(\boldsymbol{\Sigma} + \boldsymbol{A}\boldsymbol{V}\boldsymbol{A}')^{-1}\boldsymbol{A}\boldsymbol{V} = (\boldsymbol{V}^{-1} + \boldsymbol{A}'\boldsymbol{\Sigma}^{-1}\boldsymbol{A})^{-1},$$

$$\boldsymbol{R} = \boldsymbol{V}\boldsymbol{A}'(\boldsymbol{\Sigma} + \boldsymbol{A}\boldsymbol{V}\boldsymbol{A}')^{-1} = \boldsymbol{\Lambda}\boldsymbol{A}'\boldsymbol{\Sigma}^{-1}.$$

The relevance of the closed skew normal density in the context of stochastic frontiers derives from the following considerations. In model (17.1) the random component $b_i - u_{i0} - \delta_t u_{i1} + e_{it}$ is the sum of the time independent error $\varepsilon_i = b_i - u_{i0}$ and of the time dependent component $\varepsilon_{it} = e_{it} - \delta_t u_{i1}$. According to our assumptions the two components are independent in probability and are given by the difference between a normal random variable and an independent left truncated at zero normal random variable. It is well known [12] that ε_i has the following density:

$$f(\varepsilon_i) = 2\phi_1(\varepsilon_i, 0, \sigma_b^2 + \sigma_{0u}^2)\bar{\Phi}_1\left(\frac{-\sigma_{0u}^2}{\sigma_b^2 + \sigma_{0u}^2}\varepsilon_i, \frac{\sigma_{0u}^2\sigma_b^2}{\sigma_b^2 + \sigma_{0u}^2}\right).$$

It easy to see that the previous density is a $(1, 1)$ closed skew normal density.

It is also straightforward to see [12] that the joint density of the vector $\boldsymbol{\epsilon}_i$, with components $\varepsilon_{it}, t = 1, 2, \ldots, T$, is the $(T, 1)$ closed skew normal density:

$$f(\boldsymbol{\epsilon}_i) = 2\phi_1(\boldsymbol{\epsilon}_i, \mathbf{0}, \sigma_e^2 \boldsymbol{I}_T + \sigma_{1u}^2 \boldsymbol{\delta}\boldsymbol{\delta}')\bar{\Phi}_1 \left(\frac{-\sigma_{1u}^2}{\sigma_e^2 + \sigma_{1u}^2 \sum_{t=1}^T \delta_t^2} \boldsymbol{\delta}' \boldsymbol{\epsilon}_i, \frac{\sigma_{1u}^2 \sigma_e^2}{\sigma_e^2 + \sigma_{1u}^2 \sum_{t=1}^T \delta_t^2} \right).$$

From Theorem (3) of González-Farías et al. [8], it follows that the two independent random variables $\varepsilon_i, \boldsymbol{\epsilon}_i$ have a joint $(T+1, 2)$ closed skew normal density with parameters $\boldsymbol{\nu}_0 = \mathbf{0}$, $\boldsymbol{\mu}_0 = \mathbf{0}$ and

$$\boldsymbol{\Gamma}_0 = \begin{bmatrix} \sigma_{0u}^2 + \sigma_b^2 & \mathbf{0}_T' \\ \mathbf{0}_T & \sigma_e^2 \boldsymbol{I}_T + \sigma_{1u}^2 \boldsymbol{\delta}\boldsymbol{\delta}' \end{bmatrix}, \quad \boldsymbol{D}_0 = \begin{bmatrix} \frac{-\sigma_{0u}^2}{\sigma_b^2 + \sigma_{0u}^2} & \mathbf{0}_T' \\ 0 & \frac{-\sigma_{1u}^2}{\sigma_e^2 + \sigma_{1u}^2 \sum_{t=1}^T \delta_t^2} \boldsymbol{\delta}' \end{bmatrix},$$

$$\boldsymbol{\Delta}_0 = \begin{bmatrix} \frac{\sigma_{0u}^2 \sigma_b^2}{\sigma_b^2 + \sigma_{0u}^2} & 0 \\ 0 & \frac{\sigma_{1u}^2 \sigma_e^2}{\sigma_e^2 + \sigma_{1u}^2 \sum_{t=1}^T \delta_t^2} \end{bmatrix}.$$

If $\boldsymbol{L} = [\mathbf{1}_T \ \boldsymbol{I}_T]$, from Theorems 1 of González-Farías et al. [8], it follows that the T dimensional random vector $\mathbf{1}_T b_i + \boldsymbol{A}\boldsymbol{u}_i + \boldsymbol{e}_i = \boldsymbol{L}\left[\varepsilon_i, \boldsymbol{\epsilon}_i'\right]'$ with components $b_i - u_{i0} - \delta_t u_{i1} + e_{it} = \varepsilon_i + \varepsilon_{it}$ has a $(T, 2)$ closed skew normal distribution with parameters $\boldsymbol{\nu}_1 = \mathbf{0}$, $\boldsymbol{\mu}_1 = \mathbf{0}$, $\boldsymbol{\Gamma}_1 = \boldsymbol{L}\boldsymbol{\Gamma}_0\boldsymbol{L}' = \boldsymbol{\Sigma} + \boldsymbol{A}\boldsymbol{V}\boldsymbol{A}'$, $\boldsymbol{D}_1 = \boldsymbol{D}_0\boldsymbol{\Gamma}_0\boldsymbol{L}'\boldsymbol{\Gamma}_1^{-1} = \boldsymbol{R}$, $\boldsymbol{\Delta}_1 = \Delta_0 + \boldsymbol{D}_0\boldsymbol{\Gamma}_0\boldsymbol{D}_1' - \boldsymbol{D}_0\boldsymbol{\Gamma}_0\boldsymbol{L}'\boldsymbol{\Gamma}_1^{-1}\boldsymbol{L}\boldsymbol{\Gamma}_0\boldsymbol{D}_0' = \boldsymbol{\Lambda}$. From the previous result and because the parameter $\boldsymbol{\mu}$ of Definition 1 is a location parameter, the following Proposition, about the distribution of $\boldsymbol{y}_i = (\mathbf{1}_T\beta_0 + \boldsymbol{X}_i\boldsymbol{\beta}) + \mathbf{1}_T b_i + \boldsymbol{A}\boldsymbol{u}_i + \boldsymbol{e}_i$, follows.

Proposition 1. *Under the assumptions A1, A3–A7, the random vector* $\boldsymbol{y}_i = \mathbf{1}_T(\beta_0 + b_i) + \boldsymbol{X}_i\boldsymbol{\beta} + \boldsymbol{A}\boldsymbol{u}_i + \boldsymbol{e}_i$ *has a (T,2) closed skew normal distribution with density*

$$f(\boldsymbol{y}_i) = \phi_T(\boldsymbol{y}_i, \mathbf{1}_T\beta_0 + \boldsymbol{X}_i\boldsymbol{\beta}, \boldsymbol{\Sigma} + \boldsymbol{A}\boldsymbol{V}\boldsymbol{A}')\frac{\bar{\Phi}_2(\boldsymbol{R}(\boldsymbol{y}_i - \boldsymbol{X}_i\boldsymbol{\beta} - \mathbf{1}_T\beta_0), \boldsymbol{\Lambda})}{2^{-2}}$$

and moment generating function

$$E(\exp\{\boldsymbol{t}'\boldsymbol{y}_i\}) = \frac{\bar{\Phi}_2(\boldsymbol{V}\boldsymbol{A}'\boldsymbol{t}, \boldsymbol{V})}{2^{-2}} \exp\{\boldsymbol{t}'(\mathbf{1}_T\beta_0 + \boldsymbol{X}_i\boldsymbol{\beta}) + \frac{1}{2}\boldsymbol{t}'(\boldsymbol{\Sigma} + \boldsymbol{A}\boldsymbol{V}\boldsymbol{A}')\boldsymbol{t}\}.$$

When $\boldsymbol{y}_i$ is the vector of the logarithms of the outputs and $\boldsymbol{t}$ is the k-th column of the identity matrix of dimension T, the moment generating function of the previous Proposition gives the expected value of the k-th component of the vector $\exp\{\boldsymbol{y}_i\}$ of the outputs. It can be easily checked that a $(T, 1)$ closed skew normal distribution is obtained in the case of models without time invariant inefficiency or when alternatively the time dependent inefficiency is omitted. When the subject random component is omitted, the joint distribution is given by the previous results with $\sigma_b^2 = 0$. The following Proposition 2 is an immediate consequence of assumption A2 and Proposition 1.

Proposition 2. *Under the assumptions A1–A7, the log-likelihood of nT observations from model (17.1) is:*

$$L = \sum_{i=1}^{n} (\ln \phi_T(\boldsymbol{y}_i, \mathbf{1}_T\beta_0 + \boldsymbol{X}_i\boldsymbol{\beta}, \boldsymbol{\Sigma} + \boldsymbol{A}\boldsymbol{V}\boldsymbol{A}') + ln\bar{\Phi}_2(\boldsymbol{R}(\boldsymbol{y}_i - \boldsymbol{X}_i\boldsymbol{\beta} - \mathbf{1}_T\beta_0), \boldsymbol{\Lambda}) \tag{17.4}$$

which is the log-likelihood of the n independent $(T, 2)$ closed skew normal random variables $\boldsymbol{y}_i$.

17.4 Prediction of the Random Components

When $\boldsymbol{y}_i$ is the vector of the logarithms of the outputs, an important topic in applied research is forecasting the subject random components by the expected values $E(\exp\{b_i\}|\boldsymbol{y}_i)$ and the subject random inefficiencies by $E(\exp\{-\boldsymbol{u}_i\}|\boldsymbol{y}_i)$. This section shows how the relevant expected values can be computed. It is convenient to introduce the following definitions:

$$\boldsymbol{r}_i = \boldsymbol{y}_i - \boldsymbol{X}_i\boldsymbol{\beta} - \mathbf{1}_T\beta_0, \quad \tilde{\sigma}_b^2 = \sigma_b^2 - \sigma_b^4\mathbf{1}_T'\boldsymbol{\Psi}\mathbf{1}_T$$

$$\boldsymbol{\Psi} = (\boldsymbol{\Sigma} + \boldsymbol{A}\boldsymbol{V}\boldsymbol{A}')^{-1}, \quad \tilde{\boldsymbol{\Lambda}} = \boldsymbol{\Lambda} - \boldsymbol{R}\mathbf{1}_T\mathbf{1}_T'\boldsymbol{R}'\frac{\sigma_b^4}{\tilde{\sigma}_b^2}.$$

Proposition 3. *From assumptions A1, A3–A7 it follows that:*

(a) Conditionally on $\boldsymbol{y}_i$ the subject specific random component b_i has a (1,2) closed skew normal distribution with density

$$f(b_i|\boldsymbol{y}_i) = \phi(b_i, \sigma_b^2\mathbf{1}'\boldsymbol{\Psi}\boldsymbol{r}_i, \tilde{\sigma}_b^2)\frac{\bar{\Phi}_2(\boldsymbol{R}\boldsymbol{r}_i - \boldsymbol{R}\mathbf{1}_T\sigma_b^2\tilde{\sigma}_b^{-2}(b_i - \sigma_b^2\mathbf{1}_T'\boldsymbol{\Psi}\boldsymbol{r}_i), \tilde{\boldsymbol{\Lambda}})}{\bar{\Phi}_2(\boldsymbol{R}\boldsymbol{r}_i, \boldsymbol{\Lambda})}$$

(b) Conditionally on $\boldsymbol{y}_i$ the random inefficiency vector $\boldsymbol{u}_i$ is a left truncated normal random variable with density:

$$f(\boldsymbol{u}_i|\boldsymbol{y}_i) = \frac{\phi_2(\boldsymbol{u}_i, \boldsymbol{R}\boldsymbol{r}_i, \boldsymbol{\Lambda})}{\bar{\Phi}_2(\boldsymbol{R}\boldsymbol{r}_i, \boldsymbol{\Lambda})}, \quad \boldsymbol{u}_i > \mathbf{0}$$

(c)

$$E(\exp\{b_i\}|\boldsymbol{y}_i) = \frac{\bar{\Phi}_2(\boldsymbol{R}\boldsymbol{r}_i - \boldsymbol{R}\mathbf{1}_T\sigma_b^2, \boldsymbol{\Lambda})}{\bar{\Phi}_2(\boldsymbol{R}\boldsymbol{r}_i\boldsymbol{\Lambda})}\exp\{\sigma_b^2\mathbf{1}_T'\boldsymbol{\Psi}\boldsymbol{r}_i + \frac{1}{2}\tilde{\sigma}_b^2\}$$

(d)

$$E(\exp\{\boldsymbol{t}'\boldsymbol{u}_i\}|\boldsymbol{y}_i) = \frac{\bar{\Phi}_2(\boldsymbol{R}\boldsymbol{r}_i + \boldsymbol{\Lambda}\boldsymbol{t}, \boldsymbol{\Lambda})}{\bar{\Phi}_2(\boldsymbol{R}\boldsymbol{r}_i, \boldsymbol{\Lambda})} \exp\{\boldsymbol{t}'\boldsymbol{R}\boldsymbol{r}_i + \frac{1}{2}\boldsymbol{t}'\boldsymbol{\Lambda}\boldsymbol{t}\}$$

Proof. To prove (b) note that:

$$f(\boldsymbol{u}_i|\boldsymbol{y}_i) = \frac{\phi_T(\boldsymbol{y}_i, \mathbf{1}_T\beta_0 + \boldsymbol{X}_i\boldsymbol{\beta}, \boldsymbol{\Sigma} + \boldsymbol{A}\boldsymbol{V}\boldsymbol{A}')\phi_q(\boldsymbol{u}, \boldsymbol{R}\boldsymbol{r}_i, \boldsymbol{\Lambda})4}{f(\boldsymbol{y}_i)} \tag{17.5}$$

and use the result of Proposition 1 about $f(\boldsymbol{y}_i)$. Noting that $f(\boldsymbol{u}_i|\boldsymbol{y}_i)$ is the density of a left truncated at zero multi-normal random variable, (d) follows from Lemma 13.6.1 of Dominguez-Molina et al. [7]. To prove (a) we observe that:

$$f(b_i|\boldsymbol{y}_i) = \phi(b_i, \sigma_b^2\mathbf{1}'\boldsymbol{\Psi}\boldsymbol{r}_i, \tilde{\sigma}_b^2) \times \frac{\int_0^\infty\int_0^\infty \phi_2(u_0, u_1, \boldsymbol{R}\boldsymbol{r}_i - \boldsymbol{R}\mathbf{1}_T\sigma_b^2\tilde{\sigma}_b^{-2}(b_i - \sigma_b^2\mathbf{1}_T'\boldsymbol{\Psi}\boldsymbol{r}_i), \tilde{\boldsymbol{\Lambda}})\mathrm{d}u_0, \mathrm{d}u_1}{\int_0^\infty\int_0^\infty \phi_2(u_0, u_1, \boldsymbol{R}\boldsymbol{r}_i, \boldsymbol{\Lambda})\mathrm{d}u_0\mathrm{d}u_1} \tag{17.6}$$

Point (a) of the proposition follows immediately. Observing that $f(b_i|\boldsymbol{y}_i)$ is an instance of the closed skew normal density, point (c) follows from the result (17.3) on the moment generating function of a closed skew normal random variable. □

If $-\boldsymbol{t}'$ is the k-th row of the identity matrix of dimension q the result (d) of the previous Proposition gives the conditional expected value of the k-th component of the inefficiency vector $\exp\{-\boldsymbol{u}_i\}$. We stress also that conditionally on the observation $\boldsymbol{y}_i$, the subject random effect has not a normal distribution as it happens in standard random effect models.

17.5 An Alternative Specification of the Time Dependent Random Inefficiency

As we already pointed out the models TTT, FTT, TTF, FTF can be criticized because every unit shares the same set of parameters δ_t. An alternative approach to modeling time varying inefficiency is obtained if the random components $\delta_t u_{i1}$ are replaced by T independent left truncated at zero normal random variables u_{it} with null expected value and variance $\sigma_{2u}^2\psi_t$. The heteroskedasticity parameters ψ_t, $t = 1, 2, \ldots, T$ are positive and satisfy the identifiability restriction $\psi_1 = 1$. As shown in [5, 6] the results of Sect. 4 still hold with the following replacements:

$$\boldsymbol{V} = \begin{bmatrix} \sigma_{0u}^2 & \mathbf{0}_T' \\ \mathbf{0}_T & \sigma_{1u}^2\boldsymbol{\Upsilon} \end{bmatrix}, \quad \boldsymbol{A} = -[\mathbf{1}_T\ \boldsymbol{I}_T].$$

where $\boldsymbol{\Upsilon}$ denote the diagonal matrix with the parameters ψ_t on the main diagonal. The parameters σ_{0u}^2, σ_{1u}^2 and ψ_t, $t = 2, \ldots, T$ are identifiable and can be estimated by the ML method [6] even if the log-likelihood is more difficult to compute. In fact, now $f(\boldsymbol{y}_i) = \phi_T(\boldsymbol{y}_i, \mathbf{1}_T\beta_0 + \boldsymbol{X}_i\boldsymbol{\beta}, \boldsymbol{\Sigma} + \boldsymbol{A}\boldsymbol{V}\boldsymbol{A}')\frac{\bar{\Phi}_q(\boldsymbol{R}(\boldsymbol{y}_i - \boldsymbol{X}_i\boldsymbol{\beta} - \mathbf{1}_T\beta_0), \boldsymbol{\Lambda})}{2^{-q}}$ is a (q, T) closed skew normal density with $q = T + 1$ for the models TTT, FTT, and $q = T$ for the models TTF, FTF. The computational complexity of the log-likelihood comes from the dimensionality of the integrals $\bar{\Phi}_q(\boldsymbol{R}(\boldsymbol{y}_i - \boldsymbol{X}_i\boldsymbol{\beta} - \mathbf{1}_T\beta_0), \boldsymbol{\Lambda})$. The R [18] function: *SNF_maxlik*, developed by the author, can be used to compute the ML-estimates of the models of this section and of the previous section. To integrate the multinormal density the *SNF_maxlik* function resort to the Bretz-Genz [10] method, implemented in R by the *mvtnorm* package [9]. A less accurate but computationally faster approach, to the numerical integration of the multivariate distribution, is the Pandey [16] approximation which is also implemented in *SNF_maxlik*. I found this approximation particularly useful in getting good initial estimates and in dealing with the model of this section when T is large.

17.6 Examples

In the first example we consider the RICE panel data set of Battese et al. [3]. The data set is also available with the *R-package Frontier* developed by Coelli and Henningsen [4]. In this data set the *annual rice production* (in tonnes) of $n = 43$ rice producers in the Tarlac region of the Philippines is reported from 1990 to 1997. Rice production is the output and the inputs are *area planted* (hectares), *labor* used (man days) and *fertiliser* used (kg). In this application the log-linear stochastic frontier is intended to model the maximum rice production in function of the inputs. The random component b_i models unobserved differences among producers and the inefficiency components model the producer inability to reach the optimal production.

It clearly emerges from Table 17.1 (only the maximum of the log-likelihood and the estimated variances are reported) that the time dependent random inefficiency component is relevant (compare TTT with TFT) and that its omission inflates the variance of the idiosyncratic random component. On the contrary the subject random component is not significative as shown by the comparison of the model TTT with FTT. The asymptotic distribution of the log-likelihood ratio statistics to test TTF against TTT is a 0.5 mixture of a chi-squared distribution with zero degrees of freedom and a chi-squared distribution with one degree of freedom. The p-value $P = 0.029$ shows that in the model TTT the time independent inefficiency component importance is masked by the presence of the subject random component. However the relevance of the time independent inefficiency component is clearly stated by the comparison of the models FTT and FTF.

The second example examines the AIRPORT panel data set [15] about n = 38 Italian airports and the years 2005–2008. The output variable is *yearly number of aircraft movements* and the inputs are *number of runways, total area of the airport, number of check-in desks, number of luggage claim lines, number of aircraft*

Table 17.1 Skew normal stochastic frontiers for the RICE data

Model	Max. log-lik.	σ_b^2	σ_{2u}^2	σ_{1u}^2	σ_e^2
TTT	−63.181	0	0.073	0.056	0.068
TFT	−86.430	0	*0*	0.0721	0.0832
FTT	−63.181	*0*	0.073	0.056	0.068
TTF	−64.975	0.0194	0.0756	*0*	0.0689
FFT	−86.431	*0*	*0*	0.0723	0.0832
FTF	−88.101	*0*	0	*0*	0.091
TFF[a]	−88.605	0.1584	*0*	*0*	0.2901
FFF[a]	−104.907	*0*	*0*	*0*	0.3283

[a] TFF and FFF denote the subject random effect regression model and the classical regression model, respectively

Table 17.2 Skew normal stochastic frontiers for the AIRPORT data

Model	Max. log-lik.	σ_b^2	σ_{2u}^2	σ_{1u}^2	σ_e^2
TTT	28.750	0.2249	0.0112	0	0.0126
TFT	26.677	0.0493	*0*	0.3764	0.0159
FTT	27.440	*0*	0.0110	0.8802	0.0125
TTF	28.750	0.2249	0.0112	*0*	0.0016
FFT	24.107	*0*	*0*	0.6904	0.0162
FTF	−51.137	*0*	0	*0*	0.04767
TFF	25.699	0.4298	*0*	*0*	0.1261
FFF	−51.132	*0*	*0*	*0*	0.3387

parking sites. To take into account the airports heterogeneity the factor *EU category of the airport* (A: Great European Airports, B: National Airports, C: Domestic Airports and D: Regional Airports) was also introduced in the log-linear frontier. Table 17.2 reports the results obtained by applying the models of Sect. 2.

In this example the omission of the subject random component inflates the variance of the time invariant inefficiency which after controlling for subject random heterogeneity and time dependent inefficiency seems to be irrelevant. The irrelevance of the time invariant inefficiency random component is clearly shown by the comparison of TTT with TTF. Note that the asymptotic distribution of the log-likelihood ratio statistics to test FTT against TTT is a 0.5 mixture of a chi-squared distribution with zero degrees of freedom and a chi-squared distribution with one degree of freedom. The p-value is $P = 0.053$ and so model comparison enlightens that the subject specific random effect is not significant in the TTT model.

References

1. Arellano-Valle, E., Azzalini A.: On the unification of families of skew-normal distributions. Scand. J. Stat. **33**, 561–574 (2006)
2. Arellano-Valle, E., Bolfarine, H., Lachos, H.: Skew-normal linear mixed models. J. Data. Sci. **3**, 415–438 (2005)
3. Battese, G., Coelli, T., O'Donnel, C., Rao, D.: An Introduction to Efficiency and Productivity Analysis, pp. 325–326. Springer, New York (2005)
4. Coelli, T., Henningsen A.: frontier: Stochastic Frontier Analysis. R package version 0.996-6 (2010) http://www.R-project.org
5. Colombi, R.: A skew normal stochastic Frontier model for panel data. In: Poceedings of the 45th Scientific Meeting of the Italian Statistical Society, Universitádi Padova, Padova (2010) http://homes.stat.unipd.it/mgri/SIS2010/Program/contributedpaper/486-1310-1-DR.pdf
6. Colombi, R., Martini, G., Vittadini, G.: A stochastic frontier model with short-run and long-run inefficiency random effects. Aisberg, WP 012011, Università degli studi di Bergamo, Italy (2011). http://hdl.handle.net/10446/842
7. Dominguez-Molina, A., Gonzales-Farias, G., Ramos-Quiroga, R.: Skew normality in stochastic frontier analysis. In: Genton, M. (ed.) Skew Elliptical Distributions and their Applications, pp. 223–241. Chapman and Hall CRC, London (2004)
8. Gonzáles-Farías, G., Domínguez Molina, A., Gupta, A.: Additive properties of skew normal random vectors. J. Stat. Plan. Infer. **126**, 521–534 (2004)
9. Genz, A., Bretz, F., Miwa, T., Mi, X., Leisch, F., Scheipl, F., Hothorn, T.: mvtnorm - Multivariate Normal and t Distributions. R package version 0.9-9992 (2012) http://CRAN.R-project.org/package=mvtnorm
10. Genz, A., Bretz, F.: Computation of Multivariate Normal and t Probabilities. Springer, New York (2009)
11. Greene, W.: Reconsidering heterogeneity in panel data estimators of the stochastic frontier model. J. Economet. **126**, 269–303 (2005)
12. Kumbhakar, S., Lovell, K.: Stochastic Frontier Analysis. Cambridge University Press, Cambridge (2000)
13. Lee, H., Schmidt, P.: A production frontier model with flexible temporal variation in technical efficiency. In: Fried, H., Lovell, K., Scmidt, S. (eds.) The Measurement of Productive Efficiency, pp. 237–255. Oxford University Press, New York (1993)
14. Lin, T., Lee, C.: Estimation and prediction in linear mixed models with skew-normal random effects for longitudinal data. Statist. Med. **27**, 1490–1507 (2005)
15. Malighetti, P., Martini, G., Paleari, S., Redondi, R.: An empirical investigation on the efficiency capacity and ownership of italian airports. Riv. Pol. Ec. **47**, 157–188 (2007)
16. Pandey, M.: An effective approximation to evaluate multinormal integrals. Struct. Safety **20**, 51–67 (1998)
17. Pitt, M., Lee, L.: Measurement of sources of technical innefficiency in the Indonesian weawing industry. J. Dev. Econ. **9**, 43–64 (1981)
18. R Development Core Team: R: A language and environment for statistical computing. R Foundation for Statistical Computing ISBN 3-900051-07-0 (2012) http://www.R-project.org
19. Sivapulle, M.J., Sen, P.K.: Constrained Statistical Inference. Wiley, Hoboken (2005)

How Far Can Man Go?

18

Isabel Fraga Alves, Laurens de Haan, and Cláudia Neves

Abstract

In this chapter we address the question of "What is the Largest Jump at Man's reach, given today's state of the art?" To answer that question it will be used the *best from the best*, i.e., the data will be collected from the best "jumpers" from World Athletics Competitions—"Long Jump Men Outdoors" event. Our approach to the problem is based on the probability theory of extreme values (EVT) and the corresponding statistical techniques. We shall only use the top performances of World top lists. Our estimated ultimate record, i.e., the right endpoint of the jumping event, tells us what is possible to infer about the possible personal best mark, given today's knowledge, sports conditions and rules.

Keywords

Endpoint estimation in Gumbel domain of attraction • Extreme values in sports • Extreme value theory • Semi-parametric approach to statistics of extremes

---

I. Fraga Alves (✉)
CEAUL, Department of Statistics, University of Lisbon, Portugal
e-mail: isabel.alves@fc.ul.pt

L. de Haan
University of Tilburg, and Rotterdam, The Netherlands

CEAUL, University of Lisbon Portugal
e-mail: ldhaan@ese.eur.nl

C. Neves
Department of Mathematics, University of Aveiro, Portugal
e-mail: claudia.neves@ua.pt

N. Torelli et al. (eds.), *Advances in Theoretical and Applied Statistics*,
Studies in Theoretical and Applied Statistics, DOI 10.1007/978-3-642-35588-2_18,

18.1 Introduction

In this chapter we address the question of
"What is the Largest Jump that Man can achieve, given today's state of the art?"
For the purposes of our study we selected the following athletic jumping event:

- *Long Jump Men Outdoors (LJ).*

The best current mark is 8.95m from Mike Powell (USA) in Tokyo, 30 August 1991.

Our estimated ultimate record, i.e., the right endpoint of the jumping event tells us what is possible to infer about the possible personal best mark, given today's knowledge, sports conditions, and rules.

In [2], Einmahl and Magnus studied a similar problem, but LJ event could not be handled with the available methodology so far, not covering the estimation of the right endpoint of a distribution in Gumbel domain of attraction. Related contributions for this type of athletics data are [3, 5].

This contribution constitutes a first approach to the estimation of the right endpoint for a distribution in the Gumbel max-domain of attraction, a subject not yet addressed in the literature of extremes to our best knowledge. Asymptotic distributional properties of endpoint estimator are under study.

The chapter is organized as follows. In Sect. 18.2 an overview of the first and second order properties of the tail distribution and their relation to the right endpoint; special attention is devoted to the special case of finite right endpoint and Gumbel max-domain. In Sect. 18.3 the motivation for the right endpoint estimator is presented, providing consistency in accordance to asymptotic theoretical relations of Sect. 18.2. Finally, in Sect. 18.4 we apply the theory to the LJ-data.

18.2 Gumbel Domain of Attraction and Finite Right Endpoint

Let $X_1, X_2, \ldots, X_n$ be an independent, identically distributed (i.i.d.) sample from an unknown distribution function (d.f.) F. If there exist normalizing constants $a_n > 0$, $b_n \in \mathbb{R}$ and a non-degenerate d.f. G such that, for all x,

$$\lim_{n\to\infty} P\left\{a_n^{-1}\left(\max\left(X_1, \ldots, X_n\right) - b_n\right) \le x\right\} = G(x), \tag{18.1}$$

then G is, up to scale and location, an *Extreme Value* d.f., dependent on a shape parameter $\gamma \in \mathbb{R}$, and given by

$$G_\gamma(x) := \begin{cases} \exp(-(1+\gamma\, x)^{-1/\gamma}),\ 1+\gamma\, x > 0 & \text{if } \gamma \neq 0 \\ \exp(-\exp(-x)),\ x \in \mathbb{R} & \text{if } \gamma = 0 \end{cases}. \tag{18.2}$$

The EV condition (18.1) is equivalent to say that F is in the max-domain of attraction of the d.f. G_γ, i.e., $F \in \mathscr{D}_M(G_\gamma)$. For $\gamma < 0$, $\gamma = 0$ and $\gamma > 0$, the G_γ d.f. reduces to Weibull, Gumbel, and Fréchet distribution functions, respectively.

Fréchet, Weibull and Gumbel max-Domains of Attraction The shape parameter γ is closely related to the tail heaviness of the underlying distribution F: $\gamma < 0$ refers to short tails which must have finite right endpoint $x^F := \sup\{x : F(x) < 1\}$, whereas for $\gamma > 0$ the d.f. F is heavy tailed with polynomially decaying tails, for which the right endpoint must be infinite. The intermediate case $\gamma = 0$, the Gumbel domain, has been revealed very interesting for many applied sciences where extremes play the major role, and include a great variety of distributions with finite right endpoint or not. Then it is assumed that the extreme value index γ of the underlying d.f. equals 0, and statistical inference procedures concerning rare events on the tail of F, such as the estimation of high quantiles, small exceedance probabilities or return periods, bear on this assumption. The latter has been naturally imposed by the lack of attention to the estimation of the right endpoint x^F in the literature, under this Gumbel max-domain setup.

First order properties for U Consider the left-continuous inverse of $1/(1-F)$, defined by

$$U(t) := F^{\leftarrow}(1 - 1/t), \quad \text{for, } t > 1, \tag{18.3}$$

where $F^{\leftarrow}$ denotes the generalized inverse function $F^{\leftarrow}(x) = \inf\{y : F(y) \geq x\}$. The following *extended regular variation* property (see [6], for instance), denoted ERV_γ, is a well-known necessary and sufficient condition for $F \in \mathscr{D}_M(G_\gamma)$:

$$\lim_{t\to\infty} \frac{U(tx) - U(t)}{a(t)} = \begin{cases} \frac{x^\gamma - 1}{\gamma} & \text{if } \gamma \neq 0 \\ \ln x & \text{if } \gamma = 0 \end{cases}, \tag{18.4}$$

for every $x > 0$ and some positive measurable function a. For the three max-domains of attraction, the following properties on the tail of the distribution hold:

$\mathscr{D}_M(G_\gamma)_{\gamma>0}$	Heavy tails—Regularly Varying (RV) tails. No finite right endpoint, i.e., $x^F := U(\infty) = \infty$. $U \in \mathrm{RV}_\gamma$, i.e., $\lim_{t\to\infty} \frac{U(tx)}{U(t)} = x^\gamma$, for $x > 0$, with $\lim_{t\to\infty} \frac{a(t)}{U(t)} = \gamma$.

$\mathscr{D}_M(G_\gamma)_{\gamma<0}$	Short Tails—Finite right endpoint, i.e. $x^F := U(\infty) < \infty$. $U(\infty) - U \in \mathrm{RV}_\gamma$, i.e. $\lim_{t\to\infty} \frac{U(\infty)-U(tx)}{U(\infty)-U(t)} = x^\gamma$, for $x > 0$, with $\lim_{t\to\infty} \frac{a(t)}{U(\infty)-U(t)} = -\gamma$.

$\mathscr{D}_M(G_\gamma)_{\gamma=0}$	Light Tails—$U \in \Pi$-class, i.e., $\lim_{t\to\infty} \frac{U(tx)-U(t)}{a(t)} = \ln x$, for positive $a \in \mathrm{RV}_0$. The right endpoint $x^F := U(\infty)$ is finite or infinite. $U \in \mathrm{RV}_0$, i.e., $\lim_{t\to\infty} \frac{U(tx)}{U(t)} = 1$, with $\lim_{t\to\infty} \frac{a(t)}{U(t)} = 0$, for $x > 0$. If the right endpoint is finite, i.e. $x^F := U(\infty) < \infty$, then $\lim_{t\to\infty} \frac{U(\infty)-U(tx)}{U(\infty)-U(t)} = 0$, $\lim_{t\to\infty} \frac{a(t)}{U(\infty)-U(t)} = 0$, with $\lim_{t\to\infty} a(t) = 0$.

Second order properties for U The rate of convergence in (18.4) is *controlled* by the existence of a *second order* parameter $\rho \leq 0$ and a function A, possibly not changing in sign and tending to zero as $t \to \infty$, with $|A| \in \mathrm{RV}_\rho$, and such that for all $x > 0$

$$\lim_{t\to\infty} \frac{\frac{U(tx)-U(t)}{a(t)} - \frac{x^\gamma-1}{\gamma}}{A(t)} = H_{\gamma,\rho}(x) := \frac{1}{\rho}\left(\frac{x^{\gamma+\rho}-1}{\gamma+\rho} - \frac{x^\gamma-1}{\gamma}\right). \tag{18.5}$$

We then say that U is of *second order extended regular variation*, and denote this by $U \in 2\mathrm{ERV}_{\gamma,\rho}$. In (18.5), the cases $\gamma = 0$ and/or $\rho = 0$ are obtained by continuity.

In the particular case of $F \in \mathscr{D}_M(G_0)$ with $x^F < \infty$, we must have $\rho = 0$ (see Theorem B.3.6 in [6]); consequently, (18.5) becomes $H_{\gamma,\rho}(x) = H_{0,0}(x) = \frac{\ln^2 x}{2}$.

Moreover, in case $\rho = \gamma = 0$, there exists a function a_0 asymptotically equivalent to a, i.e., $a_0(t) \sim a(t)$, as $t \to \infty$, such that (see [6], pages 382–3)

$$U(t) = a_0(t) + \int_0^t \frac{a_0(s)}{s}\,\mathrm{d}s, \text{ with } a_0 \in \Pi. \tag{18.6}$$

Consequently, by (18.6)

$$U(\infty) - U(t) \sim \int_t^\infty \frac{a(s)}{s}\,\mathrm{d}s, \text{ as } t \to \infty,$$

which is slowly varying, since $U(\infty) - U(t) \in \mathrm{RV}_0$. Then

$$U(\infty) = U(t) + \int_t^\infty \frac{a(s)}{s}\,\mathrm{d}s + o(a(t)), \text{ as } t \to \infty,$$

and the following asymptotical representation holds for the right endpoint:

$$U(\infty) \sim U(t) + \tilde{a}(t), \text{ as } t \to \infty, \tag{18.7}$$

where the positive function $\tilde{a}$ is defined by

$$\tilde{a}(t) := \int_t^{\infty} \frac{a(s)}{s}\, ds = \int_1^{\infty} \frac{a(ts)}{s}\, ds = \int_0^1 \frac{a(t/s)}{s}\, ds.$$

Note that $\lim_{t\to\infty} \tilde{a}(t) = 0$ and $a(t) = o(\tilde{a}(t))$, as $t \to \infty$. Moreover, for $x > 0$,

$$\lim_{t\to\infty} \frac{\int_0^{1/t} (U(x/s) - U(1/s)) \frac{ds}{s}}{\int_0^{1/t} a(1/s) \frac{ds}{s}} = \lim_{t\to\infty} \frac{(U(tx) - U(t))\, t}{a(t)\, t} = \ln x,$$

makes possible to express $\tilde{a}(t)$ in terms of the quantile function U, which will reveal useful for estimation purposes.

This means that as $t \to \infty$ and for $x > 0$,

$$\tilde{a}(t) \sim \frac{1}{\ln x} \int_0^1 \left(U\left(\frac{t}{s} x\right) - U\left(\frac{t}{s}\right)\right) \frac{ds}{s}\,. \tag{18.8}$$

Example: Reversed Fréchet Distribution

The Reversed Fréchet model is defined for $x < x^F := U(\infty) \in \mathbb{R}$; for scale and shape parameters $\delta, \alpha > 0$ the d.f. F is given by

$$F(x) \equiv F(x; x^F, \alpha, \delta) = 1 - \exp\left\{ -\left(\frac{x^F - x}{\delta}\right)^{-\alpha} \right\}.$$

The following results are easily checked: $U(t) = x^F - \delta\, (\ln t)^{-1/\alpha}$, for $t > 1$; then $U \in \Pi(a)$, with $a(t) = \frac{\delta}{\alpha} (\ln t)^{-1/\alpha - 1}$, $t > 1$; $a \in \mathrm{RV}_0$, $\lim_{t\to\infty} a(t) = 0$ and $\lim_{t\to\infty} \frac{a(t)}{U(\infty) - U(t)} = 0$.

We conclude that $F \in \mathscr{D}_M(G_0)$, i.e., the Reversed Fréchet d.f. belongs to Gumbel max-domain of attraction with finite right endpoint x^F. Moreover, the function $\tilde{a}$ can be chosen as $\tilde{a}(t) = \delta\, (\ln t)^{-1/\alpha}$, $t > 1$, and the equality $x^F = U(t) + \tilde{a}(t)$ holds.

18.3 Estimation of the Right Endpoint in Gumbel Domain

Consider a random sample $X_1, X_2, \ldots, X_n$ of i.i.d. random variables with common d.f. $F \in \mathscr{D}_M(G_0)$, such that the right endpoint x^F is finite. Let $X_{1,n} \le X_{2,n} \le \cdots \le X_{n,n}$ be the associated ascending order statistics (o.s.). The following distributional identification will be the starting point for the estimation development:

$$\{X_{i,n}\}_{i=1}^n \stackrel{d}{=} \{U(Y_{i,n})\}_{i=1}^n, \tag{18.9}$$

denoting by $\{Y_{i,n}\}_{i=1}^n$ the o.s. pertaining to the independent random variables (r.v.) $\{Y_i\}_{i=1}^n$ with common standard Pareto d.f. $F(y) = 1 - y^{-1}$, $y \ge 1$. Let $k = k_n$ be an intermediate sequence, i.e., a sequence of positive integers such that

$$k_n \to \infty \quad \text{with} \quad k_n = o(n), \quad \text{as } n \to \infty. \tag{18.10}$$

Then, as the sample size $n \to \infty$, the intermediate sequence $Y_{n-k,n}$ converges in probability to infinity with the same rate as the ratio n/k, i.e., $(k/n)Y_{n-k,n} \xrightarrow[n\to\infty]{P} 1$, which is also equivalently denoted by the

$$Y_{n-k,n} \overset{p}{\sim} \frac{n}{k}, \quad \text{as} \quad n \to \infty. \tag{18.11}$$

The following steps provide the motivation for the proposed estimator. Consider at this stage the asymptotic relation (18.7), with n/k playing the role of t, according to (18.10); the estimator of the right endpoint $x^F = U(\infty)$ will be motivated by (18.7), through its empirical counterpart, for convenient estimators of the components:

$$\widehat{U(\infty)} := \widehat{U\left(\frac{n}{k}\right)} + \widehat{\tilde{a}\left(\frac{n}{k}\right)}. \tag{18.12}$$

From representation (18.9) together with (18.11), the estimation of the first component is immediately suggested as

$$\widehat{U\left(\frac{n}{k}\right)} := X_{n-k,n}\ .$$

For the estimation of $\tilde{a}(n/k)$, the asymptotic result (18.8) will be considered for the particular value $x = 1/2$, which has been chosen for sake of simplicity of the explicit expression of the resulting estimator

$$\tilde{a}\left(\frac{n}{k}\right) \sim \frac{1}{-\ln 2} \int_0^1 \left(U\left(\frac{n}{2ks}\right) - U\left(\frac{n}{ks}\right)\right) \frac{\mathrm{d}s}{s}\ . \tag{18.13}$$

This suggests for the estimator of $\tilde{a}\left(\frac{n}{k}\right)$ the empirical counterpart of (18.13)

$$\widehat{\tilde{a}\left(\frac{n}{k}\right)} := \frac{1}{-\ln 2} \int_0^1 \left(X_{n-[2ks],n} - X_{n-[ks],n}\right) \frac{\mathrm{d}s}{s}\ , \tag{18.14}$$

where $[z]$ denotes the integer part of z.

Since $X_{n-[ks],n}$ and $X_{n-[2ks],n}$ are identical to $X_{n,n}$ when $s \in (0, (2k)^{-1}]$, the integral $I_0 := \int_0^{1/(2k)} \left(X_{n-[2ks],n} - X_{n-[ks],n}\right) \frac{\mathrm{d}s}{s} = 0$ and is possible to rewrite the integral in (18.14) as the sum of $2k - 1$ integrals

$$\sum_{i=1}^{2k-1} I_i := \sum_{i=1}^{2k-1} \int_{i/(2k)}^{(i+1)/(2k)} \left(X_{n-[2ks],n} - X_{n-[ks],n}\right) \frac{ds}{s} .$$

After integrating (18.14) all over $\left(\frac{i}{2k}, \frac{i+1}{2k}\right]$, $i = 1, \ldots, 2k-1$ the estimator for $\tilde{a}\left(\frac{n}{k}\right)$ is finally obtained as

$$\widehat{\tilde{a}\left(\frac{n}{k}\right)} = \frac{1}{\ln 2} \sum_{i=0}^{k-1} (X_{n-2i,n} - X_{n-2i-1,n}) \ln\left(1 + \frac{1}{2i+1}\right) + \frac{1}{\ln 2} \sum_{i=1}^{k-1} (X_{n-i,n} - X_{n-2i,n}) \ln\left(1 + \frac{1}{i}\right), \tag{18.15}$$

which together with (18.12) leads to the endpoint estimator, involving $2k$ observations.

$$\hat{x}^F := X_{n-k,n} + \frac{1}{\ln 2} \sum_{i=0}^{k-1} (X_{n-2i,n} - X_{n-2i-1,n}) \ln\left(1 + \frac{1}{2i+1}\right) + \frac{1}{\ln 2} \sum_{i=1}^{k-1} (X_{n-i,n} - X_{n-2i,n}) \ln\left(1 + \frac{1}{i}\right). \tag{18.16}$$

18.4 Long Jump Men Outdoors: How Far?

For the purposes of our study we selected the following athletic jumping event:

- *Long Jump Men Outdoors (LJ)*—best current mark: 8.95 m from Mike Powell (USA) in Tokyo, 30/08/1991.

Our estimated ultimate record, i.e., the right endpoint of the jumping event, tells us what is possible to infer about the possible personal best mark, given today's knowledge, sports conditions, and rules.

LJ-Data For the LJ event the collected data have been of "the personal best" of as many of the top athletes in LJ event as it was possible. Each athlete appears only once in the data set, namely with his best own mark. Our observation period ends on December 31, 2009. The LJ data are obtained from two web sites, namely a Swedish web site compiled by Hans-Erik Pettersson, for the period up to August 2001 (source:http://hem.bredband.net/athletics/atb-m27.htm), which has been merged with the data for the period 1999–2009, from the official web site of the International Association of Athletics Federations (IAAF) (*source*: http://www.iaaf.org/statistics/toplists/). Similar to [2], only marks greater than 8 m have been collected, with an obtained top list of 720 best personal marks. Let $X_1, X_2, \ldots, X_n$

be an iid sample F and these personal bests observations associated top retained 720 order statistics $X_{n-720+1,n} \leq \cdots \leq X_{n-k,n} \leq X_{n-k+1,n} \leq \cdots \leq X_{n,n}$ be the, so that $x_{n-720+1,n} = 8$ and $x_{n,n}$ denotes the world record of 8.95, obtained by Powell. This athletics event has been previously detected by Einmahl and Magnus in [2], as a case where the Weibull Domain of attraction is questionable, since the estimated γ achieved was not clearly negative.

LJ-Data Analysis Similar to the endpoint estimation, the semi-parametric inference procedures considered in LJ-Data Analysis do not involve explicitly the value n and depend on the number k of top o.s.'s. The first LJ-Data Analysis involved the estimation of γ, which has been done according to three specific estimators, namely the Pickands, Moment and Mixed Moment estimators:

$$\hat{\gamma}^{\mathrm{P}} := (\ln 2)^{-1} \ln \frac{X_{n-k,n} - X_{n-2k,n}}{X_{n-2k,n} - X_{n-4k,n}}; \tag{18.17}$$

$$\hat{\gamma}^{\mathrm{M}} := M^{(1)} + \frac{1}{2}\left\{1 - \left(\frac{M^{(2)}}{[M^{(1)}]^2} - 1\right)^{-1}\right\}, \tag{18.18}$$

with $M^{(j)} := \frac{1}{k}\sum_{i=1}^{k}\{\ln X_{n-j+1,n} - \ln X_{n-k,n}\}^j,\ j = 1, 2;$

$$\hat{\gamma}^{\mathrm{MM}} := \frac{\hat{\varphi} - 1}{1 + 2\min(\hat{\varphi} - 1, 0)}, \tag{18.19}$$

with $\hat{\varphi} := \frac{M^{(1)} - L^{(1)}}{(L^{(1)})^2}$, where $L^{(1)} := \frac{1}{k}\sum_{i=1}^{k}\{1 - X_{n-k,n}/X_{n-j+1,n}\}$.
For details about these estimation procedures see [1,4,11]. In Fig. 18.1 are depicted the sample paths of the three γ-estimators, $\hat{\gamma}^{\mathrm{P}}$, $\hat{\gamma}^{\mathrm{M}}$ and $\hat{\gamma}^{\mathrm{MM}}$, associated with the LJ-Data against the same number k of o.s.'s involved in the estimation.

The second LJ-Data analysis consisted in performing the test for max-Gumbel domain of attraction, i.e.,

$$H_0 : F \in \mathscr{D}_M(G_0) \quad \text{versus} \quad H_1 : F \in \mathscr{D}_M(G_\gamma)_{\gamma \neq 0}\ . \tag{18.20}$$

For testing (18.20) three complementary tests statistics have been used: G (*Greenwood*), HW (*Hasofer&Wang*), and R(*Ratio*), which have been studied under semi-parametric approach in [7, 10]:

$$G := \frac{M_2}{(M_1)^2} \qquad \mathrm{HW} := \frac{1}{k(G-1)} \qquad R := \frac{X_{n,n} - X_{n-k,n}}{M_1}, \tag{18.21}$$

with $M_j := \frac{1}{k}\sum_{i=1}^{k}(X_{n-i+1,n} - X_{n-k,n})^j\ ,\ j = 1, 2$. Under the null hypothesis, the standardized versions of (18.21), $G^* := \sqrt{k/4}(G-2)$, $\mathrm{HW}^* := \sqrt{k/4}(k\mathrm{HW} - 1)$ have asymptotic Standard Normal distribution, whereas $R^* := R - \ln k$ converges to the Gumbel distribution. An overview about EV-testing is available in [8]. In Fig. 18.2-*Up* are pictured the sample paths of the three test statistics G^*, HW^* and

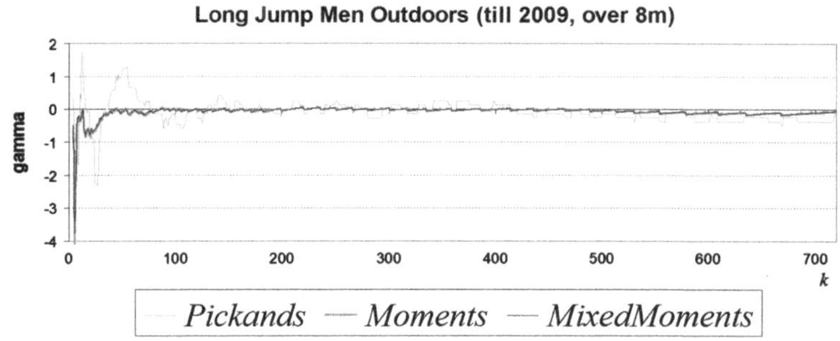

Fig. 18.1 Long Jump Men Outdoors (till 2009, over 8 m)—γ—Estimation. Sample path of γ-estimators against the number of top o.s.'s $k \leq 720$

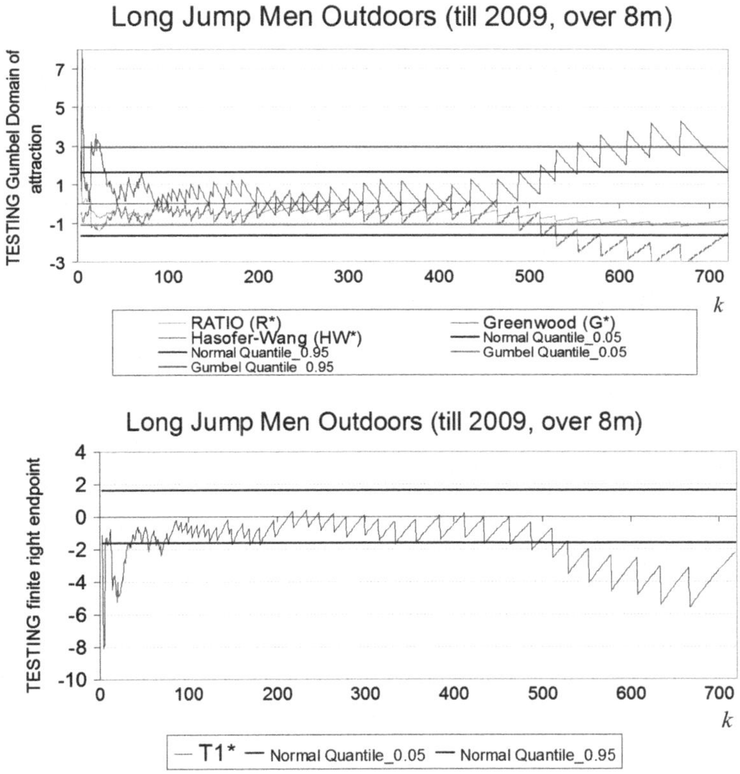

Fig. 18.2 Long Jump Men Outdoors (till 2009, over 8 m)—Testing $\mathscr{D}_M(G_0)$ and Testing finiteness of right endpoint. Sample path of the Test Statistics against the number of top o.s.'s$k \leq 720$

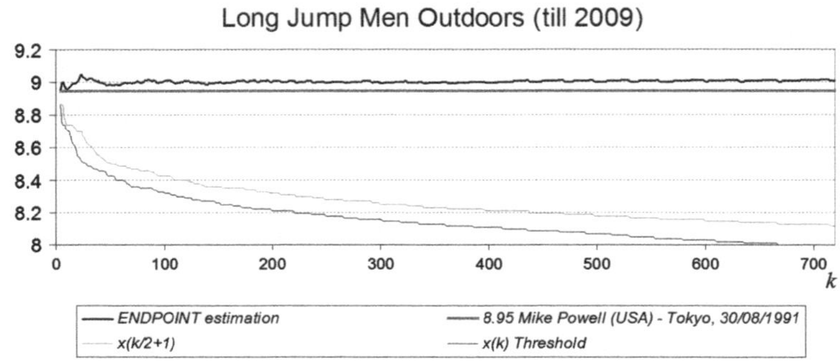

Fig. 18.3 Long Jump Men Outdoors (till 2009, over 8 m)—$\hat{x}^F$—Right Endpoint Estimation, i.e. *"ultimate record" (different thresholds,* $X_{(k/2+1)}$ *and* $X_{(k)}$ *for estimating the first and second components in* (18.12))

R^* associated with the LJ-Data, together with the respective Normal and Gumbel $(0.05, 0.95)$-quantiles. Altogether, from Figs. 18.1 and 18.2-*Up*, it seems reasonable to conclude that the distribution underlying the LJ-Data belongs to the Gumbel domain of attraction and not to the Weibull domain, which confirms the guess predicted in [2].

The third LJ-Data analysis has been testing the finiteness of the endpoint, in the same sense of Neves and Pereira in [9], i.e.,

$$H_0 : F \in \mathscr{D}_M(G_0),\ x^F = \infty \quad \text{versus} \quad H_1 : F \in \mathscr{D}_M(G_\gamma)_{\gamma \le 0},\ x^F < \infty\ . \tag{18.22}$$

through the test statistic

$$T_1 := \frac{1}{k}\sum_{i=1}^{k} \frac{X_{n-i,n} - X_{n-k,n} - T}{X_{n,n} - X_{n-k,n}}, \text{ with } T := X_{n-k,n}\frac{M^{(1)}}{2}\left(1 - \frac{\left[M^{(1)}\right]^2}{M^{(2)}}\right)^{-1}. \tag{18.23}$$

Under the null hypothesis, the standardized version of (18.23), $T_1^* := \sqrt{k}\,\ln k\, T_1$, has asymptotic Standard Normal distribution. In Fig. 18.2-*down* is pictured the sample path of T_1^* associated with the LJ-Data, together with the Normal $(0.05, 0.95)$-quantiles, from which it is not so clear that the right endpoint is finite.

This is not unexpected as T_1^*-test is conservative, as remarked in [9]. However, it yields a sample path near the lower critical point, the normal 0.05-quantile, for several values of k lying in the most stable region of the graph that can be taken as intermediate. Moreover, it seems reasonable to consider the finiteness of the ultimate record in Long Jump Men outdoors event, by the nature of the event itself.

Finally, in Fig. 18.3 are depicted the estimates of the right endpoint of the distribution underlying the LJ-data, against k.

From the LJ-Analysis we conclude that there is some space to improve the current Powell's record of 8.95 m. In fact, from this preliminary study it seems possible to attain an ultimate record of 9 m for the Long Jump Men outdoors.

Acknowledgements Research partially financed by PEst-OE/MAT/UI0006/2011 and EXTREMA-FCT/PTDC/MAT/101736/2008.

References

1. Dekkers, A.L.M., Einmahl, J.H.J., deHaan, L.: A moment estimator for an extreme-value distribution. Ann. Stat. **17**(4), 1833–1855 (1989)
2. Einmahl, J., Magnus, J.: Records in athletics through extreme-value theory. J. Am. Stat. Assoc. **103**, 1382–1391 (2008)
3. Einmahl, J.H.J., Smeets, S.: Ultimate 100M world records through extreme-value theory (July 10, 2009). CentER Discussion Paper Series No. 2009-57. Available at SSRN:http://ssrn.com/abstract=1433242
4. Fraga Alves, M.I., Gomes, M.I., de Haan, L., Neves, C.: Mixed moment estimator and location invariant alternatives. Extremes **12**, 149–185 (2009), doi:10.1007/s10687-008-0073-3
5. Gomes, M.I., Pestana, D., Rodrigues, L.H.: Athletic events and statistics of extremes: estimation of useful parameters. Tecnical Report CEAUL 06/09 (2009) http://www.ceaul.fc.ul.pt/getfile.asp?where=notas&id=250
6. de Haan, L., Ferreira, A.: Extreme Value Theory: An Introduction. Springer Series in Operations Research and Financial Engineering. Springer, New York (2006)
7. Neves, C., Fraga Alves, M.I.: Semi-parametric Approach to Hasofer-Wang and Greenwood Statistics in Extremes. TEST **16**, 297–313 (2008), doi:10.1007/s11749-006-0010-1
8. Neves, C., Fraga Alves, M.I.: Testing extreme value conditions – an overview and recent approaches. In: Beirlant, J., Fraga Alves, M.I., Leadbetter, R. (eds.) In: Statistics of Extremes and Related Fields. REVSTAT – Stat. Journal, Special issue vol. 6(1), pp. 83–100 (2008) http://www.ine.pt/revstat/pdf/rs080106.pdf
9. Neves, C., Pereira, A.: Detecting finiteness in the right endpoint of light-tailed distributions. Stat. Probabil. Lett. **80**(5–6), 437–444 (2010)
10. Neves, C., Picek, J., Fraga Alves, M.I.: Contribution of the maximum to the sum of excesses for testing max-domains of attraction. J. Stat. Plan. Infer. **136**(4), 1281–1301 (2006), doi:10.1016/j.jspi.2004.09.008
11. Pickands III, J.: Statistical inference using extreme order statistics. Ann. Stat. **3**, 119–131 (1975)

Joint Modeling of Longitudinal and Time-to-Event Data: Challenges and Future Directions

19

Dimitris Rizopoulos

Abstract

In longitudinal studies measurements are often collected on different types of outcomes for each subject. These may include several longitudinally measured responses (such as blood values relevant to the medical condition under study) and the time at which an event of particular interest occurs (e.g., death, development of a disease or dropout from the study). These outcomes are often separately analyzed; however, in many instances, a joint modeling approach is either required or may produce a better insight into the mechanisms that underlie the phenomenon under study. In this chapter we provide a general overview of the joint modeling framework, discuss its main features, and we refer to future directions.

Keywords

Dropout • Longitudinal data analysis • Missing data • Shared parameter models • Survival analysis • Time-dependent covariates

19.1 Introduction

Longitudinal studies often produce two types of outcome, namely a set of longitudinal response measurements and the time to an event of interest, such as death, development of a disease, or dropout from the study. Two typical examples of this setting are HIV and cancer studies. In HIV studies patients who have been infected are monitored until they develop AIDS or die, and they are regularly measured for the condition of the immune system using markers such as the CD4 lymphocyte

D. Rizopoulos (✉)
Department of Biostatistics, Erasmus University Medical Center, PO Box 2040, 3000 CA Rotterdam, the Netherlands
e-mail: d.rizopoulos@erasmusmc.nl

N. Torelli et al. (eds.), *Advances in Theoretical and Applied Statistics*, Studies in Theoretical and Applied Statistics, DOI 10.1007/978-3-642-35588-2_19,

count or the estimated viral load. Similarly in cancer studies the event outcome is the death or metastasis and patients also provide longitudinal measurements of antibody levels or of other markers of carcinogenesis, such as the PSA levels for prostate cancer.

These two outcomes are often separately analyzed using a mixed-effects model for the longitudinal outcome and a survival model for the event outcome. However, in mainly two settings a joint modeling approach is required. First, when interest is on the event outcome and we wish to account for the effect of the longitudinal outcome as a time-dependent covariate, traditional approaches for analyzing time-to-event data (such as the partial likelihood for the Cox proportional hazards models) are not applicable. In particular, standard time-to-event models require that time-dependent covariates are external; that is, the value of this covariate at time point t is not affected by the occurrence of an event at time point u, with $t > u$ [10, Sect. 6.3]. However, the type of time-dependent covariates encountered in longitudinal studies do not satisfy this condition due to the fact that they are the output of a stochastic process generated by the subject, which is directly related to the failure mechanism. Therefore, in order to produce valid inferences a model for the joint distribution of the longitudinal and survival outcomes is required instead. The second setting in which joint modeling is required is when interest is on the longitudinal outcome. In this case the occurrence of events causes dropout since no longitudinal measurements are available at and after the event time. When this dropout is nonrandom (i.e., when the probability of dropout depends on unobserved longitudinal responses), then bias may arise from an analysis that ignores the dropout process [12, Chap. 15]. To avoid this problem and obtain valid inferences the longitudinal and dropout process must be jointly modeled. One of the modeling frameworks that have been proposed in the missing data literature for handling nonrandom dropout is the shared parameter model [7, 28, 29], which postulates a mixed effects model for the longitudinal outcome and time-to-dropout model for the missingness process. In both settings the joint distribution of the event times and the longitudinal measurements is modeled via a set of random effects that are assumed to account for the associations between these two outcomes. Excellent overviews of this area of biostatistics and statistics research are given by [25, 32].

In this chapter we present the basics of the joint modeling framework, refer to the key features and assumptions behind these models, and provide some insight regarding future directions. To illustrate the virtues of joint modeling, we use a real data set concerning patients with primary biliary cirrhosis (PBC). The main interest lies in investigating the strength of the association between the longitudinally measured marker serum bilirubin and the time-to-death.

19.2 Joint Modeling Framework

19.2.1 Submodels Specification

In the following we will present the joint modeling framework motivated by the time-to-event point of view (i.e., in the setting in which we want to incorporate a time-dependent covariate measured with error in a survival model)—a more direct connection with the missing data framework is made in Sect. 19.3. Let T_i denote the observed failure time for the ith subject ($i = 1, \ldots, n$), which is taken as the minimum of the true event time T_i^* and the censoring time C_i, i.e., $T_i = \min(T_i^*, C_i)$. Furthermore, we define the event indicator as $\delta_i = I(T_i^* \leq C_i)$, where $I(\cdot)$ is the indicator function that takes the value 1 if the condition $T_i^* \leq C_i$ is satisfied, and 0 otherwise. Thus, the observed data for the time-to-event outcome consist of the pairs $\{(T_i, \delta_i), i = 1, \ldots, n\}$. For the longitudinal responses, let $y_i(t)$ denote the value of the longitudinal outcome at time point t for the ith subject. We should note here that we do not actually observe $y_i(t)$ at all time points but only at the very specific occasions t_{ij} at which measurements were taken. Thus, the observed longitudinal data consist of the measurements $y_{ij} = \{y_i(t_{ij}), j = 1, \ldots, n_i\}$.

Our aim is to associate the true and unobserved value of the longitudinal outcome at time t, denoted by $m_i(t)$, with the event outcome T_i^*. Note that $m_i(t)$ is different from $y_i(t)$, with the latter being the contaminated with measurement error value of the longitudinal outcome at time t. To quantify the effect of $m_i(t)$ on the risk for an event, a standard option is to use a relative risk model of the form [23]:

$$\begin{aligned} h_i(t \mid \mathcal{M}_i(t), w_i) &= \lim_{dt \to 0} dt^{-1} \Pr\{t \leq T_i^* < t + dt \mid T_i^* \geq t, \mathcal{M}_i(t), w_i\} \\ &= h_0(t) \exp\{\gamma^\top w_i + \alpha m_i(t)\}, \end{aligned} \tag{19.1}$$

where $\mathcal{M}_i(t) = \{m_i(u), 0 \leq u < t\}$ denotes the history of the true unobserved longitudinal process up to time point t, $h_0(\cdot)$ denotes the baseline risk function, and w_i is a vector of baseline covariates (such as a treatment indicator, history of diseases, etc.) with a corresponding vector of regression coefficients γ. Similarly, parameter α quantifies the effect of the underlying longitudinal outcome to the risk for an event; for instance, in the AIDS example mentioned in Sect. 19.1, α measures the effect of the number of CD4 cells to the risk for death. We should note that alternative formulations that relate other features of the longitudinal history $\mathcal{M}_i(t)$ than the current value term $m_i(t)$ can be also employed [17]. To avoid the impact of parametric assumptions, the baseline risk function $h_0(\cdot)$ is typically left unspecified. However, within the joint modeling framework [9] have recently noted that leaving this function completely unspecified leads to an underestimation of the standard errors of the parameter estimates. In particular, problems arise due to the fact that the nonparametric maximum likelihood estimate for this function cannot be obtained explicitly under the full joint modeling approach. To avoid this problem, we could either opt for the hazard function of a standard survival distribution (such as the Weibull or Gamma) or for more flexible models in which $h_0(t)$ is sufficiently approximated using step functions or spline-based approaches.

In the definition of the survival model presented above we used $m_i(t)$ to denote the true value of the underlying longitudinal covariate at time point t. However and as mentioned earlier, longitudinal information is actually collected intermittently and with error at a set of few time points t_{ij} for each subject. Therefore, in order to measure the effect of this covariate to the risk for an event we need to estimate $m_i(t)$ and successfully reconstruct the complete longitudinal history $\mathcal{M}_i(t)$, using the available measurements $y_{ij} = \{y_i(t_{ij}), j = 1, \ldots, n_i\}$ of each subject and a set of modeling assumptions. For the remaining of this chapter we will focus on normal data and we will postulate a linear mixed effects model to describe the subject-specific longitudinal evolutions. In particular, we have

$$\begin{aligned} y_i(t) &= m_i(t) + \varepsilon_i(t) \\ &= x_i^\top(t)\beta + z_i^\top(t)b_i + \varepsilon_i(t), \quad \varepsilon_i(t) \sim \mathcal{N}(0, \sigma^2), \end{aligned} \tag{19.2}$$

where β denotes the vector of the unknown fixed effects parameters, b_i denotes a vector of random effects, $x_i(t)$ and $z_i(t)$ denote row vectors of the design matrices for the fixed and random effects, respectively, and $\varepsilon_i(t)$ is the measurement error term, which is assumed independent of b_i, and with variance σ^2. When deemed appropriate, the assumption of constant variance for the error terms $\varepsilon_i(t)$ can be relaxed by postulating a suitable variance function. We should note that care should be taken in the specification of $x_i(t)$ and $z_i(t)$ in order to produce a good estimate of $\mathcal{M}_i(t)$. The main reason for this is that, as we will see later in Sect. 19.2.2, in the definition of the likelihood of the joint model the complete longitudinal history is required for the computation of the survival function, and of the risk function under the accelerated failure time formulation. Therefore, in applications in which subjects show highly nonlinear longitudinal trajectories, it is advisable to consider flexible representations for $x_i(t)$ and $z_i(t)$ using a possibly high-dimensional vector of functions of time t, expressed in terms of high-order polynomials or splines [2,4,18]. Finally, in order to complete the specification of the longitudinal submodel a suitable distributional assumption for the random effects component is required. A standard choice for this distribution is the multivariate normal distribution; however, within the joint modeling framework and mainly because of the nonrandom dropout (see also Sect. 19.3), there is the concern that relying on a standard parametric distribution may influence the derived inferences especially when this distribution differs considerably from the true random effects distribution. This motivated [22] to propose a more flexible model for the distribution of the random effects that is expressed as a normal density times a polynomial function. However, the findings of these authors suggested that the parameter estimates and standard errors of joint models fitted under the normal assumption for the random effects were rather robust to misspecification. This feature has been further theoretically corroborated by [19], who showed that as the number of repeated measurements per subject n_i increases, a misspecification of the random effects distribution has a minimal effect in parameter estimators and standard errors. Thus, here we will assume that $b_i \sim \mathcal{N}(0, D)$ and we will not further investigate this assumption.

19.2.2 Maximum Likelihood Estimation

The main estimation methods that have been proposed for joint models are (semiparametric) maximum likelihood [8,9,30] and Bayes using MCMC techniques [1, 3, 27, 31]. Moreover, [24] have proposed a conditional score approach in which the random effects are treated as nuisance parameters, and they developed a set of unbiased estimating equations that yields consistent and asymptotically normal estimators. Here we give the basics of the maximum likelihood method for joint models as the one of the more traditional approaches.

Maximum likelihood estimation for joint models is based on the maximization of the log-likelihood corresponding to the joint distribution of the time-to-event and longitudinal outcomes $\{T_i, \delta_i, y_i\}$. To define this joint distribution we will assume that the vector of time-independent random effects b_i underlies both the longitudinal and survival processes. This means that these random effects account for both the association between the longitudinal and event outcomes, and the correlation between the repeated measurements in the longitudinal process (conditional independence). Formally we have that,

$$p(T_i, \delta_i, y_i \mid b_i; \theta) = p(T_i, \delta_i \mid b_i; \theta)\ f(y_i \mid b_i; \theta) \tag{19.3}$$

$$f(y_i \mid b_i; \theta) = \prod_j f\{y_i(t_{ij}) \mid b_i; \theta\}, \tag{19.4}$$

where θ denotes the parameter vector, and y_i is the $n_i \times 1$ vector of longitudinal responses of the ith subject. Due to the fact that the survival and longitudinal submodels share the same random effects, joint models of this type are also known as shared parameter models. Under the modeling assumptions presented in the previous section, and these conditional independence assumptions the joint log-likelihood contribution for the ith subject can be formulated as

$$\log p(T_i, \delta_i, y_i; \theta) = \log \int p(T_i, \delta_i \mid b_i; \theta) \Big[\prod_j f\{y_i(t_{ij}) \mid b_i; \theta\}\Big] g(b_i; \theta)\, \mathrm{d}b_i, \tag{19.5}$$

where the likelihood of the survival part is written as

$$p(T_i, \delta_i \mid b_i; \theta) = \{h_i(T_i \mid \mathcal{M}_i(T_i); \theta)\}^{\delta_i} \mathcal{S}_i(T_i \mid \mathcal{M}_i(T_i); \theta), \tag{19.6}$$

with $h_i(\cdot)$ given by (19.1), and

$$\begin{aligned} \mathcal{S}_i(t \mid \mathcal{M}_i(t), w_i) &= \Pr(T_i^* > t \mid \mathcal{M}_i(t), w_i) \\ &= \exp\Big\{-\int_0^t h_i(s \mid \mathcal{M}_i(s); \theta)\, \mathrm{d}s\Big\}, \end{aligned} \tag{19.7}$$

$f\{y_i(t_{ij}) \mid b_i; \theta\}$ is the univariate normal density for the longitudinal responses, and $g(b_i; \theta)$ is the multivariate normal density for the random effects.

Maximization of the log-likelihood function corresponding to (19.5) with respect to θ is a computationally challenging task. This is mainly because both the integral with respect to the random effects in (19.5), and the integral in the definition of the survival function (19.7) do not have an analytical solution, except in very special cases. Standard numerical integration techniques such as Gaussian quadrature and

Monte Carlo have been successfully applied in the joint modeling framework [8, 22, 30]. Furthermore, Rizopoulos et al. [18] have recently discussed the use of Laplace approximations for joint models that can be especially useful in high-dimensional random effects settings (e.g., when splines are used in the random effects design matrix). For the maximization of the approximated log-likelihood the EM algorithm has been traditionally used in which the random effects are treated as "missing data." The main motivation for using this algorithm is the closed-form M-step updates for certain parameters of the joint model. However, a serious drawback of the EM algorithm is its linear convergence rate that results in slow convergence especially near the maximum. Nonetheless, [18] have noted that a direct maximization of the observed data log-likelihood, using for instance, a quasi-Newton algorithm [11], requires very similar computations to the EM algorithm. Therefore hybrid optimization approaches that start with EM and then continue with direct maximization can be easily employed. One of the main practical limitations for joint modeling in finding its way into the toolbox of modern statisticians was the lack of free and reliable software. The R package **JM** [16] has been developed to fill this gap to some extent. **JM** has a user-friendly interface to fit joint models and also provides several supporting functions that extract or calculate various quantities based on the fitted model (e.g., residuals, fitted values, empirical Bayes estimates, various plots, and others). At http://rwiki.sciviews.org/doku.php?id=packages:cran:jm, more information can be found.

19.3 Connection with the Missing Data Framework

As mentioned in Sect. 19.1, joint models for longitudinal and time-to-event are also used when the main interest is on the longitudinal outcome and we wish to correct for the dropout caused by the occurrence of events. Under the categorization of missing data mechanisms proposed by [21], joint models correspond to missing not at random mechanisms. To show this more clearly, we define for each subject the observed and missing part of the longitudinal response vector. The observed part $y_i^{\text{o}} = \{y_i(t_{ij}) : t_{ij} < T_i, j = 1, \ldots, n_i\}$ contains all observed longitudinal measurements of the ith subject before the observed event time, whereas the missing part $y_i^{\text{m}} = \{y_i(t_{ij}) : t_{ij} \geq T_i, j = 1, \ldots, n_i'\}$ contains the longitudinal measurements that would have been taken until the end of the study, had the event not occurred. Under these definitions and the assumptions of the joint modeling framework, we can derive the dropout mechanism, which is the conditional distribution of the time-to-dropout given the complete vector of longitudinal responses $(y_i^{\text{o}}, y_i^{\text{m}})$,

$$p(T_i^* \mid y_i^{\text{o}}, y_i^{\text{m}}; \theta) = \int p(T_i^* \mid b_i; \theta)\, p(b_i \mid y_i^{\text{o}}, y_i^{\text{m}}; \theta)\, \mathrm{d}b_i. \qquad (19.8)$$

We observe that the time-to-dropout depends on y_i^{m} through the posterior distribution of the random effects $p(b_i \mid y_i^{\text{o}}, y_i^{\text{m}}; \theta)$, which illustrates the nonrandom nature of the missing data mechanism implied by joint models. In practice, the major implication of this feature is that the observed data upon which the joint model is fitted are not a random sample of the target population, which in turn implies that the characteristics of the observed data distribution cannot be compared with the output of the joint model. Therefore, residuals plots based on the observed data alone can be deceptive regarding the fit of a joint model to the data at hand. To overcome this problem, Rizopoulos et al. [20] proposed to take dropout into account by multiply imputing the longitudinal responses that would have been observed had the event not occurred. The advantage of this approach is that after the imputation, residuals can be computed for the complete data facilitating thus the use of standard diagnostic plots can for the longitudinal [14, 26] and event time outcomes [23]. We should note however that the multiple imputation approach is merely used as a mechanism to help us investigate the fit of the model, since we are not actually interested in inferences after the event time.

19.4 Mayo Clinic Primary Biliary Cirrhosis Data

As an illustrative example of joint modeling, we consider the PBC data collected by the Mayo Clinic from 1974 to 1984 [13]. PBC is a chronic, fatal, but rare liver disease characterized by inflammatory destruction of the small bile ducts within the liver, which eventually leads to cirrhosis of the liver. Patients with PBC have abnormalities in several blood tests, such as elevated levels of serum bilirubin. In this study 312 patients are considered of whom 158 were randomly assigned to receive D-penicillamine and 154 placebo. In this analysis we are interested in the association between the longitudinal bilirubin levels and the risk for death. Due to the right skewness of the observed serum bilirubin level, we will work with the natural logarithm of serum bilirubin for the remainder of this analysis.

Patients did not return to the study centers at prespecified time points to provide serum bilirubin measurements, and thus we observe great variability between their visiting patterns. In particular, patients made on average 6.23 visits (standard deviation 3.77 visits), resulting in a total of 1945 observations. By the end of the study 140 patients had died that corresponds to 55.1 % censoring. Taking advantage of the randomization setup of the study, we fit a joint model in which we only correct for treatment. In particular, for the longitudinal process, we assume a linear mixed model, with main effects for treatment and time and their interaction for the fixed effects part, and with random intercepts and random slopes for the random-effects part. For the survival submodel, we include as a time-independent covariate the treatment effect, and as time-dependent one the true underlying effect of log serum bilirubin count as estimated from the longitudinal model. The baseline risk function is assumed piecewise constant

$$h_0(t) = \sum_{q=1}^{Q} \xi_q I(v_{q-1} < t \leq v_q),$$

Table 19.1 Parameter estimates and standard errors for the joint model analysis of the PBC data set

	Value	Std.Err	z-value	p-value
(Intercept)	0.556	0.033	16.77	< 0.0001
D-penicil	0.028	0.047	0.60	0.5461
year	0.185	0.007	26.99	< 0.0001
D-penicil×year	−0.013	0.009	−1.38	0.1690
$\log(\sigma)$	−0.963	0.018		
D-penicil	0.076	0.180	0.42	0.6743
$m_i(t)$	1.258	0.094	13.32	< 0.0001
$\log(\xi_1)$	−4.473	0.261		
$\log(\xi_2)$	−4.330	0.280		
$\log(\xi_3)$	−4.657	0.331		
$\log(\xi_4)$	−4.613	0.383		
$\log(\xi_5)$	−4.231	0.344		
$\log(\xi_6)$	−3.772	0.349		
$\log(\xi_7)$	−4.647	0.490		
D_{11}	1.027	0.088		
D_{12}	0.072	0.015		
D_{22}	0.030	0.004		

The top part contains the results for the longitudinal submodel, the middle part contains the results for the survival submodel, and the bottom part for the random-effects covariance matrix; $\xi_1, \ldots, \xi_7$ denote the parameters of the baseline risk function, D_{11} the variance for the random intercepts, D_{22} the variance for the random slopes, and D_{12} their covariance

with $Q = 7$, $v_0 = 0$, six internal knots placed at equally spaced percentiles of the observed event times, and v_7 taken larger than the largest observed time. The results are presented in Table 19.1. We observe a strongly significant effect of the log serum bilirubin levels to risk for an event. In particular, the hazard ratio for one unit increase in log serum bilirubin for patients randomized in the same treatment equals $\exp(1.258) = 3.52$, which indicates the high importance of this marker in quantifying the risk for death. To check the fit of the model we produce multiple imputation residuals for the longitudinal outcome, discussed in [20], and the Cox-Snell residuals for the survival outcome. These residual plots are shown in Figure 19.1. We observe that the fitted loess curves in the plots of the standardized marginal residuals based on the observed data alone versus the fitted values show a systematic trend (red line). Note, however, that high levels of serum bilirubin indicate a worsening of patients' condition resulting in higher death rates (i.e., dropout). Thus, the residuals corresponding to large fitted values are only based on patients with a "good" health condition, which explains why we observe the systematic trend. On the contrary the loess curves based on both the multiply imputed residuals and the residuals corresponding to y_i^o (blue line) do not exhibit any strange shapes, indicating that after taking dropout into account the fitted joint model does seem to be a plausible model for this data set.

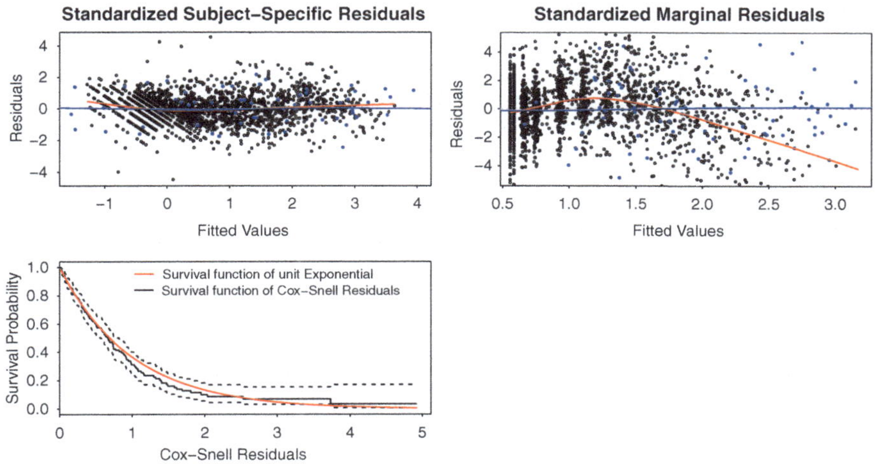

Fig. 19.1 Residuals for the joint model fitted to the PBC data: the top left panel depicts the standardized subject-specific residuals versus the subject-specific fitted values for the longitudinal outcome; the top right panel depicts the standardized marginal residuals versus the marginal fitted values for the longitudinal outcome; the bottom left panel depicts the Kaplan–Meier estimate of the survival function of the Cox-Snell residuals for the event time outcome. For the top panels the *black points* correspond to the residuals based on the observed data alone, the *blue points* to the residuals for the multiply-imputed data, the *red line* depicts the loess fit for the observed data alone, and the *blue line* the loess fit based on both the observed and multiply-imputed data

19.5 Conclusion

Joint modeling of longitudinal and time-to-event data is one of the most rapidly evolving areas of current biostatistics research, with several extensions of the standard joint model presented here already proposed in the literature. These include, among others, handling multiple failure types [5], considering categorical longitudinal outcomes [6], assuming that several longitudinal outcomes affect the time-to-event [1, 3], and associating the two outcomes via latent classes instead of random effects [15].

Furthermore, one very promising field which has emerged within the general joint modeling framework is the use of these models in personalized medicine. In particular, there is lately a great need for tools that can help physicians take better informed decisions regarding their actions for the specific patients that they treat and not for an "average" patient. Two features of joint models that allow them to become such a flexible dynamic tool is the use of random effects and their time-dependent nature. For instance, as longitudinal information is collected for patients, we can continuously update the predictions of their survival probabilities, and therefore be able to discern between patients with low and high risk for an event.

References

1. Brown, E., Ibrahim, J.: A Bayesian semiparametric joint hierarchical model for longitudinal and survival data. Biometrics **59**, 221–228 (2003)
2. Brown, E., Ibrahim, J., DeGruttola, V.: A flexible b-spline model for multiple longitudinal biomarkers and survival. Biometrics **61**, 64–73 (2005)
3. Chi, Y.Y., Ibrahim, J.: Joint models for multivariate longitudinal and multivariate survival data. Biometrics **62**, 432–445 (2006)
4. Ding, J., Wang, J.L.: Modeling longitudinal data with nonparametric multiplicative random effects jointly with survival data. Biometrics **64**, 546–556 (2008)
5. Elashoff, R., Li, G., Li, N.: A joint model for longitudinal measurements and survival data in the presence of multiple failure types. Biometrics **64**, 762–771 (2008)
6. Faucett, C., Schenker, N., Elashoff, R.: Analysis of censored survival data with intermittently observed time-dependent binary covariates. J. Am. Stat. Assoc. **93**, 427–437 (1998)
7. Follmann, D., Wu, M.: An approximate generalized linear model with random effects for informative missing data. Biometrics **51**, 151–168 (1995)
8. Henderson, R., Diggle, P., Dobson, A.: Joint modelling of longitudinal measurements and event time data. Biostatistics **1**, 465–480 (2000)
9. Hsieh, F., Tseng, Y.K., Wang, J.L.: Joint modeling of survival and longitudinal data: likelihood approach revisited. Biometrics **62**, 1037–1043 (2006)
10. Kalbfleisch, J., Prentice, R.: The Statistical Analysis of Failure Time Data, 2nd edn. Wiley, New York (2002)
11. Lange, K.: Optimization. Springer, New York (2004)
12. Little, R., Rubin, D.: Statistical Analysis with Missing Data, 2nd edn. Wiley, New York (2002)
13. Murtaugh, P., Dickson, E., Van Dam, G., Malincho, M., Grambsch, P., Langworthy, A., Gips, C.: Primary biliary cirrhosis: prediction of short-term survival based on repeated patient visits. Hepatology **20**, 126–134 (1994)
14. Nobre, J., Singer, J.: Residuals analysis for linear mixed models. Biomet. J. **6**, 863–875 (2007)
15. Proust-Lima, C., Joly, P., Dartigues, J.F., Jacqmin-Gadda, H.: Joint modelling of multivariate longitudinal outcomes and a time-to-event: a nonlinear latent class approach. Comput. Stat. Data Anal. **53**, 1142–1154 (2009)
16. Rizopoulos, D.: JM: an R package for the joint modelling of longitudinal and time-to-event data. J. Stat. Software **35**(9), 1–33 (2010)
17. Rizopoulos, D., Ghosh, P.: A Bayesian semiparametric multivariate joint model for multiple longitudinal outcomes and a time-to-event. Stat. Med. **30**, 1366–1380 (2011)
18. Rizopoulos, D., Verbeke, G., Lesaffre, E.: Fully exponential laplace approximations for the joint modelling of survival and longitudinal data. J. Roy. Stat. Soc. Ser. B **71**, 637–654 (2009)
19. Rizopoulos, D., Verbeke, G., Molenberghs, G.: Shared parameter models under random effects misspecification. Biometrika **95**, 63–74 (2008)
20. Rizopoulos, D., Verbeke, G., Molenberghs, G.: Multiple-imputation-based residuals and diagnostic plots for joint models of longitudinal and survival outcomes. Biometrics **66**, 20–29 (2010)
21. Rubin, D.: Inference and missing dat. Biometrika **63**, 581–592 (1976)
22. Song, X., Davidian, M., Tsiatis, A.: A semiparametric likelihood approach to joint modeling of longitudinal and time-to-event data. Biometrics **58**, 742–753 (2002)
23. Therneau, T., Grambsch, P.: Modeling Survival Data: Extending the Cox Model. Springer, New York (2000)
24. Tsiatis, A., Davidian, M.: A semiparametric estimator for the proportional hazards model with longitudinal covariates measured with error. Biometrika **88**, 447–458 (2001)
25. Tsiatis, A., Davidian, M.: Joint modeling of longitudinal and time-to-event data: an overview. Statistica Sinica **14**, 809–834 (2004)
26. Verbeke, G., Molenberghs, G.: Linear Mixed Models for Longitudinal Data. Springer, New York (2000)

27. Wang, Y., Taylor, J.: Jointly modeling longitudinal and event time data with application to acquired immunodeficiency syndrome. J. Am. Stat. Assoc. **96**, 895–905 (2001)
28. Wu, M., Bailey, K.: Analysing changes in the presence of informative right censoring caused by death and withdrawal. Stat. Med. **7**, 337–346 (1988)
29. Wu, M., Carroll, R.: Estimation and comparison of changes in the presence of informative right censoring by modeling the censoring process. Biometrics **44**, 175–188 (1988)
30. Wulfsohn, M., Tsiatis, A.: A joint model for survival and longitudinal data measured with error. Biometrics **53**, 330–339 (1997)
31. Xu, J., Zeger, S.: Joint analysis of longitudinal data comprising repeated measures and times to events. Appl. Stat. **50**, 375–387 (2001)
32. Yu, M., Law, N., Taylor, J., Sandler, H.: Joint longitudinal-survival-cure models and their application to prostate cancer. Statistica Sinica **14**, 835–832 (2004)

A Class of Linear Regression Models for Imprecise Random Elements

20

Renato Coppi, Maria Brigida Ferraro, and Paolo Giordani

Abstract

The linear regression problem of a fuzzy response variable on a set of real and/or fuzzy explanatory variables is investigated. The notion of *LR* fuzzy random variable is introduced in this connection, leading to the probabilization of the center and the left and right spreads of the response variable. A specific metric is suggested for coping with this type of variables. A class of linear regression models is then proposed for the center and for suitable transforms of the spreads in order to satisfy the nonnegativity conditions for the latter ones. A Least Squares solution for estimating the parameters of the models is derived, along with a goodness-of-fit measure and the associated hypothesis testing procedure. Finally, the results of a real-life application are illustrated.

Keywords

Bootstrap approach • Fuzzy random variables • Imprecise data and fuzzy methods • Linear independence test • Linear regression analysis

20.1 Introduction

In the literature a great deal of attention has been paid to the management of uncertain information. We can roughly distinguish two sources of uncertainty, namely, randomness and imprecision. In the case of randomness the information is uncertain because we do not know the (precise) outcome of a (random) mechanism. Randomness is limited to the data generation process and it can be dealt with by means of probability theory (probabilistic uncertainty). In contrast with randomness,

R. Coppi · M.B. Ferraro (✉) · P. Giordani
SAPIENZA Università di Roma, P.le A. Moro 5, 00185, Roma, Italy
e-mail: renato.coppi@uniroma1.it; mariabrigida.ferraro@uniroma1.it; paolo.giordani@uniroma1.it

N. Torelli et al. (eds.), *Advances in Theoretical and Applied Statistics*, Studies in Theoretical and Applied Statistics, DOI 10.1007/978-3-642-35588-2_20,

imprecision is connected to the uncertainty concerning the placement of an outcome in a given class and, thus, it can be seen as non-probabilistic uncertainty [2].

The different sources of uncertainty are not exclusive but can occur together. A possible way to cope with imprecision is represented by fuzzy set theory [12]. This allows us to express imprecise information in terms of fuzzy sets. When such information is also affected by randomness, the concept of fuzzy random variable (FRV) can be adopted [7, 9]. In this work we aim at investigating the linear regression problem for a very common class of FRVs, namely the *LR* family. The problem has been deeply analyzed by coping with the different sources of uncertainty in a separate way. This is the case for the possibilistic approach (see, for more details, [8, 10]) and for the least-squares approach (see, for instance, [1, 3]). Here we choose the so-called fuzzy-probabilistic (see, for instance [7]) approach that explicitly takes into account both imprecision and randomness.

The chapter is organized as follows. In the next section, some preliminary concepts are recalled. Then, Sect. 20.3 focuses on a linear regression model with *LR* fuzzy response and precise (crisp) explanatory variables and Sect. 20.4 deals with a linear regression model with *LR* fuzzy response and *LR* fuzzy explanatory variables (that is an extension of the previous one). In Sect. 20.5 we introduce a suitable determination coefficient and a linear independence test carried out by means of a bootstrap approach. Finally, the results of a real-life application are reported in Sect. 20.6 and some concluding remarks are given in Sect. 20.7.

20.2 Preliminaries

Given a universe U of elements, a fuzzy set $\widetilde{A}$ is defined through the so-called *membership function* $\mu_{\widetilde{A}} : U \to [0,1]$. For a generic $x \in U$, the membership function expresses the extent to which x belongs to $\widetilde{A}$. Such a degree ranges from 0 (complete non-membership) to 1 (complete membership).

A particular class of fuzzy sets is the LR family, whose members are the so-called *LR fuzzy numbers*. Denoting by $\mathscr{F}_{\mathrm{LR}}$ the space of the LR fuzzy numbers, the membership function of $\widetilde{A} \in \mathscr{F}_{\mathrm{LR}}$ can be written as

$$\mu_{\widetilde{A}}(x) = \begin{cases} L\left(\frac{A^m - x}{A^l}\right) & x \le A^m,\ A^l > 0, \\ 1_{\{A^m\}}(x) & x \le A^m,\ A^l = 0, \\ R\left(\frac{x - A^m}{A^r}\right) & x > A^m,\ A^r > 0, \\ 0 & x > A^m,\ A^r = 0, \end{cases} \tag{20.1}$$

where the functions L and R are particular non-increasing shape functions from $\mathbb{R}^+$ to $[0, 1]$ such that $L(0) = R(0) = 1$ and $L(x) = R(x) = 0, \forall x \in \mathbb{R} \setminus [0, 1]$, 1_I is the indicator function of a set I and A^m, A^l (≥ 0) and A^r (≥ 0) are three real-valued parameters, namely, the center, the left spread, and the right spread, respectively. $\widetilde{A}$ is a *triangular* fuzzy number if $L(z) = R(z) = 1 - z$, for $0 \le z \le 1$. Given the shape of the membership function, $\widetilde{A} \in \mathscr{F}_{\mathrm{LR}}$ can be determined uniquely in terms of the mapping $s : \mathscr{F}_{\mathrm{LR}} \to \mathbb{R}^3$, i.e., $s(\widetilde{A}) = s_{\widetilde{A}} = (A^m, A^l, A^r)$. In what follows it is indistinctly used $\widetilde{A} \in \mathscr{F}_{\mathrm{LR}}$ or (A^m, A^l, A^r).

The operations considered in $\mathscr{F}_{\mathrm{LR}}$ are the natural extensions of the Minkowski sum and the product by a positive scalar for intervals. Going into detail, the sum of $\widetilde{A}$ and $\widetilde{B}$ in $\mathscr{F}_{\mathrm{LR}}$ is the *LR* fuzzy number $\widetilde{A} + \widetilde{B}$ so that

$$(A^m, A^l, A^r) + (B^m, B^l, B^r) = (A^m + B^m, A^l + B^l, A^r + B^r),$$

and the product of $\widetilde{A} \in \mathscr{F}_{\mathrm{LR}}$ by a positive scalar γ is

$$\gamma(A^m, A^l, A^r) = (\gamma A^m, \gamma A^l, \gamma A^r).$$

Yang and Ko [11] define a distance between two *LR* fuzzy numbers $\widetilde{A}$ and $\widetilde{B}$ as follows

$$D^2_{\mathrm{LR}}(\widetilde{A}, \widetilde{B}) = (A^m - B^m)^2 + [(A^m - \lambda A^l) - (B^m - \lambda B^l)]^2 + [(A^m + \rho A^r) - (B^m + \rho B^r)]^2,$$

where the parameters $\lambda = \int_0^1 L^{-1}(\omega)\mathrm{d}\omega$ and $\rho = \int_0^1 R^{-1}(\omega)\mathrm{d}\omega$ play the role of taking into account the shape of the membership function. For instance, in the triangular case, it is $\lambda = \rho = \frac{1}{2}$. For what follows it is necessary to embed the space $\mathscr{F}_{\mathrm{LR}}$ into $\mathbb{R}^3$ by preserving the metric. For this reason a generalization of the Yang and Ko metric can be derived [4]. Given $a = (a_1, a_2, a_3)$ and $b = (b_1, b_2, b_3) \in \mathbb{R}^3$, it is

$$D^2_{\lambda\rho}(a, b) = (a_1 - b_1)^2 + ((a_1 - \lambda a_2) - (b_1 - \lambda b_2))^2 + ((a_1 + \rho a_3) - (b_1 + \rho b_3))^2,$$

where $\lambda, \rho \in \mathbb{R}^+$. $D^2_{\lambda\rho}$ will be used in the sequel as a tool for quantifying errors in the regression models we are going to introduce.

Let $(\Omega, \mathscr{A}, P)$ be a probability space. In this context, a mapping $\widetilde{X} : \Omega \to \mathscr{F}_{\mathrm{LR}}$ is an *LR* FRV if the s-representation of $\widetilde{X}$, $(X^m, X^l, X^r) : \Omega \to \mathbb{R} \times \mathbb{R}^+ \times \mathbb{R}^+$ is a random vector [9]. Much like non-FRVs, it is possible to determine the moments for an *LR* FRV. Let $\| \cdot \|^2_{\mathrm{LR}}$ be the norm associated with D^2_{LR}, the expectation of an *LR* FRV $\widetilde{X}$ is the unique fuzzy set $E(\widetilde{X})$ ($\in \mathscr{F}_{\mathrm{LR}}$) such that $(E(\widetilde{X}))_\alpha = E(X_\alpha)$ provided that $E\|\widetilde{X}\|^2_{\mathrm{LR}} = E(X^m)^2 + E(X^m - \lambda X^l)^2 + E(X^m + \rho X^r)^2 < \infty$, where X_α is the α-level of fuzzy set $\widetilde{X}$, that is, $X_\alpha = \{x \in \mathbb{R} | \mu_{\widetilde{X}}(x) \geq \alpha\}$, for $\alpha \in (0, 1]$, and $X_0 = cl(\{x \in \mathbb{R} | \mu_{\widetilde{X}} \geq 0\})$. Moreover, on the basis of the mapping s, we can observe that $s_{E(\widetilde{X})} = (E(X^m), E(X^l), E(X^r))$. The variance of $\widetilde{X}$ can be defined as $\sigma^2_{\widetilde{X}} = \mathrm{var}(\widetilde{X}) = E[(D^2_{\mathrm{LR}}(\widetilde{X}, E(\widetilde{X}))]$. In terms of s, it is $\sigma^2_{\widetilde{X}} = \mathrm{var}(\widetilde{X}) = E < s_{\widetilde{X}} - s_{E(\widetilde{X})}, s_{\widetilde{X}} - s_{E(\widetilde{X})} >_{\mathrm{LR}} = E(X^m - EX^m)^2 + E(X^m - EX^m - \lambda(X^l - EX^l))^2 + E(X^m - EX^m + \rho(X^r - EX^r))^2$. In a similar way, the covariance between two FRVs $\widetilde{X}$ and $\widetilde{Y}$ is $\sigma_{\widetilde{X}, \widetilde{Y}} = \mathrm{cov}(\widetilde{X}, \widetilde{Y}) = E < s_{\widetilde{X}} - s_{E(\widetilde{X})}, s_{\widetilde{Y}} - s_{E(\widetilde{Y})} >_{\mathrm{LR}} = E((X^m - EX^m)(Y^m - EY^m)) + E((X^m - EX^m - \lambda(X^l - EX^l))(Y^m - EY^m - \lambda(Y^l - EY^l))) + E((X^m - EX^m + \rho(X^r - EX^r))(Y^m - EY^m + \rho(Y^r - EY^r)))$.

20.3 A Regression Model with LR Fuzzy Random Response and Real Explanatory Variables

Consider a random experiment in which an *LR* fuzzy response variable $\widetilde{Y}$ and p real explanatory variables X_1, X_2,…,X_p are observed on n statistical units, $\{\widetilde{Y}_i, \underline{X}_i\}_{i=1,\dots,n}$, where $\underline{X}_i = (X_{1i}, X_{2i}, \dots, X_{pi})'$, or in a compact form $(\underline{\widetilde{Y}}, \mathbf{X})$. Since $\widetilde{Y}$ is determined by (Y^m, Y^l, Y^r), the proposed regression model concerns the real-valued random variables in this tuple. The center Y^m can be related to the explanatory variables $X_1, X_2, \dots, X_p$ through a classical regression model. However, the restriction of nonnegativity satisfied by Y^l and Y^r may entail some difficulties. We thus propose modeling a transform of the left spread and a transform of the right spread of the response through linear regressions (on the explanatory variables X_1, X_2,…,X_p). This can be represented in the following way, letting $g : (0, +\infty) \longrightarrow \mathbb{R}$ and $h : (0, +\infty) \longrightarrow \mathbb{R}$ be invertible:

$$\begin{cases} Y^m = \underline{X}' \underline{a}_m + b_m + \varepsilon_m \\ g(Y^l) = \underline{X}' \underline{a}_l + b_l + \varepsilon_l \\ h(Y^r) = \underline{X}' \underline{a}_r + b_r + \varepsilon_r \end{cases} \tag{20.2}$$

where ε_m, ε_l and ε_r are real-valued random variables with $E(\varepsilon_m|\underline{X}) = E(\varepsilon_l|\underline{X}) = E(\varepsilon_r|\underline{X}) = 0$, $\underline{a}_m = (a_{m1}, \dots, a_{mp})'$, $\underline{a}_l = (a_{l1}, \dots, a_{lp})'$, $\underline{a}_r = (a_{r1}, \dots, a_{rp})'$ are the $(p \times 1)$-vectors of the parameters related to the vector $\underline{X}$ and b_m, b_l, b_r are the intercepts. In presence of nonnegative variables, a common approach consists in transforming the constrained variable into an unconstrained one by means of g and h. Preliminary analyses showed that a good choice is given by the logarithmic transformation (that is $g = h = \log$). By this we transform the spreads into real variables without the restriction of nonnegativity.

It can be shown that the population parameters can be expressed, as usual, in terms of some moments involving the considered random variables. In fact, let $\widetilde{Y}$ be an *LR* FRV and $\underline{X}$ the vector of p real random variables satisfying the linear model (20.2), then we have that

$$\begin{aligned}
\underline{a}_m &= \{\Sigma_{\underline{X}}\}^{-1} E\left[(\underline{X} - E\underline{X})(Y^m - EY^m)\right], \\
\underline{a}_l &= \{\Sigma_{\underline{X}}\}^{-1} E\left[(\underline{X} - E\underline{X})(g(Y^l) - Eg(Y^l))\right], \\
\underline{a}_r &= \{\Sigma_{\underline{X}}\}^{-1} E\left[(\underline{X} - E\underline{X})(h(Y^r) - Eh(Y^r))\right], \\
b_m &= E(Y^m|\underline{X}) - E\underline{X}' \{\Sigma_{\underline{X}}\}^{-1} E\left[(\underline{X} - E\underline{X})(Y^m - EY^m)\right], \\
b_l &= E(g(Y^l)|\underline{X}) - E\underline{X}' \{\Sigma_{\underline{X}}\}^{-1} E\left[(\underline{X} - E\underline{X})(g(Y^l) - Eg(Y^l))\right], \\
b_r &= E(h(Y^r)|\underline{X}) - E\underline{X}' \{\Sigma_{\underline{X}}\}^{-1} E\left[(\underline{X} - E\underline{X})(h(Y^r) - Eh(Y^r))\right],
\end{aligned}$$

where $\Sigma_{\underline{X}} = E\left[(\underline{X} - E\underline{X})(\underline{X} - E\underline{X})'\right]$.

The estimators of the population parameters are based on the Least Squares (LS) criterion. In this case, using the generalized Yang-Ko metric $D^2_{\lambda\rho}$ written in vector terms, the LS problem consists in looking for $\widehat{\underline{a}}_m, \widehat{\underline{a}}_l, \widehat{\underline{a}}_r, \widehat{b}_m, \widehat{b}_l$ and $\widehat{b}_r$ in order to

$$\min \Delta^2_{\lambda\rho} = \min D^2_{\lambda\rho}((\underline{Y}^m, g(\underline{Y}^l), h(\underline{Y}^r)), ((\underline{Y}^m)^*, g^*(\underline{Y}^l), h^*(\underline{Y}^r))), \tag{20.3}$$

where $(\underline{Y}^m)^* = \mathbf{X}\underline{a}_m + \underline{1}b_m$, $g^*(\underline{Y}^l) = \mathbf{X}\underline{a}_l + \underline{1}b_l$ and $h^*(\underline{Y}^r) = \mathbf{X}\underline{a}_r + \underline{1}b_r$ are the $(n \times 1)$-vectors of the predicted values. Here we assume that the shape of the membership function of the predicted response is inherited from the observed one.

Under the assumptions of model (20.2), the solution of the LS problem is

$$\widehat{\underline{a}}_m = (\mathbf{X}^{c'}\mathbf{X}^c)^{-1}\mathbf{X}^{c'}\underline{Y}^{mc}, \quad \widehat{\underline{a}}_l = (\mathbf{X}^{c'}\mathbf{X}^c)^{-1}\mathbf{X}^{c'}g(\underline{Y}^l)^c, \quad \widehat{\underline{a}}_r = (\mathbf{X}^{c'}\mathbf{X}^c)^{-1}\mathbf{X}^{c'}h(\underline{Y}^r)^c,$$
$$\hat{b}_m = \overline{Y^m} - \overline{\underline{X}}'\hat{\underline{a}}_m, \quad \hat{b}_l = \overline{g(Y^l)} - \overline{\underline{X}}'\hat{\underline{a}}_l, \quad \hat{b}_r = \overline{h(Y^r)} - \overline{\underline{X}}'\hat{\underline{a}}_r,$$

where $\underline{Y}^{mc} = \underline{Y}^m - \underline{1}\overline{Y^m}$, $g(\underline{Y}^l)^c = g(\underline{Y}^l) - \underline{1}\overline{g(Y^l)}$, $h(\underline{Y}^r)^c = h(\underline{Y}^r) - \underline{1}\overline{h(Y^r)}$ are the centered values of the response variables, $\mathbf{X}^c = \mathbf{X} - \underline{1}\,\overline{\underline{X}}'$ is the centered matrix of the explanatory variables and, $\overline{Y^m}$, $\overline{g(Y^l)}$, $\overline{h(Y^r)}$, and $\overline{\underline{X}}$ denote, respectively, the sample means of Y^m, $g(Y^l)$, $h(Y^r)$, and $\underline{X}$. The estimators $\widehat{\underline{a}}_m, \widehat{\underline{a}}_l, \widehat{\underline{a}}_r, \widehat{b}_m, \widehat{b}_l$, and $\widehat{b}_r$ are unbiased and strongly consistent. It is worth mentioning that the significance of the regression parameters can be tested (see, for more details, [4, 5]).

20.4 A Regression Model with LR Fuzzy Random Response and Explanatory Variables

Model (20.2) can be generalized to the case in which the available information refers to an *LR* fuzzy response variable $\widetilde{Y}$ and p *LR* fuzzy explanatory variables $\widetilde{X}_1, \widetilde{X}_2, \ldots, \widetilde{X}_p$ observed on a sample of n statistical units, $\{\widetilde{Y}_i, \widetilde{X}_{1i}, \widetilde{X}_{2i}, \ldots, \widetilde{X}_{pi}\}_{i=1,\ldots,n}$. We are interested in analyzing the relationship between $\widetilde{Y}$ and $\widetilde{X}_1, \widetilde{X}_2, \ldots, \widetilde{X}_p$. This can be done by modeling the center and the spreads of $\widetilde{Y}$ by means of the centers and the spreads of $\widetilde{X}_1, \widetilde{X}_2, \ldots, \widetilde{X}_p$. However, in doing so, attention should be paid to the nonnegativity of the spreads of $\widetilde{Y}$. Analogously to the previous model, we introduce the functions g and h already defined. The linear regression model can be formalized as

$$\begin{cases} Y^m = \underline{X}'\underline{a}_m + b_m + \varepsilon_m, \\ g(Y^l) = \underline{X}'\underline{a}_l + b_l + \varepsilon_l, \\ h(Y^r) = \underline{X}'\underline{a}_r + b_r + \varepsilon_r, \end{cases} \tag{20.4}$$

where $\underline{X} = (X_1^m, X_1^l, X_1^r, \ldots, X_p^m, X_p^l, X_p^r)'$ is the $(3p \times 1)$-vector of all the components of the explanatory variables, ε_m, ε_l, and ε_r are real-valued random variables with $E(\varepsilon_m|\underline{X}) = E(\varepsilon_l|\underline{X}) = E(\varepsilon_r|\underline{X}) = 0$, $\underline{a}_m = (a^1_{mm}, a^1_{ml}, a^1_{mr}, \ldots, a^p_{mm}, a^p_{ml}, a^p_{mr})'$, $\underline{a}_l = (a^1_{lm}, a^1_{ll}, a^1_{lr}, \ldots, a^p_{lm}, a^p_{ll}, a^p_{lr})'$, $\underline{a}_r = (a^1_{rm}, a^1_{rl}, a^1_{rr}, \ldots, a^p_{rm}, a^p_{rl}, a^p_{rr})'$

are the $(3p \times 1)$-vectors of the parameters related to $\underline{X}$ and b_m, b_l, b_r are the intercepts. The generic a_{ij}^t is the regression coefficient between the component $i \in \{m, l, r\}$ of $\widetilde{Y}$ (where m, l and r refer to the center Y^m and the transforms of the spreads $g(Y^l)$ and $h(Y^r)$, respectively) and the component $j \in \{m, l, r\}$ of the explanatory variables $\widetilde{X}^t$, $t = 1, \ldots, p$, (where m, l and r refer to the corresponding center, left spread, and right spread). For example, a_{mr}^3 represents the relationship between the right spread of the explanatory variable $\widetilde{X}^3$ (X_3^r) and the center of the response, Y^m.

Also in this case, the estimation problem of the regression parameters is faced by means of the LS criterion. The LS problem consists in looking for $\underline{\hat{a}}_m$, $\underline{\hat{a}}_l$, $\underline{\hat{a}}_r$, $\hat{b}_m$, $\hat{b}_l$, and $\hat{b}_r$ such that

$$\Delta_{\lambda\rho}^2 = D_{\lambda\rho}^2((\underline{Y}^m, g(\underline{Y}^l), h(\underline{Y}^r)), ((\underline{Y}^m)^*, g^*(\underline{Y}^l), h^*(\underline{Y}^r)))$$

is minimized, where $\underline{Y}^m$, $g(\underline{Y}^l)$, and $h(\underline{Y}^r)$ are the $(n \times 1)$-vectors of the observed values and $(\underline{Y}^m)^* = \mathbf{X}\underline{a}_m + \underline{1}b_m$, $g^*(\underline{Y}^l) = \mathbf{X}\underline{a}_l + \underline{1}b_l$, and $h^*(\underline{Y}^r) = \mathbf{X}\underline{a}_r + \underline{1}b_r$ are the theoretical ones being $\mathbf{X} = (\widetilde{\underline{X}}_1, \widetilde{\underline{X}}_2, \ldots, \widetilde{\underline{X}}_n)'$ the matrix of the explanatory variables of order $(n \times 3p)$. The solution of the LS problem is

$$\widehat{\underline{a}}_m = (\mathbf{X}^{c'}\mathbf{X}^c)^{-1}\mathbf{X}^{c'}\underline{Y}^{mc}, \quad \widehat{\underline{a}}_l = (\mathbf{X}^{c'}\mathbf{X}^c)^{-1}\mathbf{X}^{c'}g(\underline{Y}^l)^c, \quad \widehat{\underline{a}}_r = (\mathbf{X}^{c'}\mathbf{X}^c)^{-1}\mathbf{X}^{c'}h(\underline{Y}^r)^c,$$

$$\hat{b}_m = \overline{Y^m} - \overline{\underline{X}}'\,\underline{\hat{a}}_m, \quad \hat{b}_l = \overline{g(Y^l)} - \overline{\underline{X}}'\,\underline{\hat{a}}_l, \quad \hat{b}_r = \overline{h(Y^r)} - \overline{\underline{X}}'\,\underline{\hat{a}}_r,$$

where $\underline{Y}^{mc}$, $g(\underline{Y}^l)^c$, $h(\underline{Y}^r)^c$, and $\mathbf{X}^c$ are defined analogously to Sect. 20.3.

20.5 Goodness of Fit

Model (20.2) is a particular case of model (20.4); so, for brevity of exposition, we consider only the latter model. Since the total variation of the response can be written in terms of variances and covariances of real random variables, by taking advantage of their properties it can be decomposed into the variation not depending on the model and that explained by the model. In particular, let $\widetilde{Y}$ and $\widetilde{X}_1, \widetilde{X}_2, \ldots, \widetilde{X}_p$ be *LR* FRVs satisfying the linear model (20.4) so that the errors are uncorrelated with $\underline{X}$, by indicating $\widetilde{Y}^{\mathrm{T}} = (Y^m, g(Y^l), h(Y^l))$, we obtain

$$E\left[D_{\lambda\rho}^2(\widetilde{Y}^{\mathrm{T}}, E(\widetilde{Y}^{\mathrm{T}}))\right] = E\left[D_{\lambda\rho}^2(\widetilde{Y}^{\mathrm{T}}, E(\widetilde{Y}^{\mathrm{T}}|\underline{X}))\right] + E\left[D_{\lambda\rho}^2(E(\widetilde{Y}^{\mathrm{T}}|\underline{X}), E(\widetilde{Y}^{\mathrm{T}}))\right]. \tag{20.5}$$

Let $\widetilde{Y}$ be the LR FRV of the linear model (20.4). Based on (20.5), it is possible to define the following determination coefficient:

$$R^2 = \frac{E\left[D_{\lambda\rho}^2(E(\widetilde{Y}^{\mathrm{T}}|\underline{X}), E(\widetilde{Y}^{\mathrm{T}}))\right]}{E\left[D_{\lambda\rho}^2(\widetilde{Y}^{\mathrm{T}}, E(\widetilde{Y}^{\mathrm{T}}))\right]} = 1 - \frac{E\left[D_{\lambda\rho}^2(\widetilde{Y}^{\mathrm{T}}, E(\widetilde{Y}^{\mathrm{T}}|\underline{X}))\right]}{E\left[D_{\lambda\rho}^2(\widetilde{Y}^{\mathrm{T}}, E(\widetilde{Y}^{\mathrm{T}}))\right]}. \tag{20.6}$$

This coefficient measures the degree of linear relationship. As in the classical case, it takes values in [0, 1].

In order to estimate the determination coefficient, it is worth introducing the decomposition of the total sum of squares. Let $\widetilde{Y}$ and $\widetilde{X}_1, \widetilde{X}_2, \ldots, \widetilde{X}_p$ be *LR* FRVs satisfying the linear model (20.4) observed on n statistical units, $\{\widetilde{Y}_i, \widetilde{\underline{X}}_{1i}, \widetilde{\underline{X}}_{2i}, \ldots, \widetilde{\underline{X}}_{pi}\}_{i=1,\ldots,n}$. The total sum of squares, *SST*, is equal to the sum of the residual sum of squares, *SSE*, and the regression sum of squares, *SSR*, that is,

$$\text{SST} = \text{SSE} + \text{SSR}, \tag{20.7}$$

where $\text{SST} = D^2_{\lambda\rho}((\underline{Y}^m, g(\underline{Y}^l), h(\underline{Y}^r)), (\overline{\underline{Y}^m}, \overline{g(\underline{Y}^l)}, \overline{h(\underline{Y}^r)}))$, $\text{SSE} = D^2_{\lambda\rho}((\underline{Y}^m, g(\underline{Y}^l), h(\underline{Y}^r)), (\widehat{\underline{Y}^m}, \widehat{g(\underline{Y}^l)}, \widehat{h(\underline{Y}^r)}))$ and $\text{SSR} = D^2_{\lambda\rho}((\widehat{\underline{Y}^m}, \widehat{g(\underline{Y}^l)}, \widehat{h(\underline{Y}^r)}), (\overline{\underline{Y}^m}, \overline{g(\underline{Y}^l)}, \overline{h(\underline{Y}^r)}))$, with $\widehat{\underline{Y}^m} = \mathbf{X}\widehat{\underline{a}}_m + \underline{1}\widehat{b}_m$, $\widehat{g(\underline{Y}^l)} = \mathbf{X}\widehat{\underline{a}}_l + \underline{1}\widehat{b}_l$, $\widehat{h(\underline{Y}^r)} = \mathbf{X}\widehat{\underline{a}}_r + \underline{1}\widehat{b}_r$ denoting the vectors of the estimated values.

Under the assumptions of model (20.4), the estimator of the determination coefficient R^2 is

$$\widehat{R}^2 = 1 - \frac{\text{SSE}}{\text{SST}} = \frac{\text{SSR}}{\text{SST}}.$$

It can be shown that $\widehat{R}^2$ is a strongly consistent estimator (see, for more details, [5, 6]).

20.5.1 Linear Independence Test

The aim of this section is to propose a procedure for testing the null hypothesis

$$H_0 : R^2 = 0 \text{ vs. } H_1 : R^2 > 0$$

on the basis of the available sample information. A suitable choice for the test statistic to be used is

$$T_n = n\widehat{R}^2 = n\frac{\text{SSR}}{\text{SST}}.$$

Since suitable generalized models for FRVs are not yet available and asymptotic tests work adequately for large size samples, a bootstrap approach can be adopted. To obtain a bootstrap population fulfilling the null hypothesis, the residual variables $Z^m = Y^m - \underline{X}'\widehat{\underline{a}}_m$, $Z^l = g(Y^l) - \underline{X}'\widehat{\underline{a}}_l$ and $Z^r = h(Y^r) - \underline{X}'\widehat{\underline{a}}_r$ must be considered. Samples of size n with replacement, $\left\{(\underline{X}^*_i, {Z^m}^*_i, {Z^l}^*_i, {Z^r}^*_i)\right\}_{i=1,\ldots,n}$, are drawn from the bootstrap population: $\left\{(X_i, Z^m_i, Z^l_i, Z^r_i)\right\}_{i=1,\ldots,n}$. For each of them the following statistic is computed:

$$T^*_n = n\frac{\text{SSR}^*}{\text{SST}^*},$$

where: $\mathrm{SSR}^* = D^2_{\lambda\rho}((\widehat{\underline{Z}^{m*}}, \widehat{\underline{Z}^{l*}}, \widehat{\underline{Z}^{r*}}), (\overline{\underline{Z}^{m*}}, \overline{\underline{Z}^{l*}}, \overline{\underline{Z}^{r*}}))$ and $\mathrm{SST}^* = D^2_{\lambda\rho}((\underline{Z}^{m*}, \underline{Z}^{l*}, \underline{Z}^{r*}), (\overline{\underline{Z}^{m*}}, \overline{\underline{Z}^{l*}}, \overline{\underline{Z}^{r*}}))$, with obvious notation.

The application of the bootstrap test is presented in the following algorithm.

Algorithm

Step 1: Compute the values $\underline{\hat{a}}_m$, $\underline{\hat{a}}_l$ and $\underline{\hat{a}}_r$ and $T_n = n\hat{R}^2$.

Step 2: Compute the bootstrap population

$$\{(X_i, Z_i^m, Z_i^l, Z_i^r)\}_{i=1,\ldots,n}.$$

Step 3: Draw a sample of size n with replacement

$$\left\{(X_i^*, Z^{m*}_i, Z^{l*}_i, Z^{r*}_i)\right\}_{i=1,\ldots,n},$$

from the bootstrap population and compute the value of the bootstrap statistic T_n^*.

Step 4: Repeat Step 3 a large number B of times to get a set of B estimators, denoted by $\{T_{n1}^*, \ldots, T_{nB}^*\}$.

Step 5: Approximate the bootstrap p-value as the proportion of values in $\{T_{n1}^*, \ldots, T_{nB}^*\}$ being greater than T_n.

It is important to note that the suggested testing procedure refers to a linear regression model in which suitable transforms (by means of functions g and h) of the spreads of the response variable are considered. It follows that the acceptation or rejection of linearity depends on the chosen scales. For more details, we refer to [5, 6].

20.6 An Application

In a recent study about the student satisfaction of a course the subjective judgements/perceptions were observed on a sample of $n = 64$ students. To formalize the problem we defined $\Omega = \{\text{set of students attending the course}\}$. Since the observations were arbitrarily chosen, P is the uniform distribution over Ω. For any student belonging to the sample, four characteristics were observed. These were the overall assessment of the course, the assessment of the teaching staff, the assessment of the course content and the average mark. The first three subjective judgements/perceptions were managed in terms of triangular fuzzy variables (hence $\lambda = \rho = 1/2$). In fact, to represent them, the students were invited to draw a triangular fuzzy number for every characteristic. The considered support went from 0 (dissatisfaction) to 100 (full satisfaction). The students were informed to place the center where they wished to represent their average judgement/perception and the lower and upper bounds of the triangular fuzzy number where they wished to represent their worst and best judgement/perception, respectively. Note that the students were informed to compute the average, minimum, and maximum values w.r.t. the variability of their subjective judgements/perceptions depending on the different course contents and/or members of the teaching staff. The average mark is a crisp variable. For analyzing the linear relationship of the overall assessment of the course ($\widetilde{Y}$) on the assessment of the teaching staff ($\widetilde{X}_1$), the assessment of the course contents ($\widetilde{X}_2$) and the average mark (X_3), the linear regression model in (20.4) was employed. To overcome the problem about the nonnegativity of spreads estimates, we used $g = h = \log$. Through the LS procedure we obtained the following estimated model

$$\begin{cases} \widehat{Y^m} = 1.08X_1^m + 0.13X_1^l - 0.07X_1^r \\ \quad -0.17X_2^m - 0.89X_2^l + 0.66X_2^r - 1.12X_3 + 34.06, \\ \widehat{Y^l} = \exp(0.01X_1^m + 0.02X_1^l + 0.02X_1^r \\ \quad +0.00X_2^m + 0.03X_2^l + 0.01X_2^r - 0.00X_3 + 0.67), \\ \widehat{Y^r} = \exp(0.00X_1^m + 0.03X_1^l - 0.02X_1^r \\ \quad -0.01X_2^m + 0.03X_2^l + 0.01X_2^r + 0.04X_3 + 1.01). \end{cases}$$

The center of $\widetilde{Y}$ is mainly related to the center of $\widetilde{X}_1$ and to X_3. This is not the case for the center of X_2. Conversely, the spreads of $\widetilde{X}_2$ remarkably affect the response Y^m. In particular, higher values of the left spreads (more imprecision on the lower values of the assessment of the course content) lead to lower values of the center of the response, whereas the opposite comment holds for the right spread. Concerning the models for the left and right spreads of the response only some of the components of $\widetilde{X}_2$ seem to be related to the transformed spreads of $\widetilde{Y}$. For instance, a positive relationship between the left spread of $\widetilde{Y}$ and the one of $\widetilde{X}_2$ was observed. We also found $\widehat{R}^2 = 0.7727$, hence approximately 77 % of the total variation of the overall assessment of the course was explained by the model. Furthermore, by applying the bootstrap procedure to test the linear independence (with $B = 1{,}000$ bootstrap samples) a p-value equal to 0 was obtained, so the null hypothesis of linear independence should be rejected. The fitted model points out the relevance of the teaching staff in the overall assessment of the course which, on the other hand, turns out to be negatively correlated with the average mark. The spreads of the overall assessment are only slightly affected by the explanatory variables.

20.7 Concluding Remarks

In this work a class of new linear regression models for *LR* fuzzy variables has been introduced, by taking into account different kinds of uncertainty, through a formalization in terms of FRVs. Furthermore, the problem of the nonnegativity of the spreads of the response has been dealt with by using reasonable transform functions. The suitability of the proposed model has been analyzed by means of a real-life application.

References

1. Coppi, R., D'Urso, P., Giordani, P., Santoro, A.: Least squares estimation of a linear regression model with LR fuzzy response. Comput. Stat. Data Anal. **51**, 267–286 (2006)
2. Coppi, R., Gil, M.A., Kiers, H.A.L.: The fuzzy approach to statistical analysis. Comput. Statist. Data Anal. **51**, 1–14 (2006)
3. Diamond, P.: Fuzzy least squares. Inform. Sci. **46**, 141–157 (1988)
4. Ferraro, M.B., Coppi, R., Gonzalez-Rodriguez, G., Colubi, A.: A linear regression model for imprecise response. Int. J. Approx. Reason. **51**, 759–770 (2010)
5. Ferraro, M.B., Colubi, A., Gonzalez-Rodriguez, G., Coppi, R.: A determination coefficient for a linear regression model with imprecise response. Environmetrics **22**, 487–596 (2011)
6. Ferraro, M.B., Giordani, P.: A multiple linear regression model for LR fuzzy random variables. Metrika **75**, 1049–1068 (2012)
7. González-Rodríguez, G., Blanco, A., Colubi, A., Lubiano, M.A.: Estimation of a simple linear regression model for fuzzy random variables. Fuzzy Sets Syst. **160**, 357–370 (2009)
8. Guo, P., Tanaka, H.: Dual models for possibilistic regression analysis. Comput. Stat. Data Anal. **51**, 253–266 (2006)
9. Puri, M.L., Ralescu, D.A.: Fuzzy random variables. J. Math. Anal. Appl. **114**, 409–422 (1986)
10. Tanaka, H., Uejima, S., Asai, K.: Linear regression analysis with fuzzy model. IEEE Trans. Syst. Man Cybernet. **12**, 903–907 (1982)
11. Yang, M.S., Ko, C.H.: On a class of fuzzy *c*-numbers clustering procedures for fuzzy data. Fuzzy Sets Syst. **84**, 49–60 (1996)
12. Zadeh, L.A.: Fuzzy sets. Inf. Contr. **8**, 338–353 (1965)

A Model-Based Dimension Reduction Approach to Classification of Gene Expression Data

21

Luca Scrucca and Avner Bar-Hen

Abstract

The monitoring of the expression profiles of thousands of genes have proved to be particularly promising for biological classification, particularly for cancer diagnosis. However, microarray data present major challenges due to the complex, multiclass nature and the overwhelming number of variables characterizing gene expression profiles. We introduce a methodology that combine dimension reduction method and classification based on finite mixture of Gaussian densities. Information on the dimension reduction subspace is based on the variation of components means for each class, which in turn are obtained by modeling the within class distribution of the predictors through finite mixtures of Gaussian densities. The proposed approach is applied to the leukemia data, a well known dataset in the microarray literature. We show that the combination of dimension reduction and model-based clustering is a powerful technique to find groups among gene expression data.

Keywords

Gene expression data • Model-based dimension reduction • SIR

L. Scrucca (✉)
Dipartimento di Economia, Finanza e Statistica, Università degli Studi di Perugia, 06100 Perugia, Italy
e-mail: luca@stat.unipg.it

A. Bar-Hen
MAP5, UFR de Mathématiques et Informatique, Université Paris Descartes, 45 rue des Saints-Pères, 75270 Paris, France
e-mail: Avner.Bar-Hen@mi.parisdescartes.fr

N. Torelli et al. (eds.), *Advances in Theoretical and Applied Statistics*, Studies in Theoretical and Applied Statistics, DOI 10.1007/978-3-642-35588-2_21,

21.1 Introduction

By monitoring the activity of many genes from a sample in a single experiment microarrays offer a unique perspective for explaining the global genetic picture of a variant, whether a diseased individual or a sample subject to whatever stressing conditions. One of the main result of an experiment with microarrays is a list of genes that are differentially expressed between two conditions, or that allow samples to be classified according to their phenotypic features. An overview of classification methods for microarray data is given in [7] while a comparison among several of them is provided in [8].

Classification of microarray data is particularly problematic due to under-resolution, i.e., the presence of a large number of features (genes) from which to predict classes compared to the relatively small number of observations (samples). Moreover the expression levels of many genes are often highly correlated. Classical procedures for dimensionality reduction are principal components analysis and factor analysis, both of which reduce dimensionality by forming linear combinations of the features. The first method seeks a lower-dimensional representation that account for the most variance of the features, while the second looks for the most correlation among the features. Alternatively, sliced inverse regression (SIR) is a dimension reduction method which exploits the information from the inverse mean function [15]. SIR directions span at least part of the dimension reduction subspace [see 5, Chap. 6], and they provide an intuitive and useful basis for constructing summary plots [22].

Dimension reduction techniques have already been studied in the microarray context. A two-stage approach based on SIR was proposed by [4], with the first stage aimed at eliminating the dimension redundancy through an eigenanalysis which identified a set of linear combinations of the original genes which retained the main variability structure. Then, the SIR method was applied on such reduced set of eigengenes [see also 16]. A final gene selection stage was proposed based on a ranking of genes in accordance to the concordance with the final estimated directions. A related approach was also discussed by [3], in which they applied both SIR and SAVE, i.e., sliced average variance estimation, to microarray data.

Clustering algorithms based on probability models offer an alternative to heuristic-based algorithms. In particular, the model-based approach assumes that the data are generated by a finite mixture of underlying probability distributions such as multivariate normal distribution. The Gaussian mixture model has been shown to be a powerful tool for many applications [1, 18, 19, 25].

In this chapter, we introduce a model-based clustering on the low dimensional space obtained by SIR projection of the high-dimensional gene expression data. SIR components allow to project the gene expression data in a lower dimension space where the direction is uncorrelated. This provided a robust framework to classify the data. On the other hand, the model-based clustering part gives the flexibility to catch the class structure within the gene expression data. The main point of this chapter is to propose a unified approach between dimension reduction and model-based clustering.

21.2 Methods

Gene expression data on p genes for n mRNA samples can be summarized by a $n \times p$ matrix $\boldsymbol{X} = [x_{ij}]$, where x_{ij} denotes the expression level of gene (variable) j in mRNA sample (observation) i. Note that in this representation the data matrix is the transpose of that adopted in the microarray literature. The expression levels might be either absolute (e.g., oligonucleotide arrays) or relative with respect to the expression levels of a suitably defined common reference sample (e.g., cDNA microarrays).

Classification methods aim at assigning a sample to one of K given classes based on information from a set of p predictor variables observed on n samples. Thus, a categorical response variable Y with K levels representing biological outcomes, such as tumors type, is also recorded along with gene expression levels. An overview of classification methods for microarray data is given in [7] while a comparison among several of them is provided in [8].

21.2.1 Dimension Reduction Methods

The general goal of a dimension reduction method is to find a $p \times d$ matrix $\boldsymbol{\beta}$, with $d \leq p$, such that $Y \perp\!\!\!\perp X \,|\, \boldsymbol{\beta}^{\top} X$, i.e., the classification response Y is independent of the predictors X given the linear reduction $\boldsymbol{\beta}^{\top} X$. The subspace of dimension d spanned by the matrix $\boldsymbol{\beta}$ is called the sufficient dimension reduction (SDR) subspace for the regression of Y on X. It always exists, since we can trivially set $\boldsymbol{\beta} = \boldsymbol{I}$, but it is not unique. The central dimension-reduction (CDR) subspace is defined as the intersection over all the SDR subspaces; under mild conditions, it exists and it is the unique minimum SDR subspace [5].

Several methods have been proposed to estimate the CDR subspace, such as sliced inverse regression [SIR; 15] and sliced average variance estimation [SAVE; 6].

SIR, which is perhaps the most commonly used DR method, estimates the basis of the CDR subspace by the first d eigenvectors of the decomposition of $\boldsymbol{\Sigma}_{X|Y} = \mathrm{Var}(\mathrm{E}(X \,|\, Y))$ with respect to $\boldsymbol{\Sigma} = \mathrm{Var}(X)$, i.e.

$$\boldsymbol{\Sigma}_{X|Y} \boldsymbol{v}_i = l_i \boldsymbol{\Sigma} \boldsymbol{v}_i, \tag{21.1}$$

with $\boldsymbol{v}_i^{\top} \boldsymbol{\Sigma} \boldsymbol{v}_j = 1$ if $i = j$ and 0 otherwise, and eigenvalues $l_1 \geq l_2 \geq \cdots \geq l_d > 0$.

SIR requires the linearity condition on the marginal distribution of the predictors, i.e., requires that $\mathrm{E}(\boldsymbol{b}^{\top} X \,|\, \boldsymbol{\beta}^{\top} X)$ is linear in $\boldsymbol{\beta}^{\top} X$ for all $\boldsymbol{b} \in \mathbb{R}^p$. This condition is satisfied if the predictors follow an elliptical symmetric distribution, such as the multivariate normal. Furthermore, this condition is not a severe restriction, since most low-dimensional projections are close to being normal [14].

Usually, a sliced version of the response variable is needed, but in classification settings Y is taken directly as the class indicator with K levels. Sample estimates are then used to compute the unknown quantities in (21.1), and directions estimated as $\widehat{\boldsymbol{\beta}}_{\text{SIR}} \equiv [\widehat{\boldsymbol{v}}_1, \dots, \widehat{\boldsymbol{v}}_d]$. In this case SIR can find at most $\min(K-1, p)$ directions. Thus, for a binary response variable SIR estimates at most a single direction.

21.2.2 Model-Based SIR

The SIR estimator is based on the information coming from the variation on slice means. Thus, the distribution of the data within any class is summarized by the class means $\boldsymbol{\mu}_k$ $(k = 1, \dots, K)$. However, it may happen that a more complicated distribution of the data is needed. Finite mixture of Gaussian densities can be used to approximate the distribution of the predictors within any class, and then the kernel matrix can be computed using the components means [23].

Assume that within each class the data can be described by the finite mixture distribution

$$f(\boldsymbol{x}|Y=k) = f_k(\boldsymbol{x}) = \sum_{c=1}^{C_k} \pi_{kc}\phi(\boldsymbol{x};\boldsymbol{\mu}_{kc}, \boldsymbol{\Sigma}_{kc}), \tag{21.2}$$

where $\boldsymbol{\mu}_{kc}$ and $\boldsymbol{\Sigma}_{kc}$ are the means and covariances for component c within class k, π_{kc} are the mixing weights such that $\pi_{kc} > 0$ and $\sum_c \pi_{kc} = 1$, and $\phi(.)$ is the multivariate Gaussian density. Geometric features, such as shape, volume, and orientation, of each component are determined by the covariance structure. [1] proposed a representation of the covariance matrix for a finite mixture model, say in general $\boldsymbol{\Sigma}_g$, in terms of its eigenvalue decomposition, $\boldsymbol{\Sigma}_g = \lambda_g \boldsymbol{D}_g \boldsymbol{A}_g \boldsymbol{D}_g{}^{\top}$, where $\boldsymbol{D}_g$ is the orthogonal matrix of eigenvectors which determines the orientation, $\boldsymbol{A}_g$ is a diagonal matrix of scales eigenvalues which determines the shape, and λ_g is a scalar determining the volume of the hyper-ellipsoid.

Consider the kernel matrix given by the covariance matrix of the between-component means

$$\boldsymbol{M} = \sum_{k=1}^{K}\sum_{c=1}^{C_k} \omega_{kc}(\boldsymbol{\mu}_{kc} - \boldsymbol{\mu})(\boldsymbol{\mu}_{kc} - \boldsymbol{\mu})^{\top}, \tag{21.3}$$

where $\omega_{kc} = \pi_{kc} \Pr(Y=k)$ $(\omega_{kc} > 0, \sum_{k,c} \omega_{kc} = 1)$, and the covariance matrix $\boldsymbol{\Sigma} = n^{-1}\sum_{i=1}^{n}(\boldsymbol{x}_i - \boldsymbol{\mu})(\boldsymbol{x}_i - \boldsymbol{\mu})^{\top}$ with $\boldsymbol{\mu} = \sum_{k=1}^{K}\sum_{c=1}^{C_k}\omega_{kc}\boldsymbol{\mu}_{kc}$. An estimate of the CDR subspace is the solution of the following constrained optimization:

$$\text{argmax}_{\boldsymbol{\beta}}\ \boldsymbol{\beta}^{\top}\boldsymbol{M}\boldsymbol{\beta}, \text{ subject to } \boldsymbol{\beta}^{\top}\boldsymbol{\Sigma}\boldsymbol{\beta} = \boldsymbol{I},$$

where $\boldsymbol{\beta} \in \mathbb{R}^{p\times d}$ is the spanning matrix, and $\boldsymbol{I}$ is the $(d \times d)$ identity matrix. This is solved through the generalized eigendecomposition

$$\boldsymbol{M}\boldsymbol{v}_i = l_i \boldsymbol{\Sigma}\boldsymbol{v}_i, \tag{21.4}$$

with $v_i^{\top} \Sigma v_j = 1$ if $i = j$ and 0 otherwise, and eigenvalues $l_1 \geq l_2 \geq \cdots \geq l_d > 0$. The eigenvectors corresponding to the first d largest eigenvalues, say $\beta_{\mathsf{MSIR}} \equiv [v_1, \ldots, v_d]$, provide a basis for the CDR subspace which shows the maximal variation between component means. There are at most $d = \min(p, C-1)$ directions which span this subspace, where $C = \sum_{k=1}^{K} C_k$ is the total number of finite mixture components.

The proposed model-based SIR (MSIR) method allows to avoid the limitations of the basic SIR procedure without imposing further conditions. In particular, MSIR is able to deal with a well-known limitation of SIR in the presence of symmetric response curves [23]. However, this additional flexibility requires that enough observations are available within any group to allow the fit of mixture models. In the case of a categorical response variable, such as tumor classes, MSIR allows to better approximate the predictors distribution within any class. Finally, note that by imposing a single mixture component for each class (i.e., $C_k = 1$ for all $k = 1, \ldots, K$) MSIR reverts to the usual SIR estimator.

21.2.3 Estimation

Preliminary filtering of genes The large number of gene expression profiles usually collected in microarray experiments can be drastically reduced since many of them exhibit near constant expression levels across samples. In our analyses we performed a preliminary filtering step by selecting the genes which have the highest ratio of between- to within-groups sums of squares [7, 22]. Filtering of genes based on the between- to within-group sums of squares is equivalent to filtering based on the between- to total-group sums of squares. This means that a gene is discarded if marginally shows no significant within-group mean differences. Therefore, it is unlikely that it will contribute useful within-class substructure when it is included jointly with other more relevant genes. Omitting this pre-filtering step basically amounts to add noise into our model, which could lead to loose power in the estimation process.

Regularized estimation of covariance matrix Usually the number of genes is larger than the number of available samples. Hence, direct application of a dimension reduction method, such as SIR, is not possible because of the singularity of the predictors covariance matrix. This fact was not recognized in [3], while [4] addressed this problem by using singular value decomposition as an intermediate step to produce the so-called eigengenes. Other approaches are based on a form of regularization for the covariance matrix. Reference [26] introduced a ridge estimator with a ridge parameter selected via bootstrap. A similar shrinkage estimator has been also employed by [22], where the proposed estimator is obtained as a convex linear combination between the full covariance matrix and the diagonal matrix having the average total variation along the diagonal; the coefficient that determines the convex linear combination is selected by cross-validation. An L2-regularization was proposed by [17] through a least-squares formulation of SIR, with the tuning parameter selected by an information criteria. Recently, [2] discussed a family of regularization procedures based on a Gaussian prior distribution, with some of the earlier proposals as special cases; however, the selection of prior parameters is still under investigation.

In this chapter we follow a simple and classical approach, estimating the predictors covariance matrix as $\widehat{\boldsymbol{\Sigma}} = \text{diag}[s_i^2]_{i=1,\ldots,p}$, where s_i^2 is the sample variance of the i-th gene. Thus, genes can have different variances but they are uncorrelated. In other words, we are more interested in their variances (within- and between-group) that in their (linear) relationships. This is arguably a strong assumption, but it should have a little effect. In fact, the (diagonal) covariance matrix defines the metric of the generalized eigendecomposition in (21.4), but it doesn't rule out that genes are between and within-group correlated. Note that some of the most successful methods in microarray data analysis assume very week dependence or even ignore correlation among genes [8, 24]. This also allows to avoid the problem of selecting any tuning parameter.

MSIR estimation Estimation of parameters for mixture models in (21.2) is obtained via the EM algorithm [20]. The number of components C_k, as well as the particular form of the covariance matrix $\boldsymbol{\Sigma}_{kc}$, may be choosen by BIC [11]. The use of BIC as model selection criterion tends to choose the simpler model which provide a good fit to the data, i.e., by penalizing the likelihood with the number of estimated parameters. This can be viewed as a form of regularization, so it avoids overfitting and unstable parameters estimates. However, given that $p \gg n$, restrictions on all the possible mixture models must be imposed. In particular, only models assuming spherical and diagonal covariance structures, obtained by imposing $\boldsymbol{D}_{kc} = \boldsymbol{I}$ in the spectral representation of $\boldsymbol{\Sigma}_{kc}$, can be fitted. Nevertheless, this appears to be sufficiently flexible to accommodate data with different characteristics.

The sample version of MSIR is then obtained by applying the eigendecomposition in (21.4) with the kernel matrix (21.3) computed from the means estimated for each mixture component, and $\widehat{\omega}_{kc} = n_{kc}/n$, the observed fraction of observations belonging to the c-th mixture component within class k.

Finally, the sample basis estimate is $\widehat{\boldsymbol{\beta}}_{\mathsf{MSIR}} \equiv [\widehat{\boldsymbol{v}}_1, \ldots, \widehat{\boldsymbol{v}}_d]$, and the projections onto such subspace are defined as $\boldsymbol{Z} = \boldsymbol{X}\widehat{\boldsymbol{\beta}}_{\mathsf{MSIR}}$.

Soft-thresholding Soft-thresholding can be used to shrink coefficients toward zero [24]. This enables to identify subsets of genes relevant for the definition of a given set of MSIR directions.

Let $\widehat{\beta}^*_{ij} = \widehat{\beta}_{ij} s_i$ be the coefficient of the j-th direction for the i-th predictor standardized to unit standard deviation. The soft-thresholding function is defined as $s(x, \Delta) = \text{sign}(x)(|x| - \Delta)_+$, where $(u)_+ = u$ if $u > 0$, and zero otherwise. For a threshold value Δ_j $(j = 1, \ldots, d)$, we define the shrunken coefficients as $\tilde{\beta}_{ij} = s(\widehat{\beta}^*_{ij}, \Delta_j)/s_i$, orthonormalized such that $\tilde{\boldsymbol{\beta}}^\top \widehat{\boldsymbol{\Sigma}} \tilde{\boldsymbol{\beta}} = \boldsymbol{I}$, where $\tilde{\boldsymbol{\beta}} = [\tilde{\beta}_{ij}] \in \mathbb{R}^{p \times d}$.

The shrunken directions $\tilde{\boldsymbol{\beta}}$ can be computed for a grid of Δ_j values. Let $\boldsymbol{X}\widehat{\boldsymbol{\beta}}$ and $\boldsymbol{X}\tilde{\boldsymbol{\beta}}$ be, respectively, the projection onto the subspace spanned by the estimated MSIR directions and the shrunken directions. The squared trace correlation [15] provides a measure of the agreement between the two subspaces. For single directions it is equivalent to the usual squared correlation coefficient. However, this measure cannot be used to select the amount of shrinkage since it is always decreasing as the threshold value is increased.

We suggest to select the amount of shrinkage applied on the basis of the maximum classes separation relative to their compactness on the projection subspace. This can be achieved by computing the Dunn index, which is given by the ratio of the smallest distance between observations not in the same class to the largest intra-class distance [9]. For a given threshold value Δ_j we compute

$$D(\Delta_j) = \min_{k=1,\ldots,K} \left(\frac{\min\limits_{i \in C_k, l \notin C_k} D_{il}}{\max\limits_{i,l \in C_k} D_{il}} \right),$$

where $D_{il} = [(\tilde{\boldsymbol{\beta}}^\top \boldsymbol{x}_i - \widehat{\boldsymbol{\beta}}^\top \boldsymbol{x}_l)^\top (\tilde{\boldsymbol{\beta}}^\top \boldsymbol{x}_i - \widehat{\boldsymbol{\beta}}^\top \boldsymbol{x}_l)]^{1/2}$ and $C_k = \{i : Y_i = k\}$ is the set of observations belonging to the k-th class ($k = 1, \ldots, K$). Then, the amount of threshold Δ_{j*} is chosen to maximize $D(\Delta_j)$. Careful study $D(\Delta_j)$ can also be useful to detect local maxima attained with smaller sets of genes.

21.3 Leukemia Data

Reference [13] analyzed data from high-density Affymetrix oligonucleotide arrays. There were 3,571 genes and 38 samples: 27 in class ALL (acute lymphoblastic leukemia) and 11 in class AML (acute myeloid leukemia). The samples in class ALL could be further splitted into B-cell and T-cell types. After the preliminary filtering a subset of 892 genes was selected. We focus on this well-known data set for pedagogical reason. Since it was intensively studied, it is easy to compare our results with results obtained from classical methods. From a practical point of view, it should be kept in mind that the classical results are strongly related to the choice of training/validation data [see 10], and that the interest of micro-array has to be compared with the predictive power of clinical data [see 21].

MSIR estimates a 3-components mixture for AML group and a 2-components mixture for the ALL group. There are four estimated directions, with corresponding eigenvalues equal to 273.3, 45.5, 40.8, and 24.3. Figure 21.1a shows a scatterplot of the samples projected along the first two MSIR directions, with marginal box-plots. The two tumor classes ALL and AML are clearly separated on the first direction. The second direction allows to split the ALL samples into B-cell (open circles) and T-cell (filled circles) types, with one unusual B-cell which is very close to the group of T-cell samples. This is interesting because MSIR estimation did not use the information on the different type of ALL tumors; nonetheless, it was able to show an underlying data structure.

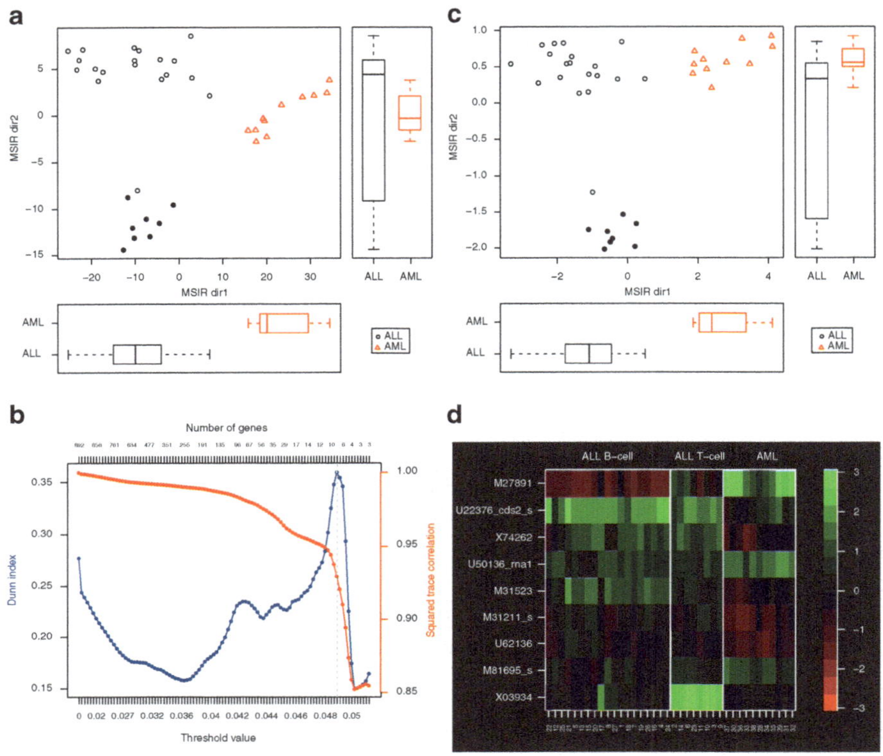

Fig. 21.1 (**a**) Scatterplot with marginal box-plots for the Golub data projected along the first two MSIR directions, with points marked by tumor classes (the ALL tumors are further identified by *open circles* for B-cell type and *filled circles* for T-cell type).
(**b**) Trace plot of Dunn index for class separation and squared trace correlation as the threshold value is increased. By increasing such value the soft-thresholding procedure selects smaller subsets of active genes.
(**c**) Scatterplot with marginal box-plots for the Golub data projected along the first two MSIR directions after applying the soft-thresholding procedure.
(**d**) Heat map of gene expression profiles for the genes selected by the soft-thresholding procedure. Values ranges from *bright green* (negative, under expressed) to *bright red* (positive, over expressed)

We also performed SIR estimation; the only direction estimable by SIR presents a large agreement with the first one estimated by MSIR (the squared correlation coefficient is equal to 0.997).

Figure 21.1b shows the results from applying the soft-thresholding procedure to the first MSIR direction. The maximal separation measure by the Dunn index is achieved for a threshold value of $\Delta_j = 0.049$; at this value 8 genes can be used to reproduce the original estimated MSIR direction to a large extent (the squared correlation coefficient is equal to 0.9292).

Applying the soft-thresholding procedure to the first two MSIR directions leads to select the previous eight genes, plus another gene (X03934) which is mainly responsible for the definition of the second direction. A plot analogous to Fig. 21.1a is shown in Fig. 21.1c; here we plotted the samples projected along the first two MSIR directions after soft-thresholding. By visual inspection we can see that the main data features are retained, but using only a small subset of genes.

Figure 21.1d shows a plot of expression levels for the selected genes. The first eight genes (read from top to bottom) allow to separate the ALL from AML tumor types, and they contribute to the definition of the first MSIR direction. The last gene is the only needed to define the second MSIR direction: as we already mentioned, the B-cell and T-cell tumors for ALL are separated along this direction. This is clearly visible in the corresponding gene profile shown at the bottom of Fig. 21.1d, where samples for ALL T-cell are over-expressed, whereas those for ALL B-cell have expression levels around zero, except for one sample (number 17) which has a level comparable to the T-cell group. Note that this is the one outlying B-cell observation which appears closer to the group of T-cell in Fig. 21.1a, c.

Finally, we report a comparison made with model-based discriminat analysis [11]. The classification errors obtained with MSIR on the train and test sets are, respectively, equal to 0 and 0.0588, using both the full set of genes and those selected by soft-thresholding. Thus, reducing the number of genes does not increase the classification error. These can be compared with the classification errors obtained directly from MCLUST [12], which are 0.0526 for the training set and 0.4118 for the test set. The large reduction in the test error for MSIR is clearly connected with the reduced set of genes employed.

Acknowledgements The authors thank the Working Group on Model-Based Clustering Summer Session (Paris, 2009) where this work was initiated. Luca Scrucca and Avner Bar-Hen conceived and collaborated in analyzing the data and writing the article.

References

1. Banfield, J., Raftery, A.E.: Model-based Gaussian and non-Gaussian clustering. Biometrics **49**, 803–821 (1993)
2. Bernard-Michel, C., Gardes, L., Girard, S.: Gaussian regularized sliced inverse regression. Stat. Comput. **19**(1), 85–98 (2009)
3. Bura, E., Pfeiffer, R.: Graphical methods for class prediction using dimension reduction techniques on DNA microarray data. Bioinformatics **19**, 1252–1258 (2003)
4. Chiaromonte, F., Martinelli, J.: Dimension reduction strategies for analyzing global gene expression data with a response. Math. Biosci. **176**, 123–144 (2002)
5. Cook, R.D.: Regression Graphics: Ideas for Studying Regressions Through Graphics. Wiley, New York (1998)
6. Cook, R.D., Weisberg, S.: Discussion of Li (1991). J. Am. Stat. Assoc. **86**, 328–332 (1991)
7. Dudoit, S., Fridlyand, J.: Classification in microarray experiments. In: Speed, T.P. (ed.) Statistical Analysis of Gene Expression Microarray Data, pp. 93–158. Chapman & Hall/CRC, Boca Raton (2003)
8. Dudoit, S., Fridlyand, J., Speed, T.: Comparison of discrimination methods for the classification of tumors using gene expression data. J. Am. Stat. Assoc. **97**, 77–87 (2002)
9. Dunn, J.: Well separated clusters and optimal fuzzy partitions. J. Cybernet. **4**, 95–104 (1974)
10. Ein-Dor, L., Kela, I., Getz, G., Givol, D., Domany, E.: Outcome signature genes in breast cancer: is there a unique set? Bioinformatics **21**(2), 171–178 (2005)

11. Fraley, C., Raftery, A.E.: Model-based clustering, discriminant analysis, and density estimation. J. Am. Stat. Assoc. **97**(458), 611–631 (2002)
12. Fraley, C., Raftery, A.E.: MCLUST version 3 for R: Normal mixture modeling and model-based clustering. Technical Report 504, Department of Statistics, University of Washington (2006)
13. Golub, T., Slonim, D., Tamayo, P., Huard, C., Gaasenbeek, M., Mesirov, J., Coller, H., Loh, M., Downing, J.R., Caligiuri, M., Bloomfield, C., Lander, E.: Molecular classification of cancer: class discovery and class prediction by gene expression monitoring. Science **286**, 531–537 (1999)
14. Hall, P., Li, K.-C.: On almost linearity of low dimensional projections from high dimensional data. Ann. Stat. **21**, 867–889 (1993)
15. Li, K.C.: Sliced inverse regression for dimension reduction (with discussion). J. Am. Stat. Assoc. **86**, 316–342 (1991)
16. Li, L., Li, H.: Dimension reduction methods for microarrays with application to censored survival data. Bioinformatics **20**(18), 3406–3412 (2004)
17. Li, L., Yin, X.: Sliced inverse regression with regularizations. Biometrics **64**(1), 124–131 (2008)
18. McLachlan, G., Basford, K.: Mixture Models: Inference and Applications to Clustering. Dekker, New York (1998)
19. McLachlan, G., Bean, R., Peel, D.: A mixture model-based approach to the clustering of microarray expression data. Bioinformatics **18**(3), 413–422 (2002)
20. McLachlan, G., Peel, D.: Finite Mixture Models. Wiley, New York (2000)
21. Michiels, S., Koscielny, S., Hill, C.: Prediction of cancer outcome with microarrays: a multiple random validation strategy. The Lancet **365**(9458), 488–492 (2005)
22. Scrucca, L.: Class prediction and gene selection for DNA microarrays using sliced inverse regression. Comput. Stat. Data Anal. **52**, 438–451 (2007)
23. Scrucca, L.: Model-based SIR for dimension reduction. Comput. Stat. Data Anal. **55**(11), 3010–3026 (2011)
24. Tibshirani, R., Hastie, T., Narasimhan, B., Chu, G.: Diagnosis of multiple cancer types by shrunken centroids of gene expression. Proc. Nat. Acad. Sci. **99**(10), 6567–6572 (2002)
25. Yeung, K.Y., Fraley, C., Murua, A., Raftery, A.E., Ruzzo, W.L.: Model-based clustering and data transformations for gene expression data. Bioinformatics **17**(10), 977–987 (2001)
26. Zhong, W., Zeng, P., Ma, P., Liu, J.S., Zhu, Y.: RSIR: regularized sliced inverse regression for motif discovery. Bioinformatics **21**(22), 4169–4175 (2005)

Exploiting Multivariate Outcomes in Bayesian Inference for Causal Effects with Noncompliance

22

Alessandra Mattei, Fabrizia Mealli, and Barbara Pacini

Abstract

A Bayesian approach to causal inference in the presence of noncompliance to assigned randomized treatment is considered. It exploits multivariate outcomes for improving estimation of weakly identified models, when the usually invoked exclusion restriction assumptions are relaxed. Using artificial data sets, we analyze the properties of the posterior distribution of causal estimands to evaluate the potential gains of jointly modeling more than one outcome. The approach can be used to assess robustness with respect to deviations from structural identifying assumptions. It can also be extended to the analysis of observational studies with instrumental variables where exclusion restriction assumptions are usually questionable.

Keywords

Bayesian statistics • Causal inference • Multivariate outcome • Noncompliance • Principal stratification

22.1 Introduction

Randomized experiments are generally considered the gold standard for causal inference. Yet even randomized experiments may suffer from a number of complications. A common complication when drawing inference about causal effects is that

A. Mattei (✉) · F. Mealli
Dipartimento di Statistica, Informatica, Applicazioni (DiSIA) "G. Parenti", Università di Firenze, Firenze, Italy
e-mail: mattei@disia.unifi.it; mealli@disia.unifi.it

B. Pacini
Dipartimento di Scienze Politiche, Università di Pisa, Pisa, Italy
e-mail: barbara.pacini@sp.unipi.it

N. Torelli et al. (eds.), *Advances in Theoretical and Applied Statistics*, Studies in Theoretical and Applied Statistics, DOI 10.1007/978-3-642-35588-2_22,

compliance is rarely perfect: noncompliance occurs when the actual treatment that units receive differs from their nominal assignment. Noncompliance is relatively frequent in experiments involving human subjects, complicating traditional theories of inference that require adherence to the random treatment assignment. A standard approach is to estimate the intention-to-treat (ITT) effect, focusing on the causal effect of assignment of treatment rather than the causal effect of receipt of treatment. Nevertheless, interest often centers on the effect of the treatment itself, and the interpretation of an ITT analysis is sometimes based on an implicit assumption, which is generally but not always plausible [5], that the effect of assignment is indicative of the effect of the treatment.

Randomized experiments with noncompliance are closely related to econometrics instrumental variable (IV) settings, where the instrument plays the parallel role of treatment assignment.

Angrist et al. [1] show that causal effects of assignment among compliers (subjects who would take treatment if offered and would not take it if not offered) are identifiable under the assumption that there are no subjects who would take the active treatment if randomized to the control arm but would not take it if assigned to the treatment arm (no-defier, monotonicity assumption) and under the additional assumption of a null effect of assignment on the outcome for those whose treatment status is not affected by assignment (exclusion restriction). The effect on compliers is usually interpreted as the effect of receipt of the treatment, under the implicit assumption that the effect of assignment is solely due to actual treatment receipt.

The stratification of units in subpopulations characterized by their potential compliance is a special case of Principal Stratification (PS, [4]), which is a more general framework to deal with *adjusting* for post-treatment variables. Principal strata are a latent classification of units and are defined by the joint values of one (or more) post-treatment variable under all possible levels of treatment assignment.

Identification of causal effects for particular strata is usually possible only under some assumptions aimed at reducing the number of strata (such as monotonicity assumptions) or at imposing certain features of the distribution of outcomes within or among strata (such as exclusion restrictions). Because the observed distributions are mixtures of distributions associated with the latent strata, these assumptions essentially allow one to uniquely disentangle these mixtures [6].

Depending on the empirical setting, some of these substantive assumptions may be questionable. For example, the exclusion restriction appears plausible in blind or double-blind experiments, but less so in open-label ones.

These structural assumptions can be relaxed by using a fully Bayesian model, thereby avoiding the need for full identification. Imbens and Rubin [7] extended the Bayesian approach to causal inference of Rubin [10] to handle simple randomized experiments with noncompliance. Hirano et al. [5] further extended the approach to handle fully observed covariates and considering the consequences of violations of exclusion restrictions by the comparison of results based on different combinations of these assumptions. Without these assumptions, inference, although straightforward in the Bayesian approach, can be imprecise even in large samples. Models may result as *weakly identified*: they have a proper posterior distribution, but do not have unique maximum likelihood estimates.

Adding information may be pursued as a general strategy for identification purposes. Covariates, for example, generally improve the precision of causal estimates (i.e., reduce posterior variability) because they improve the prediction of missing potential outcomes. In addition they can be used to achieve full identification when some of the usually invoked assumptions are relaxed. For example, Jo [8] and Frangakis [3] relax the exclusion restriction, and show how identification can be achieved by imposing some plausible behavioral hypotheses within or among groups defined by the values of the covariates, which translate into restrictions on coefficients (exclusion of some interaction terms or equality of coefficients across strata for some covariates).

Here, we investigate the role of information coming from additional outcome variables, which may help reducing the uncertainty about the treatment effects on the outcome of primary interest. We concentrate on the simple case of a binary treatment assignment, to which the units comply or do not comply (the compliance behavior is assumed to be all or none), and a bivariate continuous outcome. More general settings with multivalued treatment assignments are conceptually straightforward, although they may introduce substantial complications to inference. Also, although in principle more than one additional outcome could be used in order to reduce uncertainty about the treatment effects on the outcome of primary interest, the additional information coming from multiple outcomes would be gained at the cost of increasing model complexity and thus potential higher chances of model misspecification.

The framework we adopt uses potential outcomes to define causal effects regardless of the mode of inference. We pursue a Bayesian approach and analyze the role of jointly modeling more than one outcome; the improved classification of units eases the identification of the mixture components, thereby reducing the variability of the posterior distribution of causal estimands. It is also worth noting that, in a Bayesian setting, the effect of relaxing or maintaining assumptions can be directly addressed by examining how the posterior distributions for causal estimands change, therefore serving as a sort of sensitivity analysis.

In what follows we explain the structure of Bayesian inference in the presence of noncompliance and briefly describe methods for posterior inference based on data augmentation and MCMC algorithm (Sect. 22.2). Then we present the artificial data set we use and show some simulation results (Sect. 22.3); a brief discussion is provided in Sect. 22.4.

22.2 Bayesian Inference for Causal Estimands with Noncompliance

As a motivating example, suppose to design a social experiment to assess the efficacy of a language program for immigrants. The program is randomly offered to a sample of eligible secondary school students. Compliance is not perfect because only a subset of those students assigned to treatment (i.e., offered to participate in the program) attends the classes; access to the program is denied to those assigned to control. The primary outcome of interest is language ability measured with the score achieved in a written test; in addition to the test result, also the time to complete the test is recorded and can be considered as a secondary outcome.

Let us now introduce the potential outcome notation. Z_i is a binary treatment assignment for unit i ($Z_i = 0$ indicates the unit is assigned to the control group, whereas $Z_i = 1$ indicates the unit is assigned to the treatment group). $D_i(z)$ is the indicator of the treatment unit i would receive if assigned to treatment z ($z = 0, 1$), so that $D_i(Z_i)$ is the actual treatment received. Throughout this analysis we will make the stability assumption that there is neither interference between units nor different versions of the treatment (SUTVA; [10]). $\mathbf{Y}_i(z) = [Y_{i1}(z), Y_{i2}(z)]'$ represents a bivariate (continuous) outcome for unit i if assigned to treatment z.

G_i is an indicator (stratum membership), which partitions units in the target population into four types based on their compliance behavior: c (complier if $D_i(z) = z$ for $z = 0, 1$), n (never-takers if $D_i(z) = 0$ for $z = 0, 1$), a (always-takers if $D_i(z) = 1$ for $z = 0, 1$), d (defiers if $D_i(z) = 1 - z$ for $z = 0, 1$). In our setting a strong monotonicity of compliance assumption holds by design ($D_i(0) = 0$ for all i), which rules out the presence of defiers and always takers. This implies that the population is only composed of compliers (c) and never-takers (n). In settings with two-sided noncompliance, where there may exist also always-takers and defiers, the monotonicity assumption is not satisfied by construction anymore, but it becomes a substantive assumption that needs not always be satisfied. However, under the monotonicity (no-defier) assumption, inference is not more complicated because the likelihood structure does not substantially change.

We focus on defining and estimating two estimands: the ITT effects on the first outcome, language ability test score (Y_1), for compliers and never-takers. The ITT effect for the subpopulation of type g is defined as: $\text{ITT}_g = E[Y_{i1}(1) - Y_{i1}(0) | G_i = g]$, $g = c, n$. Note that under the exclusion restriction assumption for never-takers $\text{ITT}_n = 0$.

The observed data are: Z_i^{obs}, $D_i^{\text{obs}} = D_i(Z_i^{\text{obs}})$,$\mathbf{Y}_i^{\text{obs}} = \mathbf{Y}_i(Z_i^{\text{obs}})$, whereas the following quantities are missing: $D_i(1 - Z_i^{\text{obs}})$, $\mathbf{Y}_i(1 - Z_i^{\text{obs}})$. We observe the compliance behavior only partially, through the response to the actual assignment, D_i^{obs}. Because the type of a unit is a function of both compliance under assignment to the treatment and compliance under assignment to control, which we can never jointly observe, we generally do not know the compliance type.

Following Imbens and Rubin [7], we model the distribution of the stratum membership G_i and the conditional distribution of potential outcomes given the compliance type. Both distributions are parameterized so that, conditional on a general parameter θ, the model has an independent and identical distribution (i.i.d.) structure.

For simplicity and specificity, here we focus on commonly used outcome distributions and use conventional prior distributions, although our approach is widely applicable, also using more complex models. Specifically, we assume that language ability test scores and time to complete the test follow a bivariate normal distribution conditional on the stratum membership and treatment assignment:

$$f(\mathbf{Y}_i(Z_i)|G_i = g, Z_i = z) = \phi_i[\mu_1^{g,z}, \mu_2^{g,z}, \boldsymbol{\Sigma}^{g,z}]$$

$$\equiv \frac{1}{2\pi}\det(\boldsymbol{\Sigma}^{g,z})^{0.5}\exp\left\{-\frac{1}{2}\left(\mathbf{Y}_i(Z_i) - \begin{bmatrix}\mu_1^{g,z}\\ \mu_2^{g,z}\end{bmatrix}\right)'(\boldsymbol{\Sigma}^{g,z})^{-1}\left(\mathbf{Y}_i(Z_i) - \begin{bmatrix}\mu_1^{g,z}\\ \mu_2^{g,z}\end{bmatrix}\right)\right\},$$

$g = c, n$ and $z = 0, 1$. We use a Bernoulli distribution for the strata membership:

$$P[G_i = c] = \pi_c, \qquad P[G_i = n] = (1 - \pi_c).$$

The full parameter vector $\theta = \{\boldsymbol{\mu}^{c,0}, \boldsymbol{\mu}^{c,1}\boldsymbol{\mu}^{n,0}, \boldsymbol{\mu}^{n,1}, \boldsymbol{\Sigma}^{c,0}, \boldsymbol{\Sigma}^{c,1}, \boldsymbol{\Sigma}^{n,0}, \boldsymbol{\Sigma}^{n,1}; \pi_c\}$, containing a total of 21 parameters, characterizes the complete Likelihood function:

$$\begin{aligned} &L_{\text{comp}}(\theta|Z^{\text{obs}}, D^{\text{obs}}, \mathbf{Y}^{\text{obs}}, G) \qquad (22.1)\\ &= \prod_{i:G_i=c} \pi_c\phi_i[\boldsymbol{\mu}^{c,Z_i}, \boldsymbol{\Sigma}^{c,Z_i}] \prod_{i:G_i=n} (1-\pi_c)\phi_i[\boldsymbol{\mu}^{n,Z_i}, \boldsymbol{\Sigma}^{n,Z_i}]. \end{aligned}$$

We assume that parameters are a priori independent and use conjugate prior distributions:

$$\pi_c \sim \text{Beta}(\alpha_0, \beta_0),$$

$$\boldsymbol{\Sigma}^{g,z} \sim \text{Inv} - \text{Wishart}_{\nu_0^{g,z}}((\boldsymbol{\Lambda}_0^{g,z})^{-1}) \qquad \boldsymbol{\mu}^{g,z}|\boldsymbol{\Sigma}^{g,z} \sim N(\boldsymbol{\mu}_0^{g,z}, \boldsymbol{\Sigma}^{g,z}/k_0^{g,z}).$$

Alternative (less conventional) prior distributions could be also considered, by relaxing the conjugacy assumption and/or imposing a dependence structure in the prior distribution, using, for example, a hierarchical model. However, we expect results to point to a substantial improvement irrespective of the prior distributions.

We construct a general state-space Markov Chain that has the joint distribution of the model parameters θ and the missing membership indicators G as its unique invariant equilibrium distribution. The Markov Chain algorithm that we adopt uses the data augmentation (DA) method of Tanner and Wong [11] to impute at each step the missing membership indicators and to exploit the complete likelihood function to update the parameter distribution. At step $j + 1$ we first draw G according to $P(G|\theta^{(j)}; Z^{\text{obs}}, D^{\text{obs}}, \mathbf{Y}^{\text{obs}})$. Given our prior assumptions this conditional distribution has a simple form. Conditional on θ, Z^{obs}, D^{obs} and $\mathbf{Y}^{\text{obs}}$, the G_i are independent of $G_{i'}$, $Z_{i'}^{\text{obs}}$, $D_{i'}^{\text{obs}}$ and $\mathbf{Y}_{i'}^{\text{obs}}$ for all $i' \neq i$. By the strong monotonicity assumption $P(G_i = n|Z_i^{\text{obs}} = 1, D_i^{\text{obs}} = 0, \mathbf{Y}_i^{\text{obs}}) = 1$ and $P(G_i = c|Z_i^{\text{obs}} = 1, D_i^{\text{obs}} = 1, \mathbf{Y}_i^{\text{obs}}) = 1$. For observations with $Z_i = 0$ the following holds:

$$P(G_i = c|Z_i^{\text{obs}} = 0, D_i^{\text{obs}} = 0, \mathbf{Y}_i^{\text{obs}})$$
$$= \frac{\pi_c^{(j)} \phi_i [\boldsymbol{\mu}^{c0(j)}, \boldsymbol{\Sigma}^{c0(j)}]}{\pi_c^{(j)} \phi_i [\boldsymbol{\mu}^{c0(j)}, \boldsymbol{\Sigma}^{c0(j)}] + (1 - \pi_c^{(j)}) \phi_i [\boldsymbol{\mu}^{n0(j)}, \boldsymbol{\Sigma}^{n0(j)}]},$$

$$P(G_i = n|Z_i^{\text{obs}} = 0, D_i^{\text{obs}} = 0, \mathbf{Y}_i^{\text{obs}}) = 1 - P(G_i = c|Z_i^{\text{obs}} = 0, D_i^{\text{obs}} = 0, \mathbf{Y}_i^{\text{obs}}).$$

Once the missing G_i are drawn, the prior distribution of the parameters is updated using the complete likelihood function in (22.1). Specifically, due to the assumptions of a priori independence of the parameters and conjugacy of priors, the complete data posterior distribution of θ is the product of the following marginal posterior distributions:

$$\pi_c|G, Z^{\text{obs}}, D^{\text{obs}}, \mathbf{Y}^{\text{obs}} \sim \text{Beta}(\alpha_0 + \sum I(G_i = c); \beta_0 + \sum I(G_i = n)),$$
$$\boldsymbol{\Sigma}^{g,z}|G, Z^{\text{obs}}, D^{\text{obs}}, \mathbf{Y}^{\text{obs}} \sim \text{Inv} - \text{Wishart}_{\nu^{g,z}}((\boldsymbol{\Lambda}^{g,z})^{-1}),$$

where $I(\cdot)$ is the indicator function, and

$$\nu^{g,z} = \nu_0^{g,z} + N^{g,z}, \quad \text{where } N^{g,z} = \sum I(G_i = n) I(Z_i = z),$$
$$\Lambda^{g,z} = \boldsymbol{\Lambda}_0^{g,z} + S^{g,z} + \frac{k_0^{g,z} N^{g,z}}{k_0^{g,z} + N^{g,z}} (\bar{\mathbf{Y}}^{g,z} - \boldsymbol{\mu}_0^{g,z})(\bar{\mathbf{Y}}^{g,z} - \boldsymbol{\mu}_0^{g,z})^{\text{T}},$$

with $S^{g,z} = \sum_{i:G_i=c,Z_i=z} (\mathbf{Y}_i^{g,z} - \bar{\mathbf{Y}}^{g,z})(\mathbf{Y}_i^{g,z} - \bar{\mathbf{Y}}^{g,z})^{\text{T}}$, and

$$\boldsymbol{\mu}^{g,z}|\boldsymbol{\Sigma}^{g,z}; G, Z^{\text{obs}}, D^{\text{obs}}, \mathbf{Y}^{\text{obs}} \sim N(\boldsymbol{\mu}^{g,z}, \boldsymbol{\Sigma}^{g,z}/k^{g,z}),$$

where $k^{g,z} = k_0^{g,z} + N^{g,z}$ and $\boldsymbol{\mu}^{g,z} = \frac{k_0}{k_0 + N^{g,z}} \boldsymbol{\mu}_0^{g,z} + \frac{N^{g,z}}{k_0 + N^{g,z}} \bar{\mathbf{Y}}^{g,z}$. At the jth step of the MCMC algorithm the parameters vector $\theta^{(j)}$ are easily drawn from these complete data posteriors and then used for the DA step.

22.3 An Application to Artificial Data

Simulated data sets have been used to compare posterior inferences obtained by jointly modeling the two outcomes with those obtained working with only the first outcome.[1] Different scenarios have been considered, to take account of

[1]Consistently with our motivating example, the two outcomes are supposed to be language ability measured by the score achieved in a written test (from 0 to 100) and the time to complete the test (in minutes).

Table 22.1 Parameters' values used to generate the six data sets

	$\mu^{c,0}$	$\mu^{c,1}$	$\mu^{n,0}$	$\mu^{n,1}$	$\Sigma^{c,0}$	$\Sigma^{c,1}$	$\Sigma^{n,0}$	$\Sigma^{n,1}$
I	$\begin{bmatrix}50\\30\end{bmatrix}$	$\begin{bmatrix}70\\20\end{bmatrix}$	$\begin{bmatrix}40\\50\end{bmatrix}$	$\begin{bmatrix}30\\60\end{bmatrix}$	$\begin{bmatrix}25 & -12\\-12 & 9\end{bmatrix}$	$\begin{bmatrix}16 & -7\\-7 & 4\end{bmatrix}$	$\begin{bmatrix}36.00 & -6.75\\-6.75 & 20.25\end{bmatrix}$	$\begin{bmatrix}42.25 & -7.15\\-7.15 & 30.25\end{bmatrix}$
II	$\begin{bmatrix}50\\30\end{bmatrix}$	$\begin{bmatrix}70\\20\end{bmatrix}$	$\begin{bmatrix}40\\50\end{bmatrix}$	$\begin{bmatrix}30\\60\end{bmatrix}$	$\begin{bmatrix}25 & -12\\-12 & 9\end{bmatrix}$	$\begin{bmatrix}16 & -7\\-7 & 4\end{bmatrix}$	$\begin{bmatrix}36.00 & -21.60\\-21.60 & 20.25\end{bmatrix}$	$\begin{bmatrix}42.25 & -28.60\\-28.60 & 30.25\end{bmatrix}$
III	$\begin{bmatrix}50\\30\end{bmatrix}$	$\begin{bmatrix}70\\20\end{bmatrix}$	$\begin{bmatrix}48\\50\end{bmatrix}$	$\begin{bmatrix}30\\60\end{bmatrix}$	$\begin{bmatrix}25 & -12\\-12 & 9\end{bmatrix}$	$\begin{bmatrix}16 & -7\\-7 & 4\end{bmatrix}$	$\begin{bmatrix}36.00 & -6.75\\-6.75 & 20.25\end{bmatrix}$	$\begin{bmatrix}42.25 & -7.15\\-7.15 & 30.25\end{bmatrix}$
IV	$\begin{bmatrix}50\\30\end{bmatrix}$	$\begin{bmatrix}70\\20\end{bmatrix}$	$\begin{bmatrix}48\\50\end{bmatrix}$	$\begin{bmatrix}30\\60\end{bmatrix}$	$\begin{bmatrix}25 & -12\\-12 & 9\end{bmatrix}$	$\begin{bmatrix}16 & -7\\-7 & 4\end{bmatrix}$	$\begin{bmatrix}36.00 & -21.60\\-21.60 & 20.25\end{bmatrix}$	$\begin{bmatrix}42.25 & -28.60\\-28.60 & 30.25\end{bmatrix}$
V	$\begin{bmatrix}50\\30\end{bmatrix}$	$\begin{bmatrix}70\\20\end{bmatrix}$	$\begin{bmatrix}48\\50\end{bmatrix}$	$\begin{bmatrix}45\\60\end{bmatrix}$	$\begin{bmatrix}25 & -12\\-12 & 9\end{bmatrix}$	$\begin{bmatrix}16 & -7\\-7 & 4\end{bmatrix}$	$\begin{bmatrix}36.00 & -6.75\\-6.75 & 20.25\end{bmatrix}$	$\begin{bmatrix}42.25 & -7.15\\-7.15 & 30.25\end{bmatrix}$
VI	$\begin{bmatrix}50\\30\end{bmatrix}$	$\begin{bmatrix}70\\20\end{bmatrix}$	$\begin{bmatrix}48\\50\end{bmatrix}$	$\begin{bmatrix}45\\60\end{bmatrix}$	$\begin{bmatrix}25 & -12\\-12 & 9\end{bmatrix}$	$\begin{bmatrix}16 & -7\\-7 & 4\end{bmatrix}$	$\begin{bmatrix}36.00 & -21.60\\-21.60 & 20.25\end{bmatrix}$	$\begin{bmatrix}42.25 & -28.60\\-28.60 & 30.25\end{bmatrix}$

alternative correlation structures between outcomes for compliers and never-takers and alternative deviations from the exclusion restriction.

We present results for six data sets each containing 600 observations, generated using principal strata probabilities of 0.6 for compliers and 0.4 for never-takers, and the parameters' values for the bivariate outcome reported in Table 22.1. The simulated sample is then randomly divided into two groups, with the first assigned to treatment, and the second assigned to control.

As we can see in Table 22.1, the parameters of the outcome sub-models for compliers are set so that the true ITT effect on language ability score test for compliers is equal to 20 points, and language ability score test and time to complete the test are highly negatively correlated both under treatment and under control.

For never-takers, we consider six alternative bivariate normal distributions of language ability score test and time to complete the test. Specifically, we use three different pairs of mean vectors, which lead to ITT effects on the primary outcome for never-takers equal to $-3, -10$, and -18. These values reflect weak, strong, and very strong deviations from the exclusion restriction assumption, respectively. Each pair of mean vectors is combined with two alternative correlation structures between outcomes for never-takers: the first one implies that the correlation between outcomes is much lower for never-takers than for compliers under both the active treatment and the control treatment; the second one implies that compliers and never-takers are characterized by similar correlation structures between outcomes under both treatment arms.

We specify relatively flat prior distributions by setting the hyperparameters equal to:

$$\alpha_0 = 1, \beta_0 = 1, \nu_0^{g,z} = 5, \boldsymbol{\Lambda}_0^{g,z} = \begin{bmatrix}30 & 0\\0 & 15\end{bmatrix}, \boldsymbol{\mu}_0^{g,z} = \begin{bmatrix}50\\30\end{bmatrix},$$

$$k_0^{g,z} = 1, \; g = c, n; z = 0, 1.$$

Table 22.2 Summaries of the posterior distributions of ITT effect for compliers, c, and never-takers, n, in the univariate case (Table on the left) and in the bivariate case (Table on the right)

Univariate case

		Mean	SD	2.5 %	50 %	97.5 %
I	c	23.45	3.23	19.26	21.84	28.24
	n	−15.24	4.75	−21.92	−13.07	−8.80
II	c	22.91	3.38	18.47	21.96	27.76
	n	−15.58	5.02	−22.30	−14.52	−8.58
III	c	19.99	1.77	16.83	20.11	22.99
	n	−19.11	2.62	−23.41	−19.34	−14.37
IV	c	21.05	1.18	18.74	21.04	23.37
	n	−18.40	1.75	−21.83	−18.41	−14.93
V	c	20.88	1.53	18.05	20.91	23.63
	n	−4.12	2.29	−8.18	−4.18	0.13
VI	c	20.27	1.40	17.66	20.29	22.82
	n	−4.15	2.09	−7.91	−4.20	−0.22

Bivariate case

		Mean	SD	2.5 %	50 %	97.5 %
I	c	19.75	0.49	18.79	19.75	20.71
	n	−9.77	0.83	−11.39	−9.77	−8.16
II	c	19.17	0.50	18.18	19.17	20.15
	n	−10.00	0.85	−11.66	−10.00	−8.34
III	c	19.38	0.52	18.37	19.38	20.40
	n	−18.21	0.79	−19.76	−18.21	−16.65
IV	c	19.89	0.49	18.93	19.89	20.86
	n	−16.68	0.80	−18.26	−16.68	−15.11
V	c	20.20	0.49	19.24	20.20	21.15
	n	−3.10	0.80	−4.67	−3.10	−1.54
VI	c	19.91	0.47	18.98	19.91	20.83
	n	−3.61	0.78	−5.14	−3.61	−2.08

For each of the six generated data sets, we fitted the model running three chains from different starting values. Each chain was run for $200,000$ iterations, after a burn-in stage of 10,000 iterations. Inference is based on $600,000$ iterations, combining the three chains.

Results generally confirm our prior intuition that the simultaneous modeling of more than one outcome may improve estimation by reducing posterior uncertainty for causal estimands. Table 22.2 reports summaries of the posterior distributions of ITT effects for compliers and never-takers on the first outcome, in the univariate and bivariate case respectively. Figure 22.1 shows the histograms and 95 % posterior intervals of these posterior distributions.

The bivariate approach outperforms the univariate one in any scenario we consider, by providing considerably more precise estimates of the ITT effects for compliers and never-takers. The benefits of the bivariate approach especially arise when the mixture components of the observed likelihood are weakly overlapping, irrespective of the correlation structures between outcomes for compliers and never-takers (scenarios I and II). In such a case, jointly modeling the two outcomes reduces the width of the 95 % posterior credible intervals more than 75 %. In addition, the histograms (I) and (II) in Fig. 22.1 show that the posterior distributions of the ITT effects for compliers and never-takers are bimodal in the univariate case, but they both become unimodal in the bivariate case. This result suggests that jointly modeling the two outcomes simplifies inference; for instance the bivariate approach makes easier to choose a point estimate for the causal effects of interest.

In the overlapping scenarios, noticeable efficiency gains over the univariate models are observed when compliers and never-takers are characterized by different correlation structures (scenario III), and when deviations from the exclusion restriction assumption are smaller (scenarios V and VI). First consider scenarios III and IV, where the ITT effect for never-takers is equal to -18, implying that

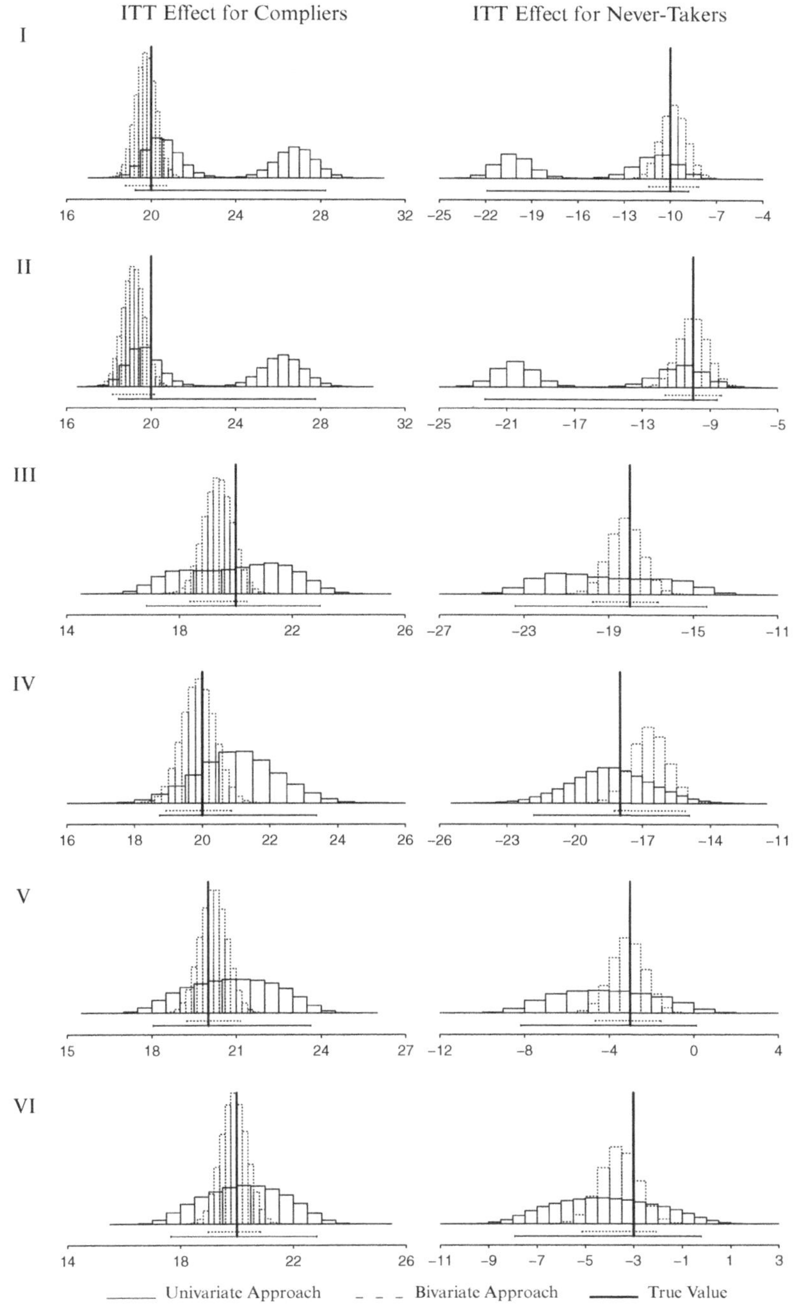

Fig. 22.1 Univariate versus bivariate approach: histograms and 95 % posterior intervals of ITT effects for compliers and never-takers

the exclusion restriction assumption is strongly unsatisfied. When compliers and never-takers are characterized by different correlation structures (scenario III), the bivariate approach clearly outperforms the univariate one leading to much less variable and more informative posterior distributions of the causal effects of interest. Actually, in the univariate approach the posterior distributions are somewhat flat (see Fig. 22.1(III)). Conversely, when compliers and never-takers are characterized by similar correlation structures (scenario IV), the univariate and bivariate approaches lead to similar results, although jointly modeling the two outcome increases the precision of the estimates. In the last two scenarios (scenarios V and VI), where deviations from the exclusion restriction assumption are smaller, jointly modeling the two outcomes reduces the posterior variability of the causal estimands, irrespective of the correlation structures between outcomes for compliers and never-takers. Moreover, the histograms (V) and (VI) in Fig. 22.1 suggest that the bivariate approach allows one to easier identify a posterior mode, which is hard to choose in the univariate case where the posterior distributions are characterized by large regions with approximately the same likelihood values.

22.4 Discussion

Although the illustration provided is necessarily limited, the approach we propose appears to be widely applicable, also in observational studies, whenever the exclusion restriction assumption for the instrument may not hold. The additional information provided by the second outcome is obtained at the cost of having to specify more complex models for the multivariate outcome, which may increase the chances of misspecification. Therefore, directions for future research include the development of specific tools for assessing model fit, such as posterior predictive checks [9] and model comparisons by marginal likelihoods and Bayes factors [2], as well as the specification of Bayesian semiparametric models using, e.g., Dirichlet processes.

Analogously to covariates, our current effort is directed to showing how imposing specific dependence structures between outcomes may allow us to achieve identification of causal effects.

References

1. Angrist, J.D., Imbens, G.W., Rubin, D.B.: Identification of causal effects using instrumental variables (with discussion). J. Am. Stat. Assoc. **91**, 444–472 (1996)
2. Chib, S., Jacobi, L.: Causal effects from panel data in randomized experiments with partial compliance. In: Chib, S., Griffiths, W.E., Koop, G. (eds.) Advances in Econometrics, vol. 23, pp. 183–215. Jai Press, Bingley (2008)
3. Frangakis, C.E.: Comment on Causal effects in the presence of non compliance: a latent variable interpretation, by A. Forcina. Metron, LXIV, pp. 8–14 (2006)
4. Frangakis, C.E., Rubin, D.B.: Principal stratification in causal inference. Biometrics **58**, 191–199 (2002)

5. Hirano, K., Imbens, G.W., Rubin, D.B., Zhou, X-H.: Assessing the effect of an influenza vaccine in an encouragement design. Biostatistics **1**(1), 69–88 (2000)
6. Imbens, G.W., Rubin, D.B.: Estimating outcome distributions for compliers in instrumental variables models. Rev. Econ. Stud. **64**, 555–574 (1997a)
7. Imbens, G.W., Rubin, D.B.: Bayesian inference for causal effects in randomized experiments with noncompliance. Ann. Stat. **25**, 305–327 (1997b)
8. Jo, B.: Estimation of intervention effects with noncompliance: alternative model specifications. J. Educ. Behav. Stat. **27**, 385–420 (2002)
9. Mattei, A., Mealli, F.: Application of the principal stratification approach to the Faenza randomized experiment on breast self-examination. Biometrics **63**, 437–446 (2007)
10. Rubin, D.B.: Bayesian inference for causal effects: the role of randomization. Ann. Stat. **6**, 34–58 (1978)
11. Tanner, M., Wong, W.: The calculation of posterior distributions by data augmentation (with discussion). J. Am. Stat. Assoc. **82**, 528–550 (1987)

Fuzzy Composite Indicators: An Application for Measuring Customer Satisfaction 23

Sergio Zani, Maria Adele Milioli, and Isabella Morlini

Abstract

Composite indicators should ideally measure multidimensional concepts which cannot be captured by a single variable. In this chapter, we suggest a method based on fuzzy set theory for the construction of a fuzzy synthetic index of a latent phenomenon (e.g., well-being, quality of life, etc.), using a set of manifest variables measured on different scales (quantitative, ordinal and binary). A few criteria for assigning values to the membership function are discussed, as well as criteria for defining the weights of the variables. For ordinal variables, we propose a fuzzy quantification method based on the sampling cumulative function and a weighting system which takes into account the relative frequency of each category. An application regarding the results of a survey on the users of a contact center is presented.

Keywords

Imprecise data and fuzzy methods • Membership function • Overall satisfaction • Quantification of ordinal variables • Weighting criteria

23.1 Introduction

The purpose of composite indicators is to measure multidimensional concepts which cannot be captured by a single variable. They are usually formed on the basis of a set of quantitative variables [11, 16]. In order to construct a composite indicator, in this chapter we suggest a method based on the fuzzy set theory [9, 17, 18, 21]. As shown in the literature, fuzzy sets permit the representation of vague concepts, e.g., poverty [5, 15], well-being [2, 7], quality of life [14], business scenarios [4],

S. Zani (✉) · M.A. Milioli · I. Morlini
Department of Economics, University of Parma, Italy
e-mail: sergio.zani@unipr.it; milioli@unipr.it; isabella.morlini@unimore.it

N. Torelli et al. (eds.), *Advances in Theoretical and Applied Statistics*, Studies in Theoretical and Applied Statistics, DOI 10.1007/978-3-642-35588-2_23,

satisfaction indices [20], etc. In Sect. 23.2, we deal with the general problem of obtaining a synthetic fuzzy measure of a latent phenomenon from a set of manifest variables measured on different scales (quantitative, ordinal, and binary). We present a few criteria to transform the values of a quantitative variable into fuzzy numbers. For ordinal variables we propose a fuzzy quantification method based on the cumulative function, and for a set of binary variables we consider the relative frequency of the symptoms of the underlying concept. In Sect. 23.3, we discuss the problem of weighting the variables and aggregating them in a composite indicator. The weights should reflect the contribution of each variable to the latent phenomenon. For ordinal variables, we suggest a new criterion taking into account the relative frequency of each category. In Sect. 23.4, we focus on the specific application of measuring customer satisfaction using ordinal (and binary) variables. The gradual transition from very dissatisfied to really satisfied customers can be captured by a fuzzy index. We apply the suggested methods to a sample of 704 respondents of a survey on the users of a contact center, in order to evaluate their satisfaction with the service. The fuzzy indicator of customer satisfaction allows us to obtain a ranking of respondents that can be compared with the traditional ones. Furthermore, the comparison between the overall satisfaction scores directly stated by the respondents and the values of the fuzzy composite indicator based on several items shows noncoherent answers for a few customers (high stated satisfaction but low value of the synthetic index, or viceversa). These units may be considered as atypical observations.

23.2 Fuzzy Transformations of the Variables

Let X be a set of elements $x \in X$. A fuzzy subset A of X is a set of ordered pairs:

$$[x, \mu_A(x)] \qquad \forall x \in X \tag{23.1}$$

where $\mu_A(x)$ is the membership function (m.f.) of x to A in the closed interval [0,1]. If $\mu_A(x) = 0$, then x does not belong to A, while if $\mu_A(x) = 1$ then x completely belongs to A. If $0 < \mu_A(x) < 1$, then x partially belongs to A and its membership to A increases according to the values of $\mu_A(x)$. Let us assume that the subset A defines the position of each element with reference to achievement of the latent concept, e.g., the well-being of a set of countries or the satisfaction of a sample of customers. In this case, $\mu_A(x) = 1$ identifies a situation of full achievement of the target (a country enjoying global well-being or a completely satisfied customer), whereas $\mu_A(x) = 0$ denotes a total failure (a country with no well-being or a very dissatisfied customer).

Consider a set of n units or elements e_i $(i = 1, 2, \ldots, n)$ and p manifest variables X_s $(s = 1, 2, \ldots, p)$ reflecting the latent phenomenon. Without loss of generality, let us assume that each variable is positively related to that phenomenon, i.e., it satisfies the property "the larger the better." If a quantitative variable X_s shows negative correlation, we substitute it with a simple decreasing function transformation, e.g.,

$f(x_{si}) = \max(x_{si}) - x_{si}$. In case of an ordinal variable, we consider it in reverse order. In order to define the m.f. for each variable it is necessary:

1. To identify the extreme situation such that $\mu_A(x) = 0$ (non-membership) and $\mu_A(x) = 1$ (full membership).
2. To define a criterion for assigning m.f. values to the intermediate categories of the variable.

Let us assume that X_s is a quantitative variable; for simplicity of notation we omit index s. For that variable X, we choose an inferior (lower) threshold l and a superior (upper) threshold u, with l and u finite, and we define the m.f. as follows:

$$\begin{cases} \mu_A(x_i) = 0, & x_i \le l \\ \mu_A(x_i) = \frac{x_i - l}{u - l}, & l < x_i < u \\ \mu_A(x_i) = 1, & x_i \ge u \end{cases} \tag{23.2}$$

In (23.2) m.f. is a linear function between the values of the two thresholds. Alternatively, we can arrange the values x_i in nondecreasing order according to i and define the following m.f.:

$$\begin{cases} \mu_A(x_i) = 0, & x_i \le l \\ \mu_A(x_i) = \mu_A(x_{i-1}) + \frac{F(x_i) - F(x_{i-1})}{1 - F[x_{i(l)}]}, & l < x_i < u \\ \mu_A(x_i) = 1, & x_i \ge u \end{cases} \tag{23.3}$$

where $F(x_{si})$ is the sampling cumulative function of the variable X and $x_{i(l)}$ is the highest value $x_i \le l$. If $l = x_1 = \min(x_i)$ and $u = x_n = \max(x_i)$, formula (23.3) corresponds to the "totally fuzzy and relative approach" suggested by Cheli and Lemmi [6]. In the literature, other specifications have been considered. If we are trying to measure the degree of achievement of a certain target, the distance $d(x)$ between the value x and the goal is an indicator of the success in achieving the target. If $d(x) = 0$, there is full membership to A, then $\mu_A(x) = 1$. If $d(x) > 0$ then $\mu_A(x) < 1$. Hence, we can write:

$$\mu_A(x) = \frac{1}{1 + d(x)} \tag{23.4}$$

In general, as highlighted by Zimmermann [21], the relationship between physical measures and perception takes an exponential form. The distance $d(x)$ can be expressed as:

$$d(x) = \mathrm{e}^{-a(x-b)} \tag{23.5}$$

so that m.f. is defined as follows:

$$\mu_A(x) = \frac{1}{1 + \mathrm{e}^{-a(x-b)}} \tag{23.6}$$

Table 23.1 Membership function to the subset of satisfied customers with reference to variable X_8

Scores (x_i)	n_i	$F(x_i)$	m.f.(x_i)
≤3	7	0.010	0
4	5	0.017	0
5	12	0.034	0
6	40	0.091	0.059
7	161	0.320	0.296
8	240	0.732	0.723
9	93	0.864	1
10	96	1	1
	704		

It is worth noting that the parameter a represents the extent of vagueness and the parameter b may be viewed as the point in which the tendency of the subject's attitude changes from rather positive to rather negative. Baliamoune-Lutz [1] uses m.f. (23.6) to measure human well-being with a fuzzy approach. If variable X_s is ordinal with k categories, a suitable choice is the following: the m.f. values of the categories up to the threshold l are equal to 0 (absence of the phenomenon) and those of the categories $\geq u$ are equal to 1 (complete presence). The intermediate values of m.f. are defined according to the formula (23.3). For example, consider the scores (on a scale 1–10) of the variable "Overall satisfaction," described in Sect. 23.4. Choosing $l = 5$ and $u = 9$, we obtain the m.f. values indicated in the last column of Table 23.1. A customer with a score equal to 5 does not belong to the set of satisfied respondents, while a customer with a score equal to 8 has a value 0.723 of the m.f. to the set of satisfied users. If variable X_s is binary, one of the categories can be considered as a symptom of the latent concept. Therefore the m.f. is a crisp function assuming only values equal to 0 (absence) and 1 (presence). Usually, we consider a set of q ($q \leq p$) binary variables reflecting several aspects of the phenomenon. In this case the m.f. can be defined as follows:

$$\mu_A(x_i) = \frac{1}{q}\sum_{s=1}^{q} z_{si} \tag{23.7}$$

where $z_{si} = 1$ if the corresponding x_{si} denotes presence of the symptom and $z_{si} = 0$ if the corresponding x_{si} denotes absence of the symptom. Definition (23.7) is consistent with interpreting membership values as the proportion of "subjects" who rate the i-th unit as an actual member of the fuzzy set A [5].

23.3 Weighting and Aggregation of the Variables

Among the steps of the construction of a crisp composite indicator, weighting and aggregation are the major ones which directly affect the quality and reliability of the results [16, 22]. For the aggregation functions in fuzzy set theory we refer, among others, to Calvo and Beliakov [3].

Let us consider the criteria for aggregating the p fuzzy variables, described in Sect. 23.2, into a fuzzy composite indicator. The simplest operations for the i-th unit are:

$$\text{intersection}: \quad {}_I\mu_A(i) = \min[\mu_A(x_{1i}), \mu_A(x_{2i}), \ldots, \mu_A(x_{pi})] \tag{23.8}$$

$$\text{union}: \quad {}_U\mu_A(i) = \max[\mu_A(x_{1i}), \mu_A(x_{2i}), \ldots, \mu_A(x_{pi})] \tag{23.9}$$

A more general aggregation function is the weighted generalized means [12]:

$$\mu_A(i) = \sum_{s=1}^{p} [\mu_A(x_{si})]^{\alpha} \cdot {w_s}^{1/\alpha} \tag{23.10}$$

where $w_s > 0$ is the normalized weight that expresses the relative importance of the variables X_s; ($\sum_{s=1}^{p} w_s = 1$). For fixed arguments and weights, function (23.10) is monotonic increasing with α; if $\alpha \rightarrow -\infty$, then formula (23.10) becomes the intersection defined in (23.8); if $\alpha \rightarrow +\infty$, then (23.10) is equal to the union (23.9). For $\alpha \rightarrow 0$ formula (23.10) becomes the weighted geometric mean. In the following, for the sake of simplicity, we will consider $\alpha = 1$, that is the weighted arithmetic mean. The weighting criteria in (23.10) may be:

- Equal weights that imply a careful selection of the variables in order to assure a balance of the different aspects of the latent phenomenon.
- Factor loadings, obtained by principal components analysis (PCA) for quantitative variables or by nonlinear PCA for ordinal variables; this method of weighting is valid if the first component accounts for a high percentage of the total variance and the weights (loadings) of the variables are proportional to their correlation with the first component (factor) reflecting the underlying concept.
- Obtained from expert judgements, e.g., using focus groups.
- Determined by an Analytic Hierarchy Process [13].

We suggest a criterion for the determination of the weights, considering for each variable X_s the fuzzy proportion $g(X_s)$ of the achievement of the target:

$$g(X_s) = \frac{1}{n} \sum_{i=1}^{n} \mu(x_{si}) \tag{23.11}$$

If X_s is binary, formula (23.11) coincides with the crisp proportion and in general it may be seen as an index of the proportion of the units having (totally or partially) the latent phenomenon [6]. The normalized weights may be determined as an inverse function of $g(X_s)$, in order to give higher importance to the rare features in the n units. To avoid excessive weights to the variables with low value of $g(X_s)$ we choose [5]:

$$w_s = \ln\left[\frac{1}{g(X_s)}\right] / \sum_{s=1}^{p} \ln\left[\frac{1}{g(X_s)}\right] \tag{23.12}$$

Using (23.12), it is possible to attach to each variable a weight sensitive to the fuzzy membership.

23.4 A Fuzzy Indicator of Customer Satisfaction

Customer satisfaction may be defined as the degree of happiness that a customer experiences with a product or a service and is a personal function of the gap between expected and perceived quality. The expected degree of happiness is a random vague concept. This latent phenomenon is here considered for all individuals and analyzed with fuzzy method [8, 10]. We apply the methods of the previous sections to the results of a survey on customer satisfaction of the users of the Contact Center of the Municipality of Parma, considering a sample of 704 respondents calling for information (see [19] for a complete description). The questions on the degree of satisfaction of the users are:

X_1 = Contact at the first call (no = 0, yes = 1).
X_2 = Waiting time (too long = 1, normal = 2, fairly short = 3).
X_3 = Courtesy of the operator.
X_4 = Skill of the operator.
X_5 = Quality of the provided information.
X_6 = Speed of the information.
X_7 = Complete answer (no = 1, partially = 2, yes = 3).
X_8 = Overall satisfaction with the service, with scores from 1 to 10.

All the variables whose categories are not specified are measured on Likert scale (very dissatisfied = 1, dissatisfied = 2, neither satisfied nor dissatisfied = 3, satisfied = 4, very satisfied = 5). The cumulative function of the variables X_1 - X_7 and the corresponding m.f. according to formula (23.3) are presented in Table 23.2 and for variable X_8 in the previous Table 23.1. For the variables on Likert scale, the inferior threshold is "dissatisfied" and the superior threshold is "very satisfied." We consider the following weighting systems, with and without the variable X_8, which indicates overall satisfaction with the service (see Table 23.3):

1. Equal weights of the variables (W_1).
2. Normalized factor loadings obtained by standard (linear) PCA. The first PC accounts for 49.7 % of the total variance with X_8 and for 50.4 % without X_8 (W_2).
3. Normalized factor loadings applying PCA on τ rank correlation matrix. The first PC explains 46.67 % of the total variance with X_8 and 46.44 % without X_8 (W_3).
4. Normalized weights as inverse function of the fuzzy proportion of each variable, according to formula (23.12) (W_4).

The least important variable is always X_1 and could be deleted. The correlation coefficient between W_2 and W_3 is 0.964 considering X_8 and 0.970 without X_8, but the correlation coefficients of W_4 with the previous weights are in the interval [0.791, 0.909]. Therefore, the last weighting criterion is slightly different from the others.

Table 23.2 Cumulative function F of the categories of each item in the sample and corresponding membership function to the subset A of satisfied customers

X_1			X_2			X_3			X_4			V_5			X_6			X_7		
	F	$m.f.$		F	$m.f.$		F	$m.f.$		F	$m.f$		F	$m.f.$		F	$m.f.$		F	$m.f.$
0	0.08	0	*1*	0.03	0	*1*	0	0	*1*	0	0	*1*	0.01	0	*1*	0.01	0	*1*	0.05	0
1	1	1	*2*	0.29	0.27	*2*	0.01	0	*2*	0.03	0	*2*	0.04	0	*2*	0.04	0	*2*	0.13	0.08
			3	1	1	*3*	0.05	0.05	*3*	0.14	0.12	*3*	0.13	0.09	*3*	0.12	0.08	*3*	1	1
						4	0.46	0.45	*4*	0.60	0.59	*4*	0.57	0.55	*4*	0.58	0.56			
						5	1	1	*5*	1	1	*5*	1	1	*5*	1	1			

Table 23.3 Values of W_2, W_3 and W_4

	With X_8			Without X_8		
	W_2	W_3	W_4	W_2	W_3	W_4
X_1	6.97	6.93	3.59	7.99	7.83	4.38
X_2	9.21	8.84	10.57	10.62	10.05	12.88
X_3	12.29	13.67	13.65	14.17	15.64	16.63
X_4	14.76	15.50	16.18	16.95	17.71	19.71
X_5	15.61	16.08	16.43	17.97	18.25	20.02
X_6	17.73	16.02	16.22	18.13	18.32	19.76
X_7	12.27	10.67	5.43	14.17	12.20	6.62
X_8	13.16	12.29	17.93	–	–	–
Tot	100	100	100	100	100	100

Table 23.4 Frequency distribution of the fuzzy composite indicators with different weights

	With X_8				Without X_8			
Classes	W_1	W_2	W_3	W_4	W_1	W_2	W_3	W_4
0.0 ⊢ 0.1	4	7	7	8	5	7	7	7
0.1 ⊢ 0.2	8	12	12	12	7	12	12	14
0.2 ⊢ 0.3	14	13	16	20	15	15	15	18
0.3 ⊢ 0.4	36	36	37	39	30	34	33	35
0.4 ⊢ 0.5	30	31	33	26	35	28	29	30
0.5 ⊢ 0.6	49	49	46	120	26	27	37	81
0.6 ⊢ 0.7	109	158	151	103	91	205	195	142
0.7 ⊢ 0.8	134	96	106	82	161	68	68	69
0.8 ⊢ 0.9	105	77	66	99	122	69	69	66
0.9 ⊢ 1.0	129	136	144	109	42	69	69	72
1.0	86	86	86	86	170	170	170	170
Total	704	704	704	704	704	704	704	704

Table 23.4 shows the frequency distribution of the values of the fuzzy composite indicators with the mentioned weighting criteria. None of the respondents can be regarded as completely dissatisfied, since the values of the composite indicators are at least 0.02. Even clients experiencing dissatisfaction with most indicators

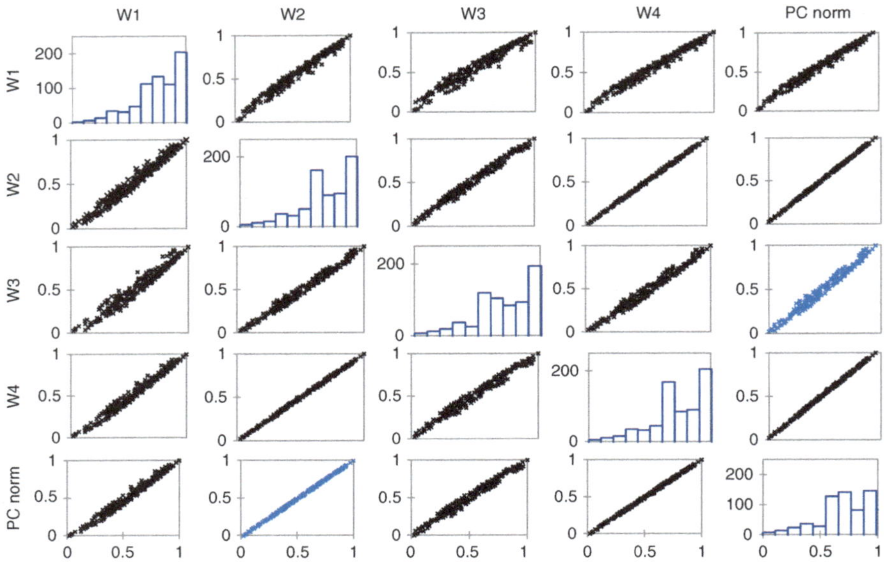

Fig. 23.1 Scatter plot of the fuzzy indicators (obtained with inclusion of X_8) and of the crisp indicator (obtained normalizing the scores of the first principal component)

are found to be not completely dissatisfied with other indicators. On the other hand, 86 respondents (with X_8) and 170 (without X_8) belong to the subset of completely satisfied customers. In both analyses, the use of the weights increases the frequencies of small values (less than 0.4) and decreases the frequencies in the classes $[0.7 - 0.8)$ and $[0.8 - 0.9)$.

Figure 23.1 displays the scatter plot matrix of the fuzzy indicators obtained with inclusion of X_8 and of the crisp indicator obtained normalizing the scores of the first principal component. As outlined previously, fuzzy and crisp indices show high pairwise correlations (the correlation coefficient of each off-diagonal panel of the scatter plot matrix is nearly 0.99). With $l = 1$ the distribution of the values is more symmetric. In order to compare the fuzzy indicators computed without X_8 with the mostly used crisp indicators of customer satisfaction, we normalize both the values of X_8 (overall satisfaction) and the scores of the first principal component computed on the τ rank correlation matrix, to lie in the interval $[0, 1]$. Figure 23.2 presents boxplots of the fuzzy and crisp indices. While all composite indicators show the presence of outlying values (due to small size), the single variable X_8 does not reveal this presence. Moreover, all fuzzy indicators show that 75 % of the respondents belong to the set of satisfied customers, with m.f. values higher than 0.6. Figure 23.3 shows the distribution of the fuzzy index (computed without X_8 and with weight W_4) for each category of the variable X_8. Note that none of the respondents has a category smaller than 3 (i.e., no one is completely dissatisfied with the service). Boxplots show that the distribution of the values of the m.f. to the subset

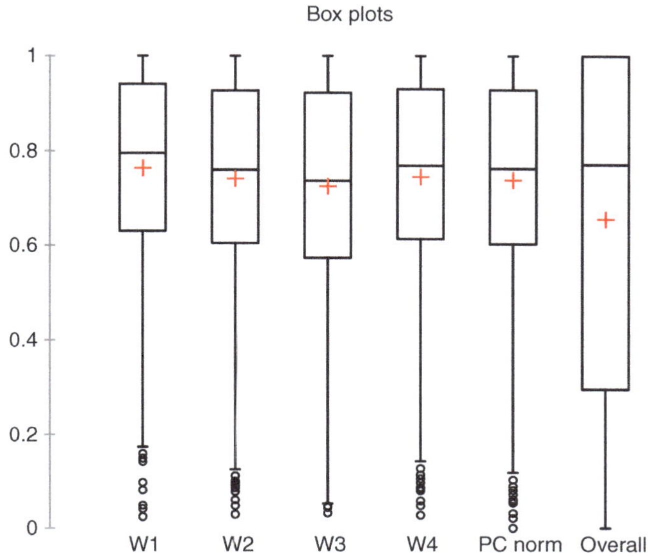

Fig. 23.2 Boxplots of the fuzzy indicators computed without X_8, of the normalized values of variable X_8 (Overall) and of the scores of the first principal component (Pcnorm)

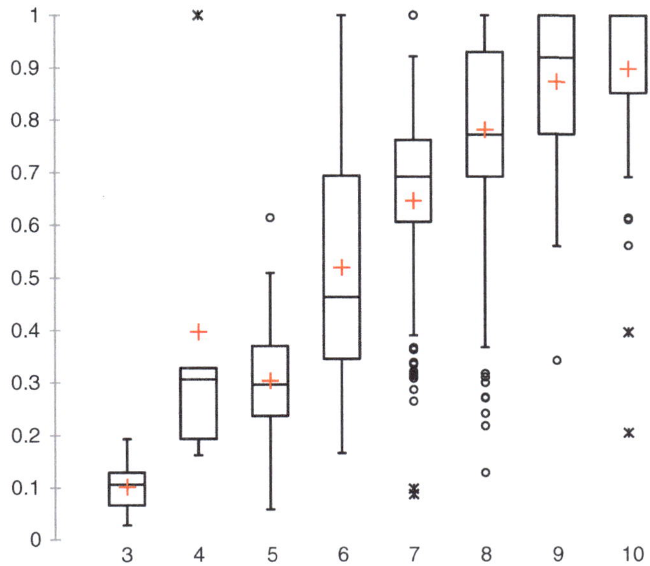

Fig. 23.3 Boxplots of the fuzzy indicators with weights W_4 for each value of the variable X_8 (overall satisfaction for the service)

of satisfied customers is increasing in the median, with respect to the score given to the overall satisfaction (X_8). It also reveals the presence of few incoherent responses. For example, two customers that have a score equal to 4 in the overall satisfaction (they are neither satisfied nor very dissatisfied) are however quite satisfied with all the specific items that have been considered in the questionnaire. On the other hand, some clients that have a scores equal to 8, 9, or 10 in the overall satisfaction indicate a low degree of satisfaction in nearly all the items. We point out that the comparison between the direct question on overall satisfaction and the values of a fuzzy index based on the items reveals the customers with inconsistent answers who may be considered as atypical observations.

References

1. Baliamoune-Lutz, M.: On the measurement of human well-being: Fuzzy set theory and Sen's capability approach. In: UNU-WIDER, Helsinki (2004)
2. Baliamoune-Lutz, M., McGillivray, M.: Fuzzy well-being achievement in Pacific Asia. J. Asia Pacific Econ. **11**, 168–177 (2006)
3. Calvo, T., Beliakov, G.: Aggregation functions based on penalties. Fuzzy Sets Syst. **161**, 1420–1436 (2010)
4. Castillo, C., Lorenzana, T.: Evaluation of business scenarios by means of composite indicators. Fuzzy Econ. Rev. **15**, 3–20 (2010)
5. Cerioli, A., Zani, S.: A fuzzy approach to the measurement of poverty. In: Dagum, C., Zenga, M. (eds.) Income and Wealth Distribution, Inequality and Poverty, pp. 272–284. Springer, Berlin (1990)
6. Cheli, B., Lemmi, A.: A totally fuzzy and relative approach to the multidimensional analysis of poverty. Econ. Notes **24**(1), 115–134 (1995)
7. Chiappero Martinetti, E.: A multidimensional assessment of well-being based on Sen's functioning approach. Rivista Internazionale di Scienze Sociali **108**(2), 207–239 (2000)
8. Chien, C.-J., Tsai, H.-H.: Using fuzzy numbers to evaluate perceived service quality. Fuzzy Sets Syst. **116**, 289–300 (2000)
9. Coppi, R., Gil, M.A., Kiers, H.A.L.: A fuzzy approach to statistical analysis. Comput. Stat. Data Anal. **51**, 1–14 (2006)
10. Darestani, A.Y., Jahromi, A.E.: Measuring customer satisfaction using a fuzzy inference system. J. Appl. Sci. **9**(3), 469–478 (2009)
11. JRC: An information server on composite indicators and ranking systems. http://composite-indicators.jrc.ec.europa.eu. (2013)
12. Klir, G.J., Folger, T.A.: Fuzzy Sets, Uncertainity and Information, p. 61. Prentice-Hall Int., London (1988)
13. Kwong, C.K., Bai, H.: A fuzzy AHP approach to the determination of importance weights of customer requirements in quality function deployment. J. Intell. Manufact. **13**, 367–377 (2002)
14. Lazim, M.A., Osman, M.T.A.: A new Malaysian quality of life index based on Fuzzy sets and hierarchical needs. Soc. Indicat. Res. **94**(3), 499–508 (2009)
15. Lemmi, A., Betti, G. (eds.): Fuzzy Set Approach to Multidimensional Poverty Measurement. Springer, New York (2006)
16. OCDE: Handbook on Constructing Composite Indicators. OCDE Publications, Paris (2008)
17. Smithson, M., Verkuilen, J.: Fuzzy Sets Theory: Applications in the Social Sciences. Sage, London (2006)
18. Zadeh, L.A.: Fuzzy sets. Inform. Contr. **8**, 338–353 (1965)
19. Zani, S., Berzieri, L.: Measuring customer satisfaction using ordinal variables: an application in a survey on a contact center. Stat. Applicata-Italian J. Appl. Stat. **20**(3–4), 331–351 (2008)

20. Zani, S., Milioli, M.A., Morlini, I.: Fuzzy methods and satisfaction indices. In: Kenett, R.S., Salini, S. (eds.) Modern Analysis of Customer Surveys, pp. 439–455. Wiley, Chichester (2012)
21. Zimmermann, H.J.: Fuzzy Sets Theory and Its Applications, 4th edn. Kluwer, Boston (2001)
22. Zhou, P., Ang, B.W., Zhou, D.Q.: Weigthing and aggregation in composite indicator construction: a multiplicative optimization approach. Soc. Indicat. Res. **96**, 169–181 (2010)

Part IV

Survey Methodology and Official Statistics

Symmetric Association Measures in Effect-Control Sampling

24

Riccardo Borgoni, Donata Marasini, and Piero Quatto

Abstract

In order to measure the association between an exposure variable X and an outcome variable Y, we introduce the effect-control sampling design and we consider the family of symmetric association measures. Focusing on the case of binary exposure and outcome variables, a general estimator of such measures is proposed and its asymptotic properties are also discussed. We define an allocation procedure for a stratified effect-control design, which is optimal in terms of the variance of such an estimator. Finally, small sample behavior is investigated by Monte Carlo simulation for a measure belonging to the family, which we believe particularly interesting as it possess the appealing property of being normalized.

Keywords

Association measures • Optimal sampling • Sample design and estimation

24.1 Introduction

In survey sampling theory, the control sample has been introduced to measure the association between two dichotomous variables, denoted by X and Y [2].

In this context, X may be the indicator variable for the exposure to a possible cause of an effect, represented by an outcome variable Y. The aim is to measure the strength of the assumed causal relation.

To exemplify, suppose that one wants to identify potential causes of students drop out of the University. If one restricts his attention only on a sample of those who dropped out, it may happen that a high percentage of students interviewed says that

R. Borgoni (✉) · D. Marasini · P. Quatto
Dipartimento di Economia, Metodi Quantitativi e Strategie di Impresa, Università degli Studi di Milano-Bicocca, Piazza Ateneo Nuovo, 1 Milano 20126, Italy
e-mail: riccardo.borgoni@unimib.it; donata.marasini@unimib.it; piero.quatto@unimib.it

N. Torelli et al. (eds.), *Advances in Theoretical and Applied Statistics*,
Studies in Theoretical and Applied Statistics, DOI 10.1007/978-3-642-35588-2_24,

they withdrew because the University studies were very difficult. This suggests the study difficulty as a cause of the drop out. However, if a control sample of students having continued their studies is considered together with the drop out sample, one may find a similar proportion of students believing that the university studies are very difficult among the controls too. In this case, the study difficulty can be hardly considered a cause of drop out. Clearly, many other variables can have an impact on individuals' decision to withdraw, such as being a student–worker, the school of provenience, and so on. For these variables, i.e., the potential causes, we need to calculate the association with the effect under study. As there may be other variables that affect the association between the effect and its potential causes, one may want to control for their influence in order to prevent confounding. In this sense, such variables can be considered as auxiliary variables and can be used to divide the population into suitable strata (e.g., gender and age).

In this chapter we introduce a family of association measures which can be used in effect-control sampling. Effect-control designs are introduced in Sect. 24.2. The proposed family of association measures is described in Sect. 24.3 and a class of estimators is introduced in Sect. 24.4 along with the homogeneity hypothesis. Section 24.5 discusses the issue of optimal designs, and in Sect. 24.6 we focus on the estimation procedure of a normalized measure of the family, which finite sample behavior is evaluated by Monte Carlo simulation in Sect. 24.7.

24.2 Effect-Control Sampling

Let Y be the outcome variable, i.e., Y is the dichotomous variable taking value 1 if the effect is present and 0 otherwise. The target population U is assumed to be finite and it is divided into two subpopulations according to Y. Let N and M be the number of units for which $Y = 1$ and $Y = 0$, respectively. The N units for which Y equals 1, identify the effect group, whereas the M for which Y equals 0 identify the control group.

The two groups are partitioned in H strata according to the values of a set of auxiliary variables, coded by a univariate variable S, identifying homogeneous profiles.

Let N_h and M_h be the size of the stratum h $(h = 1, 2, \ldots, H)$ in the two groups with

$$N = \sum_{h=1}^{H} N_h$$

and

$$M = \sum_{h=1}^{H} M_h.$$

The effect and control samples, say e and c, are drawn in two steps. First a sample, say e_h, is taken within the units in stratum h of the effect group; then a sample c_h of size m_h is taken from stratum h of the control group, in order to guarantee that effect and control units are similar in terms of the stratification variables. Clearly, we obtain that

$$e = \bigcup_{h=1}^{H} e_h$$

and

$$C = \bigcup_{h=1}^{H} C_h$$

with size

$$n = \sum\nolimits_{h=1}^{H} n_h$$

and

$$m = \sum\nolimits_{h=1}^{H} m_h,$$

respectively. The design allocates a positive probability to the samples of size $n_h + m_h$ that can be obtained from the $N_h + M_h$ units in stratum h, assuming a simple random sampling without replacement. Every $n_h + m_h$-tuple is supposed to have the same probability

$$1 \Big/ \binom{N_h}{n_h}\binom{M_h}{m_h}$$

$h = 1, \ldots, H$ to occur. It can be noted that the effect-control sampling design may be considered as a particular stratified sampling design in which the population is partitioned into $2H$ sub-populations according to the total number H of strata, which rises from the product of the values each auxiliary variable can assume, and to the two values of the dichotomous outcome variable.

24.3 Symmetric Association Measures

As mentioned above, let X be a dichotomous exposure variable and let Y be a dichotomous outcome variable possibly associated with X. We consider the population U consisting of $N + M$ units of which N are characterized by $Y = 1$ and M are characterized by $Y = 0$.

Assume that N' units out of N and M' units out of M are exposed ($X = 1$). Thus, the odds ratio (OR) parameter is defined by

$$\theta = \frac{N'}{N - N'} \frac{M - M'}{M'} \tag{24.1}$$

In particular, an odds ratio of 1 indicates that X and Y are independent, an odds ratio greater than 1 indicates positive association, and an odds ratio less than 1 indicates negative association. Unlike other measures of association for binary data (such as the relative risk), the odds ratio treats the two variables X and Y symmetrically.

In general, according to [3], a symmetric association measure between X and Y is a function of

$$\frac{N'}{N}, \frac{M'}{M}$$

or, alternatively, of

$$\frac{N'}{N'+M'}, \frac{N-N'}{N+M-(N'+M')}$$

such that these alternative measures are equal. Hence it is of the form $f(OR)$, where f is a strictly increasing function. Examples of measures belonging to the family of strictly increasing functions of odds ratio are given by $\log(OR)$, Yule's coefficient of association and Yule's coefficient of colligation [3].

Moreover, it can be proved by solving a functional equation that all the symmetric and scale-invariant measures of association are strictly increasing function of OR, where, according to [4], a measure of association

$$\alpha(N', M', N-N', M-M')$$

is called symmetric and scale invariant if

$$\alpha(N', M', N-N', M-M') = \alpha(N', N-N', M', M-M')$$

and

$$\alpha(N', M', N-N', M-M') = \alpha(\lambda N', \lambda M', \mu N-\mu N', \mu M-\mu M')$$

for any $\lambda, \mu > 0$.

With respect to the stratification, we restrict our attention to the family of the association measures defined by $f(\theta_h)$ herein,

where f is a strictly increasing function and

$$\theta_h = \frac{N'_h}{N_h - N'_h} \frac{M_h - M'_h}{M'_h}, \tag{24.2}$$

is the stratum specific version of (24.1), assuming that N'_h units out of N_h and M'_h units out of M_h are exposed in the stratum $h = 1, 2, \ldots, H$.

24.4 Homogeneity Hypothesis

Cross-classifying the N units of the population according to the response variable, Y, the exposure variable, X, and the auxiliary variable used for stratification, S produces a $2 \times 2 \times H$ three-way contingency table. A relevant hypothesis for a contingency table of this form often considered in practical applications [6,7] is that the H stratum-specific odds ratios are all equal, that is to say:

$$\theta_1 = \ldots = \theta_H = \theta.$$

This hypothesis of homogeneity of odds ratios amounts to assume that there is not a three way interaction amongst Y, X and S. The homogeneity condition can then be stated equivalently as

$$f(\theta_1) = \ldots = f(\theta_H) = f(\theta) \tag{24.3}$$

for every strictly increasing function f.

Under such a hypothesis, the parameter $f(\theta)$ can be estimated by the plug-in estimate

$$\sum_{h=1}^{H} w_h f(\hat{\theta}_h), \tag{24.4}$$

once a suitable weight system w_h $h = 1, \ldots, H$ has been identified. In Eq. (24.4)

$$\hat{\theta}_h = \frac{n'_h}{n_h - n'_h} \frac{m_h - m'_h}{m'_h}$$

is the usual estimate of (24.2), assuming that n'_h units are exposed amongst the n_h included in the sample for which $S = h$ and $Y = 1$, whereas m'_h units are exposed amongst the m_h units in the sample for which $S = h$ and $Y = 0$.

Traditionally, statistical inference for contingency tables has relied heavily on large-sample approximations for the sampling distribution of test statistics. Under the homogeneity condition it is straightforward to prove that a chi-square with $H - 1$ degree of freedom asymptotic distribution can be employed to approximate the p-value of a suitable test statistic whatever the transformation $f(\theta)$ is.

More recently exact inference for small samples has become of paramount importance given an increasing number of studies with small sample size. In addition even when the sample size is quite large, large-sample approximations can be very poor when the contingency table contains both small and large expected frequencies [5]. Large sparse tables present typically such a problem. Agresti [1], amongst others, reviews these methods.

Zelen [13] presents an exact test for the homogeneity of the odds ratio in a $2 \times 2 \times H$ contingency table. Since f in (24.3) is a strictly monotone function of the odds ratio, the Zelen's test for odds ratio homogeneity is also the one for rejecting the null hypothesis in (24.3) independently of f.

In the presence of large tables, computational advantages are obtained by resorting to appropriate algorithms such as network algorithm [10] which can be used in conjunction with Monte Carlo techniques for sampling the contingency tables [11].

24.5 Optimal Designs

For the sake of simplicity, in this section the hypergeometric distributions of n'_h and $m'_h (h = 1, \ldots, H)$ are approximated by suitable binomial distributions (according to [9], p. 114).

By applying the delta method [9, p. 315], it can be proved that the estimator corresponding to (24.4) is approximately normal with mean $f(\theta)$ and variance given by

$$\sum_{h=1}^{H} \left[w_h f'(\theta_h)\theta_h\right]^2 \left[\frac{M_h^2}{m_h M_h'(M_h - M_h')} + \frac{N_h^2}{n_h N_h'(N_h - N_h')}\right] \tag{24.5}$$

where f' denotes the derivative of f.

Then, from Slutsky's theorem [9, p. 283], it follows that (24.5) can be consistently estimated by

$$\sum_{h=1}^{H} \left[w_h f'(\hat{\theta}_h)\hat{\theta}_h\right]^2 \left(\frac{1}{m_h'} + \frac{1}{m_h - m_h'} + \frac{1}{n_h'} + \frac{1}{n_h - n_h'}\right). \tag{24.6}$$

Moreover, optimal weights can be obtained by minimizing (24.5) under the constraint

$$\sum_{h=1}^{H} w_h = 1.$$

For this purpose, the method of Lagrange multipliers provides the solution of the optimization problem

$$w_h = \frac{\frac{1}{\sigma_h^2}}{\sum_{h=1}^{H} \frac{1}{\sigma_h^2}}, \tag{24.7}$$

where

$$\sigma_h^2 = [f'(\theta_h)\theta_h]^2 \left[\frac{M_h^2}{m_h M_h'(M_h - M_h')} + \frac{N_h^2}{n_h N_h'(N_h - N_h')}\right] \tag{24.8}$$

$(h = 1, 2, \ldots, H)$.

It should be noted that, under hypothesis (24.3), the optimal weights (24.7) turn out to be independent of the choice of transformation f.

Hence, the optimal weights (24.7) can be approximated by

$$\hat{w}_h = \frac{\frac{1}{\hat{\sigma}_h^2}}{\sum_{k=1}^{H} \frac{1}{\hat{\sigma}_k^2}}, \tag{24.9}$$

where the consistent estimate

$$\hat{\sigma}_h^2 = \frac{1}{m_h'} + \frac{1}{m_h - m_h'} + \frac{1}{n_h'} + \frac{1}{n_h - n_h'} \tag{24.10}$$

[7, p. 27] corresponds to the variance stabilizing transformation

$$f(OR) = \log(OR).$$

In addition, we observe that such a choice of f leads to the classical estimate suggested by [12]

$$\sum_{h=1}^{H} \hat{w}_h \log(\hat{\theta}_h).$$

Finally, optimal allocation can be obtained by finding the minimum of (24.5) subject to the constraint

$$\sum_{h=1}^{H}(n_h + m_h) = n + m = t,$$

where t represents the total sample size.

Under homogeneity hypothesis (24.3), the method of Lagrange multipliers provides the optimal sample sizes which turn out to be independent of f.

$$\begin{cases} n_h = \dfrac{w_h/\eta_h}{\sum_{h=1}^{H} w_h \left(\dfrac{1}{\eta_h} + \dfrac{1}{\gamma_h}\right)} t \\ m_h = \dfrac{w_h/\gamma_h}{\sum_{h=1}^{H} w_h \left(\dfrac{1}{\eta_h} + \dfrac{1}{\gamma_h}\right)} t \end{cases} \tag{24.11}$$

where

$$\begin{cases} \gamma_h = \sqrt{\frac{M'_h(M_h - M'_h)}{M_h^2}} \\ \eta_h = \sqrt{\frac{N'_h(N_h - N'_h)}{N_h^2}} \end{cases} \tag{24.12}$$

$(h = 1, 2, \ldots, H)$.

In order to realize the optimal allocation (24.11), it would be desirable to conduct a pilot survey for estimating the unknown parameters (24.12) or to use previous surveys or other statistical sources, if available.

24.6 A Normalized Symmetric Measure of Association

In this section we consider the normalized OR suggested by [8],

$$f(OR) = \frac{OR}{1 + OR} = \frac{(M - M')N'}{M'(N - N') + (M - M')N'} \in [0,\ 1]. \tag{24.13}$$

In particular, the value taken on by (24.13) is equal to 0.5 when the variables X and Y are independent, is greater than 0.5 in the presence of positive association, and is less than 0.5 in the presence of negative association.

Putting

$$\lambda_h = f(\theta_h) = \frac{\theta_h}{1+\theta_h} = \frac{(M_h - M'_h)N'_h}{M'_h(N_h - N'_h) + (M_h - M'_h)N'_h}$$

$(h = 1, 2, \ldots, H)$, the homogeneity hypothesis (24.3) becomes

$$\lambda_1 = \ldots = \lambda_H = \lambda.$$

So, the estimate (24.4), the variance (24.5), and its estimate (24.6) turn out to be, respectively,

$$\sum_{h=1}^{H} w_h \hat{\lambda}_h, \tag{24.14}$$

$$\sum_{h=1}^{H} [w_h \lambda_h (1-\lambda_h)]^2 \left[\frac{M_h^2}{m_h M'_h (M_h - M'_h)} + \frac{N_h^2}{n_h N'_h (N_h - N'_h)} \right] \tag{24.15}$$

and

$$\sum_{h=1}^{H} \left[w_h \hat{\lambda}_h (1-\hat{\lambda}_h) \right]^2 \left(\frac{1}{m'_h} + \frac{1}{m_h - m'_h} + \frac{1}{n'_h} + \frac{1}{n_h - n'_h} \right),$$

where

$$\hat{\lambda}_h = \frac{(m_h - m'_h)n'_h}{m'_h(n_h - n'_h) + (m_h - m'_h)n'_h}$$

$(h = 1, 2, \ldots, H)$.

Under homogeneity hypothesis (24.3), the parameter λ can be estimated by means of (24.14), in which the optimal weights specified by (24.7) and (24.8) can be approximated using (24.9) and (24.10), and the sample sizes are provided by the proportional allocation

$$\begin{cases} n_h = \dfrac{N_h}{N+M} t \\ m_h = \dfrac{M_h}{N+M} t \end{cases} \tag{24.16}$$

$(h = 1, 2, \ldots, H;\ t$ being the total sample size).

Finally, the variance (24.15) could further be reduced by the optimal allocation (24.11), in which the parameters (24.12) may be estimated on the basis of the preceding sample survey.

Table 24.1 Results of the Monte Carlo study in the case A of proportional allocation and in the case B of optimal allocation

OR	λ	p	A bias	B bias	B mse/A mse
4	0.8	0.05	0.011	−0.005	0.87
		0.1	0.007	−0.003	0.90
		0.2	0.002	−0.001	0.83
		0.3	0.002	−0.001	0.99
2	0.66	0.05	0.007	−0.002	0.76
		0.1	0.002	0.000	0.80
		0.2	−0.001	0.000	0.83
		0.3	0.001	−0.001	0.86
1	0.5	0.05	−0.004	−0.002	0.67
		0.1	0.001	−0.002	0.77
		0.2	0.001	0.001	0.75
		0.3	0.000	0.001	0.73
0.5	0.33	0.05	−0.008	0.002	0.69
		0.1	−0.007	0.000	0.71
		0.2	−0.003	0.002	0.77
		0.3	−0.001	0.001	0.62
0.25	0.2	0.05	−0.014	0.006	0.80
		0.1	−0.008	0.003	0.71
		0.2	−0.003	0.001	0.71
		0.3	−0.001	0.001	0.75

24.7 A Simulation Study

In order to evaluate the small sample behavior of the estimate (24.14) with the optimal coefficients (24.7) approximated by (24.9), an extensive simulation study have been performed. Results are reported in Table 24.1.

The simulation design is as follows. A finite population consisting of 4,000 units has been considered assuming $N/M = 0.6$. This population was stratified in two groups ($H = 2$). The values of the proportion of cases and controls within the two categories of an exposure binary variable X have also been chosen in advance. This corresponds to fix the value of λ. A grid of possible values of this parameter has been considered to explore the performance of alternative sampling procedures under different scenarios. More precisely the population was sampled 1,000 times for each value of λ using both the proportional and the optimal allocation for the strata size. Calculating the optimal allocation (24.11) requires a preliminary estimate of the parameters (24.12), a problem that can be addressed by a pilot survey in real situations. In order to implement the optimal design in the simulation study presented herein, we have first extracted a pilot sample from the simulated population using a simple random sample without replacement inside each stratum. This dataset has been used to estimate the unknown parameters (24.12) which, in

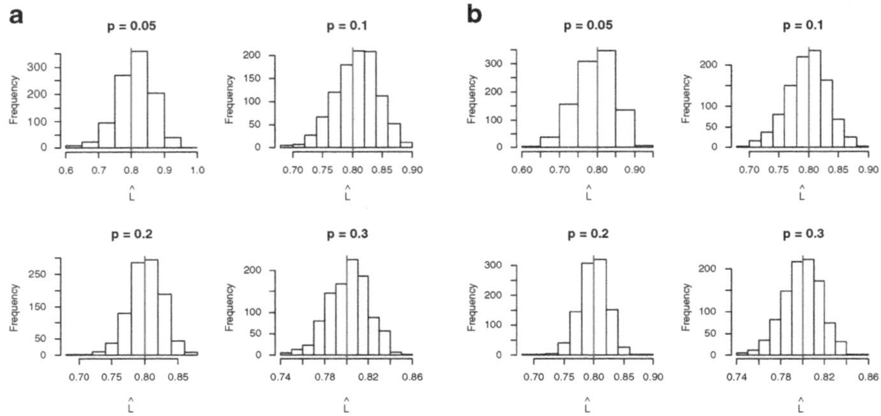

Fig. 24.1 Monte Carlo sampling distribution of $\hat{\lambda}$ under proportional (**a**) and optimal (**b**) sample size allocation for different sampling fractions. *Vertical line* represents the "true" value of λ used to simulate samples

turn, permitted to calculate the optimal sample size of each stratum. Then a second sample was selected from the population and, finally, employed to estimate λ. Note that only cell counts are necessary to calculate (24.14), hence the entire Monte Carlo experiment has been efficiently implemented using a routine to generate from the hypergeometric distribution.

The simulation has been repeated for different sampling fractions (p) in order to evaluate whether the performance of the estimator is influenced by the overall sample size and the velocity at which its distribution converges to normality. The Monte Carlo (MC) bias has been reported in Table 24.1 for the two schemes together with the ratio between the MC MSE obtained by the optimal allocation (B) and the MC MSE obtained by the proportional allocation (A). Hence, a value smaller than 1 of this ratio suggests a better performance of the optimal allocation.

It can be observed that the MC MSE in the case of the optimal allocation is always smaller than the MC MSE obtained using a proportional allocation and quite often much smaller. Broadly speaking, a bigger advantage was found when the sampling fraction is small for a value of λ bigger than 0.5. For λ smaller than 0.5 the optimal allocation produces, roughly, a constant reduction of the MSE in the case of the proportional allocation whatever the sampling fraction is.

As far as the bias is concerned results are less clear as both designs tend to have a good performance. This was somehow expected as the estimator is asymptotically unbiased under both allocation schemes. However, the optimal allocation has a bias which is generally smaller than the proportional allocation, suggesting that the optimal scheme seems preferable also in this respect.

Finally, Fig. 24.1 shows that the distribution of the considered estimator approaches normality quickly under both designs as it resembles the Gaussian density, also in the presence of small samples. The graphs are drown using samples simulated under values of λ equal to 0.8. Similar results were found for all the other values of λ considered in the simulation.

References

1. Agresti, A.: A survey of exact inference for contingency tables. Stat. Sci. **7**(1), 131–153 (1992)
2. Borgoni, R., Marasini, D., Quatto, P.: Effect-control sampling and odds ratio, ITACOSM 2009, Siena, (2009)
3. Edwards, A.W.F.: The measure of association in 2 × 2 table. JRSS A **126**(1), 109–114 (1963)
4. Good, I.J., Mittal, Y.: The amalgamation and geometry of two-by-two contingency tables. Ann. Stat. **15**(2), 694–711 (1987)
5. Haberman, S.J.: A warning on the use of chi-squared statistics with frequency tables with small expected cell counts. J. Am. Stat. Assoc. **83**, 555–560 (1988)
6. Jewell, N.P.: Statistics for Epidemiology. Chapman&Hall/CRC, New York (2004)
7. Lachin, J.M.: Biostatistical Methods: The Assessment of Relative Risks. Wiley, New York (2000)
8. Landenna, G.: Fondamenti di statistica descrittiva. il Mulino, Bologna (1994)
9. Lehmann, E.L.: Elements of Large-Sample Theory. Springer, New York (2001)
10. Mehta, C.R., Patel, N.R., Gray, R.: Computing an exact confidence interval for the common odds ratio in several 2 by 2 contingency tables. J. Am. Stat. Assoc. **80**, 969–973 (1985)
11. Mehta, C.R., Patel, N.R., Senchaudhuri, P.: Importance sampling for estimating exact probabilities in permutational inference. J. Am. Stat. Assoc **83**, 999–1005 (1988)
12. Woolf, B.: On estimating the relation between blood group and disease. Ann. Hum. Genet. **19**, 251–253 (1955)
13. Zelen, M.: The analysis of several 2 × 2 contingency tables. Biometrika **58**, 129–137 (1971)

Spatial Misalignment Models for Small Area Estimation: A Simulation Study

25

Matilde Trevisani and Alan Gelfand

Abstract

We propose a class of misaligned data models for addressing typical small area estimation (SAE) problems. In particular, we extend hierarchical Bayesian atom-based models for spatial misalignment to the SAE context enabling use of auxiliary covariates, which are available on areal partitions non-nested with the small areas of interest, along with planned domains survey estimates also misaligned with these small areas. We model the latent characteristic of interest at atom level as a Poisson variate with mean arising as a product of population size and incidence. Spatial random effects are introduced using either a CAR model or a process specification. For the latter, incidence is a function of a Gaussian process model for the spatial point pattern over the entire region. Atom counts are driven by integrating the point process over atoms. In the proposed class of models benchmarking to large area estimates is automatically satisfied. A simulation study examines the capability of the proposed models to improve on traditional SAE model estimates.

Keywords

Atom-based models • Hierarchical Bayesian methods • Small area estimation • Spatial misalignment

M. Trevisani (✉)
Department of Economics, Business, Mathematics and Statistics, University of Trieste, 34100 Trieste, Italy
e-mail: matildet@econ.units.it

A. Gelfand
Department of Statistical Science, Duke University, Durham, NC 27708-0251, USA
e-mail: alan@stat.duke.edu

N. Torelli et al. (eds.), *Advances in Theoretical and Applied Statistics*,
Studies in Theoretical and Applied Statistics, DOI 10.1007/978-3-642-35588-2_25,

25.1 Introduction

This chapter considers small area estimation (SAE) when multiple data sources, possibly misaligned with small areas, are available for inference. Supplementary sources of relevant information may consist either in auxiliary data (from census or administrative archives) or in survey data collected on planned or major domains. Both these types of data may be available at a level of spatial resolution incompatible or misaligned with that of small areas. Hence, developing models to combine most appropriately and efficiently any source of potentially important information becomes of crucial concern for research in SAE.

Under the hierarchical Bayesian (HB) framework, misaligned data models have been recently proposed to address the issue of making inference for domains (target zones) different from the ones (source zones) for which data are available (for a review see [1, 7]; some applied recent works are [3, 6]). In particular, we consider the so-called *atom-based models* and extend them to originally address typical SAE questions: (a) combining auxiliary covariates which are available on non-nested areal partitions (misaligned areal regression problem); (b) providing small area estimates by using planned domain data (misaligned areal interpolation problem).

We offer two versions of the proposed class of atom-based models. The first one, which we will refer to as *Small*$\circlearrowleft$*Large area estimation model* (S$\circlearrowleft$LAE for short), introduces area-specific effects (not accounted for by auxiliary covariates) as heterogeneous and/or spatial random effects (these last, e.g., by a conditionally autoregressive prior specification), as is customarily done in mixed effects models. The second, the so-called *Small*$\odot$*Large area estimation model* (S$\odot$LAE), fits over the entire region, an intensity surface which drives a latent spatial point pattern of locations within atoms representing the "cases" of interest (e.g., of unemployment as in our example). Expected atom counts are obtained by integrating the intensity over atoms.

To illustrate our approach we consider the problem of estimating the number of unemployed at Local Labour Market (LLM) area level by using two misaligned data sources: auxiliary information available on different administrative partitions; reliable estimates of unemployed for Labour Force Survey (LFS) planned domains. A simulation study examines the capability of the proposed models to improve on traditional SAE model estimates.

25.2 Regression-Only Atom Models for SAE

Before proposing a general methodology to address SAE problems, we consider a more familiar setting. Namely, we start with a traditional SAE model and suitably extend it in order to perform a regression analysis also on covariates available over areal partitions misaligned with the small area set.

Suppose that $i = 1, \ldots, I$ index the small areas (target zones) on which we want to estimate the characteristic θ of interest. Here, driven by the motivating example, we consider θ to be a total or a count. A crude but still challenging HB model takes the form

$$d_i|\theta_i \sim N(\theta_i, \sigma_i) \tag{25.1}$$

$$\theta_i|\mu_i \sim \text{Po}(S_i\mu_i), \quad \log(\mu_i) = \mathbf{x}_i^{'}\boldsymbol{\beta} + \nu_i \tag{25.2}$$

$$\nu_i|\tau \sim N(0, \tau) \tag{25.3}$$

plus an hyperprior on $(\boldsymbol{\beta}, \tau)$. In sampling model (25.1), d_i and σ_i are a direct estimate of θ_i and, respectively, a design variance estimate (of d_i) which is imputed as known (but, in model extensions, it can be made model dependent). In linking model (25.2), S_i is a regional synthetic estimate (of θ_i) which takes the place of an expected count as in standard Poisson mean formulations. Moreover, $\mathbf{x}_i^{'}$ and ν_i are auxiliary data available for each area i and, respectively, area-specific effects not accounted for by the observed auxiliary variables. Hyperparameters $\boldsymbol{\beta}$ and τ are (regional) regression coefficients and area effect variance respectively. The form (25.1)–(25.3) is an unmatched model developed according to the general hierarchical perspective [data| process, parameters][process| parameters][parameters] (see for [13, 14]).

Now, let $j = 1, \ldots, J$ index the areas (source zones) on which a possibly important covariate $\mathscr{X}$ is available. Then, some sort of realignment is needed to carry out a regression analysis using the X's to produce better small area estimates θ_i.

A number of techniques exist to obtain estimates for misaligned data at the required area level (usually called the *modifiable areal unit problem* in statistical literature, or *areal interpolation* in geography): the "pycnophylactic" density surface estimation [12], the "areal weighting" interpolation [8], the "intelligent" areal interpolation [4, 5] as well as other methods implemented in *Geographical Information System* environment. None of these methods allows for a fully inferential approach to the problem of areal interpolation. We follow instead a fully inferential philosophy and adopt a HB approach allowing for a full uncertainty integration and propagation through the model. In particular we adopt the so-called atom-based models borrowed from the most recent HB literature on spatial misalignment [1, 10]. The strategy behind atom-based models is to realign the misaligned data onto a common "intersection–partition" whose areal unit is the "atom," then conveniently set the regression model at the atom level, and, finally, build the estimates of interest by a suitable aggregation over atoms.

Then, let $k = 1, \ldots, K$ index the intersection areas (atoms) arising from crossing each target zone with each source zone. Attractively, for us, there are no "edge" atoms; the target and source zones partition the same region. Accommodating the general case is straightforward (see [1] for modeling edge atoms).

In particular, with reference to our motivating example, we address the case in which both the small area characteristic of interest θ and the possibly important covariate $\mathscr{X}$ are count data (e.g., number of unemployed and of unemployment list enrolled respectively). Thus, the skeleton of an atom-based model, as we have framed it in the SAE context, is as follows. At atom level, small area quantity and non-nested auxiliary variable are modeled as

$$\theta_k|\eta_k \sim \text{Po}(S_k e^{\eta_k}), \qquad X_k|\omega_k \sim \text{Po}(N_k e^{\omega_k}) \tag{25.4}$$

$$\eta_k = \beta_0 + \beta_1 X_k + \beta_2 z_k + \gamma_{i(:\ni k)}, \qquad \omega_k = \zeta_0 + \zeta_1 w_k + \psi_{j(:\ni k)} \tag{25.5}$$

where θ_k, X_k and (z_k, w_k) are, respectively, the latent (or unknown) total of interest, the latent auxiliary variable and further known covariates on atom-k; S_k and N_k are known synthetic estimate and, respectively, population count for atom-k; e^{η_k} and e^{ω_k} represent the incidence rate respectively associated with θ and X on atom-k; γ_i and ψ_j are random effects inherited—by atom-k—from small area-i and, respectively, auxiliary source area-j (which atom-k belongs to, indicated in (25.5) by "$\ni k$" notation); β's and ζ's are unknown regression coefficients.

Then, on regional partitions, θ-model and $\mathscr{X}$-model consist in

$$d_i|\theta_i \sim \text{Norm}(\theta_i, \sigma_i) \tag{25.6}$$

$$\theta_i = \sum \theta_{k:k\in i}, \quad X_j = \sum X_{k:k\in j} \tag{25.7}$$

which amount to imposing the sum-constraints (25.7) on each atom set making up either a small area or an auxiliary source zone. Besides, likewise above—see (25.1), we assume a customary sampling model (25.6) for small area direct survey estimates d_i's.

Note 1. Our methodology assumes Poisson counts at the atom level for both the unknown of interest and the latent auxiliary variable. In other cases, these variables may have other distributions, e.g., Bernoulli trials with probit or logit links, or normal variates.

Note 2. In principle, the linking model for θ might be set at small area level. In this case, atom values X_k should be aggregated so as to reconstruct small area value X_i (i.e., $X_i = \Sigma X_{k:k\in i}$) which is to be imputed into the regression equation set for θ_i. However, a regression analysis is preferably carried out at atom or finer level whenever further covariates are available on regional partitions finer than the small area set. In (25.5), z_k and w_k denote known values of auxiliary variables whose source is nested within atoms.

Note 3. Linking models (25.5) are defined on the log-linear scale although other functional forms, such as the identity-linear or the mixed types, can be considered. For instance, if we adopt a linear in covariates form such that arising from a mixing (identity- and log-) link, θ-model (25.4)–(25.5) is replaced by

$$\theta_k|\eta_k \sim \text{Po}(S_k \eta_k) \quad \text{with } \eta_k = (\beta_0 + \beta_1 X_k + \beta_2 z_k)e^{\gamma_{i(:\ni k)}}$$

(or $\eta_k = \beta_0(1 + \beta_1 X_k + \beta_2 z_k)e^{\gamma_{i(:\ni k)}}$ in the version of [10]). The use of linear forms is more opportune when there are regressors thought to produce additive rather than multiplicative effects (see e.g. [2]). Moreover, an attractive feature of additive models may be that, regardless of aggregation level, a regression slope maintains the same interpretation (that is why we turn to them in the simulation study below).

Note 4. In the regression analysis for θ or X, random effects, generically denoted by γ_i and ψ_j in (25.5), can be set either as spatially structured (e.g., by specifying a Gaussian conditionally autoregressive (CAR) prior) or simply as heterogeneous effects (by customarily assuming an exchangeable Gaussian prior), or perhaps a sum of such effects.

25.3 A Point Level Version

An alternative to the class of misaligned data models proposed in Sect. 25.2 is based on Gaussian process specifications. This approach envisions an underlying intensity surface which is block-averaged to the areal units relevant to the study. Process model priors are commonly used for explaining spatial point patterns when the patterns are aggregated to areal units. They are interpreted as intensity surface for the unobservable spatial point pattern of cases of interest. For us, these cases are the unknown point locations for the unemployed individuals. Here the surface is assumed to be a realization of a log Gaussian process model.

Suppose we revise the model for the expected incidences in (25.4) with (25.5) using integration of latent intensity surfaces over atoms. Likewise (25.4), latent counts of both the quantity of interest, θ_k, and the misaligned covariate, X_k, are modeled as Poisson variates with mean arising as the product of the population size (or a function of this) and the incidence rate. But now the mean arises as

$$\int_{A_k} S(s)e^{\eta(s)}\,ds, \qquad \int_{A_k} N(s)e^{\omega(s)}\,ds, \tag{25.8}$$

for process-θ and X respectively, where s denotes a spatial point and A_k indicates the atom-k (with area $|A_k|$). Atoms are assumed small enough to approximate $S(s)$ and $N(s)$ by a constant function over atom, $S(s) \approx S_k/|A_k|$, $N(s) \approx N_k/|A_k|$. Moreover, $\eta(s) = \beta_0 + \beta_1 X_k + \beta_2 z_k + \gamma(s)$ and $\omega(s) = \zeta_0 + \zeta_1 w_k + \psi(s)$ where, differently from (25.5), random effects $\gamma(s)$ and $\psi(s)$ are given Gaussian process priors. Being intensively measured, X_k, z_k, and w_k values are unvaryingly assigned to point level covariates. Hence, the expected incidences (25.8) become

$$S_k e^{\beta_0+\beta_1 X_k+\beta_2 z_k} \int_{A_k} e^{\gamma(s)}/|A_k|, \qquad N_k e^{\zeta_0+\zeta_1 w_k} \int_{A_k} e^{\psi(s)}/|A_k| \tag{25.9}$$

respectively. Note the difference between these expressions and those in (25.4) with (25.5). In the latter $\gamma_{i(:\ni k)} = \int_{A_k} \gamma(s)\mathrm{d}s/|A_k|$ and $\psi_{j(:\ni k)} = \int_{A_k} \psi(s)\mathrm{d}s/|A_k|$. But $\exp(\gamma_{i(:\ni k)}) \neq \int_{A_k} \exp(\gamma(s))\mathrm{d}s/|A_k|$ and if $\gamma(s)$ is a Gaussian process, the resulting aggregated random effect is not normally distributed, similarly for $\psi_{j(:\ni k)}$. Under this construction, we see that (25.4) with (25.5) induces an ecological fallacy (see [16] for reference).

Note 5. Following Note 3, the log-link can be replaced by the mixed form, thus maintaining the log Gaussian process prior specification for the spatial random effects.

Note 6. Initially, spatial processes $\gamma(s)$ and $\psi(s)$ are set independent, but in principle could be made dependent by using coregionalization (see [1] for reference).

This version of misaligned data models fits an incidence surface over the region, obviously a more flexible model than the "step" surface modeled through a finite set of random effects as for the previous version. With the new approach we can have local adjustment to these steps at every location allowing richer understanding of the latent surface. Moreover, working at the point level achieves the highest spatial resolution, thus avoiding the dependence of the prior model on the data collection procedure, i.e., the number, shapes, and sizes of the areal units chosen in the particular study. Still, it replaces the specification of a proximity matrix, which spatially connects the subregions, with a covariance function, which directly models dependence between arbitrary pairs of locations (and induces a covariance between arbitrary subregions using block averaging). Lastly, it avoids an ecological fallacy. But, in (25.9), we see that the desired relationships with X_k, z_k, and w_k, respectively, are unaffected in this regard, and so we can fit the simpler areal level model without much concern. The simulation study below will distinguish the case in which this relationship is unaffected by ecological fallacies from the one in which it is affected.

25.4 SAE via S℧LAE/S⊙LAE Atom Models

We proceed by addressing the problem of the additional integration of large area estimates, i.e., direct estimates evaluated on survey planned domains. It is reasonable that small area estimates will further improve given that large estimates are generally reliable. We will conveniently refer to the final model as *Small℧Large* area estimation model—S℧LAE for short—for the discrete-space version (℧ symbol just means a "circular influence" between small and large areas), as S⊙LAE for the continuous-space approach of Sect. 25.3 (the "dot" within the circle depicts a spatial point within an area) .

To this aim we generalize the spatial structure of the foregoing section by adding a further layer, i.e., large areas $p = 1, \ldots, P$. Accordingly, atom product space is updated integrating this further zonal system into it. Then, survey information related to large areas is conveyed by an additional stage, $d_p|\theta_p \sim \text{Normal}(\theta_p, \sigma_p)$, that is by assuming an appropriate sampling model with the large area estimates subject to the sum-constraint $\theta_p = \sum \theta_{k:k\in p}$, as θ_i's are in (25.7). d_p and σ_p have the customary interpretation of direct estimates and, respectively, sampling error variances for survey planned domains (e.g., for provinces in our application).

We point out that inference coherency synthesized by the so-called benchmarking propriety (small area estimates sum up to large area estimates) is automatically satisfied (since $\sum_k \theta_k = \sum_i \sum \theta_{k:k\in i} = \sum_i \theta_i = \sum_p \sum \theta_{k:k\in p} = \sum_p \theta_p$) in the final model version of the present section.

25.5 A Simulation Study

In this section we present a subset of results from a larger plan of simulation studies. First we give some details of the simulation experiments.

Survey data simulation was carried out on the basis of the Italian LFS design. In particular, a number of LFS samples was generated for the sole Veneto Region using 2001 census data on population counts available at municipality level, and following a two-stage stratified sampling scheme with varying probability at stage-1 and simple random sampling without replacement at stage-2. Then, $\mathscr{X}$ (unemployment list enrolled) and θ (unemployed) numbers were model-based generated at municipality level. Setting the synthetic data generation at this level—the highest spatial resolution available from census data—allows for a straightforward survey data simulation. Besides, data generation at atom level would not be realistic. More importantly, it allows to distinguish the scenario (a) (see below) in which the relationship with covariates is not affected by ecological fallacies from the one (b) in which it is.

In particular, $\mathscr{X}$ and θ numbers were generated according to the following models: (a) $X_l \sim \text{Po}(N_l\omega_l)$ with $\omega_l = \zeta_0(1 + \zeta_1 w_l)e^{\phi_j}$ and $Y_l \sim \text{Po}(S_l\eta_l)$ with $\eta_l = \beta_0(1 + \beta_1 x_l + \beta_2 z_l)e^{\nu_i}$; (b) $X_l \sim \text{Po}(N_l e^{\omega_l})$ with $\omega_l = \zeta_0 + \zeta_1 w_l + \phi_j$ and $Y_l \sim \text{Po}(S_l e^{\eta_l})$ with $\eta_l = \beta_0 + \beta_1 x_l + \beta_2 z_l + \nu_i$. In both scenarios l is the municipality index; ζ_0, ζ_1, β_0, β_1 and β_2 were set to the estimates found out from fitting the foregoing regression models on real data; w_l and z_l are covariates available at municipality level (number of employed in industry and, respectively, of registered immigrants); moreover, ϕ_j and ν_i have been specified as heterogeneous effects. For more information on the real context of the simulation study see [13].

Notice that in scenario (a) the mixing (additive in covariates, multiplicative in random effects) link was chosen for its property of maintaining the same form when aggregated on higher levels (see Note 3). Thus, first, it makes the comparison between traditional (25.1)–(25.3) and atom-based models "fair" (otherwise possible limits of a traditional model might depend on how synthetic data have been generated, while might not on its intrinsic capability). Second, such a choice allows us to avoid ecological bias issues. Whilst, the customary log-link chosen for scenario (b) can induce ecological bias issues (the same functional form may

Table 25.1 Simulation study: scenario (a)

	Bias				Accuracy				Efficiency		Reliability	
Estimator	$\overline{\text{rb}}$	rb^M	$\overline{\text{arb}}$	arb^M	$\overline{\text{are}}$	are^M	$\overline{\text{rE}}$	rE^M	$\overline{\text{eff}}$	eff^M	$\overline{\text{rel}}$	rel^M
T without $\mathscr{X}$	4	2	26	22	30	28	33	32	13	13	57	50
non-na	2	2	23	21	26	25	31	30	15	14	60	59
T with $\mathscr{X}$ plug-in(1)	3	2	20	18	23	21	28	28	13	13	61	60
non-na	2	2	18	16	20	19	27	28	14	14	66	64
T with $\mathscr{X}$ plug-in(2)	1	1	11	9	17	16	23	20	14	14	80	75
non-na	1	0	10	8	16	15	22	21	15	15	89	91
RA	0	1	9	7	14	13	17	16	15	15	96	96
non-na	0	0	8	6	13	11	16	15	17	16	104	108
S℧LAE	0	0	5	3	12	11	15	14	16	15	102	103
non-na	0	0	3	3	11	10	14	13	18	16	110	109
S⊙LAE	−1	0	5	2	11	11	15	14	16	15	101	104
non-na	0	0	3	3	10	11	15	13	17	16	106	108

Overlined measures are mean values, whereas $_M$ superscript denotes median values
T is the traditional model (25.1)–(25.3)
⁻ Without $\mathscr{X}$: the misaligned covariate X is not included as a regressor
⁻ With $\mathscr{X}$ plug-in(1): with X plugged-in by "population weighting" interpolation
⁻ With $\mathscr{X}$ plug-in(2): with X plugged-in by using aggregated atom-based model estimates
RA is the regression-only atom-based model (Sect. 25.2)
S℧LAE is the discrete-space version of the atom-based model (Sects. 25.2 and 25.4)
S⊙LAE is the continuous-space version of the atom-based model (Sects. 25.3 and 25.4)
non-na rows refer to measures calculated leaving out survey non-sampled areas.

not be maintained at any level). So, we are given the opportunity for distinguishing the performance of S℧LAE from that of S⊙LAE.

To compare the performance of $\hat{\theta}_i^{HB}$ estimators, we compute a series of standard measures out of $R = 100$ simulated samples: relative bias (rb) and absolute relative bias (arb), absolute relative error (are) and relative root mean squared error (rE), efficiency (eff) and reliability (rel) of the standardly used efficiency indicator. Details for their computation can be found in [14].

From the foregoing, the results shown here (Tables 25.1 and 25.2) refer to scenarios where both the misaligned variable $\mathscr{X}$ and the nested covariates at finer area level (z and w) are important auxiliary sources of information. Then, an SAE traditional model, such as (25.1)–(25.3), might be unfit since it neglects the auxiliary information $\mathscr{X}$ or, at best, includes it by merely using a "plug-in" method. Moreover, the unit of analysis of a traditional model is set to a level not finer than that defined by small areas. Last but not least, it fails to integrate direct large area information.

In the tables best performance values have been framed though if misleading the box is gray-filled. The other values are gray-colored where intensity grows with worsening performance (by dividing, for each measure, the range of expectation/median values into five equal intervals).

Table 25.2 Simulation study: scenario (b)

	Bias				Accuracy				Efficiency		Reliability	
Estimator	$\overline{\text{rb}}$	rb^M	$\overline{\text{arb}}$	arb^M	$\overline{\text{are}}$	are^M	$\overline{\text{rE}}$	rE^M	$\overline{\text{eff}}$	eff^M	$\overline{\text{rel}}$	rel^M
T without $\mathscr{X}$	5	3	27	23	31	29	34	33	14	14	56	49
non-na	4	3	24	22	27	26	32	31	16	15	59	58
T with $\mathscr{X}$ plugged-in(1)	4	3	23	21	26	24	31	31	14	14	58	57
non-na	3	3	21	19	23	22	30	31	15	15	63	61
T with $\mathscr{X}$ plugged-in(2)	2	2	16	14	22	21	28	25	15	15	75	70
non-na	2	1	15	13	21	20	27	26	16	16	84	86
RA model	1	2	13	11	19	18	22	21	16	16	92	91
non-na	1	1	12	10	18	16	21	20	18	17	104	105
S↺LAE	−1	−1	9	7	16	15	19	18	18	17	103	104
non-na	0	0	7	7	15	14	18	17	20	18	109	108
S⊙LAE	0	0	4	2	11	11	15	14	18	18	102	103
non-na	0	0	2	3	10	11	15	13	19	18	106	107

Overlined measures are mean values, whereas M superscript denotes median values
T is the traditional model (25.1)–(25.3)
⁻ Without $\mathscr{X}$: the misaligned covariate X is not included as a regressor
⁻ With $\mathscr{X}$ plug-in(1): with X plugged-in by "population weighting" interpolation
⁻ With $\mathscr{X}$ plug-in(2): with X plugged-in by using aggregated atom-based model estimates
RA is the regression-only atom-based model (Sect. 25.2)
S↺LAE is the discrete-space version of the atom-based model (Sects. 25.2 and 25.4)
S⊙LAE is the continuous-space version of the atom-based model (Sects. 25.3 and 25.4)
non-na rows refer to measures calculated leaving out survey non-sampled areas

Main outcomes of the selected simulation experiments are concisely listed below, with possible explanations according to Authors' opinion *in italic*. (For reading see the table legend.)

- The T model without/with $\mathscr{X}$ plugged-in(1), *as expected*, gets the worst performance scores.
- The T model with $\mathscr{X}$ plugged-in(2) gets far better scores than the coarser T models (see bias), *since the areal unit of analysis is set to a finer level than small area*, yet it performs worse than RA model (see either accuracy or reliability), *since it fails to properly account for the uncertainty related to this integration.*
- Both the S↺LAE and S⊙LAE models have generally the best performance, *since*, differently from the RA version, they *integrate large area direct estimates as well.*
- S↺LAE and S⊙LAE models show equivalent results in scenario (a) (Table 25.1) *since here information is available on a spatial resolution not "substantively" finer than atoms* (thanks to a property of the mixing link, see Note 3).
- The S⊙LAE model improves on S↺LAE one in scenario (b) (Table 25.2) as it does not run into ecological bias issues here arising *for being information available on finer scale than atoms* (the log-link fails to maintain the functional form across aggregated levels)

25.6 Concluding Remarks

We conclude by pointing out some relevant innovative aspects of the proposed class of misaligned data models (in both versions) with respect to a traditional model for SAE.

The proposed HB framework builds small area estimates (θ_i) through aggregation (over θ_k's), hence similar to the way survey estimates are generally obtained. However, it integrates any source of potentially important information within a full probability model, thus formally combining the multiple sources of variation within a global analysis. Sources of information are those traditionally incorporated into a SAE model, such as small area direct estimates (d_i), auxiliary variables (z_k), (possibly) synthetic estimates (S_k). Furthermore, the atom-based models here proposed allow for misaligned supplementary variables (X_k) and large area direct estimates (d_p) being incorporated as well, properly accounting for the uncertainty related to this integration.

Finally, traditional model-based estimators do not generally benchmark to reliable direct survey estimates for large areas (the construction of benchmarked estimators is currently an active thread of SAE research, see e.g. [9, 11, 15]). While the proposed atom-based models have the desirable property of automatic benchmarking to large area estimates.

References

1. Banerjee, S., Carlin, B.P., Gelfand, A.E.: Hierarchical Modeling and Analysis for Spatial Data. Chapman & Hall/CRC, Boca Raton (2004)
2. Best, N.G., Ickstadt, K., Wolpert, R.L.: Spatial Poisson regression for health and exposure data measured at disparate resolutions. J. Am. Stat. Assoc. **95**, 1076–1088 (2000)
3. Chakraborty, A., Gelfand, A.E., Wilson, A.M., Latimer, A.M., Silander Jr., J.A.: Modeling large scale species abundance with latent spatial processes. Ann. Appl. Stat. **4**, 1403–1429 (2010)
4. Flowerdew, R., Green, M.: Developments in areal interpolation methods and GIS. Ann. Reg. Sci. **26**, 67–78 (1992)
5. Flowerdew, R., Green, M.: Areal interpolation and types of data. In: Fotheringam, S., Rogerson, P. (eds.) Spatial Analysis and GIS, pp. 121–145. Taylor and Francis, London (1994)
6. Gelfand, A.E., Silander Jr., J.A., Wu, S., Latimer, A.M., Lewis, P., Rebelo, A.G., Holder, M.: Explaining species distribution patterns through hierarchical modeling. Bayesian Anal. **1**, 42–92 (2005)
7. Gelfand, A.E., Diggle, P., Fuentes, M., Guttorp, P. (eds.): Handbook of Spatial Statistics. Chapman & Hall/CRC, Boca Raton (2010)
8. Goodchild, M.F., Lam, N.S.-N.: Areal interpolation: a variant of the traditional spatial problem. Geo-Processing **1**, 297–312 (1980)
9. Lu, L., Larsen, M.D.: Small area estimation in a survey of high school students in Iowa. In: Proceedings of the Survey Research Methods Section, American Statistical Association, pp. 2627–2634 (2007)
10. Mugglin, A.S., Carlin, B.P., Gelfand, A.E.: Fully model-based approaches for spatially misaligned data. J. Am. Stat. Assoc. **95**, 877–887 (2000)
11. Pfeffermann, D., Tiller, R.: Small-area estimation with state-space models subject to benchmark constraints. J. Am. Stat. Assoc. **101**, 1387–1397 (2006)

12. Tobler, W.R.: Smooth pycnophylactic interpolation for geographical regions. J. Am. Stat. Assoc. **74**, 519–530 (1979)
13. Torelli, N., Trevisani, M.: Labour force estimates for small geographical domains in Italy: problems, data and models. Int. Rev. Soc. Sci. **4**, 443–464 (2008)
14. Trevisani, M., Torelli, N.: Hierarchical Bayesian models for small area estimation with count data. DiSES Working Papers 115, University of Trieste (2007)
15. Ugarte, M.D., Militino, T.G., Goicoa, T.: Benchmarked estimates in small areas using linear mixed models with restrictions. Test, 18, 342–364 (2009)
16. Wakefield, J.C.: Disease mapping and spatial regression with count data. Biostatistics **8**, 158–183 (2007)

Scrambled Response Models Based on Auxiliary Variables

26

Pier Francesco Perri and Giancarlo Diana

Abstract

We discuss the problem of obtaining reliable data on a sensitive quantitative variable without jeopardizing respondent privacy. The information is obtained by asking respondents to perturb the response through a scrambling mechanism. A general device allowing for the use of multi-auxiliary variables is illustrated as well as a class of estimators for the unknown mean of a sensitive variable. A number of scrambled response models are shown and others discussed in terms of the efficiency of the estimates and the privacy guaranteed to respondents.

Keywords

Privacy protection • Regression estimator • Sensitive variable

26.1 Introduction

Posing direct questions on private and confidential topics such as gambling, alcoholism, sexual behavior, abortion, drug taking, tax evasion, illegal income, and so on can cause embarrassment or fear of social disapproval to respondents. Even when the interviewers do their best to guarantee confidentiality, subjects can be skeptical and may be reluctant to supply truthful answers.

P.F. Perri (✉)
Department of Economics, Statistics and Finance, University of Calabria, Cubo OC, 87036 Arcavacata di Rende (CS), Italy
e-mail: pierfrancesco.perri@unical.it

G. Diana
Department of Statistical Sciences, University of Padova, Via Cesare Battisti 241, 35121 Padova, Italy
e-mail: giancarlo.diana@unipd.it

N. Torelli et al. (eds.), *Advances in Theoretical and Applied Statistics*, Studies in Theoretical and Applied Statistics, DOI 10.1007/978-3-642-35588-2_26,

To reduce the rate of nonresponse and minimize the underreporting of embarrassing, threatening, or even incriminating behaviors, survey statisticians have developed a number of procedures which ensure interviewee anonymity or, at least, a high degree of confidentiality. When sensitive topics are surveyed, researchers often use self-administered questionnaires with paper and pencil, computer-assisted self-interviews, or Internet-delivered interviews. An entirely different approach for eliciting information on sensitive data while ensuring respondent privacy protection is the interviewing procedure pioneered by Warner [21] and known as the Randomized Response (RR) technique. The rationale of the technique is simple: respondents are instructed to give partially misclassified (or perturbed) responses by using a randomization device, such as a dice, coin, spinner, or a deck of cards. For instance, let us consider a population divided into two mutually exclusive groups: one group with a stigmatizing attribute A (e.g., tax evader) and the other without such a characteristic. Then, under Warner's procedure a deck of cards is used as randomization device to collect information. A proportion $p \neq 0.5$ of cards is marked with the statement "I posses the sensitive attribute A", the others are marked with the complementary statement "I do not posses the sensitive attribute A". The respondents are requested to draw a card, note its mark unseen by the enquirer and say "yes" if the statement on the card matches their true status, "no" if there is mismatch. Since the interviewer does not know to which type of card the "yes" or "no" refers to, the use of the mechanism ensures that respondents cannot be identified on the basis of their answers.

Different studies have assessed the validity of RR methods and shown that these can significantly outperform conventional data collection methods (see, e.g., [13, 20]). In the Netherlands, there is a well-established tradition of the use of RR data collection in official surveys mainly concerned with measuring the prevalence of fraud, especially in the area of disability benefits (see [14]). However, many authors use RR techniques in a wide range of empirical studies. Among others, Lara et al. [12] have recently estimated the prevalence of abortion in Mexico; van der Heijden and Böckenholt [19] discussed applications of the methodology in e-commerce; Ostapczuk et al. [15] and Krumpal [11] treated the issue of xenophobia and anti-semitism in Germany, while Arnab and Singh [1] measured the impact of AIDS infection in Botswana.

Standard RR methods are used primarily in surveys which require a "yes" or "no" response to a sensitive question, or a choice of responses from a set of nominal categories. The purpose is to estimate the proportion of people bearing a sensitive characteristic (see, e.g., [5]). Nevertheless, there is a growing amount of theoretical literature dealing with situations where the response to a sensitive question results in a quantitative variable and the interest is on the estimation of the mean (total) of the sensitive variable. Consider, for instance, the problem of estimating the average personal income or the number of fraudulent acts and see Gjestvang and Singh [9] for a recent application to the estimation of average GPA of students at St Cloude State University. In these cases, people are asked to scramble the true response algebraically by means of a coding mechanism.

In the remainder of this chapter, we discuss a general approach for scrambling the responses which takes into account the possibility of using multi-auxiliary variables, and we show a class of estimators for the mean of a sensitive variable of interest. The problem of evaluating the degree of privacy protection is also considered.

26.2 A Class of Estimators for the Sensitive Mean

Let $P = \{1, 2, \ldots, N\}$ be a finite population of identifiable individuals and $Y \geq 0$ be a quantitative sensitive variable with mean and variance, $\mathbb{E}(Y) = \mu_y$ and $\mathbb{V}(Y) = \sigma_y^2$, assumed to be unknown.

In RR theory, many devices have been implemented to estimate μ_y. Most of these use a coding mechanism of the response on Y in the sense that respondents are requested to perturb the true value of Y algebraically through one or more random numbers generated from known scrambling distributions. For instance, Pollock and Bek [16] considered the additive/multiplicative models which involve the respondents adding/multiplying the answer to the sensitive question by a random number from a known distribution. The multiplicative model was later taken up and investigated in depth by Eichhorn and Hayre [8]. Gupta et al. [10] introduced an optional randomized response technique which is more efficient than Eichhorn and Hayre's scrambled device. Bar-Lev et al. [2] generalized Eichhorn and Hayre's device by introducing a design parameter which is used for randomizing the responses. Hereafter, we refer to coding procedures for quantitative variables as Scrambled Response (SR) models.

Most of the SR models in the literature do not consider the possibility of getting more accurate estimates of μ_y by using multi-auxiliary information directly at the estimation stage. Here we discuss this possibility in a design-based set-up when a k-vector of nonsensitive auxiliary variables, say $\mathbf{X} = (X_1, \ldots, X_k)'$, correlated to Y, is available. Alternatively, a model-assisted approach could be followed as, for instance, in Chaudhury and Roy [3]. Auxiliary variables of this type are not unusual in many areas of social, clinical, and medical research where the researcher could consider social, economic, or demographic information coming from administrative sources or which are in the public domain. For instance, on studying people's income and tax evasion, we may consider variables such as the type of car, house size, and neighborhood which are certainly nonsensitive but connected with the people's living standards.

Suppose that the vector $\boldsymbol{\mu}_x$ of the auxiliary means is known and that the values of Y [$\mathbf{X}$] can be perturbed algebraically by means of positive scrambling variables W and U [$\mathbf{T}$ and $\mathbf{H}$] whose (marginal) distributions are known. The scrambling variables are assumed to be mutually independent and also independent of Y [$\mathbf{X}$]. Finally, let $S = \varphi(W, U, Y)$ and $\mathbf{R} = \psi(\mathbf{T}, \mathbf{H}, \mathbf{X})$ denote the coded responses on Y and $\mathbf{X}$, respectively, according to the *scrambling functions* φ and ψ induced by the coding device. See Diana and Perri [6, 7] and Table 26.1 for examples.

In our approach, to estimate μ_y, a sample of n individuals is selected from the population and each respondent is asked to perform a Bernoulli trial with known probability of success p. If this is successful, the respondent then gives the true

Table 26.1 Selected models from the generic scheme

ψ	φ	p	c	h
$\mathbf{X}$	$Y + U$	$[0, 1]$	$(1 - p)\mu_u$	1
$\mathbf{X}$	YW	$[0, 1]$	0	$(1 - p)\mu_w + p$
$\mathbf{X} + \mathbf{H}$	$Y + U$	$[0, 1]$	$(1 - p)\mu_u$	1
$\mathbf{X} \odot \mathbf{T}$	YW	$[0, 1]$	0	$(1 - p)\mu_w + p$
$\mathbf{X}$	$W(Y + U)$	$[0, 1]$	$(1 - p)\mu_w\mu_u$	$p + (1 - p)\mu_w$
$\mathbf{X}$	$W[\alpha U + (1 - \alpha)Y]$	$[0, 1]$	$(1 - p)\alpha\mu_w\mu_u$	$(1 - p)(1 - \alpha)\mu_w + p$
$\mathbf{X}$	$\xi(Y + U) + (1 - \xi)WY$	$[0, 1]$	$(1 - p)\xi\mu_u$	$(1 - p)[\xi + (1 - \xi)\mu_w] + p$
$\mathbf{T} \odot (\mathbf{X} + \mathbf{H})$	$W(Y + U)$	$[0, 1]$	$(1 - p)\mu_w\mu_u$	$p + (1 - p)\mu_w$
$\mathbf{T} \odot [\alpha\mathbf{H} + (1 - \alpha)\mathbf{X}]$	$W[\alpha U + (1 - \alpha)Y]$	$[0, 1]$	$(1 - p)\alpha\mu_w\mu_u$	$(1 - p)(1 - \alpha)\mu_w + p$
$\xi(\mathbf{X} + \mathbf{H}) + (1 - \xi)\mathbf{T} \odot \mathbf{X}$	$\xi(Y + U) + (1 - \xi)WY$	$[0, 1]$	$(1 - p)\xi\mu_u$	$(1 - p)[\xi + (1 - \xi)\mu_w] + p$

$\odot$ Denotes the Hadamard product while α and ξ are design parameters in the unit interval

values of Y and $\mathbf{X}$; in case of failure the values S and $\mathbf{R}$. Obviously, the interviewer does not know the outcome of the Bernoulli experiment. Thus, the distribution of the responses is

$$(Z, \mathbf{V}) = \begin{cases} (Y, \mathbf{X}), & \text{with probability } p \\ (S, \mathbf{R}), & \text{with probability } 1 - p. \end{cases} \tag{26.1}$$

Under this generic scheme, we consider the following class of unbiased estimators of μ_y based on a SRSWOR sample $\{(z_1, \mathbf{v}_1), (z_2, \mathbf{v}_2), \ldots, (z_n, \mathbf{v}_n)\}$ of n responses

$$\hat{\mu}_g = \frac{\bar{z}_d - c}{h}, \quad (h \neq 0) \tag{26.2}$$

where $\bar{z}_d = \bar{z} + (\boldsymbol{\mu}_v - \bar{\mathbf{v}})'\mathbf{b}$, $\bar{z} = \sum_{j=1}^{n} z_j/n$ and $\bar{\mathbf{v}} = \sum_{j=1}^{n} \mathbf{v}_j/n$ being unbiased estimators of μ_z and $\boldsymbol{\mu}_v$, respectively; c and h are real constants such that $c + h\mu_y = \mathbb{E}(\bar{z})$ that depend exclusively on the scrambling device adopted for Y; $\mathbf{b} = (b_1, \ldots, b_k)'$ is a vector of constants linked to the efficient use of the auxiliary variables. In fact, setting $\mathbf{b} = \mathbf{0}$ excludes the possibility of using auxiliary information at the estimation stage and, thus, for fixed c and h, the class boils down to the specific estimator (with $\bar{z}_d = \bar{z}$) generated by the adopted SR method. Therefore, equation (26.2) has the merit of including, under different coding procedures, both classical estimators without supplementary information and possible estimators based on auxiliary variables. Some simple estimators belonging to the class are discussed in Sect. 26.4 as well as in Diana and Perri [6, 7].

Let $\gamma = (N - n)/n(N - 1)$, $\boldsymbol{\Sigma} = \{\sigma_{v_i v_j}\}_{i,j=1}^{k}$ and $\boldsymbol{\Omega} = \{\sigma_{zv_i}\}_{i=1}^{k}$, $\sigma_{..}$ being the covariance between the variables indicated in the subscript. Then, the variance of $\hat{\mu}_g$ is given by (see, e.g., [4])

$$\mathbb{V}(\hat{\mu}_g) = \gamma \frac{\sigma_z^2}{h^2}(\sigma_z^2 + \mathbf{b}'\boldsymbol{\Sigma}\mathbf{b} + 2\mathbf{b}'\boldsymbol{\Omega}). \tag{26.3}$$

Minimization of (26.3) with respect to $\mathbf{b}$ is achieved for $\mathbf{b} = \boldsymbol{\Sigma}^{-1}\boldsymbol{\Omega} = \{\beta_{zv_i}\}_{i=1}^{k}$, where β_{zv_i} denotes the *partial regression coefficient* of Z on V_i. With this choice, the minimum variance bound is given by

$$\mathbb{V}(\hat{\mu}_g)_{\min} = \gamma \frac{\sigma_z^2}{h^2}(1 - \rho_{z.\mathbf{v}}^2) \tag{26.4}$$

where $\rho_{z.\mathbf{v}}^2 = \boldsymbol{\Omega}'\boldsymbol{\Sigma}^{-1}\boldsymbol{\Omega}/\sigma_z^2$ is the *squared multiple correlation coefficient* of Z on $\mathbf{V}$.

Expression (26.4) well emphasizes the role of the auxiliary variables in improving the accuracy of the estimates. As regard this, $\gamma\sigma_z^2/h^2$ is the variance of the estimator which does not employ the k-vector $\mathbf{X}$ and $(1 - \rho_{z.\mathbf{v}}^2)$ denotes the reduction in the variance due to the use of $\mathbf{X}$ through the traditional *regression estimator*. Therefore, the use of one or more auxiliary variables is certainly profitable.

We note that the optimum coefficients β_{zv_i} are generally unknown and need to be estimated. The ordinary least squares method provides consistent estimates that can always be computed since they are based on the observed scrambled responses provided by the sampled respondents. Obviously, OLS estimates based on scrambled responses are less efficient than those obtained if the true values of the sensitive variable are surveyed exactly. On this see Singh et al. [18].

The class we have introduced in a general framework is highly flexible since it can be easily adapted to whatever perturbing response mechanism the researcher decides to use for the sensitive problem at hand. Once the choice of the model has been made, the coding functions φ and ψ assume an explicit form and the constant c and h are defined. Table 26.1 summarizes a number of examples concerning different specifications of the general SR model and gives the expressions for c and h to estimate μ_y from (26.2). Note that the choice $p = 0$ excludes any possibility of giving a direct response on Y, while for $p \in (0,1)$ responses on Y can be partially perturbed. To evaluate the performance of a particular scrambling scheme in terms of efficiency of the estimates we need to rewrite the following expressions

$$\mu_z = p\mu_y + (1-p)\mu_s, \quad \mu_{v_i} = p\mu_{x_i} + (1-p)\mu_{r_i} \tag{26.5}$$

$$\sigma_z^2 = p\mu_{2,y} + (1-p)\mu_{2,s} - \mu_z^2, \quad \sigma_{v_i}^2 = p\mu_{2,x_i} + (1-p)\mu_{2,r_i} - \mu_{v_i}^2 \tag{26.6}$$

$$\sigma_{v_i z} = p\sigma_{x_i y} + (1-p)\sigma_{r_i s} + p(1-p)(\mu_y - \mu_s)(\mu_{x_i} - \mu_{r_i}) \tag{26.7}$$

$$\sigma_{v_i v_j} = p\sigma_{x_i x_j} + (1-p)\sigma_{r_i r_j} + p(1-p)\left[\mu_{x_i}(\mu_{x_j} - \mu_{r_j}) + \mu_{r_i}(\mu_{r_j} - \mu_{x_j})\right] \tag{26.8}$$

where $\mu_{2,.}$ denotes the second moment of the variable indicated in the subscript. The details have been omitted since most of these expressions, as well as the guidelines for their derivation, can be found in Diana and Perri [6, 7].

26.3 Respondent Privacy Protection

One relevant aspect, often underrated in RR theory, is the measure of respondent privacy protection. An index used to measure privacy should indicate how closely the original values of the perturbed sensitive variable Y can be estimated. The closer these values, the higher the privacy disclosure. SR devices which ensure high performance in terms of efficiency are usually less protective. Therefore, a fair comparison of alternative scrambling strategies should necessarily take into account both considerations of efficiency and privacy protection. Different ways of measuring confidentiality are discussed in the literature and a brief review is reported in Diana and Perri [7].

Starting from the proposal by Zaizai et al. [22], we first consider the measure $\Delta = \mathbb{E}(Q - Y)^2 = \mathbb{E}_d\mathbb{E}_r(Q - Y)^2$ where $\mathbb{E}_d$ and $\mathbb{E}_r$ denote, respectively, the expectations with respect to the sampling design and the scrambling device, and Q is a randomized response unbiased estimator of the true value of Y. The magnitude of deviation of the unbiased coded response Q_i from the true undiscovered value Y_i determines the *degree of protection* to the ith respondent induced by the adopted SR device. If $\Delta = 0$ then privacy is completely jeopardized, in other words respondent privacy has not been protected. Focusing, for simplicity, on scrambling methods allowing us to obtain only coded sensitive responses (i.e., $p = 0$ and $Z = S$ in (26.1)), it follows that $Q = (S - c_0)/h_0$ where c_0 and h_0 denote the expressions for c and h when $p = 0$. Therefore, the Δ measure can be expressed as

$$\Delta = \mathbb{E}\left(\frac{S - c_0}{h_0} - Y\right)^2 = \mathbb{E}_p\mathbb{E}_r\left(\frac{S - c_0}{h_0} - Y\right)^2. \qquad (26.9)$$

After some algebra, we get the form

$$\Delta = \frac{h_0^2\mu_{2,y} - 2h_0\mu_{ys} + 2c_0h_0\mu_y + \mu_{2,s} - 2c_0\mu_s + c_0^2}{h_0^2} \qquad (26.10)$$

which can be adapted to a given coding method by means of an appropriate specification of the constants c_0 and h_0, as well as the moments involved.

26.4 Comparisons

For illustrative purposes, we consider the following simple SR models which are taken as a starting point for the analysis of more complex devices

Author	Model	Coded response **Z**	Estimator
Pollock and Bek [16]	Additive	$Y + U$	$\hat{\mu}_{\text{PB}} = \bar{z} - \mu_u$
Eichhorn and Hayre [8]	Multiplicative	YW	$\hat{\mu}_{\text{EH}} = \bar{z}/\mu_w$
Saha [17]	Mixed 1	$W(Y + U)$	$\hat{\mu}_1 = (\bar{z}/\mu_w) - \mu_u$
Diana and Perri [6]	Mixed 2	$W[\alpha U + (1 - \alpha)Y]$	$\hat{\mu}_2 = (\bar{z} - \alpha\mu_w\mu_u)/(1 - \alpha)\mu_w$
Diana and Perri [6]	Mixed 3	$\xi(Y + U) + (1 - \xi)WY$	$\hat{\mu}_3 = (\bar{z} - \xi\mu_u)/[\xi + (1 - \xi)\mu_w]$

At this stage, we assume that no auxiliary variable is involved. Considerations on the performance of these models are detailed in Diana and Perri [6] where it is shown that, when $\mu_{2,y} > \sigma_u^2 C_w^{-2}$, with $C_.$ denoting the coefficient of variation of the variable indicated in the subscript, and for ξ minimizing $\mathbb{V}(\hat{\mu}_3)$, the following ordering is valid

$$\mathbb{V}(\hat{\mu}_1) \geq \mathbb{V}(\hat{\mu}_2|\alpha) > \mathbb{V}(\hat{\mu}_{\mathrm{EH}}) > \mathbb{V}(\hat{\mu}_{\mathrm{PB}}) > \mathbb{V}(\hat{\mu}_3) \quad \text{for } 0 < \alpha \leq 0.5$$

$$\mathbb{V}(\hat{\mu}_2|\alpha) > \mathbb{V}(\hat{\mu}_1) > \mathbb{V}(\hat{\mu}_{\mathrm{EH}}) > \mathbb{V}(\hat{\mu}_{\mathrm{PB}}) > \mathbb{V}(\hat{\mu}_3) \quad \text{for } 0.5 < \alpha < 1.$$

For the above models, the Δ measure takes the form

$$\Delta_{\mathrm{PB}} = \sigma_u^2, \quad \Delta_{\mathrm{EH}} = \mu_{2,y} C_w^2, \quad \Delta_1 = \mu_{2,u} l_w + \mu_{2,y}(l_w - 1) - \mu_u^2$$

$$\Delta_2 = \Delta_1 + \frac{(2\alpha - 1)a + 2\alpha(1-\alpha)b}{(1-\alpha)^2}\mu_u, \quad \Delta_3 = \frac{\mu_{2,y}\sigma_w^2\sigma_u^2}{\mu_w^2\sigma_u^2 + \mu_{2,y}\sigma_w^2}$$

where $a = \mu_u(l_u l_w - 1)$ and $b = \mu_y(l_w - 1)$ with $l_u = \mu_{2,u}/\mu_u^2$, $l_w = \mu_{2,w}/\mu_w^2$. Comparing the measures we find that the multiplicative model is more protective than the additive one if the condition $\mu_{2,y} > \sigma_u^2 C_w^{-2}$ holds, while Δ_1 is always greater than Δ_{EH} since $\Delta_1 - \Delta_{\mathrm{EH}} = l_u l_w - 1$. Similarly, it can be demonstrated that $\Delta_{\mathrm{PB}} > \Delta_3$. Therefore, the following ordering holds: $\Delta_1 > \Delta_{\mathrm{EH}} > \Delta_{\mathrm{PB}} > \Delta_3$. Note that the variances are ordered in the same manner but with an opposite meaning: the more protective the method, the less efficient the estimator. The analysis for the privacy protection induced by the mixed 2 model is not so straightforward. It can be proved, however, that Δ_2 is an increasing function of α in $(0, 1)$ with $\lim_{\alpha \to 1^-} \Delta_2 = \infty$ and $\Delta_2 = \Delta_{\mathrm{EH}}$ for $\alpha = 0$. Then $\Delta_2 > \Delta_{\mathrm{EH}} > \Delta_{\mathrm{PB}}$ for $\alpha \in (0, 1)$. Moreover, a value of α exists, say $\alpha_0 = (a + b - \sqrt{a^2 + b^2})/2b$ in the interval $(0, 0.5)$, such that $\Delta_2 = \Delta_1$ and $\Delta_2 > \Delta_1$ $[\Delta_2 < \Delta_1]$ for $\alpha > \alpha_0$ $[\alpha < \alpha_0]$. From efficiency considerations, it follows that for $\alpha \in (\alpha_0, 0.5)$ the mixed 2 model outperforms mixed 1 model both in efficiency and privacy protection.
We note that the additive and multiplicative models are perfectly equivalent in terms of efficiency and privacy protection when $\mu_{2,y} = \sigma_u^2 C_w^{-2}$.

The results given in (26.9) and (26.10) can be easily extended to the case where $p \in (0, 1)$ and, thus, a direct response on Y is possible. In this situation we find

$$\Delta = (1 - p)\mathbb{E}\left(\frac{S - c_0}{h_0} - Y\right)^2$$

which well emphasizes the (unsurprising) fact that privacy is less protected when a direct response is allowed.

When the k-vector $\mathbf{X}$ of auxiliary variables is used, the variance of the estimators decreases according to expression (26.4), while the confidentiality of the respondents is expected to decrease as well. This aspect is not captured by the Δ measure. To overcome this drawback, we extend here the measure proposed by Diana and Perri [7] and consider the following

$$\tau = 1 - \rho_{y.\mathbf{vz}}^2, \tag{26.11}$$

$\rho_{y.\mathbf{vz}}$ being the multiple correlation coefficient of Y on the coded response variable and the coded auxiliary variables. This is a normalized and intuitive measure since, according to the literature on the topic, it indicates how closely the true values of the sensitive variable can be estimated by the perturbed ones. Maximum privacy protection is attained when $\tau = 1$ since $\rho^2_{y.vz} = 0$. On the contrary, as τ approaches zero, privacy protection declines and respondents are expected to become less collaborative. Besides the elements given in (26.5) and (26.6), the determination of the τ measure requires additional calculations for the covariance terms, namely

$$\sigma_{yv_i} = p\sigma_{x_i y} + (1-p)\sigma_{yr_i}, \quad \sigma_{yz} = p\sigma^2_y + (1-p)\sigma_{ys}.$$

When no auxiliary information is used, from (26.11) we get $\tau = 1 - \rho^2_{yz}$ which reduces to

$$\tau = 1 - h^2 \frac{\sigma^2_y}{\sigma^2_z}$$

if the variable Y is linearly perturbed whatever $p \in (0,1)$. The considered models meet this requirement of linearity. Therefore, from (26.4), the more efficient the estimates, the less protective the coding procedure.

For illustrative purposes, let us suppose information on X is directly collected and the sensitive response scrambled by the aforementioned SR models ($p = 0$ and $Z = S$). Then, after some algebra, we get

$$\tau = 1 - \frac{\rho^2_{xy} + \rho^2_{ys} - 2\rho_{xy}\rho_{ys}\rho_{xs}}{1 - \rho^2_{xs}} = 1 - \frac{a + (1-2a)\rho^2_{xy}}{1 - a\rho^2_{xy}} \tag{26.12}$$

where $a = h_0\sigma^2_y/\sigma^2_s$.

Finally, when a direct response on Y is allowed with probability $p \in (0,1)$, the privacy can be still measured through τ in (26.12) by considering h instead of h_0 in a. Additionally, privacy can be better guaranteed by means of a suitable choice of p. For instance, the response model

$$Z = \begin{cases} Y, & \text{with probability } p \\ S = Y + U, & \text{with probability } 1-p \end{cases}$$

is more protective (in the τ sense) than the simple model $Z = Y + U$ when $p < 2 - l_u$. If the scrambling variable U is chosen such that $l_u > 2$, then the condition is never met and respondent confidentiality can be increased by avoiding direct questioning.

26.5 Discussion and Final Remarks

The idea of releasing algebraically perturbed data rather than the exact values of a quantitative variable can be particularly useful to combat deliberate misreporting or untruthful responses when people are surveyed on sensitive issues. The chapter shows how to treat scrambled response models in a general framework in order to estimate the mean of a sensitive variable when auxiliary variables are available and are used directly at the estimation stage provided that correlation with the study character can be assumed. The superiority of estimators based on auxiliary information is clearly shown, whatever the adopted scrambling procedure.

One relevant aspect when dealing with coding methods is the degree of privacy protection assured to the respondents. Methods which produce highly efficient estimates tend to be more intrusive and jeopardize privacy to a much greater extent. The key issue, when choosing a model, is to find a right trade-off between privacy protection and efficiency in the estimates. In practical situations, the choice among different models is not always easy and may deserve special attention since it depends to a greater extent on the problem at hand. See on this the simulation studies given in Diana and Perri [6, 7]. However, the issue of how to measure the degree of privacy protection correctly still remains a matter of concern. Different measures are discussed in the literature, two of which are mentioned in the chapter. One important problem is that the measures can yield different conclusions and do not always allow us to capture certain relevant aspects such as the effect of the auxiliary information. Moreover, even if they can provide the researcher with an indication of privacy disclosure, it is questionable whether the respondents themselves are able to measure the extent of privacy disclosure offered by a scrambling design. Whatever the privacy measure, respondents usually have a subjective impression regarding disclosure which can be higher or lower than that measured by an index. In this sense, an empirical study would certainly be useful to ascertain and quantify the difference between the perceived and measured degree of respondent privacy protection. Finally, since respondents cannot fully appreciate the level of privacy protection ensured by a scrambled response method, researchers may be more inclined towards designs allowing more efficient estimates rather that ones that offer high protection in theory. On the other hand, simple coding mechanisms should be preferred since it may not be easy to understand how to perturb the responses algebraically. This may be a limitation for the application of the scrambled methods since their correct execution can only be performed by highly educated people unless computer support is provided.

To conclude we observe that no complete investigation on SR models can leave two standard survey issues out of consideration. The first refers to the assessment of the accuracy of the estimates via variance estimation. The second is related to the extension of existing models to more complex designs such as unequal probability sampling, multistage sampling, stratified random sampling, and small area estimation. We hope that future investigation can shed light on these key issues.

References

1. Arnab, R., Singh, S.: Randomized response techniques: an application to the Botswana AIDS impact survey. J. Stat. Plann. Inference **140**, 941–953 (2010)
2. Bar-Lev, S.K., Bobovitch, E., Boukai, B.: A note on randomized response models for quantitative data. Metrika **60**, 255–260 (2004)
3. Chaudhury, A., Roy, D.: Model assisted survey sampling strategies with randomized response. J. Stat. Plann. Inference **60**, 61–68 (1997)

4. Diana, G., Perri, P.F.: Estimation of finite population mean using multi-auxiliary information. Metron **LXV**, 99–112 (2007)
5. Diana, G., Perri, P.F.: Estimating a sensitive proportion through randomized response procedures based on auxiliary information. Stat. Pap. **50**, 661–672 (2009)
6. Diana, G., Perri, P.F.: New scrambled response models for estimating the mean of a sensitive quantitative character. J. Appl. Stat. **37**, 1875–1890 (2010)
7. Diana, G., Perri, P.F.: A class of estimators for quantitative sensitive data. Stat. Pap. **52**, 633–650 (2011)
8. Eichhorn, B., Hayre, L.S.: Scrambled randomized response methods for obtaining sensitive quantitative data. J. Stat. Plann. Inference **7**, 307–316 (1983)
9. Gjestavang, C.R., Singh, S.: An improved randomized response model: estimation of mean. J. Appl. Stat. **36**, 1361–1367 (2009)
10. Gupta, S., Gupta, B., Singh, S.: Estimation of sensitive level of personal interview survey questions. J. Stat. Plann. Inference **100**, 239–247 (2002)
11. Krumpal, I.: Estimating the prevalence of xenophobia and anti-semitism in Germany: a comparison of randomized response and direct questioning. Soc. Sci. Res. **41**, 1387–1403 (2012)
12. Lara, D., García, S.G., Ellertson, C., Camlin, C., Suaréz, J.: The measure of induced abortion in Mexico using random response technique. Sociol. Methods Res. **35**, 279–301 (2006)
13. Lensvelt-Mulders, G.J.L.M., Hox, J.J., van der Heijden, P.G.M., Maas, C.J.M.: Meta-analysis of randomized response research: thirty-five years of validation. Sociol. Methods Res. **33**, 319–348 (2005)
14. Lensvelt-Mulders, G.J.L.M., van der Heijden, P.G.M., Laudy, O., van Gils, G.: A validation of a computer-assisted randomized response survey to estimate the prevalence of fraud in social security. J. R. Stat. Soc. Ser. A **169**, 305–318 (2006)
15. Ostapczuk, M., Musch, J., Mashagen, M.: A randomized-response investigation of the education effect in attitudes towards foreigners. Eur. J. Soc. Psychol. **39**, 920–931 (2009)
16. Pollock, K.H., Bek, Y.: A comparison of three randomized response models for quantitative data. J. Am. Stat. Assoc. **71**, 884–886 (1976)
17. Saha, A.: A simple randomized response technique in complex surveys. Metron **LXV**, 59–66 (2007)
18. Singh, S., Joarder, A.H., King, M.L.: Regression analysis using scrambled responses. Aust. J. Stat. **38**, 201–211 (1996)
19. van der Heijden, P.G.M., Böckenholt, U.: Applications of randomized response methodology in e-commerce. In: Jank, W., Shmueli, G. (eds.) Statistical Methods in e-Commerce Research, pp. 401–416. Wiley, New York (2008)
20. van der Heijden, P.G.M., van Gils, G., Bouts, J., Hox, J.J.: A comparison of randomized response, computer-assisted self-interview, and face-to-face direct questioning. Sociol. Methods Res. **28**, 505–537 (2000)
21. Warner, S.L.: Randomized response: a survey technique for eliminating evasive answer bias. J. Am. Stat. Assoc. **60**, 63–69 (1965)
22. Zaizai, Y., Jingyu, W., Junfeng, L.: An efficiency and protection degree-based comparison among the quantitative randomized response strategies. Commun. Stat. Simul. Comput. **38**, 400–408 (2009)

Using Auxiliary Information and Nonparametric Methods in Weighting Adjustments

27

Emilia Rocco

Abstract

Weighting adjustments are commonly used in survey estimation to compensate for unequal selection probabilities, nonresponse, noncoverage, and sampling fluctuations from known population values. Over time many weighting methods have been proposed, mainly in the nonresponse framework. These methods generally make use of auxiliary variables to reduce the bias of the estimators and improve their efficiency. Frequently, a substantial amount of auxiliary information is available and the choice of the auxiliary variables and the way in which they are employed may be significant. Moreover, the efficacy of weighting adjustments is often seen as a bias–variance trade-off. In this chapter, we analyze these aspects of the nonresponse weighting adjustments and investigate the properties of mean estimators adjusted by individual response probabilities estimated through nonparametric methods in situations where multiple covariates are both categorical and continuous.

Keywords

Bias–variance trade-off • Nonresponse • Quasi-randomization inference

27.1 Introduction

Estimation in sample surveys is carried on mainly by attaching weights to observed data and then computing the weighting summation. The weights are generally developed in a series of stages to compensate for unequal selection probabilities, nonresponse, noncoverage, and sampling fluctuations from known population

E. Rocco (✉)
Department of Statistica, Informatica, Applicazioni "G. Parenti", University of Florence, Viale Moragni 59, 50134 Firenze, Italy
e-mail: rocco@disia.unifi.it

N. Torelli et al. (eds.), *Advances in Theoretical and Applied Statistics*, Studies in Theoretical and Applied Statistics, DOI 10.1007/978-3-642-35588-2_27,

values. In the first stage of weighting for unequal selection, the base weights are usually readily determined from selection probabilities. Conversely, all their subsequent adjustments involve models or assumptions of some kind, either explicit or implicit, and make use of auxiliary information. Thus their efficacy in terms of bias and/or variance reduction depends on the validity of the assumed model and the choice of auxiliary variables.

In the remainder of the chapter we just consider the adjustments for nonresponse. Their essence is to increase the weights of the respondents so that they represent the nonrespondents. The respondent base weights are increased by a multiplying factor given by the reciprocal of the estimated response probabilities. A critical step in this procedure is the estimation of the response probabilities. This step is usually carried out under the assumption that, after conditioning on a set of covariates, the response mechanism is missing at random (MAR). Therefore, the response probabilities are estimated under explicit or implicit models linking the response occurrences to the auxiliary variables. The most common method used to estimate them is the weighting within cell (WWC) method, where respondents and nonrespondents are classified in adjustment cells based on covariate information known for all the sample units, and for all cases in a cell the response weight is calculated as the reciprocal of the response rate in that cell. Another popular way to estimate the individual response probabilities is by fitting a parametric model, such as logistic, probit, or exponential. An alternative, less-known approach is through kernel regression method. The kernel type-smoothing method in the nonresponse context was first proposed by Giommi [5] as an alternative to the WWC method when no clearly identifiable "weighting cells" exist. He suggested turning the idea of homogeneity within "weighting cells" into the idea that pairs of units with similar values of an auxiliary variable have similar response probabilities. For each unit in the sample a group of nearest neighbors is identified and the response probability is estimated as the proportion of response units in the group. The resulting estimator of the response function is a running mean of the response indicators but it can also be viewed as a kernel smoother in the form of the Nadaraya–Watson estimator. Giommi [6] extends this estimator using more general Kernel functions, and his results have been further investigated by Niyonsenga [17, 18]. Both authors used the estimated response probabilities to adjust the regression mean estimator. More recently, Da Silva and Opsomer [3] used the response probabilities estimated through the Nadaraya–Watson estimator to adjust the Horvitz–Thompson mean estimator and the ratio-adjusted mean estimator (Hàjek estimator), and obtained several properties of these two estimators. Da Silva and Opsomer [4] extended these results to the estimation of the response probabilities by local polynomial regression. In this chapter, for the three response probability estimation methods quoted above (WWC, parametric estimation (PE), kernel regression estimation (KRE)), we first discuss the characteristics that a covariate used in the weighing adjustment process must have so that the adjustments are effective in reducing both the bias and the variance of the unweighted estimator. Then, the three methods are empirically contrasted for different scenarios, some using a unique continuous covariate and others using two mixed type covariates. However our main interest is to evaluate the nonparametric approach in the multiple covariate case since it has not been investigated so far.

27.2 Framework for the Sampling Design and Response Model

Let U be a finite population of N units labeled $i(i = 1, \ldots, N)$ and let the population mean $\overline{Y} = \sum_{i \in U} y_i / N$ the parameter of interest, where y_i is the value of the study variable for unit i. A sample s of n units is drawn from U according to a sampling design $p(s)$; let π_i be the inclusion probability of unit i for all $i \in U$. When nonresponse occurs the y_i's values are only observed on a subset $r \subseteq s$ of $n_r (n_r \leq n)$ respondents, and to account for the information lost in the estimation of the parameters of interest, it becomes necessary to model the response process. To this end, a response indicator, R_i, is defined assuming value one if the unit i responds and zero otherwise. The distribution of the vector $(R_i : i \in s)$ is called response mechanism and needs to be modeled. We will keep the response mechanism quite general. More specifically, we assume that the response mechanism is MAR given a vector of q auxiliary variables $\mathbf{x}$ correlated both to nonresponse and to y and fully observed throughout the sample. Moreover, we assume that, given the sample, the response indicators are independent random variables with: $\Pr(R_i = 1 | i \in s, y, \mathbf{x}) = \phi(\mathbf{x}_i) \equiv \phi_i$, for all $i \in U$ where the exact form of $\phi(.)$ is unspecified, but it is assumed to be a smooth function of $\mathbf{x}_i$, with $\phi(.) \in (0, 1]$. If all the response probabilities were known, the theory of two-phase sampling would lead us to use the following as possible mean estimators:

$$\hat{\bar{y}}_\pi = \sum\nolimits_{i \in S} y_i \pi_i^{-1} \varphi_i^{-1} R_i / N \quad \text{or}$$
$$\hat{\bar{y}} = \sum\nolimits_{i \in S} y_i \pi_i^{-1} \varphi_i^{-1} R_i / \sum\nolimits_{i \in S} \pi_i^{-1} \varphi_i^{-1} R_i \tag{27.1}$$

which are the Horvitz–Thompson and the Hàjek estimators adjusted to compensate for nonresponse [14]. The second, although not unbiased, is usually preferred as it is location-scale invariant and generally more efficient. Actually, both formulas in (27.1) are unfeasible as the response probabilities are unknown.

So we need to replace ϕ_i with their estimates $\hat{\phi}_i$, satisfying $0 < \hat{\phi}_i \leq 1$. The resulting estimators are:

$$\hat{\bar{y}}_\pi = \sum\nolimits_{i \in S} y_i \pi_i^{-1} \hat{\varphi}_i^{-1} R_i / N \quad \text{or}$$
$$\hat{\bar{y}} = \sum\nolimits_{i \in S} y_i \pi_i^{-1} \hat{\varphi}_i^{-1} R_i / \sum\nolimits_{i \in S} \pi_i^{-1} \hat{\varphi}_i^{-1} R_i . \tag{27.2}$$

In order to implement the expressions in (27.2), it is necessary to estimate the response probabilities ϕ_i. Since the WWC method and the PE method are well known and extensively present in literature (among others [9, 15]), we only describe the KRE method here to estimate the response probability. When auxiliary information consists of a scalar continuous variable x, the Kernel regression estimator (or local constant regression (LCR) estimator) of response probability assumes the following expression:

$$\hat{\phi}_i = \sum_s K\left(\frac{x_j - x_i}{h}\right) R_j / \sum_s K\left(\frac{x_j - x_i}{h}\right) \tag{27.3}$$

where $K(.)$ denotes the kernel function and h is referred to as the bandwidth. The possibility of generalizing this approach in the case of a vector of q auxiliary variables lies in the use of a "generalized product kernel function":

$$K_1\left(\frac{x_{j1} - x_{i1}}{h_1}\right) \times K_2\left(\frac{x_{j2} - x_{i2}}{h_2}\right) \times \ldots \times K_q\left(\frac{x_{jq} - x_{iq}}{h_q}\right) \tag{27.4}$$

i.e., the product of q univariate kernel function. Each one may be a continuous data kernel function, a categorical ordered data kernel function, or a categorical unordered data kernel function. Moreover, Da Silva and Opsomer [4] generalized the kernel regression estimator in (27.3) with the estimation of the response probability by local polynomial regression (general pth order). In other words, they suggested a more general estimator based on minimizing the following expression:

$$\min_{\{b_0, b_1, \ldots b_p\}} \sum_{j=1}^{n} \left(R_j - b_0 - b_1(x_j - x_i) - \ldots - b_p(x_j - x_i)^p\right)^2 K\left(\frac{x_j - x_i}{h}\right) \tag{27.5}$$

We refer to [4] for the detailed description of this estimator. Its expression is already cumbersome in the case of only one continuous variable. For polynomial of order 1, that is, for local linear regression (LLR), Li and Racine [11] in a more general context suggested the extension to the case of more mixed continuous and discrete variables. The extension is based on the generalized product kernel function, and it actually treats the continuous regressors in a local linear fashion and the discrete regressors in a local constant one. We refer to [11] and [12, Chap. 4] for a description. Their idea is applied here to estimate nonresponse probability in the presence of mixed discrete and continuous covariates.

27.3 Weighting Nonresponse Adjustments and Bias–Variance Trade-Off

A widespread view in the survey sampling literature is that the nonresponse weighting methods aim at reducing nonresponse bias at the expense of an increase in variance. Kalton e Kasprzyk [8] and Kish [10], among others, define the effectiveness of nonresponse weighting adjustments as a trade-off between bias reduction and variance increase. Kish also gives a measure of this loss of precision that is $\mathrm{F} = 1 + \mathrm{CV}^2(\mathrm{w_i})$, where $\mathrm{CV}^2(\mathrm{w_i})$ is the coefficient of variation of the weights $w_i = \phi_i^{-1}$. The measure F represents the multiplying factor that is applied to the variance of a survey estimate due to the variability in the weights in the situation where equal weights are optimal. According to Little and Vartivarian [15], this view is an oversimplification, *nonresponse weighting can in fact lead to a reduction in variance as well as in bias*. The fact that nonresponse weighting can reduce variance is also implicit in [16], and noted in [13] and in Holt and Smith [7]. Some other authors, including Rosenbaun [19] and Beaumont [2], noted that the weighted adjusted estimator using the estimated response probability may be more efficient than the estimator using the true response probability. This last result is known as "super efficiency." Actually, the question concerns the relation of the auxiliary information used in the estimation of the response probability with both the nonresponse mechanism and the outcome of interest. We address this question explicitly here for the WWC, PE, and KRE response probability estimation methods.

Kim and Kim [9] show for the PE method that when the response probability is estimated using a parametric maximum likelihood method, not only the bias of the weighted estimator is reduced but its efficiency is also generally improved more than that of an estimator which uses the true response probability, since it incorporates the additional information contained in the auxiliary variables used in the response model.

For the WWC method, Little and Vartivarian [15] propose an equation that captures the bias and variance components if the adjustment cell variables are related to the study variables and if they are not related. From said equation it is evident that:

1. A substantial bias reduction requires adjustment cell variables that are related to both nonresponse and outcome variables.
2. Cell variables only related to the nonresponse increase the variance by the Kish multiplying factor without any reduction in bias.
3. Cell variables only related to the study variable tend not to have any impact on the bias but reduce variance as far as they are good predictors of the outcome variable.
4. If the adjustment cell variables are good predictors of both the nonresponse mechanism and outcome variables, weighting tends to reduce both bias and variance.

The negative effect on the variance of the weights variability is also present in the situations defined in bullet points 3 and 4, but it is generally less than the positive effect of a good predictor of the study variable.

The leading idea is that some of the considerations listed in points 1 - 4 can be extended to the case of overlapping classes that are the starting point of KRE methods.

As regards point 1, we can argue as follows: suppose that for each respondent a group of neighbor units is identified on the basis of a covariate x. In this case, if x_i is the value of x for the respondent i, the group of units centered on i contain all units $j \in s$ for which $|x_j - x_i| \leq h$ with h fixed. We let n_i and n_{ir} be the number of sampled and respondent units in that group respectively. Then, assuming that units are selected by simple random sampling, the weighted and the unweighted (or naive) mean estimators are respectively:

$$\bar{y}_w = \sum\nolimits_{i \in r} \frac{y_i}{n}\frac{n_i}{n_{ir}} = \sum\nolimits_{i \in r} \frac{y_i}{n_r}\frac{n_i}{n_{ir}}\frac{n_r}{n} \quad \text{and}$$

$$\bar{y}_r = \sum\nolimits_{i \in r} \frac{y_i}{n_r} = \sum\nolimits_{i \in r} \frac{y_i}{n}\frac{n}{n_r} \tag{27.6}$$

The construction of the groups of neighbors with respect to variable x is equivalent to sorting the units with respect to x and allows us to consider the terms n_{ir}/n_i as a running mean of the response indicators. If we assume that the conditional distribution of y given $x = x_i$ and $R_i = 1$ has mean μ_{ir}, then, conditional to $D = \{n, nr, (n_i, n_{ir})i \in r\}$, the means of $\bar{y}_w$ and $\bar{y}_r$ are:

$$E[\bar{y}_w|D] = \sum\nolimits_{i \in r} \frac{\mu_{ir}}{n}\frac{n_i}{n_{ir}} = \tilde{\mu_r} \text{ and } E[\bar{y}_r|D] = \sum\nolimits_{i \in r} \frac{\mu_{ir}}{n_r} = \mu_r \tag{27.7}$$

Moreover, by neglecting the difference between the weights and their expectation, the bias reduction may be expressed as:

$$b(\bar{y}_r) - b(\bar{y}_w) = \sum\nolimits_{i \in r} \frac{\mu_{ir}}{n_r}\left(1 - \frac{n_i}{n_{ir}}\frac{n_r}{n}\right) \tag{27.8}$$

and this difference is zero if either nonresponse is unrelated to x (in this case $n_i/n_{ir} \approx (n_r/n)^{-1}$) or y is unrelated to x (in this case $\mu_{ir} \approx \mu_r \forall_i$ and as $\sum_{i \in r} \frac{n_i}{n_{ir}} \approx n$ then $b(\bar{y}_r) - b(\bar{y}_w) = \mu_r - \frac{\mu_r}{n}\sum_{i \in r} \frac{n_i}{n_{ir}} \approx 0$). Thus, also in the case of overlapping classes, a substantial bias reduction requires that the variable used to identify the neighborhoods be related to both the nonresponse mechanism and the outcome variable.

With regard to the effect of weighting on the variance, even if detailed analytical proof is not shown here, it can be noted that also in the case of overlapping classes as shown by Little and Vartivarian [15] for those not overlapping, the effect of weighting on the variance may be measured as $V(\bar{y}_r) - V(\bar{y}_w)$, and this difference may be expressed as the sum of two terms: V_1 equal to the difference among the variances of the conditional expectations of the two estimators and V_2 equal to the difference among the expected values of their conditional variances. With a few simplifying assumptions, including the equality between the weights and their expectation and the constant variance (σ^2) of the conditional distribution of y, it is possible to write: $V_1 = \sum_{i \in r} \frac{1}{n_r}(\mu_{ir} - \mu_r)^2 \frac{n_r}{n} - \sum_{i \in r} \frac{1}{n}(\mu_{ir} - \tilde{\mu}_r)^2 \frac{n_{ir}}{n_i}$ and $V_2 = -\lambda\sigma^2$, where λ is the population analog of the variance of the response weights n_i/n_{ir}. If the covariate x is only related to the nonresponse $V_1 \approx 0$ and $V_2 < 0$ and therefore the variance of the weighted estimator increases. If the covariate x is related to the study variable, V_1 is positive, and it is more likely for its size to compensate the negative effect of V_2 ($V_2 \leq 0$ and tends to zero when x is not related to the nonresponse), therefore the variance of the weighted estimator decreases. In the next section the empirical results already show these properties: the KRE method, as well as the logistic method and WWC method, tends to reduce both the bias and variance of the mean if it uses covariates that are good predictors of both the nonresponse mechanism and the outcome variable.

27.4 Simulation Study

Several Monte Carlo experiments have been performed in order to compare the LCR and the LLR estimators of response probabilities with the two more common response probability estimation methods: WWC and the logistic regression method. In the first setting a unique continuous covariate $x_1 \sim Uniform(0,1)$ is considered and two outcome variables are generated: $y_1 = 20 + 60x_1 + \varepsilon_1$ and $y_2 = 25 + 135(x_1 - 0.5)^3 + \varepsilon_2$, where $\varepsilon_1 \sim N(0,16)$ and $\varepsilon_2 \sim N(0,6.25)$. For both populations, nonresponse is generated according to two response functions of x_1 : $\phi_1(x_1) = -\beta_0 x_1^2 + \beta_1 x_1 + \beta_2$ and $\phi_2(x_1) = \exp(\beta_3 + \beta_4 x_1)(1 + \exp(\beta_3 + \beta_4 x_1))^{-1}$ where the coefficients $\beta_0, \ldots, \beta_4$ are chosen so that the response rate is equal to approximately 0.7.

In the second setting we consider three different pairs of covariates: (1) $x_1 \sim Uniform(0,1)$ plus an indicator variable x_2; (2) $x_1 \sim Uniform(0,1)$ plus a categorical variable x_3 with *four* categories; (3) $x_1 \sim Uniform(0,1)$ plus a discrete variable x_4 assuming four values. In case 1, two outcome variables are generated: y_3 as a linear function of x_1 with different parameters in the two subpopulations identified by x_2, and y_4 as the quadratic function of x_1 with different parameters in the two subpopulations identified by x_2. In case 2, an outcome variable y_5 is generated as a linear function of x_1 with four different sets of parameters corresponding to the four categories of x_3. In case 3, an outcome variable y_6 is generated as a linear function of $x_1 \times x_4$. For all four populations, nonresponse is generated according to two response functions ϕ_3 and ϕ_4 equivalent to ϕ_1 and ϕ_2 respectively but with different parameters corresponding to the two categories of x_2 or to two groups of categories of x_3 or of x_4. In all cases the response rate for the entire population is equal to approximately 0.7 and varies at the most between 0.65 and 0.75 for the two subpopulations corresponding to different parameters of the response function.

For each combination of the population model and response mechanism, the simulation procedure consists of the following steps:

1. Select a simple random sample of 300 units (population size is $N = 3000$)
2. Perform a Bernoulli trial for each unit $i \in s$ with probability $\phi_i (i = 1,2,3,4)$ for "success" (response) and $(1 - \phi_i)$ for "failure" (nonresponse)
3. Compute the second expression in (27.2) on the set of respondents to obtain the following weighted estimators of the mean:
   T_0 : adjusted with true response probabilities
   T_1 : adjusted with response probabilities estimated through LCR
   T_2 : adjusted with response probabilities estimated through LLR

Table 27.1 Percentage biases of weighting-adjusted mean estimators

Response functions	Outcomes	T_0	T_1	T_2	T_3	T_4	T_5
ϕ_1	y_1	− 0.015	0.458	− 0.042	0.272	0.050	2.664
ϕ_1	y_2	− 0.005	0.374	− 0.040	0.249	− 0.051	1.822
ϕ_2	y_1	− 0.009	− 0.981	− 0.073	− 0.017	− 0.076	− 7.432
ϕ_2	y_2	− 0.005	− 0.795	− 0.041	− 0.011	− 0.093	− 5.076
ϕ_3	y_3	0.093	0.994	0.022	0.166	0.673	7.717
ϕ_3	y_4	0.021	0.658	0.203	− 0.237	0.167	4.405
ϕ_3	y_5	0.042	0.793	0.191	− 0.078	1.316	5.255
ϕ_3	y_6	0.074	1.908	1.060	0.565	2.063	8.734
ϕ_4	y_3	− 0.010	− 1.102	− 0.218	0.729	− 0.458	− 7.297
ϕ_4	y_4	− 0.028	− 0.452	0.026	0.073	− 0.320	− 3.122
ϕ_4	y_5	− 0.019	− 0.735	− 0.135	0.313	− 1.378	− 4.519
ϕ_4	y_6	− 0.011	− 0.731	0.113	− 0.965	− 1.723	− 5.011

T_3 : adjusted with response probabilities estimated through a logistic model
T_4 : adjusted with response probabilities estimated with the WWC method
T_5 : adjusted with $\hat{\phi}_i = 1, i \in s$ (naive estimator)

4. Repeat steps 1–3 10,000 times.

For the first two populations, all the four response estimation methods use the unique covariate x_1. For y_3 and y_4, the pair (x_1, x_2) is used. For y_5 and y_6, the pairs (x_1, x_3) and (x_1, x_4) are used respectively. In the local regression methods, x_2 and x_3 are considered as factors, whereas, for x_4, an ordered kernel function is used. For the WWC method the cells are identified in the univariate case by the deciles of the distribution of x_1 and in the bivariate case by crossing x_1 (first categorized using the quartiles) with x_2, x_3, and x_4, respectively. In the KRE methods other choices in estimator settings concern the kernel functions and the selection criterion of the bandwidths: the Epanechikov kernel function is used for x_1, the Aitchison and Aitken kernel function [1] is used for x_2 and x_3, and the Wang and van Ryzin kernel function [20] is used for x_4, in all cases the bandwidths are selected through the least-squares cross-validation method.

The experimental results are reported in Tables 27.1 and 27.2: for each combination of population and response mechanism Table 27.1 gives the relative bias which is the bias divided by the population mean percent. Table 27.2 shows the corresponding empirical mean square errors (MSE). It is immediately evident that in all the scenarios each weighting adjustment method successfully reduces the bias and the MSE of the unweighted (naive) estimator. In the single covariate scenarios: (a) all the weighted estimators using estimated response probabilities also show a lower MSE than T_0; (b) moreover, among these, T_1 is the worst in both bias and MSE, T_2 and T_4 are almost equivalent, both in the bias and in the MSE, and T_3 shows better bias performance than T_2 and T_4 for response mechanism ϕ_2, while it is worse for response mechanism ϕ_1, and for both response mechanisms it is almost equivalent to T_2 and T_4 in the MSE. For scenarios with two covariates: (a) for all populations, when the response mechanism is ϕ_2, T_2 appears the best both in the

Table 27.2 Empirical mean square errors of weighting-adjusted mean estimators

Response functions	Outcomes	T_0	T_1	T_2	T_3	T_4	T_5
ϕ_1	y_1	1.596	1.022	0.946	0.937	0.944	3.031
ϕ_1	y_2	0.262	0.173	0.154	0.157	0.161	0.396
ϕ_2	y_1	1.582	1.227	0.952	0.955	0.943	14.903
ϕ_2	y_2	0.254	0.208	0.162	0.178	0.160	1.802
ϕ_3	y_3	1.488	1.112	0.856	0.880	0.964	13.596
ϕ_3	y_4	0.588	0.522	0.387	0.387	0.377	6.525
ϕ_3	y_5	1.016	0.848	0.645	0.621	1.193	9.410
ϕ_3	y_6	1.040	1.282	0.877	0.739	1.374	12.346
ϕ_4	y_3	1.392	1.137	0.851	0.977	0.894	12.278
ϕ_4	y_4	0.637	0.456	0.378	0.394	0.413	3.456
ϕ_4	y_5	0.975	0.814	0.629	0.678	1.282	7.117
ϕ_4	y_6	1.192	0.848	0.746	0.864	1.167	4.580

bias and the MSE; (b) when the response mechanism is ϕ_1, T_2 still appears the best in two scenarios, in one scenario T_3 is the best and in another the best is T_4; (c) for all scenarios the MSE of T_2 and T_3 is lower than that of T_0, whereas this behavior is not always valid for T_1 and T_4.

Another simulation experiment, not reported here, considers a situation with two covariates z_1 and z_2 and two outcomes, one a function of z_1 and the other a function of z_2 and a response mechanism which depends on both z_1 and z_2. If the response probability is estimated only using z_1, all the weighting estimators work "well" both in bias and in MSE for the outcome function of this covariate, but they are more biased and more variable than the unweighted estimator for the other outcome variable.

27.5 Conclusions

Both the theoretical considerations in Sect. 27.3 and the empirical results in Sect. 27.4 show that the most important aspect in weighting methods is the choice of auxiliary information. This is to some extent an expected result, but here we also shed light on several features than could at times be hidden in the shadow of the real-life applications. Two main results stem from our simulation study. First, when the set of auxiliary information is related to both the nonresponse mechanism and the outcome variable, all the examined weighting estimation methods are effective in reducing both the bias and the variance of the basic and unweighted estimator of the mean. Moreover, they may also be more effective in reducing MSE than the method based on the true response probability. Therefore we must admit that there is no evidence of a bias–variance trade-off from using or not using a weighting method. Second, given the set of auxiliary variables, the bias–variance trade-off may regard the choice among different weighting methods since an inverse relationship between the bias reduction and the variance reduction can sometimes be observed. When there are several outcome variables and the same weights are applied to them all, the problem can become more complicated. Some variables may be unrelated to the response mechanisms and for these variables the unweighted estimator should be better. Moreover, to include all the relevant information in a weighting estimation method may be "difficult" and produce more unstable response probability estimations. With respect to the comparison between nonparametric and other estimation methods, the empirical results show that the LLR method could be an appealing alternative, especially in the multivariate case. Nevertheless, further research is needed since when the number of auxiliary variables increases, the application of the LLR may become very complex. Finally, we cannot rule out the possibility that other nonparametric models such as generalized additive models, not explored here, could give rise to more effective results.

References

1. Aitchison, J., Aitken, C.G.G.: Multivariate binary discrimination by the kernel method. Biometrica **68**, 301–309 (1976)
2. Beaumont, J.F.: Calibrated imputations in survey under a model-assisted approach. J. R. Stat. Soc. B **67**, 445–458 (2005)
3. Da Silva, D.N., Opsomer, J.D.: A kernel smoothing method of adjusting for unit nonresponse in sample survey. Can. J. Stat. **34**, 563–579 (2006)
4. Da Silva, D.N., Opsomer, J.D.: Nonparametric propensity weighting for survey nonresponse through local polynomial regression. Surv. Meth. **35**, 165–176 (2009)
5. Giommi, A.: A simple method for estimating individual response probabilities in sampling from finite populations. Metron **42**, 185–200 (1984)
6. Giommi, A.: Nonparametric methods for estimating individual response probabilities. Surv. Meth. **13**, 127–134 (1987)
7. Holt, D., Smith, T.M.F.: Post stratification. J. R. L. Stat. Soc. B **142**, 33–46 (1979)
8. Kalton, G., Kasprzyk, D.: The treatment of missing survey data. Surv. Meth. **12**, 1–16 (1986)
9. Kim, J.K., Kim, J.J.: Nonresponse weighting adjustment using estimated response probabilities. Can. J. Stat. **35**, 501–514 (2007)
10. Kish, L.: Weighting for unequal P_i. J. Offic. Stat. **8**, 183–200 (1992)
11. Li, Q., Racine, J.S.: Cross validated local linear nonparametric regression. Statistica Sinica **14**, 485–512 (2004)
12. Li, Q., Racine, J.S.: Nonparametric Econometrics: Theory and Practice. Princeton University Press, Princeton, NJ (2006)
13. Little, R.J.A.: Survey nonresponse adjustment. Int. Stat. Rev. **54**, 139–157 (1986)
14. Little, R.J.A., Rubin, D.B.: Statistical Analysis with Missing Data. Wiley, New York (2002)
15. Little, R.J.A., Vartivarian, S.: Does weighting for nonresponse increase the variance of survey mean. Surv. Meth. **31**, 161–168 (2005)
16. Oh, H.L., Scheuren, F.S.: Weighting adjustments for unit nonresponse. In: Madow, W.G., Olkin, I., Rubin, D.B. (eds.) Incomplete Data in Sample Surveys, 2, Theory and Bibliographies, pp. 143–184. Academic Press, New York (1983)
17. Niyonsenga, T.: Nonparametric estimation of response probabilities in sampling theory. Surv. Meth. **20**, 177–184 (1994)
18. Niyonsenga, T.: Response probability estimation. J. Stat. Plann. Infer. **59**, 111–126 (1997)
19. Rosenbaum, P.R.: Model-based direct adjustment. J. Am. Stat. Assoc. **82**, 387–394 (1987)
20. Wang, M.C., van Ryzin, J.: A class of smooth estimators for discrete distributions. Biometrica **68**, 301–309 (1981)

Open Source Integer Linear Programming Solvers for Error Localization in Numerical Data

28

Gianpiero Bianchi, Renato Bruni, and Alessandra Reale

Abstract

Error localization problems can be converted into Integer Linear Programming problems. This approach provides several advantages and guarantees to find a set of erroneous fields having minimum total cost. By doing so, each erroneous record produces an Integer Linear Programming model that should be solved. This requires the use of specific solution softwares called Integer Linear Programming solvers. Some of these solvers are available as open source software. A study on the performance of internationally recognized open source Integer Linear Programming solvers, compared to a reference commercial solver on real-world data having only numerical fields, is reported. The aim was to produce a stressing test environment for selecting the most appropriate open source solver for performing error localization in numerical data.

28.1 Introduction

In many statistical data collection, data can be affected by errors, i.e. alterations of some of the values. Errors can be due to the original answer or introduced at any later stage of data conversion or processing. Automatic procedures for finding and correcting errors are nowadays necessary, especially in the case of large datasets. Data are generally organized into conceptual units called *records*. A record has the formal structure of a set of n fields $R = (f_1, \ldots, f_n)$, and by giving each field f_i a value v_i we obtain a record instance, or, simply, a record $r = (v_1, \ldots, v_n)$.

G. Bianchi (✉) · A. Reale
Istat, DCCG, Via A. Ravà 150, Roma, Italy
e-mail: gianbia@istat.it; reale@istat.it

R. Bruni
Univ. Roma "Sapienza", DIS, Via Ariosto 25, Roma, Italy
e-mail: bruni@dis.uniroma1.it

N. Torelli et al. (eds.), *Advances in Theoretical and Applied Statistics*,
Studies in Theoretical and Applied Statistics, DOI 10.1007/978-3-642-35588-2_28,

In general, fields can be numerical or categorical. Throughout this chapter, we will restrict our attention to numerical data.

The above data cleaning tasks can be performed by following several different approaches, each of which having its own features. A main one is based on the use of rules, called *edits*, that each record must respect in order to be declared exact. Records not respecting such rules are declared erroneous. A seminal paper on the subject is due to Fellegi and Holt [9]. After the detection of erroneous records, for each of them one needs to determine which are its erroneous fields. Such operation is called *error localization*. This is generally done by assigning each field f_i a cost c_i, and by searching for the set E of fields having minimum total cost such that, if their values are changed, edit compliance for that record can be restored. Fields belonging to such set E are assumed to be erroneous, and are the ones that should be modified during a subsequent error correction phase.

Finding the above-mentioned set of fields E is a nontrivial combinatorial optimization problem in general (since one needs to choose a subset among many possible subsets of the ground set, see e.g. [14] for details and [12] for the complexity of those problems) and can be performed by solving Integer Linear Programming models (see e.g. [14] for an introduction to the field) having the structure described in Sect. 28.2. This allows to overcome the limits of other methodologies (see e.g. [3, 17]) based on the Fellegi Holt approach, and guarantees the individuation of the set of fields having minimum total cost. This has been first done within the data Editing and Imputation software system DIESIS [5], and, subsequently, in other works such as those described in [8, 16].

Integer Linear Programming models may however be quite time demanding to be solved. Efficient software solvers are therefore needed. This has been a very active field of research in the last 50 years, and performance improvements were impressive. A well-known and very fast solver, based on advanced branch-and-cut techniques [14], is Cplex (http://www-01.ibm.com/software/integration/optimization/cplex), from the international software company ILOG. It has been successfully used in the mentioned DIESIS system. However, such solver is a commercial and closed source software, so there are many applications for which such choice would not be admissible.

Luckily, open source Integer Linear Programming solvers are also available. An overview of the most internationally recognized and used among these softwares is given in Sect. 28.3. They too are generally based on branch-and-cut techniques. However, there are many different algorithmic aspects that may vary within such techniques, and that may produce very sensible variations in the resulting time performance. Other fundamental issues for such kind of solvers are numerical precision and stability, which also may vary greatly. Therefore, for solving efficiently the described error localization problems, a careful selection of the solver, and of the settings that the solver should use, is needed.

In order to make such a selection, a group of *candidate* open source solvers has been initially identified. This was done after fixing the requirements and surveying the literature for the best ones. Among them, by performing preliminary tests, the group of *suitable* solvers has been selected. These solvers were Cbc (Coin- or branch and cut), from the COIN-OR (COmputational INfrastructure for Operations Research) project of the International Business Machines Corporation, and SCIP (Solving Constraint Integer Programs) from the research institute Konrad- Zuse-Zentrum für Informationstechnik Berlin. They were compared to a reference solver that was decided to be the mentioned Cplex, on error localization problems arising from the processing of real-world data having numerical fields, as reported in Sect. 28.4. Such data present wide numerical excursions in the data values, and constitute a stressing test environment for selecting the most appropriate open source solver for performing error localization in numerical data.

28.2 The Structure of the Integer Linear Programming Problems

We encode error localization problems into a specific form of optimization problems, namely Integer Linear Programming (ILP), even though solving ILP models is known to be a difficult problem in general [14]. This is because, unfortunately, easier models could not represent the full complexity of localization problems: a problem with a certain degree of complexity can be correctly converted only into another one having the same degree of complexity. Indeed, if a conversion into an easier one is attempted, a part of the problem's features would be lost [12]. Integer Linear Programming models are constituted by a linear objective function and by constraints expressed in the form of linear inequalities (or, equivalently, linear equations) over a set of variables bounded to be integer. This particular structure have been intensively studied, and very powerful solvers are available today, probably more efficient than those available for any other class of problems sharing the same complexity level.

Therefore, the basis of the adopted modelling technique consists in encoding the edits by means of linear inequalities (or, equivalently, linear equations). Any other form, though correct, would cause the impossibility of using the above-mentioned powerful solvers.

Edits expressing linear relationships among fields are immediately convertible into linear inequalities using n variables z_i representing the values of the different fields of the data records. In the numerical cases considered in this work, such variables are bound to be non-negative integer, so we assume their domain to be over Z_+ (even if this is not strictly required in general) up to a maximum value U. More complex edits, possibly expressing also logical conditions, can be converted into linear inequalities by using also a suitable set of binary variables x_j. Hence, the variables used for expressing the edits, and their respective domains, are the following:

$$z_{\rm i} \in \{0, \ldots, U\}; \quad x_{\rm j} \in \{0, 1\} \tag{28.1}$$

We give now some examples of conversion from edit into linear inequalities. Consider a very simple mathematical rule expressing that, for every record, the value of field h should be greater or equal to 100 plus the value of field k.

This would easily become the following linear inequality, whose meaning is clear enough.

$$z_h \geq 100 + z_k \text{ that is } z_h - z_k \geq 100$$

In case a logical condition appears in the rule, such as the following, we need also one binary variables x.

when field j is strictly greater then 10, the value of field h should be greater or equal to the value of field k, otherwise (= when value of field j is less or equal to 10) it should be the opposite (the value of field k should be greater or equal to the value of field h).

This edit becomes a set of linear inequalities, as follows.

$$z_j - 10 \leq Ux \tag{28.2}$$

$$11 - z_j \leq U(1 - x) \tag{28.3}$$

$$U(1 - x) + z_h \geq z_k \tag{28.4}$$

$$z_h \leq z_k + Ux \tag{28.5}$$

This holds because, when $z_j > 10$ (and recall it's integer), value $(z_j - 10)$ is positive and value $(11 - z_j)$ is less than or equal to 0. Therefore, $x \in \{0, 1\}$ is forced by (28.2) to have value 1 so that, multiplied by the greatest value U, inequality (28.2) is satisfied. On the contrary, (28.3) does not force any value for x. Being $x = 1$, we have that $U(1 - x)$ is 0 and so $z_h \geq z_k$ must hold in order to satisfy (28.4). On the contrary, (28.5) is always satisfied for any value of z_h and z_k by the presence of Ux.

On the other hand, when $z_j \leq 10$, value $(z_j - 10)$ is less than or equal to 0 and value $(11 - z_j)$ is positive. Therefore, $x \in \{0, 1\}$ is forced by (28.3) to have value 0 so that $U(1 - x)$ has value U and inequality (28.3) is satisfied. On the contrary, (28.2) does not force any value for x. Being now $x = 0$, we have that Ux is 0 and so $z_h \leq z_k$ must hold in order to satisfy (28.5). On the contrary, (28.4) is always satisfied for any value of z_h and z_k by the presence of $U(1 - x)$. So, all the above rule is enforced when the values of the variables satisfy inequalities (28.2), (28.3), (28.4), (28.5).

Similar techniques are used in order to convert each edit into one or mode liner inequalities (or equations, if that is the case) [7]. Afterwards, by using elementary algebraic transformations, all equations can be converted into couples of inequalities, and all inequalities can be converted into $\geq$ form. All in all, we obtain a system of linear inequalities that can be denoted using the following compact notation.

$$A'x + A''z \geq b \tag{28.6}$$

The generic erroneous record g correspond to a set of values $(g_1, \ldots, g_n)$ for these z variables. By assigning each field f_i a cost c_i, our aim is to find the set E of fields having minimum total cost such that, if their values are changed, edit compliance for that record can be restored. The total cost of a set of fields is the sum of their individual costs c_i. In order to represent the changes, we introduce n binary variables $w_i \in \{0, 1\}$ meaning

$$w_i = \begin{cases} 1 & \text{if we change } g_i \\ 0 & \text{if we keep } g_i \end{cases}$$

The objective of our Integer Linear Programming model is therefore:

$$\min \sum\nolimits_{i=1,\ldots,n} c_i w_i \tag{28.7}$$

whereas restoring edit compliance means respecting the above system (28.6). One could observe that the variables appearing in the two parts (28.6) and (28.7) of the model are not the same. However, the following relation can be identified between the w and the z variables:

$$w_{\mathrm{i}} = \begin{cases} 0 & \text{if } z_i = g_i \\ 1 & \text{if } z_i < g_i \text{ or } z_i > g_i \end{cases}$$

This can be used to link those variables, as described in greater detail in [5, 6], obtaining so, for each erroneous record, an Integer Linear Programming model representing the error localization problem for that record.

Generally, in real-world surveys, a large number of records should be processed, and, consequently, a large number of such models should be solved. Their complexity usually increases exponentially with the number of variables of each model. Therefore, the optimal solutions can be obtained only by using specific softwares called Integer Linear Programming solvers.

28.3 Overview on Open Source Solvers

The group of candidate open source Integer Linear Programming solvers has been selected after a phase of study of the solver features and of literature survey. The most relevant and internationally recognized solvers declaring to possess the required features have been chosen. In particular, the candidate group was composed by the following solvers:

1. GLPK (GNU Linear Programming Kit [13]), from the GNU operating system project
2. Cbc (Coin-or branch and cut [10]), from the COIN-OR (COmputational INfrastructure for Operations Research) project
3. Symphony [15], again from the COIN-OR project
4. SCIP (Solving Constraint Integer Programs [1]), from an Integer Linear Programming and Constraint Programming integration project

Moreover, in order to check the correctness of the obtained solutions, a reference solver has been considered. This reference was chosen to be Cplex, from the international software company ILOG, currently owned by IBM (International Business Machines Corporation). This is a commercial solver, not an open source one, and was chosen because it is deemed to be the best Integer Linear Programming solver nowadays available. Moreover, it is the one currently used by the data Editing and Imputation software system DIESIS [5].

GLPK (GNU Linear Programming Kit, available from http://www.gnu.org/software/glpk/) is an implementation of branch-and-cut techniques, primal and dual simplex, interior- point methods [14], written in ANSI C by a research group headed by Prof. Andrew Makhorin, from the Department of Applied Informatics of the Moscow Aviation Institute (http://www.mai.ru/english/). This solver can be used as free software under the General Public License (GNU GPL).

Cbc (Coin-or branch and cut, available from https://projects.coin-or.org/Cbc) is a quite complete implementation of branch-and-cut techniques [14] written in C++ by a research group headed by Dr. John J. Forrest, from the Watson Research Center (http://www.watson.ibm.com/index.shtml) of IBM, within a joint project among IBM, Maximal and Schneider called COIN-OR (COmputational INfrastructure for Operations Research, http://www.coin-or.org/index.html). This solver can be used as open source code under the Common Public License (CPL). Cbc requires the use of another solver for solving the liner relaxations of the problems. The default one for this task is Clp (Coin-or linear programming, available from https://projects.coin-or.org/Clp), which is a complex implementation of primal and dual simplex and barrier methods [14], written in C++ by the same research group of Cbc, and can be used under the same licence.

Symphony (available from https://projects.coin-or.org/SYMPHONY) is a flexible implementation of branch-and- cut techniques [14] that can be customized for specific classes of problems, written in C by a research group headed by Prof. Ted Ralphs, from the COR@L research laboratory (Computational Optimization Research at Lehigh, http://coral.ie.lehigh.edu/index.html) of the Department of Industrial and Systems Engineering at Lehigh University, again within the described COIN-OR project. This solver can be used as open source code under the Common Public License (CPL). Symphony also requires the use of another solver for solving the liner relaxations of the problems, and the default one for this task is again the described Clp.

SCIP (Solving Constraint Integer Programs, available from http://scip.zib.de) is an implementation of branch-and-cut techniques [14] integrating also constraint programming capabilities, written in ANSI C by Dr. Tobias Achterberg, from the Optimization Department of the Scientific Calculus Division of the Konrad-Zuse Zentrum für Informationstechnik of Berlin (http://www.zib.de) within a project of integration between Integer Linear Programming and Constraint Programming. This solver can be used under the ZIB Academic License only for academic or non-commercial institutions. SCIP also requires the use of another solver for solving the liner relaxations of the problems, and can use the described Clp, even if the default one is SoPlex (Sequential object-oriented simplex, available from http://soplex.zib.de/) which is also an implementation of primal and dual simplex [14].

Integer Linear Programming solvers generally accept a number of parameters for setting the algorithmic choices that should be used during the solution process (i.e. cuts applied, branching strategy, backtracking policy, etc.) and the numerical precision (i.e. integer tolerance, feasibility tolerance, scaling technique, etc.). This is necessary for handling successfully heterogeneous problems, and setting such kind of parameters was allowed by all the described solvers.

Those choices basically affect the evolution of the solution algorithm in reaching the numerical solution of the optimization model, and therefore they impact (but generally in a random way) on which one of the solutions is obtained for the cases of problems having more than one optimal solution. However, it is impossible to draw a direct correspondence between the above parameter choices and the statistical quality of the data values corresponding to the obtained optimal solution.

Note, finally, that all of the above solvers are the results of several years of development, and all of them are currently active projects, in the sense that new releases, revised and updated, are periodically issued.

28.4 Computational Experience

Two phases of tests have been performed over the described open source solvers:

1. Preliminary tests, with the aim of selecting the group of suitable solvers within the group of the candidate ones
2. Final tests, with the aim of selecting the most appropriate solver within the group of the suitable ones

Note that the above tests are focused on evaluating the solvers from the computational performance point of view, and not the statistical qualities of the obtained data. The preliminary tests were performed under Linux operating system Red Hat EL Server ver. 5.4 64 bit. Features required to the solvers for being considered suitable were not only speed, but also robustness (in the sense of producing acceptable performances even in unplanned conditions) and numerical precision. A C++ code for artificially generating series of Integer Linear Programming models having the same characteristics of those described in Sect. 28.2, and passing them to the solvers, was developed. A generally positive behaviour of all the considered solvers was experienced during this phase. However, some negative points also emerged. In particular, GLPK and Symphony are not suitable for solving all in one run, one after another, a large number of models of the described type. They suffer in fact from lack of stability and occasionally poor performance, and this could compromise and even interrupt the sequential automatic solution of the whole set of models. The group of suitable solvers was therefore composed only by Cbc and SCIP.

The final tests were again performed under Linux operating system Red Hat EL Server ver. 5.4 64bits by generating 8,017 files in standard LP format (Linear Program) encoding the error localization problem for 8,017 records of data from agricultural production each having 211 fields, all numerical (see [4] for further details). Such data were obtained from the Italian sample survey on Farm Structure and Production of 2005 (Indagine campionaria su Struttura e Produzioni delle

Table 28.1 Aggregate results of Cbc, SCIP and Cplex

	Cbc	SCIP	Cplex
Number of records processed	8,017	8,017	8,017
Number of records solved	7,994	7,291	8,017
Percentage of records solved	99.7%	90.9%	100%
Total processing time	1 h 20 min	1 h 30 min	30 min
Average time per record	0.6 s	0.7 s	0.2 s

Table 28.2 Algorithmic and numerical settings chosen for Cbc and SCIP

Feature affected	Cbc settings	SCIP settings
Preprocessing	−preprocess on, −presolve on	ppmethods = –
Scaling	−scaling automatic,	scaling = TRUE
Integer tolerance	−integerT 1e-20	–
Feasibility tolerance	–	feastol = $1e^{-06}$
Branching strategy	−strat 1, −trust 1	strategy = R, priority = −1000
Cutting planes used	clique, flow, knapsack, mixed, probing, two	Default
Linear relax. Solver	Clp simplex algorithm	Clp simplex algorithm

Aziende Agricole), which contained random errors. Those data exhibit wide numerical excursions in the data values, and were chosen because they constitute a stressing test environment for the described solvers, and also because they can be viewed as good representative of other types of numerical data.

The 8,017 model files have, on average, 7,681 variables (∼97% binary, 3% integers) and 8,437 constraints. They were solved by using Cbc, SCIP and Cplex, obtaining the aggregate results reported in the following Table 28.1. SCIP was used by solving the linear relaxations by means of the more performing Clp [11] instead of the default SoPlex.

As observable, the reference solver Cplex numerically computes the optimal solution in the totality of the cases. Quite the opposite, the two open source solvers are not able to correctly compute a certain (small) percentage of the considered models. This happens notwithstanding a considerable effort spent for choosing the algorithmic and numerical settings (reported in Table 28.2) more fitting to solving all in one run, one after another, the whole set of considered models, so as to fully automatize the correction process. This is not surprising: numerical precision is a very delicate issue in many real-world cases. Just as an example, the solver used by the quite known software Banff [2], from Statistics Canada, running on the same 8,017 error localization problems, could not find a solution for 199 records, with a total running time of 69 h, as reported in [4].

The cases of unsolved problems are analysed in detail in the following Table 28.3. They are split in the two following categories:

1. Problems not solved because of strongly erroneous solution values (for instance, an objective value of 0, whereas all the considered models should produce an

Table 28.3 Differences among the open source solvers and the reference solver

	Cbc	SCIP
Strongly erroneous solution values (28.2)	0	111
Numerically incorrect solution values (28.3)	23	615
Total number of not solved	23	726
Percentage of not solved	0.3%	9.1%

Table 28.4 Optimal values of the objective function for Cbc, SCIP, Cplex

	Cbc	SCIP	Cplex
Number of problems with obj. = 1	4,890	4,459	4,902
Number of problems with obj. = 2	2,100	1,902	2,123
Number of problems with obj. = 3	592	430	599
Number of problems with obj. = 4	258	247	267
Number of problems with obj. >= 5	154	253	126
Total	7,994	7,291	8,017

objective value strictly positive and integer) due mainly to numerical instability and ill-conditioning of the corresponding models

2. Problems not solved because of (slightly) numerically incorrect solution values (for instance, fractional values smaller than 1 in the objective value) mainly due to scaling inaccuracies, effect of the wide numerical excursions in the data values

It is not traceable a correlation among problems not solved by Cbc and those not solved by SCIP: the two sets are only partially overlapping. It is worth to note that even the reference solver Cplex may in general suffer from numerical errors, even if this did not happen in the present test. Moreover, the mentioned wide numerical excursions in the data values produce also some differences among the optimum values obtained for the objective function by the different solvers. Recall that, in each model, the value of the objective function is the weighted sum of the changes that should be applied to the record values in order to restore edit compliance. Note also that, for the considered tests, the cost of all fields were set to value 1. Therefore, the objective should assume only integer values.

Differences are reported in the following Table 28.4, by considering the number of problems producing the value respectively of 1, 2, 3, 4 and 5 or more for the objective function. One can observe that SCIP, and in smaller amount Cbc, occasionally tend to compute optimal objective values higher than those computed by the reference solver Cplex. Therefore, among the suitable solvers, the most appropriate open source solver for Error localization problems in numerical data appears to be Cbc.

Moreover, the Integer Linear Programming problems solved in these tests have a quite general structure (though a set covering [14] predominance, see also [6]). Therefore, there is some evidence that a similar solver behaviour would be repeated when solving other Integer Linear Programming problems of general structure having wide numerical excursions in the values of the constraint coefficients.

28.5 Conclusions

Error localization problems can be converted into Integer Linear Programming problems. By doing so, each erroneous record produces an Integer Linear Programming model that should be solved. This kind of problems are in general computationally demanding. However, easier models could not represent the full complexity of error localization problems, and if a conversion into some easier class of problems is attempted, a part of the error localization problem's features would be lost. Luckily, this has been a very active field of research in the last decades, and performance improvements were impressive: solvers for Integer Linear Programming probably are today more efficient than the solvers available for any other class of problems sharing the same complexity level. On the other hand, in many statistical applications, treatment should be performed by means of open source software. Efficient and open source Integer Linear Programming solvers are therefore needed. A study on the performance of internationally recognized open source Integer Linear Programming solvers, compared to the reference commercial solver Cplex on data from agricultural production having only numerical fields, is reported in this chapter. Such data present wide numerical excursions in their values, and constitute so a stressing test environment for selecting the most appropriate open source solver for performing error localization in numerical data.

Cbc resulted to be the fastest, more flexible, stable and robust among the tested open source solvers. Moreover, it has a powerful modelling interface allowing the dynamic generation of models. For these reasons, it appears to be the most appropriate for solving the described problems. Cbc inevitably suffers from some numerical imprecision, like any other scientific computing software, but to a degree lower than the other open source solvers tested, even if probably higher than the reference commercial software Cplex. Note, finally, that Cbc allows the use of several numerical settings, so that a better precision on single model solving could be achieved by tuning such settings specifically for that model.

References

1. Achterberg, T.: SCIP - a framework to integrate Constraint and Mixed Integer Programming. Technical Report 04-19, Zuse Institute Berlin (2004)
2. Banff Support Team: Functional Description of the Banff System for Edit and Imputation System. Statistics Canada, Quality Assurance and Generalized Systems Section Tech. Rep. (2003)
3. Bankier, M.: Canadian Census Minimum change Donor imputation methodology. In: Proceedings of the Workshop on Data Editing, UN/ECE, Cardiff, United Kingdom (2000)
4. Bianchi, G., Manzari, A., Reale, A., Salvi, S.: Valutazione dell'idoneità del software DIESIS all'individuazione dei valori errati in variabili quantitative. Istat - Collana Contributi Istat n. 1-2009 (2009)
5. Bruni, R.: Discrete models for data imputation. Discrete Appl. Math. **144**(1), 59–69 (2004)
6. Bruni, R.: Error Correction for Massive Data Sets. Optimization Methods and Software **20** (2-3), 295–314 (2005)

7. Bruni, R., Bianchi, G.: A formal procedure for finding contradictions into a set of rules. Appl. Math. Sci. **6**(126), 6253–6271 (2012)
8. De Waal, T.: Computational Results with Various Error Localization Algorithms. UNECE Statistical Data Editing Work Session, Madrid, Spain (2003)
9. Fellegi, I.P., Holt, D.: A systematic approach to automatic edit and imputation. J. Am. Stat. Assoc. **71**, 17–35 (1976)
10. Forrest, J., Lougee-Heimer, R.: CBC User Guide. Online available at http://www.coin-or.org/Cbc (2005)
11. Forrest, J., De la Nuez, D., Lougee-Heimer, R.: CLP User Guide. Online available at http://www.coin-or.org/Clp (2005)
12. Garey, M.R., Johnson, D.S.: Computers and Intractability: A Guide to the Theory of NP-Completeness. W.H. Freeman and Company, San Francisco (1979)
13. Makhorin, A.: GLPK Reference Manual. Online at http://www.gnu.org/software/glpk/
14. Nemhauser, G.L., Wolsey, L.A.: Integer and Combinatorial Optimization. Wiley, New York (1999)
15. Ralphs, T.: Symphony Documentation. Online available at https://projects.coin-or.org/SYMPHONY
16. Riera-Ledesma, J., Salazar-Gonzalez, J.-J. New Algorithms for the Editing and Imputation Problem. UNECE Statistical Data Editing Work Session, Madrid, Spain (2003)
17. Winkler, W.E.: State of Statistical Data Editing and current Research Problems. In: Proceedings of the Workshop on Data Editing, UN/ECE, Rome, Italy (1999)

Integrating Business Surveys

29

Guidelines and Principles Based on the Belgian Experience

Maria Caterina Bramati

Abstract

In the past few years the modernization of business surveys has been extensively discussed by several NSI's in Europe. The MEETS program (Modernisation of European Enterprise and Trade Statistics) has been recently launched for the years 2008–2013 by EUROSTAT for encouraging the reform process at the European level in order to identify new areas for business statistics, to enhance the integration of data collection and treatment, and to improve the harmonization of methods and concepts in business statistics. At Statistics Belgium the debate has been brought especially through a revision of concepts and methods in business surveys with the aim of reducing the survey costs for the Administration and the response burden for the enterprises. In the present contribution, the issue of integration of business surveys is tackled with a variable-oriented approach and using classification techniques.

Keywords

Business surveys • Classification • Data integration • Survey design

29.1 Introduction: Why Integration?

The issue of survey integration has been object of several discussions in many Statistical Institutes of European and non-European countries.

The process of integrating surveys has the aim of reducing costs and response burdens by a complete reorganization of the currently existing surveys into a unique and coherent structure. Therefore, the horizontal complexity of the stove-pipe

M.C. Bramati (✉)
Dipartimento di Metodi e Modelli per l'Economia, il Territorio e la Finanza, Sapienza University of Rome, Via del Castro Laurenziano 9, 00161 Rome, Italy
e-mail: mariacaterina.bramati@uniroma1.it

N. Torelli et al. (eds.), *Advances in Theoretical and Applied Statistics*, Studies in Theoretical and Applied Statistics, DOI 10.1007/978-3-642-35588-2_29,

statistical production process currently in use, in which each surveying step is led independently from the others, is replaced by a vertical increase in complexity.

Many key-points have been addressed as advantages coming with integration, such as a reduction in costs, gain in efficiency, statistical productivity and data quality from the Statistical Institute viewpoint, and reduction of the burdens from the respondent side.

The aim of this study is to shed a light on the advantages of integration of business surveys, measuring the current response burden for enterprises and suggesting the criteria and the structure on which the Belgian *Integrated Business Survey* (IBS) is based.

The experience of integration is not new in the current practice of some National Statistical Offices (NSO) such as the Centraal Bureau voor de Statistiek (NL), the Office for National Statistics (UK), and Statistics Canada among others.

Going towards integration in the business surveys is more than a need for Statistics Belgium, since it provides a solution to many problematic issues raised by the current way of surveying businesses, such as the harmonization of the definitions (economic, statistical, etc.), of the statistical methods (like sampling, calibration, construction of indexes), enhancing synergies and cooperation between statisticians for improving the statistical procedures and the data treatment, in a systematic approach to data quality and documentation of the surveying process. Furthermore, too many and different tools for survey management and web surveys are often employed which requires ad hoc IT implementation and intervention for each survey.

Though on the one hand, integration boosts the improvement of the statistical practices, allowing for harmonization of concepts, definitions and methods, for synergies in the use of statistical tools, for monitoring and coordination of the statistical production process enhancing data quality, on the other hand it involves some increase in complexity of the process at both theoretical and practical level. This in turns implies some non-negligible investment in terms of human capital and of IT tools. However, one of the most challenging target is to act at the conceptual level, enlarging the current notion of many single parallel surveys to one coherent variable-oriented structure able to satisfy the whole end-users' demand.

Section 29.2 describes the *variable-oriented* approach suggested for the reorganization of business surveys, whereas in Sect. 29.3 are presented the criteria for clustering the existing surveys. Some classifications are proposed and compared by means of a logistic regression.

29.1.1 Business Response Burdens in Belgium

A reform in the surveying strategy for businesses has been a priority in the past few years at Statistic Belgium. In order to quantify the response burden for enterprises and therefore the benefit of a reorganization of business surveys into one coordinated and integrated, the distribution of the *number of potential surveys for an establishment in a year* has been derived from the population frames for

Table 29.1 List of business surveys entering the IBS with their sampling frame size as percentage of the universe of firms ($N = 796230$) in 2005

Survey	Acronym	% of the sampling frame
Structural Business Survey (SBS)	SBS	88.9
Wastes from industry	Afval-ind	91.4
Wastes from agriculture	Afval-Landbouw	8.6
Unskilled labor costs	Arbeidkost	5.6
Composition and structure of salaries	Losa	7.9
Evolution of salaries	Loon	1.4
Retail trade	Dethand (Detailhandel)	13.2
Consumer credit	Kred	0.04
Continuing Vocational Training Survey	CVTS	6.7
Industrial production (PRODCOM)	Prod-ind	6.5
Producer price index	Prod-prijs	1.4
Tourism	Tour	0.5
Information and Communication Technology	ICT	8.9
Cinema	Cine	0.02
Census for agriculture	Landbouwtelling	8.1
Vegetables production	Groenten (Groenteteelten)	0.4
Stock of wine	Wijn	0.3
Stock of cereals	Graan (Graanvoorraden)	0.2
Fruits exploitation and production	Fruit (Fruitteelten)	0.3
Harvest	Oogsten (Ramingen van de oogst)	6.5
Milk production	Melk	0.02
Slaughtered animals	Slacht (Geslachte dieren)	0.1
Short census for agriculture in November	Novenq	6.5

each of the 24 business surveys produced at Statistics Belgium. In particular, the frequency count is in terms of number of establishments receiving potentially x_i, $i = 1, 2, \ldots$ questionnaires in a year according to the time frequency of each survey (monthly, quarterly, bi-annual, yearly, etc.) and to the specific population frame of the survey considered. In 2005 each local unit could potentially be surveyed 4.47 times per year in average, with a maximum of 44 questionnaires. Almost for a 10 % of cases (probably the largest firms in manufacturing) the response burden is very high, over 13 questionnaires in a year and it attains the 25.5 questionnaires for the last percentile of the distribution.

The business surveys here considered are 23 over the 24 running at Statistics Belgium (the survey on transports of goods is not taken into account). They are listed in Table 29.1, where in the last column is reported the size of the target populations for each survey with their percentage on the total number of non-missing (valid) entries for the local units. For example, in the case of the Structural Business Survey (SBS), over the 796,230 valid entries (i.e., those for which the information is available, which we call the reference universe), 708,212 are the units composing the SBS universe, i.e., the 88.9 % of the reference universe.

29.2 The Variable-Oriented Approach

The logic of the integration goes further beyond the simple summation of the existing surveys into one. The *IBS* has a new *variable-oriented* structure which is not directly related to the structure of each single survey. The starting point of the integration process consists of the required final outcome of the surveying process. This means that one needs to look at the output side of the problem, which is represented by a set of variables (say *output* variables) to "produce" for some final users. This output-based approach here suggested implies a variable-oriented structure of the IBS. The integration process should be driven mainly at four levels: the *architecture* of the statistical production process, the statistical *methods*, the *data flows* (collection, storage, rule-based processing, metadata), and the *revision* of the economic variables involved in the surveys. The objectives of the methodological framework are the *standardization of methods*, the *improvement of quality*, the *transparency*, and the *IT-standardization* by linking methods to tools. In the next sections the main guidelines and principles at the basis of the methodological framework are explained.

From the viewpoint of the final outcome to deliver in terms of *output-variables* v_{o} contained in the set V_{o}, the basic assumption on which the following analysis relies is that there exists a one-to-one mapping $f_{\mathrm{o}} : \boldsymbol{\omega}_{\mathrm{o}} \rightarrow V_{\mathrm{o}}$, where the domain $\boldsymbol{\omega}_{\mathrm{o}} = \mathrm{ID}_{\mathrm{o}} \times \mathrm{SU}_{\mathrm{o}} \times \mathbf{T}_{\mathrm{o}} \times \mathbf{TP}_{\mathrm{o}}$ is the cartesian product of sets $\mathrm{ID}_{\mathrm{o}} = \{\text{identification of the variable}\}$, $\mathrm{SU}_{\mathrm{o}} = \{\text{sampling unit}\}$, $\mathbf{T}_{\mathrm{o}} = \{\text{frequency, delivering time}\}$, $\mathbf{TP}_{\mathrm{o}} = \{\text{target population}\}$.[1] In other words, the output required by the final users of business surveys is a set of variables which are uniquely identified by four attributes: the *kind* of variable (added value, number of full-time equivalent employees, etc.), the reference *sampling unit* (establishment, statistical unit or group of enterprises), the *time* (which might be an array of information concerning time, like the survey frequency, the delivering time…), the *target population* (i.e., information concerning the breakdown details of the variable, like NACE[2] class or size class in terms of number of employees). For instance, an output-variable i is a combination of those characteristics exogenously[3] given, i.e.

$$v_{\mathrm{o}}^{i} = f_{\mathrm{o}}(\mathrm{id}_{\mathrm{o}}^{i}, \mathrm{su}_{\mathrm{o}}^{i}, \mathbf{t}_{\mathrm{o}}^{i}, \mathbf{tp}_{\mathrm{o}}^{i}).$$

[1] Boldface symbols are used to distinguish sets and variables characterized by more complex objects, like arrays.

[2] Nomenclature of economic activity.

[3] Not chosen by the statistician, but defined by the end-user.

Those attributes are defined by the requirements of the statistical authorities and more in general by recipients like Eurostat (see 1. and 2.), the National Bank of Belgium (see 4.), etc. (indicated in the diagram by R). Therefore, a survey j from the output side is a collection of output-variables $V_o^j \subseteq tV_o$ subset of the class of all variables. In this view and under the hypothesis that no redundant variables are present across the current business surveys,[4] the current survey structure can be seen as a way of partitioning the output variables in V_o into clusters (which are currently 24).

One of the advantages of this approach is that the structure of the survey is flexible to changes in the criteria defining the target population. To those changes would correspond new output variables to produce. Now, ascending in the process to the *input-side* of the variable "production," i.e., the *collection strategy*, similar assumptions to those of the output-side can be made.

An *input-variable* v_{I}^i is an intermediate product in the survey process, a piece of information concerning the output-variable which is either collected (part of a questionnaire), or obtained by external sources (administrative records or other sources). In symbols

$$v_{\mathrm{I}}^i = f_{\mathrm{I}}(\mathrm{id}_o^i, \mathrm{su}_{\mathrm{I}}^i, \mathbf{t}_{\mathrm{I}}^i, \mathbf{tp}_{\mathrm{I}}^i, \mathbf{es}^i),$$

where $f_{\mathrm{I}} : \boldsymbol{\omega}_{\mathrm{I}} \rightarrow V_{\mathrm{I}}$ is a one-to-one mapping with domain given by the cartesian product $\boldsymbol{\omega}_{\mathrm{I}} = \mathrm{ID}_o \times \mathrm{SU}_{\mathrm{I}} \times \mathbf{T}_{\mathrm{I}} \times \mathbf{TP}_{\mathrm{I}} \times \mathbf{ES}$, where the subscript I indicates the input-side of the process.

The input-variables are the result of the combination of five attributes, the variable *identification*, the reference *sampling unit*, the *time*, the *target population*, and the *external sources*. Those attributes in this case are defined by the statistician under some constraints determined by the outcome to deliver. The constraints which are generally respected in the current business surveys are

$$\mathrm{su}_{\mathrm{I}}^i = \mathrm{su}_o^i \quad \mathbf{t}_{\mathrm{I}}^i = \mathbf{t}_o^i \quad \mathbf{tp}_{\mathrm{I}}^i = \mathbf{tp}_o^i \quad \forall v_{\mathrm{I}}^i \in V_{\mathrm{I}}, \quad v_o^i \in V_o,$$

where it is assumed an input/output correspondence between the input variable v_{I}^i and the output variable v_o^i. With those constraints it is clear that the statistician at present does not dispose of a lot of flexibility in the effort of optimizing and simplifying the business survey structure. Basically the statistician is allowed to act on the external source component only. A more intensive use of the administrative records was indeed the policy adopted by many statistical institutes in the past few years in view of a more efficient surveying process and of a reduction of the statistical burdens for enterprises. However, external sources are not always such a flexible tool for the statistician, since their use is constrained to the delivering time, the sampling units to which they refer (very often the legal unit) and the target population coverage. Therefore, several transformations and approximations are often required. The transformation of the input variable into the output variable

[4] This is to ensure the invertibility of mapping f_o.

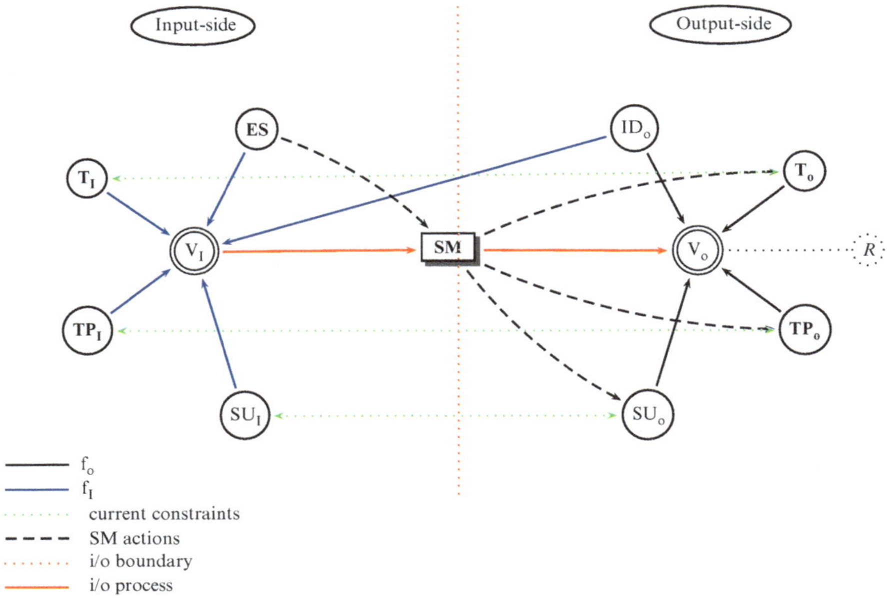

Fig. 29.1 The *variable-oriented* approach

is operated by a set of techniques, i.e., statistical methods. An appropriate and extensive use of those methods may help in relaxing some constraints under which the current business survey structure runs (Fig. 29.1).

So far, once that redundancies in V_I are eliminated, the main objective of the IBS project is to act on the input-side of the survey process, relaxing some of the constraints imposed by the output-variables by means of the statistical methods and the economic knowledge. This can entail a reduction in the number of input-variables.

For instance, for some variables the time constraint could be relaxed combining the information coming from external sources and estimation techniques. Also the target population constraint could be partially relaxed finding convenient estimators of the total (with properties like the *stratification-equivariance/invariance*). As for the sampling unit constraint, conversion methods from one definition to another would be helpful.

29.3 Classification Issues

As mentioned in the previous section, surveys can be considered as clusters of variables. Since variables are characterized by attributes like time, target population, sampling unit, etc., variable classification can be done by one single attribute or by a combination of them. The classification is a requirement for grouping surveys according to their "similarities." There is no evidence of the advantage of keeping the current partitioned structure of surveys in the sampling design, collection, and treatment of data. Of course, to respond to institutional obligations the output (at the dissemination step) should be converted by an interface into the usual disaggregated survey structure. Several clustering criteria might be adopted. Basically, four main grouping criteria can be considered:

1. *Economic* criterion. It is based on the category of variables of interest in a given survey. For instance, surveys can be clustered by groups of variables which summarize the economic structure of a business. These groups can be the *activity*, the *expenditure*, the *investment*, the *employment*, and the *innovation* of the firm (see 3.). A definition of "proximity" between activities needs to be introduced in order to reflect similarities between their distributions.
2. *Overlapping population* criterion. Surveys are grouped according to their target population. Survey clusters should be formed on a quantitative basis, i.e., computing and comparing the overlapping population rates for surveys.
3. *Enterprise-based* criterion. This criterion is *business-centric* in the sense of adopting the viewpoint of the business. To each business is attributed a code according to the survey for which it enters the target population. In this way every business can be identified by a series of numbers which represents the list of surveys to which it is concerned. Then, enterprises can be grouped by the survey code obtaining clusters of *similar* businesses in terms of survey subject. A ranking of the clusters can be made (in principle if the surveys are 24, there are 2^{24} possible clusters for the enterprises, but only a subset of them currently existing) and descriptive statistics can be easily calculated (like the mode cluster). Not only this classification might lead to a new way of classifying enterprises (according to the *survey-similarity*) but also it provides crucial insights on the way in which surveys are likely to naturally cluster between them. This topic should be object of further discussion and research.
4. *Geographical* criterion. Geographical dimension is an important component for understanding economic interactions of businesses. For instance, territorial proximity can be used as a measure of *structural similarity* regarding some economic aspects. Little attention has been payed until now to the impact of the locational component for businesses to the sectoral economic activity and to the geographical agglomeration (industrial districts). It would be interesting to increase the efforts in this direction.

All of these criteria present both advantages and disadvantages. However, the *enterprise-based* criterion might be privileged (at least as a starting point for research in classification) because it entails a new clustering approach in line with the philosophy of the IBS project. For instance, the classification criterion of surveys by *overlapping population* is equivalent to a classification of variables by the target population attribute. The classification according the *enterprise* criterion is also a classification of variables by the target population attribute. Indeed, it can be seen as the dual of the overlapping population criterion.

Also classification by the time attributes is possible. For example, variables can be grouped by the time frequency, i.e., monthly, quarterly, biannual, yearly, etc.

Table 29.2 Clusters of surveys obtained with Ward's method

Group nr.	List of surveys (LW measure)	Surveys (SE distance)
1	SBS, Afval-ind, Dethand	SBS, Afval-ind
2	Arbeidk, Losa, ICT, CVTS	Arbeidk, Losa, ICT, CVTS
3	Prod-ind, Lonen	Dethand
4	Prod-prijs, Toer, Cine, Kredieten, Wijn, Graan, Melk, Slacht, Groente, Fruit	Prod-prijs, Toer, Cine, Prod-ind, Lonen Kredieten, Wijn, Graan, Melk, Slacht, Groente, Fruit
5	Oogsten, Novenq, Afval-land, Landbouwtelling	Oogsten, Novenq, Afval-land, Landbouwtelling

variables. In the same way, classification by delivering time is possible, or by the sampling unit attribute, or even by kind of variables (*economic* criterion).

Clearly, each of these classification methods has the limit to consider only one of the variable attributes at time. Of course, a classification by a combination of the variable attributes could be a more efficient solution. In particular, two ways of classifying variables are considered: by the overlapping population criterion and by the enterprise-based criterion. In the first approach, target populations for each surveys are constructed using the NACE class and the size (turnover and number of employees) requirements. To make comparisons feasible, the statistical unit used in this exercise is the establishment and the time unit is the year.

29.3.1 Classification by the Overlapping Population Criterion

The aim of this classification is to form clusters of surveys by means of similarities between the sampling units belonging to each survey universe. Of course, some natural clusters can be made on the basis of the overlap of the population frames. For example, we observe that Wastes from Industry, SBS, and ICT are subsamples nested one into the other, whereas SBS has the largest population frame overlap with all the other non-agricultural surveys, which ranges between 87 % and 100 %. In what follows, surveys are grouped by means of the hierarchical cluster approach, using Ward's method with the Squared Euclidean distance (SE, see 5.) and the Lance and Williams measure (LW based on non-metric multidimensional scaling, see 6.).

The squared Euclidean distance for a binary variable takes value $d_{ijk}^2 = 0$ if cases i and j both have attribute k "present" or both "absent," or 1 if attribute k is "present" in one case and "absent" in the other case. In Table 29.2 are displayed the survey clusters (the chain reported ranges from three to eight groups, we selected the scenario for five groups) obtained by applying the Lance and Williams measure and the squared Euclidean distances, respectively.

The two classification are quite close. It is clear that in both cases "agricultural" surveys are grouped together; the SBS is grouped with the Wastes from industry (in some cases with the Retail Trade). The surveys concerning the human capital and technology are also clustered. When the survey populations are considered in terms of year-equivalent units, the previous results are partially confirmed. Adding the time constraint in the classification analysis does not produce conclusive results (at least in this approach).

29.3.2 Classification by the Business Criterion

Establishments are clustered by combination of survey questionnaires they might receive. Again, we observe that more than half of the establishments are subjected to be surveyed by SBS and Wastes from Industry, with few possible combinations of surveys concerning more than 300 establishments at time.

Two step cluster is used. This cluster method is designed to handle very large datasets. It has two steps: (1) pre-cluster the cases into many small sub-clusters and (2) cluster the sub-clusters resulting from the pre-cluster step into a number of clusters automatically selected. If the partition into five groups of units is considered, the following clusters are made. Clusters of establishments by surveys are

1. Harvest, short census for agriculture in November, wastes from agriculture, census for agriculture, vegetables production, fruits exploitation
2. SBS (69.2 %), wastes from industry (67.4 %)
3. SBS (11.1 %), wastes from industry (13.4 %), unskilled labor costs (18.6 %), composition and structure of salaries (21.6 %), retail trade (4.1 %), consumers credit (78.4 %), producer prices (84.8 %), industrial production (3.8 %), continuing vocational training (7.3 %), tourism (83.4 %), ICT (37.4 %), cinema (68.6 %), stock of wine (94.9 %), stock of cereals (93.9 %), milk production (64.9 %), slaughtering (85.7 %)
4. SBS (12.7 %), wastes from industry (12.4 %), retail trade (85.9 %)
5. SBS (7 %), wastes from industry (6.8 %), retail trade (10.1 %), unskilled labor costs (81.4 %), composition and structure of salaries (78.4 %), continuing vocational training (92.7 %), evolution of salaries (100 %), industrial production (96.2 %), consumers credit (21.6 %), producer prices (15.6 %), tourism (16.6 %), ICT (62.6 %), cinema (31.4 %), stock of wine (5.1 %), stock of cereals (6.1 %), milk production (35.1 %), slaughtering (14.3 %)

In parenthesis are indicated percentages of those establishments between the ones entering potentially the survey, which are selected for the given group. For example, group 1 is formed by the 69.2 % of establishment potentially surveyed by SBS and by the 67.4 % of establishments potentially participating to the survey on industrial wastes. Of course, in this group there might be also establishments which enter potentially both surveys, and therefore contained at the same time in the 69.2 % of SBS and 67.4 % of wastes from industry. Therefore, it is confirmed the separation in different clusters of the *agriculture* versus the *non-agriculture* surveys which are clustered with the SBS.

29.4 Conclusions

So far, some classification issues have been addressed, especially focusing on the *target population* component of an output-variable set. Of course, there are many other ways of classifying variables, using other variable components (such as time and sampling units) or a combination of them. One could also set a *multi-step* classification, using a priority list of criteria.

Classification by target population criteria, in particular by the enterprise-based approach, has the main advantage to allow the construction of criteria which include each new-born business into the appropriate universe for the target variables. On the other hand, this system would not be enough flexible with respect to changes in the target population criteria. Suppose for instance that for some variables a different industry breakdown is required by the statistical authority. Then a new classification by target population is required for all businesses to create the universes for the variables. If the creation of new variables or changes in their breakdown industry happen often, the enterprise-based criterion require several updates and moreover does not allow for comparisons across the periods of change. From the classification study, we observe that the main discriminant factors in clustering surveys are the *NACE* class, which helps in separation between the agricultural versus non-agricultural surveys, the industrial surveys concerning transformation and manufacturing (such as food in milk, wine, slaughtering, producer prices), services, businesses concerned by retail trade survey. Also, the employment/labor surveys are naturally clustered, stressing the relevance of the *economic* classification criterion.

The advantages of integration for both businesses and the NIS have been pointed out throughout the chapter. It is clear that the integration process involves also some operational costs for the administration, which need to be evaluated according to the survey domain.

Acknowledgements The author acknowledges the anonymous referee and colleagues at Statistics Belgium for support and many insightful discussions. All errors are my own.

References

1. Aucremanne, L., Druant, M.: Price-setting behaviour in Belgium: what can be learned from an ad hoc survey? ECB Working Paper Series n. 448 (2005)
2. Butzen, P., Fuss, C.: The interest rate and credit channels in Belgium: an investigation with micro-level firm data. ECB Working Paper Series n. 107 (2001)
3. Lallemand, T., Plasman, R., Rycx, F. Wage structure and firm productivity in Belgium. NBER Working Paper Series n. 12978 (2007)
4. Van Gastel, G.: Small and medium sized enterprises in business tendency surveys: some critical remarks. Internal discussion paper, National Bank of Belgium (1999)
5. Ward, J.H.J.: Hierarchical grouping to optimize an objective function. J. Am. Stat. Assoc. **58**, 236–244 (1963)
6. Williams, W.T., Lance, G.N.: Computer programs for monothetic classification. Comput. J. **8**, 246–249 (1965)

Part V

Social Statistics and Demography

A Measure of Poverty Based on the Rasch Model

30

Maria Cristiana Martini and Cristiano Vanin

Abstract

Asking questions about income is often a hard task: nonresponse rates are high, and the reliability of the surveyed answers is sometimes poor. In this contribution we propose a measure of the economic status based on a set of wealth-related items. This measure is based on the Rasch model, which allows to estimate different relevance weights for each item: difficulty parameters indicate the severity of each of the situations described by the items, while ability parameters indicate poverty level of each unit. This poverty measure is computed on survey data collected on a sample of 2,465 households living in Veneto. Some analyses conducted on the estimated poverty measure indicate a good consistency with the expected relationships, and confirm the possibility to get this kind of estimates from indirect measures.

Keywords

Composite indicators • Income • Measuring poverty and inequality • Rasch model

30.1 A Rasch Model to Measure Poverty

Income questions are among the most threatening questions in general population surveys, and often produce inadequate information. Nonresponse rates are frequently quite high, and family income tends to be underestimated because

M.C. Martini (✉)
Università di Modena e Reggio Emilia, Italy
e-mail: cmartini@unimo.it

C. Vanin
Regione Veneto, Italy
e-mail: cristiano.vanin@regione.veneto.it

N. Torelli et al. (eds.), *Advances in Theoretical and Applied Statistics*,
Studies in Theoretical and Applied Statistics, DOI 10.1007/978-3-642-35588-2_30,

respondents are likely to report earning after taxes and deductions, or to forget about less common sources of income. Reluctance to report income data and difficulty to collect and summarize information from different family income sources has led to a number of strategies to collect data and to correct estimates. In this chapter we propose a method to estimate wealth levels (or poverty levels) which does not require to ask threatening questions about income; this method is based on the Rasch model.

The Rasch model was developed by Georg Rasch in 1960, aimed at objectively assessing the school achievement of Danish soldiers. Then, the most natural applications of the Rasch model are in psychometrics and education, but its flexibility and simplicity makes it suitable for other applications as well, such as quality measurement and social sciences in general. Broadly speaking, this model is a good instrument to measure latent variables.

The Rasch model can also be used to study poverty, but in literature this use is not frequent. Among the most recent works we can remind Fusco and Dickes [5], while one of the first examples comes from the French Gailly and Hausman [6].

To apply this psychometric model, we consider poverty as a latent variable. The Rasch model properties make it suitable to study poverty: the specific objectivity allows to obtain an objective measure of poverty, seen as a unidimensional trait in the model hypotheses; then, the possibility to rank items and grade them in a hierarchical way grants a definition of economic disease as the sum of different factors.

In the following application, the original and simpler version of the Rasch model is used, with items coded by dichotomous variables. The starting matrix is given by a set of units (households) observed in the values of a set of dichotomous items, regarding the ownership of some goods, or the manifestation of some situations, considered as deprivation indicators. The positive response $x_{ij} = 1$ indicates a state of deprivation for the i-th family with regard to the good or characteristic expressed by the j-th item, while the null response $x_{ij} = 0$ indicates absence of deprivation. The aim is to transform the answers in a synthetic measure that allows to place the latent variable on a continuous scale.

The peculiarity of this model consists in considering two factors to measure the latent variable: the individual ability and the item difficulty. So, the latent variable is evaluated for each individual not only on the basis of how many questions he answered positively, but also based on the degree of difficulty of each question: the answer x_{ij} given by the i-th respondent depends on his relative deprivation, i.e., his intrinsic poverty combined with the level of deprivation indicated by the j-th item. This method produces two separate rankings: one ranks individuals from the most to the least needy (controlling for the items difficulty parameters), the second orders questions according to the level of poverty they imply, taking into account how many individuals lack that good or are in that situation. The Rasch model represents these two factors by means of two families of parameters: k parameters of difficulty for the items (β_j, $j = 1, \ldots, k$), and n ability parameters for the individuals ($\theta_i, i = 1, \ldots, n$).

The parameter of difficulty β_j is renamed "parameter of severity," since it corresponds to the idea of inequality: the corresponding ranking shows which are the goods whose lack mainly contribute to generate poverty, and describes the inequalities in the goods distribution among families. The higher the severity parameter, the more widespread is the ownership of the good, and the higher is the inequality degree corresponding to its lack. The underlying hypothesis is that lacking a good is more weighty if its ownership is common in the reference population [10]. In this sense, we refer to a relative poverty approach, since the inequality scales are not built in absolute terms, but in comparison with other households.

The ability parameter θ_i directly refers to poverty, and is renamed "position parameter." Its meaning is clear, and it allows to rank families from the most to the least poor. In fact, the higher the position parameter, the harsher its deprivation and poverty situation. Moreover, once we know the position parameter of a family and the severity parameter for a certain good, we can calculate the probability for that family to suffer deprivation with respect to the item.

To formalize, the relationship between the latent variable and the observed items corresponds to the probability $P(X_{ij} = 1|\theta_i, \beta_j)$ of a positive answer to the j-th item for the i-th individual. The formula resembles that of a logistic regression model, with parameters belonging to two families:

$$P(X_i = 1|q_i, b_j) = \frac{e^{(\beta_j - \theta_i)}}{1 + e^{(\beta_j - \theta_i)}}, \; i = 1, \ldots, n; \; j = 1, \ldots, k$$

Compared with the psychometric applications, the use of Rasch models to study poverty has to square with non-experimental survey designs. In fact, while psychometrics and pedagogy studies are based on questionnaires especially designed to present batteries of valid and reliable tested items, all measuring the same construct, in the studies on economic disease the data are often collected according to a different rationale, and data must be recoded to satisfy the basic assumptions.

30.2 Application to Survey Data

The Rasch model has been applied to the data coming from a survey conducted on a sample of 2,465 households living in Veneto. The survey, carried out from December 2004 to April 2005, has been conducted via computer assisted telephone interviews by the Statistics Department of the University of Padua, on behalf of the Local Centre for Household Documentation and Analysis. The Rasch model analysis responds to the demand for a variable expressing the situation of material poverty, meant as lack of goods or monetary shortage. In fact, the questionnaire includes a question about household income, but the wide response classes proposed[1] and the problem of tailoring these classes to the family size, do

[1]The four wide classes are: "Less than 1,500 € net per month," "Between 1,500 € and 3,000 € net per month," "Between 3,000 € and 6,000 € net per month," and "More than 6,000 € net per month."

not allow to estimate income in an easy way. Then, in order to build a measure of material poverty, we consider a set of items that refer, to some extent, to the economic sphere.

Unlike Gailly and Hausman [6], who consider 32 objective and subjective indicators of disease, we try to disentangle the material and nonmaterial components of poverty, therefore the Rasch model is fitted on the basis of a set of items all concentrated on different aspects of material deprivation, in order to obtain a unique principal latent trait. The selected items, dichotomized to indicate forms of material deprivation, are those reported in Table 30.1. All the variables take value 1 for households in a situation of deprivation with regard to the item.

In order to asses the internal consistency of the selected items, we calculate the Cronbach α [3] and we perform an exploratory factor analysis to test the existence of a unique common latent factor. The Cronbach α measures the internal consistency of the items by measuring their correlations. In our case the value is 0.67, acceptable even if not high.[2] The factor analysis is performed on the described items with principal components analysis and varimax rotation [7], and the plot of eigenvalues (scree-plot) shows a unique latent factor, interpretable as a deprivation factor. The first factor, corresponding to the first eigenvalue ($\lambda_1 = 3.5$), explains only 15% of the total variance. However, the reliability of the factorial procedure, expressed by Kaiser measure [8], equals 0.73, which indicates an "average" goodness according to the indications found in literature.

The algorithm used to select the final set of items is a backward procedure. First we apply the Rasch model to the whole matrix X, then we estimate the severity parameters β_j referred to the items, and the position parameters θ_i for the families through the conditional maximum likelihood method [1, 13].[3] Then, we check the

[2]Nunnaly [12] defines 0.70 an acceptable value for Cronbach α; 0.67 is quite close to that threshold.

[3]Among different possible estimation methods for the Rasch model, conditional maximum likelihood is the most frequent. The method here used maximizes the log-likelihood function conditional to the total score of each individual $r_i = \sum_{j=1}^{J} x_{ij}$:

$$L = -\sum_{j=1}^{J} s_j \beta_j - \sum_{r=1}^{I-1} f_r \log \gamma(r, \beta) \text{ with } s_j = \sum_{i=1}^{I} x_{ij}$$

where f_r indicates the number of individuals achieving a score equal to r, J is the number of considered items and I the number of individuals, and $\gamma(r, \beta)$ is the following symmetric function:

$$\sum_{(x)|r} \exp\left(-\sum_{j=1}^{J} x_{ij}\beta_j\right).$$

where $\sum_{(x)|r}$ is a summation over all possible vectors $(x)_{ij}$ such that the total score is r. Due to the property of sufficiency of the r_i scores, the maximum only depends on the severity parameters, which are the only elements in the estimate equations, resolvable through a Newton–Raphson procedure. Once the severity parameters have been estimated, their values can be used to estimate the position parameters, through a separate procedure.

Table 30.1 Severity parameters estimates and standard errors, for the Rasch model fitted on the selected items. Ranking of the deprivation items according to their relevance

Item	Severity parameter	S.E.	Outfit MSQ	Infit MSQ
Eviction	5.77	0.30	0.23	0.82
Use of savings or loans in the last year	3.83	0.12	1.84	0.95
No TV in the household	3.41	0.10	0.91	0.92
Insufficient housing for the household needs	3.24	0.10	1.64	1.01
Need for economic support in the last year	3.12	0.09	1.28	0.97
No cars in the household	2.45	0.07	0.47	0.77
Lives in a building with 10 flats or more (vs. independent house)	2.34	0.07	1.30	0.99
Rental house/apartment	2.33	0.07	1.35	1.01
No holidays in the last year for economic reasons	1.79	0.06	1.02	0.99
Savings consumption in the last year	1.48	0.06	1.54	1.17
Lives in a building with less than 10 flats (vs. independent house)	1.14	0.05	1.10	1.08
Household income less than 1,500 € per month	1.08	0.05	0.64	0.75
No videotape recorder in the household	1.03	0.05	0.71	0.79
Pension as the main household income	0.70	0.05	0.81	0.83
Unable to save money in the last year	0.27	0.05	1.58	1.37
No DVD player in the household	−0.26	0.05	0.72	0.80
No Internet connection in the household	−0.39	0.05	0.61	0.69
No motorcycle in the household	−1.30	0.05	0.98	1.01
No satellite TV in the household	−1.55	0.06	0.82	0.90
No ADSL connection in the household	−1.83	0.06	0.60	0.81
No laptops in the household	−2.16	0.07	0.71	0.88
No paid domestic help in the household	−2.36	0.07	1.62	1.02
No flat screen TV in the household	−3.15	0.09	0.63	0.85

The standard errors are computed in the following way:

$$SE(\beta_j) = \left[1/\sum_{r=1}^{I} n_r p_{rj}(1-p_{rj})\right]^{1/2}$$

where p_{rj} is the probability estimate to correctly answer the j-th item, for an individual with score r, and n_r is the number of individuals scoring r.

item-fit indices of infit and outfit[4]: the item corresponding to the highest fit index is removed, and the Rasch model re-estimated. The procedure is repeated until

[4]Outfit and infit statistics are residual based measures of accord between the model and the data [16]. Outfit are unweighted mean-square outlier-sensitive fit statistics, while in infit statistics each observation is weighted by its statistical information (model variance), yielding an inlier-pattern-sensitive fit statistic. Their expected values are 1.0, and the mean-squares show the amount of

we obtain a set of items for which the model assumptions hold.[5] After the initial estimate, only the parameter corresponding to the variable "no computers in the household" is removed from the model. The second (and final) estimate of the Rasch model produces the parameters of severity and the fit indices shown in Table 30.1.

Some outfit indices are higher than 1.5, suggesting the presence in the dataset of outlying individual answer pattern; this is not totally unexpected, given the content of the items; however, the corresponding infit indices do not force the item removal.

We can note that the scale range is quite wide (ranging from –3.15 to 5.77), making the set of items quite informative; in fact, when the items have similar severity parameters, the different response profiles tend to have the same probability, not all the units are measured precisely, and cases of zero and perfect scores are likely to occur. Table 30.1 reports the items ordered according to the severity parameter: the ones with higher estimates of β_j mainly concur to generate inequalities, and can increase the probability of material poverty.

The first remark regards the ownership of material goods: these characteristics usually do not generate poverty, and are placed at the bottom of the ranking, with the only exceptions of the possess of car and TV. This confirms a predictable result, e.g., the ownership of rare goods such as satellite TV, motorcycle, DVD players, or Internet connection does not discriminate between poor and nonpoor households, while the lack of common goods as a car or a TV indicates economic deprivation. In fact, not having the first group of goods depends more on cultural factors and life style, since they are more frequent among young, highly educated households, while the elderly rarely hanker these devices.

Among the most relevant items we can notice eviction and recourse to savings and loans; in fact, they are quite direct indicators of economic poverty and monetary trouble. Items concerning housing conditions are also quite relevant in the ranking, especially living in a rental house/apartment and in an apartment (as opposed to independent house): this confirms the importance of the housing expenses to determine the economic situation. Moreover, Italian families traditionally aspire to buy their house, even at the cost of great economic sacrifice. Finally, the parameter corresponding to the monthly income is placed in a medium ranking position; one could expect a higher position for a variable, income, which is usually a very good indicator of economic deprivation, but this position is due to the extremely poor quality of the variable as collected in this survey, as seen in the beginning of Sect. 30.2.

---

distortion of the measurement system. Values less than 1.0 indicate that observations are too predictable, i.e. redundancy and model overfit. Values greater than 1.0 indicate unpredictability, i.e. unmodeled noise and model underfit.

[5]The threshold used here is 2.0, in agreement with a "rule of thumb" often mentioned by practitioners: items with infit and outfit indices larger than 2 distort or degrade the measurement system, while items with indices between 1.5 and 2.0 are unproductive for measurement construction, but not degrading, items with indices from 0.5 to 1.5 are productive for measurement, and items with indices lower than 0.5 are too predictable but do not contradict the model [15].

Fig. 30.1 Percentage of households who perceive a poverty situation corresponding to each estimated value of the position parameters θ_i

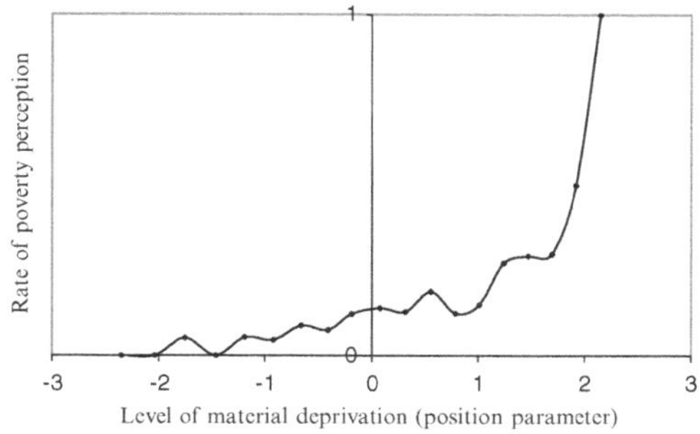

30.3 Validation of the Poverty Scale

The variable resulting from the Rasch model fit should be a measure of poverty. In order to assess its validity, we must check if it behaves as expected. First of all, we find the poverty measure consistent with a variable on poverty self-perception asked in the same survey: families who perceive themselves in economic trouble have significantly higher levels of the new poverty measure (Fig. 30.1).

Several studies show that single person, mono-parental and large households are more prone to the risk of poverty [11]; the Rasch score is consistent with the expectations (Table 30.2), except for large families, but when asked about general and economic disease, large families in this sample report lower rates of problems than smaller families [4]. Moreover, results from recent surveys on poverty in Italy show that large families are severely at risk in Southern Italy, but not as much in Veneto, where poverty is more related to old age, job uncertainty, and household disruption [2].

Human capital is known to be a powerful resource immunizing from poverty [9], and the Rasch poverty score consistently decreases for higher levels of education (Table 30.2). Also occupational status is obviously expected to be related to poverty [11], and in fact the poverty score is significantly higher for unemployed heads of household, families with unemployment problems and in general for heads of household in nonworking positions.

Moreover, we model a regression tree where the (quantitative and continuous) criterion variable is the estimated position parameter resulting from the Rasch model. The regression tree is estimated by maximizing a criterion function $\Phi(s, t)$, giving a measure of diversity between the child nodes generated by partitioning the corresponding parent node t. In our application, the criterion function is Fisher's η^2 index, derived from the heterogeneity function $i(t)$ measuring the lack of homogeneity of a group with respect to the dependent variable. The $i(t)$ function is the deviance of node t:

$$i(t) = \sum_i |y_i - \bar{y}_t|^2$$

Table 30.2 Average Rasch score for some household characteristics, and p-value of the ANOVA mean comparison

Household characteristics	Rasch score
Number of family members (p-value < 0.001)	
1	0.60
2	0.22
3–4	−0.21
5 or more	−0.28
Familiar structure (p-value < 0.001)	
Single person household	0.60
Couple	0.22
Nuclear family	−0.24
Extended family	−0.19
Single parent with children 0–5	0.31
Other	0.10
Head of household's education (p-value < 0.001)	
University or postgraduate	−0.22
High school	−0.07
Compulsory education	−0.03
Lower education	0.45
Head of household's occupational status (p-value < 0.001)	
Employed	−0.25
Unemployed	0.45
Retired	0.36
Other (invalid, student, housewife, etc.)	0.43
Presence of unemployed (p-value < 0.001)	0.29
General average	0.02

where $\bar{y}_t$ is the mean value y in the parent node t. Thus, we can estimate the mean value of y in each node in order to characterize the group. The tree building process is very similar to a multiple regression model where the variables are selected through a forward procedure. Additionally, the regression tree allows to detect combinations of the predictive variables with interaction effects, and to insulate and describe groups of families particularly subjected to economic deprivation.

The regression tree is fitted by means of the library *tree* of software R [14], devoted to Classification And Regression Trees (CART) analysis. The halt rules are set as follows: the minimum dimension of groups is set at 100 units to avoid the making of tiny groups; the maximum number of final nodes is set at 14 to avoid too complex branching; the minimum total deviance of a parent node is set at 1% of the starting deviance of y to avoid further partitioning of homogeneous nodes.

The analysis of the regression tree (Fig. 30.2) also shows results which are consistent with what we expect from a poverty measure. The traditional family with an employed head of household is the wealthier type of family. These are couples, couples with children and extended families, whose principal income derives from

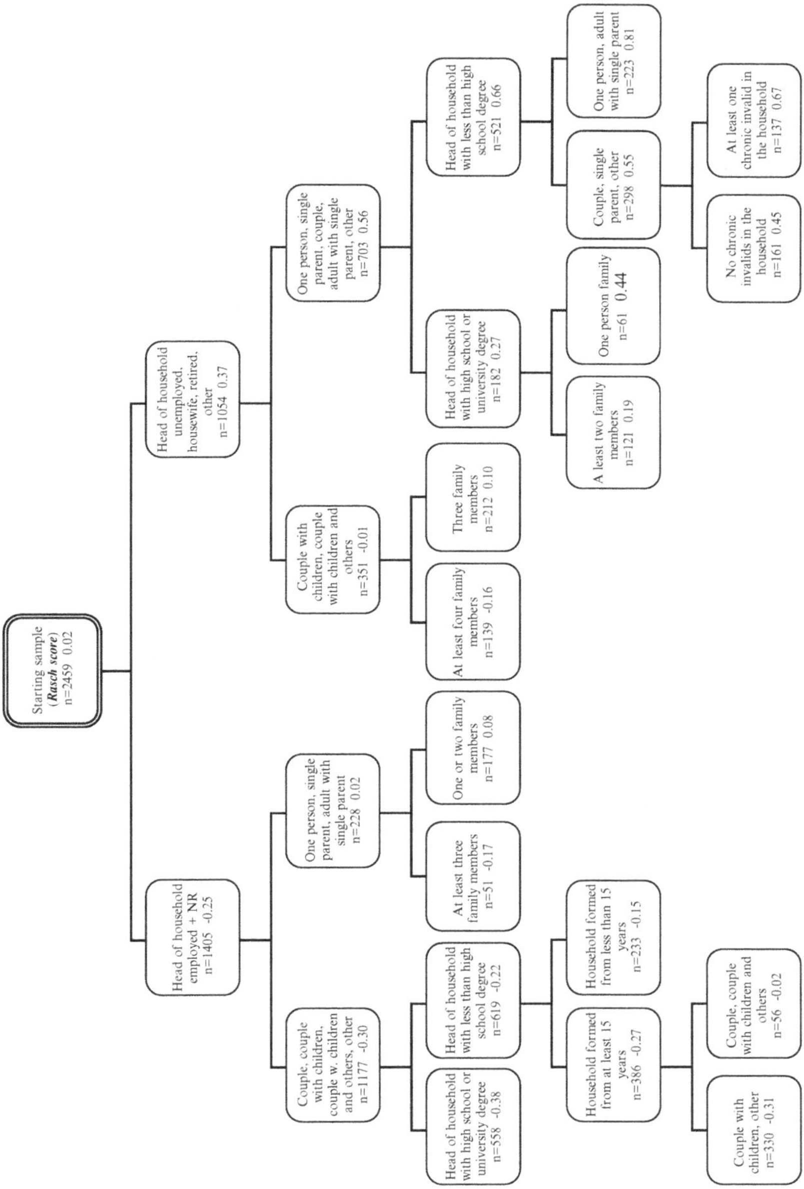

Fig. 30.2 Regression tree for the poverty measure resulting from the Rasch model

employment or self-employment. Among these, the most protected from poverty are households with an average or high level of education. Also for households with lower education levels, steady families are less vulnerable to poverty, both when they have been formed from longer periods and when they have a stable structure: in this sense, the traditional couple with children is the less economically frail. A traditional and stable familiar structure seems to prevent from poverty independently on the working situation of the head of the household. In any case, for each type of family structure, children are a factor that protects from poverty; probably their presence is a stimulus to face toward critical situations and assure them a certain future.

Families whose main income comes from pensions or other sources have higher poverty levels. The traditional household structure (couple with children or couple with children and other members) is again less vulnerable to poverty. Moreover, the existence of more than one child further protects from material deprivation; since these are mostly old families with a retired head of the household, more children probably imply more incomes. The main difference, compared with the traditional families in the left side of the tree, concerns the couples without children: when the head of the household is not employed (hence, in most cases, old), this type of family is more frail to poverty. Finally, chronic invalids in the household correspond to higher levels of poverty, probably because they require medical attention and care expenses, in addition to the constant assistance of family members, thus subtracted to other activities.

30.4 Final Remarks and Future Perspectives

The Rasch model allows to estimate a poverty measure, also in default of a reliable and accurate measure of income. The necessary variables for the estimate pertain the economic area and the ownership of durable goods, and are therefore much easier to survey than income, and less prone to definition complexity, nonresponses, and reluctant respondents. A comparison between the new poverty measure and the perceived poverty, as well as the analysis of the relationship with some household variables likely to influence the poverty level, confirm the adequacy of this indicator. The lack of a reliable income variable in the survey data used for the application prevents from performing a complete validation of the obtained measure. The application could be replicated on a different dataset which includes an income variable, in order to compute an alternative measure of poverty independent from the income. The comparison of the two (direct and indirect) measures of poverty will also shed light on the possible reasons of discrepancy, and suggest future improvements.

References

1. Andrich, D.: Rasch models for measurement. Sage University Paper series on Quantitative Applications in the Social Sciences, 07–068. Sage Publications, Beverly Hills (1988)
2. Caritas Italiana – Fondazione E. Zancan: Famiglie in salita. Rapporto 2009 su povertà ed esclusione sociale in Italia. Il Mulino, Bologna (2009)
3. Cronbach, L.J.: Coefficient alpha and the internal structure of tests. Psychometrika **16**, 297–334 (1951)
4. Fabbris, L., Martini, M.C., Vanin, C.: Le categorie di famiglie a rischio di povertà. In: Fabbris, L. (a cura di): Le famiglie venete a rischio di disagio, pp. 65–89. Cleup, Padova (2007)
5. Fusco, A., Dickes, P.: The Rasch model and multidimensional poverty measurement. In: Kakwani, N., Silber, J. (eds) Quantitative Approaches to Multidimensional Poverty Measurement, pp. 49–62. Palgrave Macmillan, New York (2008)
6. Gailly, B., Hausman, P.: Des Désavantages Relatifs à une Mesure Objective de la Pauvreté. In: Sarpellon, G. (eds.) Understanding Poverty, pp. 192–216. FrancoAngeli, Milano (1984)
7. Kaiser, H.F.: The Varimax criterion for analytic rotation in factor analysis. Psychometrika **23**, 141–151 (1958)
8. Kaiser, H.F.: An index of factorial simplicity. Psychometrika **39**, 31–36 (1974)
9. Micheli, G.A.: Cadere in povertà. Le situazioni a rischio, i processi, i terreni di coltura dell'impoverimento. Franco Angeli, Milano (1999)
10. Micheli, G.A., Laffi, S., (a cura di): Derive. Stati e percorsi di povertà non estreme. FrancoAngeli, Milano (1995)
11. Nanni, W., Vecchiato, T., (a cura di): La rete spezzata. Rapporto su emarginazione e disagio nei contesti familiari (Caritas Italiana – Fondazione E. Zancan). Feltrinelli, Milano (2000)
12. Nunnaly, J.: Psychometric Theory. McGraw-Hill, New York (1978)
13. Rasch, G.: Probabilistic models for some intelligence and attainment tests. Danish Institute for Educational Research, Copenaghen (1960). New edition, The University of Chicago Press, Chicago (1960/1980)
14. Venables, W.N., Ripley, B.D.: Modern Applied Statistics with S-PLUS. Springer, New York (1999)
15. Wright, B.D., Linacre, J.M., Gustafson, J.-E., Martin-Löf, P.: Reasonable mean-square fit values. Rasch Meas. Trans. **8**, 370 (1994)
16. Wright, B.D., Masters, G.N.: Rating Scale Analysis. MESA Press, Chicago (1982)

Chronic Poverty in European Mediterranean Countries

31

Daria Mendola and Annalisa Busetta

Abstract

This chapter investigates the characteristics of chronic poverty among youth in Southern European countries. These countries have the highest levels of poverty in Europe and welfare systems unable to smooth social inequalities effectively. The main aim of this chapter is to investigate on how long-lasting poor among the Mediterranean youth differs in socio-demographic and economic characteristics. The intensity of poverty over time is measured by a very recent longitudinal poverty index based on the rationale of cumulative hardship. We tested the effects of many covariates on the incidence and intensity of chronic poverty, controlling for main socio-demographic and economic characteristics, using a ZINB model. The peculiar mix between the lack of effective state institutions and a strong presence of families explains why many factors, which usually are significantly related to poverty, do not have an important role here. A strong inertial effect is due to the intensity of poverty at the start of the time-window. Italian youth has the worst performance.

Keywords

Chronic poverty index • Initial conditions • Young adults • ZINB model

31.1 Introduction

There is a great turnover in the stock of people living around the poverty line. Indeed many of those who have left poverty return back relatively quickly, while only a minority of people, poor in a certain year, will experience a long-lasting poverty.

D. Mendola (✉) · A. Busetta
Department of Economics, Business and Statistics, University of Palermo, Italy
e-mail: daria.mendola@unipa.it; annalisa.busetta@unipa.it

N. Torelli et al. (eds.), *Advances in Theoretical and Applied Statistics*, Studies in Theoretical and Applied Statistics, DOI 10.1007/978-3-642-35588-2_31,

Only around 60% of young Europeans are never poor in 7 consecutive years with relevant differences among countries [22].

The interest goes therefore to longitudinal patterns of poverty and to the factors related to the process of impoverishment, and marginalization. In particular, which individual and household characteristics influence poverty persistence? Do economics and labour market-related initial conditions prevail over socio-demographic ones?

This chapter focuses on young people living in four countries in Southern Europe (Greece, Italy, Portugal, and Spain) where chronic poverty is widely spread. Indeed, analysing only this group of countries allows for controlling those confounding factors related to the interaction among individual behaviours, family models, labour market characteristics, and social protection schemes, which are all the features of the "Mediterranean welfare typology" [12]. We use a synthetic measure of chronic poverty based on a minimum and fully nonparametric set of assumptions. This chronic poverty index is then modelled to spot factors associated with long-lasting poverty, with particular attention to the initial conditions problem.

The chapter is structured as follows. In Sect. 31.2 we present the literature on measurement of longitudinal poverty. In Sect. 31.3 we introduce data and describe chronic poverty among young people in Southern European countries. The last two sections (Sects. 31.4 and 31.5) sum up the main results of the model and our conclusions.

31.2 Theoretical Background

Poverty is characterized by a strong state dependence (since the experience of poverty in the past can influence the occurrence of poverty in the future) and affects, with different strength, specific subpopulations. There is a wide literature on the determinants of poverty persistence that uses longitudinal data, although part of this refers to papers on income dynamics from which, however, information about poverty can be derived. These models separate the true state dependence from the propensity to experience poverty due to observable and unobservable factors. Among the most successful models in the literature we highlight (a) transition probability models, (b) covariance components models, and (c) longitudinal poverty pattern models (see [18]).

The first category is the most applied and deals with transitions in and out of poverty over time as a function of individual characteristics which may be fixed, time-variant, but also lagged covariates (see the seminal paper of [2] and also [26]).

The second class introduces stochastic time-series structures in the error component (usually autoregressive terms of order 1 or greater) and includes a further error term that may account for individual heterogeneity, persistent effect of random shocks, and unobserved individual specific variables.[1]

[1] A drawback of these two categories of models is that they usually do not take into account that there is a strong correlation among *multiple* spells of poverty or non-poverty for the

Finally the last category embraces models with a dependent variable accounting for the complete longitudinal sequence of incomes observed for each individual. This kind of models (see among others [14, 17]) requires that all the covariates refer to the first year of the sequence if they are time-varying, since they have to be related with a response variable that is a simple fixed measure synthesizing all the individual poverty and non-poverty spells. Nevertheless we think that these "simpler" models, from which it is much easier to derive poverty persistence predictions, could give a positive pay-off. In addition, according to us, they could be improved by using as response variables more sophisticated measures, and this could represent an interesting alternative way to study state dependence in poverty while connecting it to household and individual characteristics. In this perspective it is noteworthy to say that between 2009 and 2011 many papers proposed new indices of longitudinal (or chronic) poverty which account for poverty over time.[2] These indices take into account the nonrandom patterns in the sequences of poverty (multiple) spells and acknowledge, in some way, correlation structure of poverty and nonpoverty spells without making any parametric assumption. The model proposed in Sect. 31.4 uses just one of these indices inside the longitudinal poverty pattern models framework.

31.3 Data and Methods

Data come from the European Community Household Panel (ECHP). The sample includes people aged 16–29 at the first wave, living in Greece, Italy, Portugal, or Spain, with complete information about household income along the seven waves of the panel (balanced panel from 1994 to 2001). The sample is of 6,201 individuals. The poverty status is assessed according to the Eurostat criterion: the poverty line is set at the 60% of the median net equivalized income, for each country and each year using the OECD modified equivalence scale; the households are classified as poor or not poor, and all their members accordingly. We calculate the measure of chronic poverty, explained in the following, upon six waves ranging from wave 2 to wave 7; while the supposed predictive factors (covariates) are all recorded at the first wave, but four variables related to changes in the observation time-span.

The intensity of poverty over time is measured *via* a reduced form of the individual longitudinal poverty index *PPI* in [21], which focuses here simply on how poverty and poverty spells follow each other, ignoring both the "decay effect" and the intensity weights. This reduced *PPI* is a rational number in the [0,1] interval, where 0 corresponds to an individual who is never poor or is poor only once in

same individual. In particular, in transition models, *spells in the two alternating states (poverty and non-poverty) for the same individual are also assumed to be uncorrelated* [...] *However, when individuals differ in unobserved terms like ability, effort, tastes, and these unobservables remain constant over the individual's lifetime, the assumption of uncorrelated spells might be inappropriate* [8].

[2]See among others [5, 6, 9, 13, 15, 21, 23, 27].

Table 31.1 Longitudinal poverty in Southern European countries

	Spain	Italy	Greece	Portugal	Southern Europe
Longitudinal poverty (from 1st to 7th wave)					
Sample size (percentage of the sample)	1,560 (25.15)	2,291 (36.95)	1,115 (17.98)	1,235 (19.92)	6,201 (100.00)
% Never longitudinally poor	73.27	68.31	74.80	82.11	73.47
Ever poor	2.05	5.50	1.97	2.02	3.31
Longitudinal poverty index among who experience at least two years of poverty (from 2nd and 7th wave)					
Median PPI	0.21	0.41	0.29	0.29	0.29
Mean PPI	0.33	0.46	0.36	0.39	0.40
Standard Deviation	0.29	0.33	0.29	0.32	0.32
Percentage of years in poverty	52.04	63.36	48.04	48.87	55.92
Mean PG along six waves	0.19	0.29	0.18	0.21	0.24

the observation interval (named "longitudinally never poor" from now on) and 1 corresponds to an individual who is always poor. The more numerous and close the years of poverty, the higher the intensity of poverty in a longitudinal perspective (hypothesis of cumulative hardship). Note that *PPI* accounts for duration of poverty and it refers to the spells approach, but here the emphasis is not simply on the number of spells of poverty in the temporal sequence but also on the length of the recovery spells. We refer the reader to the original paper for technical details and properties of the index.

Let us now describe our sample (see Table 31.1). Comparing the situation of chronic poverty in Mediterranean countries it is evident that Italian young people are the most exposed to long-lasting poverty (only around 68% of them are never longitudinally poor), while Portuguese young people are the least exposed (Table 31.1). At the same time, the percentage of individuals who never escaped poverty along the seven observed years ranges between around 2% in Greece, Portugal, and Spain to more than 5% in Italy.

A synthetic display of the *PPI* distribution across the four Mediterranean countries is given in Table 31.1. If we concentrate only among individuals who experimented at least 2 years in poverty, again Italian young people are in the worst situation also in terms of intensity of chronic poverty (highest median, mean); while the best situation is lived by Spanish young adults. Moreover the intensity of chronic poverty in Italy shows a median which is almost double of the Spanish one. Indeed Greece and Spain show quite similar distributions of long-lasting poverty, but the median value of the index is considerably higher among Greek young people.

Among those who experienced at least 2 years of poverty (i.e. the longitudinally poor people, according to the duration cut-off adopted in this chapter), Italy is the country with the highest mean permanence in poverty with an average number of years spent in poverty of more than 63% of the entire time interval. An important feature of poverty is its severity. Again if we concentrate only on longitudinally poor individuals, the intensity of longitudinal poverty (measured strictly by the country-level mean of the individual poverty gaps) is substantially around 0.20, but in Italy where it reaches 0.28.

As announced, the main goal of this chapter is studying the effect of household and individual socio-economic characteristics on the intensity of poverty persistence and on the process determining an individual lives permanently out of poverty. Note that in this case study, due to the shortness of the observation period (we evaluate longitudinal poverty over six waves), the index takes 29 different values only (apart from 0) and the distribution is over-dispersed. This feature drives us to consider the suitability of a negative binomial model.[3]

Indeed the most common motivation for using a negative binomial model is in terms of unobserved heterogeneity, but in this case we find more appealing the idea of the "contagion model" suggested by Eggenberger and Pólya in 1923, as reported in [19]: "contagion occurs when individuals with a given set of x's initially have the same probability of an event occurring, but this probability changes as events occur [...] thus contagion violates the independence assumption of Poisson distribution". This seems to fit perfectly the case of individual poverty sequences. In fact transitions from poverty to non-poverty do not have the same probability of the reversed ones, and, in addition, the longer the permanence in poverty the lower the probability to escape. Since, longitudinally never poor people represent nearly 75% of the distribution (see Table 31.1), this suggests the use of a zero-inflated negative binomial model (ZINB).[4] In particular firstly a logit model is estimated for predicting whether or not a young individual is never longitudinally poor along the six waves ($PPI = 0$): it shows the effect (on the logit scale) of the covariates (referring to initial conditions and changes occurred during the observation period) on being never poor. Secondly, a negative binomial model is generated for predicting the values of *PPI*: it shows the magnitude and the direction of covariates' effect on

[3]Note that here, we use rounded values of the *PPI*. In particular, we multiplied the *PPI* by 1,000 and then rounded to the nearest integer. This transformation allows us to keep 30 different values also in the transformed values. Indeed in presence of a longer panel a more appropriate model would be a mixture model allowing zero-inflation and a suitable model for asymmetric data (looking at the distribution observed a reasonable model seems to be the Beta distribution with parameters lower than 1).

[4]In our case part of the inflation of zeroes is due to the fact that the *PPI* is zero also for individuals with a single spell of poverty. Indeed people who are poor only once in the 6 years of observation are 11.7% of our sample. The likelihood ratio test and the Vuong test confirm the appropriateness of the chosen functional form compared to a standard negative binomial distribution.

the longitudinal poverty intensity. These two models (simultaneously estimated) are then combined. Note that the two models in the ZINB show opposite signs for the same covariates. In the negative binomial model the estimated coefficients with a positive sign indicate the amount of increase in longitudinal poverty index value that would be due to an increase (or to a change in state) in the explanatory variables, while the negative sign shows the opposite. On the other side, in the logit model the estimated coefficients with a positive sign indicate the amount of increase in the predicted probability to be never poor during the following 6 years (on the logit scale). The higher the coefficient, the higher the predictive (and protective) effect of the covariate.

31.4 Chronic Poverty Among Young Southern Europeans

As mentioned above, we focus our attention on Southern European youth poverty; this is not only because these countries present the highest scores of chronic poverty in Europe [16] but mainly because they represent an interesting object of study giving that they share many common features, not only in the social welfare organization, but more broadly in the interplay of state institutions and families which generates a peculiar welfare mix (see [11]). Indeed, as highlighted in [20], all these countries are characterized by "the marginal role of social assistance and the absence of minimum income programmes [. . .]. It is argued that the 'patchiness' of safety nets in Southern Europe is due to a unique set of constraints, the most relevant of which are the role of families and the 'softness' of state institutions".

In addition note that young people in these countries have very similar patterns of transition to adulthood: they tend to postpone leaving of parental home, take a longer time to complete education, entry late into the labour market, postpone the setting of a new family and the choice to have children. In the first wave of our sample, 77% of young people lives with their parents; less than 25% is married or lives with a partner; only 15% has a child. The reasons for this delayed transition to adulthood are quite controversial and actually broadly studied [3, 4]. All this helps to understand why, as we will see in the following, most of the demographic characteristics which usually explain poverty does not play an important role in explaining chronic poverty among Southern European youth, whereas economic characteristics and the changes during the observation window still reveal their importance.

Let us now introduce the main results from the ZINB model for Southern European countries (see Table 31.2). At this stage, it is worthwhile saying that, as mentioned in Sect. 31.2, due to the nature of our response variable, all covariates needed to be either fixed or measured at the start of the sequence. However we also built up four dummy variables accounting for changes, occurred at some, even if not-specified point, in the observation period. We believe that these could add an interesting interpretative value to the overall model.

Among the initial conditions considered, a key role is played by the intensity of poverty experience at the first wave (poverty gap, PG in the following). This

Table 31.2 ZINB model: estimates for Southern European countries

	Negative binomial model		Logit model	
Personal Characteristics	Coef.	Std. Err.	Coef.	Std. Err.
– Woman	n.s.	–	n.s	–
– Age at first wave	n.s.	–	n.s.	–
– Squared age at first wave	n.s.	–	n.s.	–
Socio-economic characteristics (at 1st wave)				
– Poverty gap (PG)	1.21	0.51	−3.10	0.99
– Upper Secondary School (ref. Tertiary School)	n.s.	–	n.s.	–
Below the Upper Secondary School	n.s.	–	−0.68	0.19
– Upper Secondary School * PG	n.s.	–	n.s.	–
Below the Upper Secondary School* PG	n.s.	–	n.s.	–
– Student (ref. worker >15 hours)	n.s.	–	n.s.	–
Unemployed or nonlabour forces	n.s.	–	n.s.	–
– Saving habits	−0.38	0.09	0.87	0.13
– Saving habits * PG	0.70	0.29	−2.76	0.85
Changes from 2nd to 7th wave				
– Out of parental home for at least 3 years	n.s.	–	−0.35	0.12
– Student/nonpaid trainer for at least 3 years	0.23	0.07	−0.33	0.12
– Unemployed or nonlabour forces for at least 3 years	0.30	0.05	−0.60	0.11
– 1 step up in education level	n.s.	–	n.s.	–
Household characteristics (at 1st wave)				
– Out of parental home and without kids (ref. live in parental home)	n.s.	–	n.s.	–
Out of parental home and with kids	n.s.	–	n.s.	–
– Number of kids in the household	0.12	0.04	−0.64	0.11
– Number of kids in the household * PG	0.10	0.04	−0.52	0.11
– Material deprivation	0.07	0.02	−0.48	0.04
– Material deprivation* PG	0.07	0.01	−0.29	0.04
Country's effects				
– Portugal (ref. Italy)	−0.19	0.07	1.09	0.12
Greece	−0.27	0.06	0.59	0.12
Spain	−0.30	0.05	0.23	0.10
– Constant	6.16	0.80	2.91	1.41

Vuong test $z = 17.15$, P-value 0.0000; n.s.= not statistically significant at 5%

last has a relevant direct effect in explaining both the increase of the intensity of poverty in the next six waves (beta = 1.21) and the decrease of probability to be never longitudinally poor (−3.10 on the logit scale) as expected under the state dependence hypothesis. Moreover PG interacts, sometimes in a very strong way, with the effect of most of the other covariates.

Having a higher education at the first wave does not give any advantage in reducing the intensity of long-lasting poverty (whatever the intensity of poverty -PG- in the first wave), on the contrary it shows to have some effect on the probability to never experience longitudinal poverty. In particular the coefficients estimated for education levels attained at the first wave in the binary model show that (compared to who completed the tertiary school) the probability to be never poor is significantly reduced by having an education below the upper secondary school (−0.68). Even in case young achieves a better level of education, during the 6 years considered, this does not imply an improvement of her condition, maybe also due to the shortness of time-window. Being unemployed or out of the labour force[5] (vs. being a worker), again recorded at the first wave, does not act neither on the chance to be never poor nor on the intensity of poverty duration. The lack of significance of these key variables (both education and activity status) could be explained considering the peculiarities of Mediterranean welfare state and labour market. Indeed in Southern European countries there is a very low economic return to education [25], increasing difficulties in entering the labour market, and a large diffusion of underpaid and precarious employment for young adults. Indeed "transitions to stable employment can be long. In Europe, it takes, on average, almost two years to find a first job after leaving school", and it takes even more in the South [24].

Moreover, being a student or unemployed have no significant impact both on the magnitude of the longitudinal poverty index and on the slipping into poverty, but the condition of lasting as a student or in the unemployment (measured as more, or less than 50% of time spent in these conditions) represents a risk factor for the experience of high intensity poverty and of being poor more than 1 year.

It is a well-known fact that the median age of leaving home in Mediterranean countries is higher than in any other European country. Recent studies [1, 22] suggest that delaying the transition out of the parental home can be a strategy for young individuals to reduce their risk of entering poverty. In this context living with parents, as well as with a partner/spouse (with or without children), does not show any effect on poverty. This is probably due to the fact that in all Southern European countries young adults tend to stay at parental home until they achieve economic independence and/or find a stable employment, and in any case until the completion of their studies (nearly 80% of the individuals in our sample live with their parents at the first wave). This concurs also to explain the absence of a gender effect.

As expected an important influence can be attributed to the number of kids (aged less than 16 years) living in the household: the higher the number of kids, the higher the probability to experience more severe forms of longitudinal poverty and, similarly, the lower the probability to be never poor (betas respectively equals to 0.12 and −0.64 for the negative binomial and the binary model). This effect is amplified as long as the poverty gap at the first wave increases.

---

[5]Please note that the classification of the activity status is made through three categories, mutually exclusive, derived from the shrinking of the main activity declared by respondents. So a young individual is a "worker" if she is employed for at least 15 h a week, or a "student"/"unemployed or out of the labour force" even if doing a small/odd paid job but for less than 15 h per week.

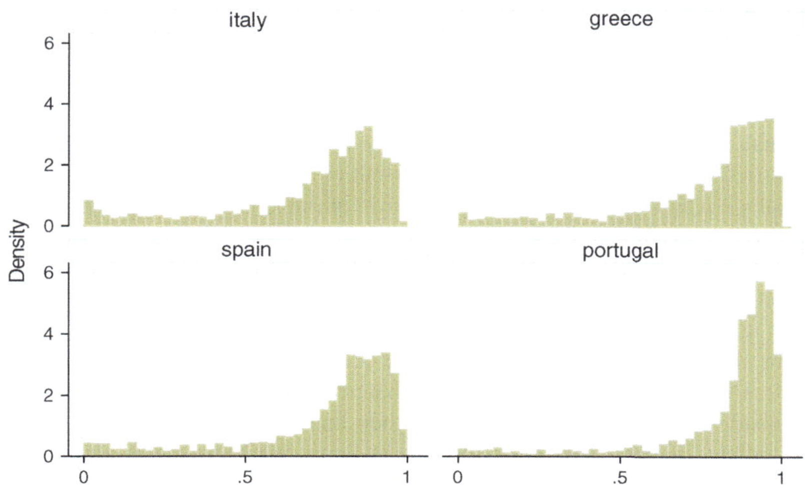

Fig. 31.1 Distribution of predicted individual probabilities to be never longitudinally poor

Neat and expected is also the effect of the economic covariates in explaining chronic poverty, even though with different degrees. The model shows that in the Mediterranean countries the material deprivation (measured as the number of not possessed items among the 7 indicators listed in [10]) slightly, but significantly, increases the intensity of chronic poverty (beta = 0.07) and decreases the probability to be never poor (beta = − 0.48 on the logit scale). This effect shows significant interaction with poverty gap at first wave. Moreover, the model shows that the individual ability to save up for the future protects them in terms of both reducing the magnitude of longitudinal poverty and increasing the probability to not experience poverty. But this protective power is reversed with the increasing of poverty gap at the first wave.

In addition, the model takes into account the existence of country peculiarities among the Mediterranean ones, including dummy variables for each country (assuming Italy as reference category). These dummies act as a sort of control for residual unobserved heterogeneity (in terms of legislation, labour market regulation, and demographic behaviour not captured elsewhere). The beta estimates are all statistically significant both in the negative binomial and in the logit model, and show that Italian welfare system is the least able to smooth inequalities in terms of long-lasting poverty. A more effective view can emerge from predicted probabilities to be never longitudinally poor (Fig. 31.1), where it is evident that the higher probability to be never poor is expected for Portuguese young adults where the Italians show the lower one.

31.5 Conclusion

The literature on longitudinal poverty debated a lot on how to take into account the distortion introduced by omitting information on the characteristics of the individuals. Most of the papers propose more and more sophisticated models that imply strict assumptions on the error distribution. These models fully exploit the potential of panel data (accounting both for initial conditions and time-varying variables) but usually do not take into account that there is a strong correlation among multiple spells of poverty or non-poverty for the same individual. Recently some scholars proposed to model the initial poverty conditions and non-random attrition in addition to the modelling of poverty transitions [7].

In an alternative way in this chapter we deal with these issues starting from a synthetic measure of sequences of states (that accounts for the closeness of poverty spells inside the observed sequence) and then including it as response variable in a zero inflated negative binomial model. As far as we know there are no papers which include indices of poverty persistence as response variable so that this chapter represents a contribution, though not ambitious, in this direction.

The results of the zero inflated negative binomial model show that economic characteristics are crucial in explaining the highest levels of chronic poverty. In particular among the initial conditions considered a key role is played by the intensity of poverty experience at the first wave which has a direct effect both on the intensity of poverty in the next six waves and on the probability to be never longitudinally poor. Moreover it interacts, sometimes in a very strong way, with the effect of most of the other covariates (e.g., material deprivation and saving habits).

On the contrary socio-demographic characteristics seem to have a less relevant impact on chronic poverty. In a context where families are called upon to play a role in smoothing the negative effects of the delayed transition into adulthood, living with a partner/spouse (with or without children) does not show any effect on poverty. This result is likely due to the fact that young Mediterranean individuals tend to stay in parental home until they achieve fully economic independence, and this concurs to explain the non-relevance of different living arrangements which usually are associated with poverty. As far as countries' differences, Italy is the least able to smooth inequalities in terms of incidence and duration of long-lasting poverty.

As a consequence Italian young adults suffer more severe forms of longitudinal poverty. The wider difference among the four countries is between Italy and Portugal: the odds to be never poor for Portuguese young adults is three time greater than the odds for Italian people.

Acknowledgements Although this chapter is the result of the joint effort of both authors, who share the full responsibility for the content, Sects. 31.1, 31.2, and 31.3 are attributable to D. Mendola, while Sects. 31.4 and 31.5 to A. Busetta. The authors gratefully acknowledge the financial support provided by University of Palermo [grant no. 1267929730 - CORI] under the responsibility of professor A.M. Milito, and IRISS at CEPS/INSTEAD for providing the ECHP data under the FP6 (Trans-national Access contract RITA 026040).

References

1. Aassve, A., Iacovou, M., Mencarini, L.: Youth poverty and transition to adulthood in Europe. Demogr. Res. **5**(2), 21–50 (2006)
2. Bane, M.J., Ellwood, D.T.: Slipping into and out of poverty: the dynamics of spells. J. Hum. Resour. **21**(1), 1–23 (1986)
3. Billari, F.C.: Becoming an adult in Europe: a macro(/micro)-demographic perspective. Demogr. Res. Spec. Collection **3**(2), 15–43 (2004)
4. Billari, F.C., Rosina, A.: Italian "latest-late" transition to adulthood: an exploration of its consequences. Genus **69**, 71–88 (2004)
5. Bossert, W., Chakravarty, S.R., D'Ambrosio, C.: Poverty and time. J. Econ. Ineq. **10**(2), 145–162 (2012)
6. Calvo, C., Dercon, S.: Chronic poverty and all that the measurement of poverty over time. In: Addison, T., Hulme, D., Kanbur, R. (eds.) Poverty Dynamics. Interdisciplinary Perspectives, pp. 29–58. Oxford University Press, Oxford (2009)
7. Cappellari, L., Jenkins, S.P.: Modelling low income transitions. J. Appl. Econ. **19**, 593–610 (2004)
8. Devicienti, F.: Estimating poverty persistence in Britain. Empir. Econ. **40**(3), 657–686 (2011)
9. Dutta, I., Roope, L., Zank, H.: On intertemporal poverty measures: the role of affluence and want. Soc. Choice Welf. (2012). doi:10.1007/s00355-012-0709-8
10. Eurostat: Material deprivation in the EU. Stat. Focus **21**, 1–11 (2005)
11. Ferreira, L.V.: Persistent poverty: Portugal and the southern European welfare regime. Eur Soc **10**(1), 49–71 (2008)
12. Ferrera, M.: The "southern model" of welfare in social Europe. J. Eur. Soc. Policy **6**(1), 17–37 (1996)
13. Foster, J.E.: A class of chronic poverty measures. In: Addison, T., Hulme, D., Kanbur, R. (eds.) Poverty Dynamics. Interdisciplinary Perspectives, pp. 59–76. Oxford University Press, Oxford (2009)
14. Hills, M.S., Jenkins, S.P.: Poverty among children: chronic or transitory? In: Bradbury, B., Jenkins, S.P., Micklewright, J. (eds.) The Dynamics of Child Poverty in Industrialised Countries, pp. 174–195. Cambridge University Press, Cambridge (2001)
15. Hoy, M., Zheng, B.: Measuring lifetime poverty. J. Econ. Theory **146**(6), 2544–2562 (2011)
16. Iacovou, M., Aassve, A.: Youth Poverty in Europe. Joseph Rowntree Foundation, York (2007) (Available at http://www.jrf.org.uk/sites/files/jrf/2121-poverty-youth-europe.pdf)
17. Jalan, J., Ravaillon, M.: Transient poverty in postreform rural China. J. Comp. Econ. **26**(2), 338–357 (1998)
18. Jenkins, S.P.: Modelling household income dynamics. J. Popul. Econ. **13**(4), 529–567 (2000)
19. Long, J.S.: Regression Models for Categorical and Limited Dependent Variables. Sage, Thousand Oaks (1997)
20. Matsaganis, M., Ferrera, M., Capucha, L., Moreno, L.: Mending nets in the south: anti-poverty policies in Greece, Italy, Portugal and Spain. Soc. Policy Adm. **37**(6), 639–655 (2003)
21. Mendola, D., Busetta, A.: The importance of consecutive spells of poverty: a path-dependent index of longitudinal poverty. Rev. Income Wealth **58**(2), 355–374 (2012)
22. Mendola, D., Busetta, A., Aassve, A.: What keeps young adults in permanent poverty? A comparative analysis using ECHP. Soc. Sci. Res. **38**(4), 840–857 (2009)

23. Mendola, D., Busetta, A., Milito, A.M.: Combining the intensity and sequencing of the poverty experience: a class of longitudinal poverty indices. J. R. Stat. Soc. A **174**, 953–73 (2011)
24. Quintini, G., Martin, S.: Starting Well or Losing their Way? The position of youth in the labour market in OECD countries. OECD Social, Employment and Migration Working Papers, No. 39 (2006). doi:10.1787/351848125721
25. Reyneri, E.: Sociologia del mercato del lavoro. Il Mulino, Bologna (2005)
26. Stevens, A.H.: Climbing out of poverty, falling back in: measuring the persistence of poverty over multiple spells. J. Hum. Resour. **34**(3), 557–588 (1999)
27. Zheng, B.: Measuring chronic poverty: a gravitational approach. Seminar paper presented at University of Colorado, Denver (2012)

Do Union Formation and Childbearing Improve Subjective Well-being? An Application of Propensity Score Matching to a Bulgarian Panel

32

Emiliano Sironi and Francesco C. Billari

Abstract

The link between childbearing, union formation and subjective well-being is still under-investigated. A key problem is disentangling causal effects, a challenge when the interplay between life course pathways and states of mind is investigated. Here we use propensity score matching estimates applied to panel data to demonstrate how the birth of a first child or entry into union increase individuals' psychological well-being and reduce disorientation in Bulgaria, a transition country with lowest-low fertility and postponement of union formation. Sensitivity analyses confirm the robustness of our findings to heterogeneous levels of hidden bias.

Keywords

Union Formation • Childbearing • Propensity Score

32.1 Introduction

After the fall of communism in 1989 and the subsequent economic crisis that spread in the former Iron Curtain, several changes in demographic behaviour arose as those countries became engaged in the difficult transition to a market economy.

Bulgaria has been no exception to this scenario. The Total Fertility Rate (TFR) reached "lowest low" fertility levels, i.e. lower than 1.3. Moreover, the mean age at

E. Sironi (✉)
Department of Statistical Sciences, Catholic University and P. Baffi Centre, Bocconi University, Milan, Italy
e-mail: emiliano.sironi@unibocconi.it

F.C. Billari
"Carlo F. Dondena" Centre for Research on Social Dynamics, Department of Decision Sciences and IGIER, Bocconi University, Milan, Italy
e-mail: francesco.billari@unibocconi.it

N. Torelli et al. (eds.), *Advances in Theoretical and Applied Statistics*, Studies in Theoretical and Applied Statistics, DOI 10.1007/978-3-642-35588-2_32,

marriage has become higher than the mean age at first birth [12], with an increase of out-of-wedlock births. Since the early 1980s, a decrease in the relative risks of marriage and a simultaneous increase in the rate of entry in cohabitation have modified the common patterns of union formation [10].

Given the crisis, one might ask with [11] "Why do individuals in developed countries continue to form union and have children?" Most studies discuss the costs and benefits of having children and entering partnerships. This approach assumes that individuals derive utility from life course events such as childbirth and marriage/cohabitation. The "utility" deriving from these events cannot be clearly expressed in monetary terms: a possible measure of the benefit associated with life course pathways and events is subjective well-being.

In fact, the demographic literature has also underlined how ideational dimensions (including proxies of subjective well-being) can produce a causal effect on individuals' intentions and subsequent life course decisions. In particular, Philipov et al. [15] focus on *anomie* measures as determinants of childbearing decisions (the concept of *anomie* was developed by Durkheim [5, 6] and later by Merton [14] to define the loss of prescriptive power of norms in a society). The weakened power of norms causes a state of uncertainty and disorientation on people's state of mind that may affect happiness and subjective well-being. Under the influence of *anomie* people become indecisive, due to feelings such as disorientation, uncertainty and psychological discomfort. This situation is widespread in transitional societies like Bulgaria, where sudden changes can lead to rapid increases in material well-being as well as entry into poverty status.

After this brief introduction the remainder of this chapter is organized as follows: Sect. 32.2 presents the data and some introductory descriptive statistics on the outcome of interest. Section 32.3 illustrates estimation strategy, whereas Sect. 32.4 gives empirical results of the analysis. Section 32.5 describes robustness checks and Sect. 32.6 concludes.

32.2 Data

This study on Bulgaria is based on data from the recent survey "The Impact of Social Capital and Coping Strategies on Reproductive and Marital Behavior", sponsored by the Max Planck Institute for Demographic Research in Rostock, Germany. The first wave of the survey was carried out in 2002. The sample includes 10,003 men and women aged 18–34. The second wave took place in the winter of 2005/2006 and contains the interviews of 7,481 subjects already present in the first wave; one of the aims of the second wave is to capture changes in intentions, behaviours, values and social characteristics through a panel design.

The survey investigates *anomie* using several dimensions: powerlessness, lack of orientations (that are included in a unique factor named *disorientation*[1]), loneliness

[1]This variable is obtained after a principal factor analysis on three items that ask to respondents whether they agree or disagree to the following statements: "I have little influence over my fate";

Table 32.1 Change in anomie factors in the presence or less of the birth of the first child between the waves

	First child between waves	Without children between waves	Difference
Psychological well-being	0.088	0.055	0.033
Disorientation	−0.136	−0.158	−0.022

*** $p < 0.01$; ** $p < 0.05$ * $p < 0.1$

Table 32.2 Change in anomie factors in the presence or less of a new union formation between the waves

	Union formation between waves	Not in union between waves	Difference
Psychological well-being	0.308	0.031	0.277***
Disorientation	−0.212	−0.128	−0.084**

*** $p < 0.01$; ** $p < 0.05$ * $p < 0.1$

(labelled as *psychological well-being*[2]). These dimensions measured in terms of factor scores are subject to a process of adaptation on new life course paths similar to what happens for attitudes and value orientations. In this framework we follow [11] approach and test whether childbearing and marriage/cohabitation contribute to increase subjective well-being and to reduce individual *anomie*.

In order to reach this purpose, it is pivotal to define more precisely the outcome of the research as the variation in factor scores for each of the *anomie* dimensions analysed. However, this chapter does not limit the analysis on the cases of marriage and non-marital cohabitation, but aims at extending its focus also on the effects of childbearing on *anomie*.

Tables 32.1 and 32.2 preliminary offer summary statistics about the variation of *anomie* indices in the presence or less of two crucial life course events between the two waves of the interview. The first life course event analysed in Table 32.1 is the birth of a first child that is supposed to increase well-being for the subset of individuals that experimented the first childbearing. The first column displays the variation in factor scores between the two waves for the subset of individuals that have had the first child in that period, while the second column shows the variations for the childless sample.

When we compare the two groups of individuals, a larger increase in psychological well-being and disorientation has been displayed for people that have had the first child, even if the difference appears as not significant. Similarly, Table 32.2 compares the subset of individuals that entered into a union between the two waves

"One can hardly find his/her orientation in life nowadays" and "No one cares what happens to other people".

[2]This variable is obtained from a factor extracted from answers to two questions where respondents had to choose over a five-point scale ranked from Completely Disagree to Strongly Agree. The wording of the items was: "During the past month have you ever felt very lonely or remote from other people?" and "During the past month have you ever felt depressed or very unhappy?"

with a subset of individuals that remain single. On a descriptive point of you, Table 32.2 emphasizes larger benefits than in childbirth's case, both in increasing psychological well-being and in reducing disorientation for people that entered in a union.

Although descriptive analyses support the idea that childbearing has not any effect in modifying *anomie* measures while a union formation increases individual subjective well-being, the identification of a possible effect of life course events on *anomie* shows statistical problems that are not present in [11], who examine a sample of Danish twins. Indeed, estimates may easily be polluted by problems of selection bias. For example, the estimation of a positive effect of childbearing on individuals' subjective well-being may be polluted by the presence of external confounders like the gender of respondent, economic well-being, age and pre-pregnancy level of satisfaction towards his/her own life. These factors may affect both the probability of having a child and post-childbirth subjective well-being, with the effect of overestimating the positive effect of childbearing. An analogous phenomenon might arise for union formation.

32.3 Estimation Strategy

The aim of our empirical analysis is twofolded. First, we want to investigate the causal effect of bearing a first child on individuals' measures of anomie. Second, the purpose is to detect the effect of the entry into union on *anomie* for individuals that were not in union after the first wave. To measure the impact of childbearing or union formation, we are interested in the difference of the factor scores in the presence of childbearing/union or not between the two waves. Because of the binary typology (yes/no) of these life course events, the problem can be analysed in a treatment evaluation setting. Rosenbaum and Rubin [16] suggested to focus on the quantity $E[Y_{i1} - Y_{i0}|T_i = 1]$, defined as the Average Treatment Effect on Treated (*ATT*). Here, Y_{i1} is the outcome for individuals who experienced a union or childbearing ("treatments") between the waves; conversely Y_{i0} is the potential outcome for individuals who did not experience any union formation or fertility event ("controls").

Many problems deal with the identification of *ATT*. The first is that only one among the potential outcomes is observed, making a direct comparison impossible; consequently, in order to estimate *ATT* one needs to identify the quantity $E[Y_{i0}|T_i = 1]$. This is impossible because the outcome in the case of absence of pregnancy/union for an individual that has had a child or started a union instead is not observable. A possible solution to overcome the problem is to use an estimator of $E[Y_{i0}|T_i = 0]$ for $E[Y_{i0}|T_i = 1]$. Nevertheless, these estimators are biased in the presence of observable confounders. A possible solution is conditioning on pre-treatment covariates $\boldsymbol{X}$: $ATT(\mathbf{X}_i) = E[Y_{i1} - Y_{i0}|T_i = 1, \mathbf{X}_i]$.

In this case identification of *ATT* is feasible if we impose mean independence, i.e.

$$Y_{i1}, Y_{i0} \perp T_i|\mathbf{X}_i \qquad (32.1)$$

However, this solution is not conclusive, because a second problem dealing with *ATT* estimates consists in the difficulty to find individuals with identical values in the vector $\boldsymbol{X}$, when the number of covariates is high or when some of them are continuous.

Rosenbaum and Rubin [16] proposed matching based on propensity score, which is the conditional probability of receiving the treatment given $\boldsymbol{X}$,

$$p(\mathbf{X}_i) = \Pr[T_i = 1|\mathbf{X}_i]. \tag{32.2}$$

Rosenbaum and Rubin [16] demonstrated that matching individuals with the same propensity score is equivalent to compare them on the basis of $\boldsymbol{X}$ with the advantage that an estimate of propensity score is easy to obtain through a simple logistic regression. This result reduces the dimensionality of the problem, as we only need to condition on a one-dimensional variable. Then, *ATT* can be formalized as follows:

$$\begin{aligned} ATT &= E_{p(\mathbf{X}_i)}[ATT(\mathbf{X}_i)] = E_{p(\mathbf{X}_i)}[E[Y_{i1} - Y_{i0}|T_i = 1, p(\mathbf{X}_i)]] = \\ &= E_{p(\mathbf{X}_i)}[E[Y_{i1}|T_i = 1, p(\mathbf{X}_i)] - E[Y_{i0}|T_i = 0, p(\mathbf{X}_i)]|T_i = 1] \end{aligned} \tag{32.3}$$

Several matching algorithms can be used to estimate (32.3); in such case we will have that

$$A\hat{T}T = \frac{1}{N_1} \sum_{i \in \{T_i = 1\}} \left[Y_{1i} - \sum_{j \in \{T_i = 0\}} w_{ij} Y_{0j} \right] \tag{32.4}$$

where N_1 indicates the number of units that experienced childbearing or union formation in the considered time interval; w_{ij} represents a sample weight for control units used in matching procedure. In this framework, it is useful to implement nearest neighbour within caliper matching, consisting in matching each treated unit i with the closest control unit in terms of propensity score. More in detail, treated and controls are matched within a tolerance level that is settled at 0.05 in terms of propensity score. The use of a fairly strict caliper jointly with the imposition of the common support[3] requires a high degree of observational similarity between treatment and control cases.

Propensity score matching (*PSM*) presents also a strong disadvantage: its estimates give robust results when individuals are matched on observables but fail in the presence of unobservable confounding factors. When in presence of unobservables affecting both assignment into treatment and the outcome variable simultaneously, hidden bias might arise. In order to preserve estimates from irregular assignment to treatment in logistic regression caused by hidden bias, we implement different strategies.

[3]Estimate results showed in Table 32.3 and 32.4 are invariant with respect to the imposition of common support, because very few units were excluded from the analysis after its imposition.

First, the nature of our survey allows us for adding variables related to ideational factors that limit the influence of omitted or unobservable variables and are proxies for individuals' religiosity, attitudes and family values.

Secondly, the presence of a longitudinal design gives us the opportunity of measuring the outcome both before and after the treatment. This allows comparing the mean change of *anomie* from the first wave t and the second wave $t + \Delta t$. After replacing the expected values with the sample means, the so-called difference in difference estimator can be obtained for the quantity:

$$ATT_{\mathrm{DD}} = E_{p(X_i)}\{E[\Delta_{1i}|p(\mathbf{X}_i), T_i = 1] - E[\Delta_{0i}|p(\mathbf{X}_i), T_i = 0]\} \tag{32.5}$$

where Δ_i represents the difference in factor scores for the individual i between the waves. An estimator of (32.5) is called combined *PSM-DD* estimator.

An important advantage of *DD* estimator is that controls for selection due to unobservables that are supposed to be time invariant. Thus, it is possible to combine *DD* with propensity score matching estimator, summing the advantages of both the approaches.

Obviously, even this assumption might be violated, if some time varying confounders exist.

Rosenbaum [17] developed a bound method for sensitivity analysis with the aim to assess the effect of unobservable factors on the computation of hypothesis test from the family of sign score statistics that includes the nonparametric statistics Wilcoxon's signed rank test used in presence of continuous outcomes. In the line below we offer a summary of the theoretical procedure as formulated by Di Prete and Gangl [4] and Guo and Fraser [8].

Suppose there is an unobservable source of bias U_1 that varies across individuals and that affects the assignment to treatment. Let p_i be the probability that the i-th unit is treated, then the treatment assignment rule is described as follows:

$$\ln\left[\frac{p_i}{1 - p_i}\right] = f(\mathbf{X}_i) + \gamma U_i \tag{32.6}$$

Rosenbaum proved that a transformation of (32.6) implies that the ratio of the odds for treated and untreated units considered in each of the N_1 pairs of treated and untreated units is bounded:

$$\frac{1}{\Gamma} \le \frac{p_{1i}(1 - p_{1i})}{p_{2i}(1 - p_{2i})} \le \Gamma \tag{32.7}$$

where $\Gamma = \mathrm{e}^{\gamma}$, p_{1i} is the probability that treated unit i receives the treatment, p_{2i} is the probability that the control unit matched to i receives also a treatment. The sensitivity analysis goes further to assume that this odds ratio for units with the same value of propensity score was at most some number $\Gamma \geq 1$. By the above definitions, if $\Gamma = 1$, then $p_{1i} = p_{2i}$, so the study would be free of hidden bias and the odds ratio of receiving a treatment for matched individual with identical values for propensity score computed with observable covariates is obviously equal to one, i.e. $\gamma = 0$. For example, if $\Gamma = 2$, then the two units that appear with closest value of propensity score could differ in their odds ratio of receiving the treatment by as much as a factor of 2, so one unit might be twice as likely as the other to receive treatment. In other words, as [17] pointed out, Γ is a measure of the degree of departure from a study that is free of hidden bias.

Sensitivity analysis consists in evaluating the sensitivity of estimates of *ATT* for different values of Γ; from the equation (32.7) Γ represents the bound for the odds ratio that units *1i* (the treated) and *2i* (the control) might be selected to form a matched pair [8].

Rosenbaum [17] developed a procedure that uses the distribution of the Wilcoxon's signed-rank statistic (see [4] for analytical aspects) under the null hypothesis that *ATT* is zero: the distribution of the Wilcoxon's signed-rank is function of Γ, i.e. of the odds ratio explained in (32.7). If *ATT* maintains significant for high values of Γ, it means that there exists a significant causal effect of treatment also in the presence of hidden sources of bias. Our table gives the critical value for the odds ratio of receiving the treatment Γ for matched and control pairs for that the confidence interval for *ATT* at 90% starts to include the 0. Literature has not discussed in detail critical Γ for considering causal effects robust to hidden bias. However, Di Prete and Gangl [4] stated that an optimal level depends on the research question, even if a value of Γ close to 1.5 would represent a rather reliable threshold to not question the significance of *ATT*.

32.4 Results

The empirical analysis is carried using the Stata module written by Leuven and Sianesi [13] and is divided into two parts. In the first part we address the effect of the birth of the first child on *anomie*. In the second part we analyse the effect of marriage or cohabitation on the same dimensions.

According to the results in Table 32.3, childbearing increases psychological well-being for women, whereas no effect appears for men. Results are consistent with the ones obtained by Kohler et al. [11], who show how the birth of first child increases subjective well-being in a sample of Danish twins. Focusing now on disorientation, childbearing seems to have no effect for both women and men. However, the combined *PSM-DD* estimator is reliable only in the presence of time-invariant sources of bias. As we mentioned above, we implemented a sensitivity analysis, using Rosenbaum's [2002] procedure for bounding the treatment effect estimates in previous tables. Results are displayed in the last column of Table 32.3. This procedure provides evidence for the robustness of estimates at the maximum-tolerated level of hidden bias expressed in terms of the odds ratio of differential treatment assignment due to an unobservable covariate. In the case of the impact of a first child on females' psychological well-being, the positive effect is questioned in the presence of an unobservable covariate with odds ratios of differential treatment assignment of about 1.4.

Table 32.3 Causal effect of first childbirth on anomie—Subjective well-being

		No. of treated	ATT	t-stat	Γ critical
Psychological well-being	Men	247	−0.003	−0.03	1
	Women	286	0.272**	2.49	1.4
Disorientation	Men	246	−0.008	−0.09	1
	Women	283	0.065	0.86	1

*** $p < 0.01$; ** $p < 0.05$ * $p < 0.1$

Table 32.4 Causal effect of union formation on anomie—Subjective well-being

		No. of treated	ATT	t stat	Γ critical
Psychological well-being	Men	234	0.277***	2.86	1.35
	Women	251	0.428***	3.85	1.6
Disorientation	Men	234	−0.065	−0.85	1
	Women	250	−0.233***	−3.07	1.4

*** $p < 0.01$; ** $p < 0.05$ * $p < 0.1$

Table 32.4 shows how psychological well-being increases after marriage or cohabitation: estimates are higher for women, but they are also significantly positive for men. The estimates are also more reliable with respect to unobservable confounders in the females' sub-sample. Rosenbaum bounds suggest that selection due to unobservables would have to attain level larger than 1.5 for odds ratio of receiving the treatment for treated units and controls to undermine the significance of *ATT*; this confirms the results as relatively robust to hidden bias, in the approach of [4]. The effect of disorientation for females' sub-sample also seems to be significantly reduced after a union formation; the finding is quite robust to unobservable heterogeneity, showing a critical value for Γ equal to 1.4. Finally, Rosenbaum bounds approach does not allow for returning remarkable results for estimates that were initially not significant with Γ equal to one. In these cases, for any value of Γ , confidence interval estimates for *ATT* include 0.

32.5 Matching Details and Robustness Checks

The choice of the variables used for estimating the propensity score applied in Table 32.3 is based on both statistical and demographic reasons. Following [9], omitting important variables can increase bias in matching results. Even though [18] are in favour of parsimony in choosing the regressors, the response is unanimous in suggesting the inclusion of covariates that are supposed to be correlated with the treatment and the outcome. Previous demographic surveys such as [15] identified possible determinants for childbearing between the two waves of the same interview; age of respondents, number of siblings, number of children at first wave, marital status, household income, labour market position (working in private or state market), respondents' and parents' education (primary, secondary or higher education), religiosity are included in this survey as decisive regressors for the treatment. Also economic theory (see for example [7]) gives to some of the variables listed above a central role in affecting the subjective well-being, which represents the outcome of our research.

The list of *PSM* confounders includes also positive and negative attitudes towards childbearing as estimated in [2] and that are significantly connected to the life course events we have analysed. Finally, intentions of having a child appear in the model as direct antecedents of the related behaviour, following the theory of planned behaviour [1].

With respect to the model implemented in Table 32.4, a partially different list of covariates is used, because of the different treatment (union formation instead of childbearing): age of respondents, number of siblings, number of children at first wave, household income, labour market position, respondents' and parents' education, religiosity are confirmed also as determinants also of entry into union. Due to the specificity of the matter, the presence of past unions is a strong predictor of the probability of receiving this kind of treatment, whereas attitudes towards partnership specularly replace attitudes towards childbearing here, such as intentions of living with a partner appear in the list of covariates instead of the intentions of having children.

Finally, as suggested in literature, *PSM* results are more robust adding in both the models the starting levels of *anomie* indicators (Disorientation and Psychological Wellbeing) that represent the outcome of our analysis registered at the time of the first interview.

However, the quality of matching depends not only on the variables used for estimating the propensity score but also on the satisfaction of the assumption of the balance property and of the reduction of average absolute standardized bias, according to the thresholds suggested in literature [3]: all the covariates described above are not different in means in treated and untreated groups at 1%, 5% and 10% levels without the need to introduce interactions between covariates and higher order polynomial terms. Therefore, treatments and controls are sufficiently homogeneous and can be successfully compared in order to reduce selection bias due to observables. In order to conclude the discussion around the robustness of *PSM* estimates, literature suggests also a brief discussion on the matching procedure. In fact, the goodness of estimates depends also on the matching method implemented in the analyses. Nearest neighbour matching results are compared with those obtained via radius and kernel matching, joining very close results in terms of causal effect estimates.

32.6 Conclusions

Our results confirm the presence of a process of anomie reduction in the presence of childbearing and new partnerships. In particular, the loneliness dimension of anomie is summarized by the psychological well-being factor that seems to be more sensitive to a process of adaptation, especially for women, who increase well-being when they start a new cohabitation or when they bear a child. Disorientation dimension of anomie shows a more controversial result: it seems to be reduced by partnership only in the females' sub-sample.

References

1. Ajzen, I.: Attitudes, Personality and Behavior. Open university Press, Milton Keynes (1988)
2. Billari, F.C., Philipov, D., Testa, M.R.: Attitudes, norms and perceived behavioural control: explaining fertility intentions in Bulgaria. Eur. J. Popul. **25**(4), 439–465 (2009)
3. Caliendo, M, Kopeinig, S.: Some practical guidance for the implementation of propensity score matching. IZA DP N. 1588 (2005)
4. Di Prete, T., Gangl, M.: Assessing bias in estimation of causal effects: Rosenbaum bounds on matching estimators and instrumental variables estimation with imperfect instruments. Socio. Meth. 35 (2005)
5. Durkheim, E.: The Division of Labour in Society. Free Press, New York (1893)
6. Durkheim, E.: Suicide. Free Press, New York (1897)
7. Easterlin, R.A.: Income and happiness: towards a unified theory. Econ. J. **111**, 465–484 (2001)
8. Guo, A., Fraser, M.K.: Propensity Score Analysis. Sage Publications, London (2010)
9. Heckman, J., Ichimura, H., Todd, P.: Matching as an econometric evaluation estimator: evidence from evaluating a job training programme. Rev. Econ. Stud. **64**, 605–654 (1997)
10. Hoem, J.H., Kostova, D.: Early traces of the second demographic transition in Bulgaria: a joint analysis of marital and non-marital union formation, 1960–2004 Popul. Stud. **62**(3), 259–271 (2008)
11. Kohler, H.P., Behrman, J.R., Skytthe, A.: Partner + children = happiness? An assessment of the effect of fertility and partnerships on subjective well-being in Danish twins. Popul. Dev. Rev. **31**(3), 407–445 (2005)
12. Koytcheva, E.: Social-Demographic Differences of Fertility and Union Formation in Bulgaria Before and After the Start of the Societal Transition. Universitat Rostock (2006)
13. Leuven, E., Sianesi, B.: PSMATCH2: Stata Module to Perform Full Mahalanobis and Propensity Score Matching, Common Support Graphing and Covariate Imbalance Testing. Statistical Software Components S432001 Boston College Department of Economics (2003)
14. Merton, R.: Social Theory and Social Structure. Free Press, New York (1968)
15. Philipov, D., Spéder, Z., Billari, F.C.: Soon, later, or ever? The impact of anomie and social capital on fertility intentions in Bulgaria (2002) and Hungary (2001). Popul. Stud. **60**(3), 289–308 (2006)
16. Rosenbaum, P.R., Rubin, D.: The central role of the propensity score in observational studies for causal effects. Biometrika **70**(1), 41–55 (1983)
17. Rosenbaum, P.R.: Observational Studies. Springer, New York (2002)
18. Rubin, D., Thomas, N.: Matching using estimated propensity scores: relating theory to practice. Biometrics **52**, 249–264 (1996)

Health and Functional Status in Elderly Patients Living in Residential Facilities in Italy

33

Giulia Cavrini, Claudia Di Priamo, Lorella Sicuro, Alessandra Battisti, Alessandro Solipaca, and Giovanni de Girolamo

Abstract

Over the last 50 years the aging of the population in Italy has been one of the fastest among developed countries, and healthcare professionals have witnessed a rapid increase in the complexity of the case mix of older patients. In Italy in 2006, Residential Facilities (RFs) cared for 230,468 people aged 65 and over. Due to the increase in the overall proportion of the aged in the general population (particularly in those over 85) and the sharp decline in the number of extended families (with the consequent reduction in informal support), the probability of an increase in the number of RF residents in future years is very high. The objective of this work is twofold. Firstly, to report on the availability of institutional care in Italy by analysing the territorial distribution of residential facilities, rehabilitation centres and the hospital structure with the intent of gathering both quantitative and qualitative data. Secondly, to examine the health conditions of the elderly in these institutions. Functional status, multiple pathology and medical conditions requiring care have been evaluated in 1,215 elderly subjects living in Residential Facilities across five Italian regions.

Keywords

Aging and Health • Care needs • Residential Facilities

G. Cavrini (✉)
Department of Statistics, University of Bologna, Italy
e-mail: giulia.cavrini@unibo.it

C. Di Priamo · L. Sicuro · A. Battisti · A. Solipaca
Italian Information system on disability, Istat, Italy
e-mail: dipriamo@esterni.istat.it; sicuro@esterni.istat.it; albattis@istat.it; solipaca@istat.it

G. de Girolamo
IRCSS Fatebenefratelli, Brescia, Italy
e-mail: gdegirolamo@fatebenefratelli.it

N. Torelli et al. (eds.), *Advances in Theoretical and Applied Statistics*,
Studies in Theoretical and Applied Statistics, DOI 10.1007/978-3-642-35588-2_33,

33.1 Introduction

The national and international debate on demographic aging in recent years has been focused on a number of aspects. The main problem addressed has been the effect of aging on the welfare and health care systems in various countries. Italy has the highest proportion of aged people in Europe. In 2009, 20.1% of the population was aged 65 and over, corresponding to about 12 million individuals. Furthermore, this proportion is predicted to increase over the next few decades, rising to a figure of 32.5% in 2050 [7]. However, the condition of being elderly does not necessarily imply the loss of autonomy, and health improvements among the elderly have led to an increase in disability-free life expectancy. Nevertheless, the rapid ageing of the population has caused an increasing level of disability and the consequent need for long-term care. All OECD countries are agreed on the general policy direction of maintaining disabled older people in their homes where possible rather than in residential institutions. This tendency may guarantee a better quality of life by providing assistance to the elderly without breaking down the social network. According to the international literature, supply of long- term care at home is not always the most appropriate, particularly when highly qualified and specialized health care is required. Therefore, these findings indicate that residential institutions are likely to play a key role if home care or hospitalization is inappropriate [1]. In the latter case, in fact, a prolonged hospital stay frequently gives rise to a loss of residual capabilities and a consequent worsening of quality of life [2]. In Italy, the number of beds in elderly residential institutions is lower than other European countries, with a territorial heterogeneity. The south of Italy is in fact totally inadequate in the supply of residential care. The lack of statistics on the health conditions of the institutionalized elderly hinders the assessment of the effects of some illnesses. As a result, information on need for care, therapy and rehabilitation is deficient.

The purpose of this study is twofold. Firstly, to analyse the supply of residential services taking into account the territorial distribution of social and health institutions, rehabilitation centres ex article 26[1] and hospital structures (particularly the long-stay wards). Secondly, to analyse the health conditions of the institutionalized elderly. For this purpose, the Italian Ministry of Health has funded an elderly residential care survey addressing five Italian regions. This project is called PROGRES-Older people (PROGetto strutture RESidenziali per Anziani - Residential Facilities for Older People Project). The five regions are Veneto in the north-east, Umbria in the centre, Calabria in the south, Sicily and Sardinia,

[1]Former art. 26 Law 833/78: the health benefits direct to the functional and social recovery of people with physical, mental or sensory impairments, employees from any cause, are provided by local health care through their services. The Local Health Authority may not be able to provide the service directly; in this case, it shall make an agreement with existing institutions in the region where a person lives or even in other regions. The recognition of institutions or rehabilitation centers was conducted in 1998 with an ad hoc model, more complex and specific sections on the structural characteristics, the activities and personnel. (www.salute.gov.it).

both Italian islands conventionally linked to southern Italy. The data analysed was in reference to the demographic and clinical characteristics of residents, staffing arrangements and discharge rates. Further to this, a representative sample of institutions and residents was studied in greater detail [3].

33.2 Data and Methods

This work analysed two different kinds of data:

1. The national statistical data provided by the National Institute of Statistics (*Istat*) and the Ministry of Health. In particular, the Residential Facilities Survey (2004), Hospital Discharge Files and the Survey on Rehabilitation Centres ex art 26 were used.
2. The regional data obtained from a specific project on residential facilities (*PROGRES*).

33.2.1 Italian Context

The Residential Facilities Survey, conducted annually by Istat since 1999, gathers information on admissions to residential facilities and their residents, producing data on the structure of the institutions, staff, recipients and economic information. The aim of the survey is to report on the supply of residential facilities from both a qualitative and quantitative perspective. The Hospital Discharge Files, established in 1991 under a specific law, is the administrative instrument that gathers information on general admissions and outpatients across the national territory. This data source was used to identify long-stay hospitalization. Finally, the third source allows for the integration of data regarding public hospital beds dedicated to rehabilitation ex article 26. This survey also provides data on users, period of care and type of staff.

Descriptive statistical analyses were conducted to show the distribution of residential institutions in Italy and to describe the health conditions of the institutionalized elderly. The available data enabled the health conditions of the institutionalized elderly to be demonstrated, which included autonomy in mobility, in daily activities and in the presence of severe mental (cognitive and behavioural) disturbances. In addition to this, a factorial analysis was applied taking into account the supply of alternative services and aid (living conditions, informal aid, home care and inappropriate and long-stay hospitalization).

33.2.2 The PROGRES Project

In terms of the second point to further the analysis of the five above-mentioned regions, we used the PROGRES project, initiated by the National Health Institute, which includes three phases. Phase 1 was aimed at gathering accurate data on RFs, RF features, demographic and clinical characteristics of residents, staffing arrangements and discharge rates. The main aim of Phase 2 and 3 was to assess a representative sample of facilities (Phase 2) and residents (Phase 3) in greater detail. The selection of facilities to be used in the study was made from regional RF registers which are prepared and periodically updated with the aim of authorizing and regulating facility functioning. Facilities with less than four beds were excluded. Furthermore, since most RFs have no strict age limits for admission, the 23 facilities (3% of all RFs selected) which had more than 50% of residents under 65 years were also excluded from the analyses. At the end of Phase 1 (November 2003), we identified and surveyed 754 RFs in five regions. Information about RFs, their staff and management characteristics as well as suicidal behaviour which occurred has been published elsewhere [3].

For Phase 2, we randomly selected a stratified sample of RFs surveyed in Phase 1 in order to obtain a fair representation in the sample of all categories of RFs (e.g. those with the highest number of beds) which otherwise may have been lost during the sampling procedure. For each RF surveyed, the sample size of the elderly to be studied was estimated according to the distribution of older people living in each of the five regions, to the size of each RF and to the financial resources needed to cover the cost of the individual assessment of patients. Based on these criteria, we estimated a number of 280 subjects to be assessed in each region as a fair representative of the regional population of older people, as well as of the residential population. Each resident was assessed with a "Patient Form" covering socio-demographic, clinical and treatment-related variables. The prevalence of multiple pathologies was evaluated and some specific conditions requiring skilled care were investigated, which were: more severe impairment in mobility, double incontinence, severe disturbances in speech and communication and severe decline in sight and/or hearing. A physician or a nurse of the facility who was in contact with the patients and was aware of their clinical history and present status was involved in the patient assessment. The "Patient Form" included the Resident Assessment Instrument (in Italian, *Valutazione Anziano Ospite di residenza-VAOR* [4]). Trained research assistants interviewed patients using the Mini Mental State Examination (MMSE) [5] to define global cognitive impairment, and caregivers using the Neuropsychiatric Inventory (NPI) [6] to assess the behavioural and psychological symptoms (BPS) of patients. Functional status was assessed using the Index of Independence in Activities of Daily Living (ADLs). The MMSE is a widely used neurocognitive test measure by means of 16 items including: orientation, language, verbal memory, attention, visuospatial function and mental control with scores ranging from 30 (no impairment) to 0 (maximum impairment). The total score of the MMSE was corrected for age and years of education. The NPI was used to measure the BPS in elderly people with or without dementia. It is a valid and reliable inventory to assess 12 neuropsychiatric dimensions such as delusion, hallucination, agitation, depression, anxiety, euphoria/elation, apathy/indifference, lack of inhibition, irritability, aberrant motor behaviour, night-time disturbances and appetite/eating patterns. An informant rated the frequency and severity of each of these dimensions, and the multiplication of the two scores was used as a final codification. The score for each dimension ranges from 0 to 12 with a maximum total score of 144 in the 12-item version. A product equals 0 when there is no symptom. If the score is between 1 and 3 the symptom is not clinically relevant. A score of 4 or more means that the symptom is clinically relevant, it probably deserves further clinical evaluation and adequate treatment (pharmacological and/or non-pharmacological).

The Multiple Correspondence Analysis (MCA) was used to highlight correspondence between some variables displaying the social habits of elderly people living in RFs. In this analysis we also added some passive variables, such as the size of the RF and its type of administration (Health Authority, private, religious or mixed).

33.3 Results

33.3.1 General Overview

The descriptive analyses showed heterogeneity in the availability of residential institutions across the Italian regions: the north of Italy seems to be able to meet the demand the most appropriately.

The activity rate of residential institutions and hospital structures was higher in the northern regions. Since rehabilitation centres were more widespread in the south than in the north, it seems likely that these services are substitutes for residential care. However, some central and southern regions such as Umbria, Calabria, Sicily and Basilicata had the lowest level of all care services. Focusing on social and health institutions, territorial heterogeneity was shown, particularly for nursing homes, and a lack of these services in southern Italy was observed. The distribution of non self-sufficient elderly recipients of social and health institutions revealed a high prevalence of totally dependent elderly patients, particularly in the north. Mobility difficulties were territorially homogeneous and more frequent than other dependencies: 40% of recipients were mobile with the help of personnel, while 35% were not at all mobile. The islands and the south had a lower value of residents with autonomous movement.

In regards to cognitive and behavioural mental disorders, 45% of patients had little or no disturbance, 27% moderate and 29% had a severe disorder. Territorial variability was also displayed in the higher proportion in the northern regions of those with severe disorders. It was also found that the health conditions of the institutionalized elderly were consistent with the Italian supply of residential services. The territorial distribution of residential institutions might be partly affected by northern Italy having the highest share of very old people and elderly with severe health conditions who need more care, as showed in Fig. 33.1.

The north-south gradient is confirmed by the principal component analysis. It showed that the northern regions had more residential care facilities that provide both social and health services, and also an older population with a bad health status than in the southern regions. Nevertheless, the south was characterized more by informal care, private services and inappropriate hospital admissions. It seems like

Elderly patients in Residential Facilities (Rate per 10,000 inhabitants aged 65-years and over)	Elderly patients in Rehabilitation Residences ex art. 26 (Rate per 10,000 inhabitants aged 65-years and over)	Long-term discharge from hospitals (Rate per 10,000 inhabitants aged 65-years and more)

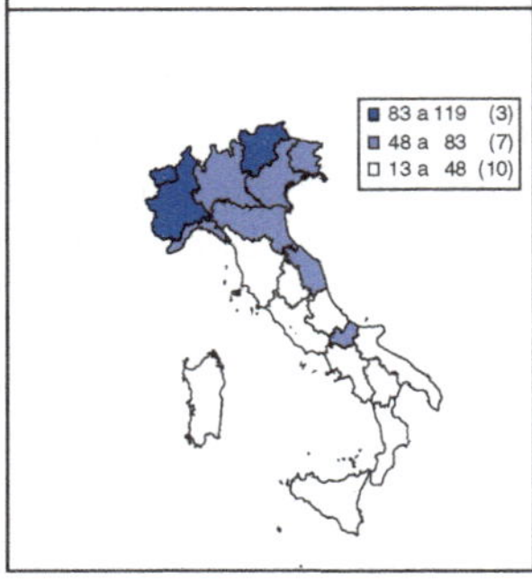

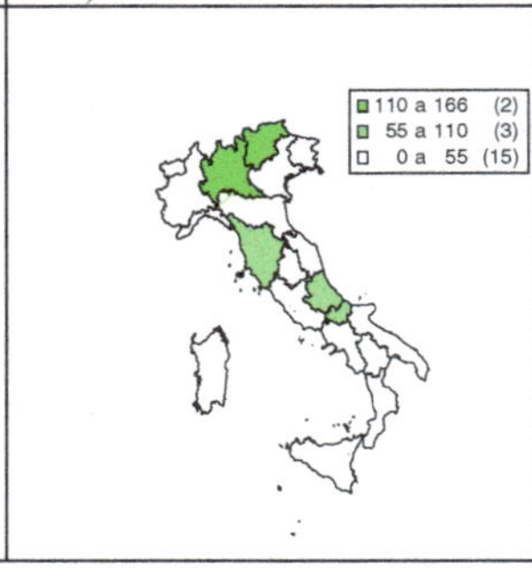

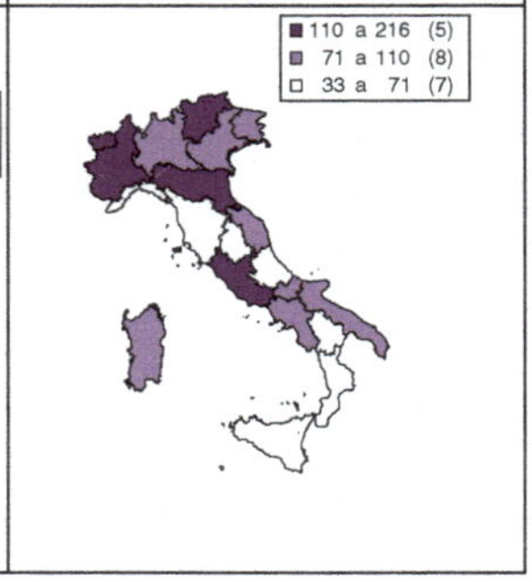

Fig. 33.1 Distribution of elderly by region and type of residential care. Year 2004

that in the south of Italy the lack of specific services was balanced by the family and by personnel paid by the family. Furthermore, when the family is no longer able to help their elderly, the hospital becomes the last resort, also if in an inappropriate way.

33.3.2 PROGRES Project

With respect to the PROGRES project, Phase 1 was completed in November 2003. 76 (8.9%) of the 853 RFs in the five regional registers did not reply, despite several requests to do so. The response rate differed across regions, with a 100% rate in Veneto, Umbria and Sardinia, 89.9% in Calabria and 65.6% in Sicily. Twenty-three facilities, 3% of the participating RFs, hosted more than 50% of residents under the age of 65. Therefore, only 754 of the 777 eligible RFs were further examined, and all analysis of patients presented in this chapter is based on this sample. The 754 surveyed RFs represented about 17% of the 4,304 RFs for older people operating in Italy, as identified by the Italian Institute of Statistics as of December 31, 2000. There was a mean number of 198 residential beds per 10,000 people aged 65 years and older, with great variability in the availability of beds across the five different regions, e.g. the Veneto-Calabria bed ratio was 4.3:1. 132 (17.5%) RFs had more than 100 residential beds, and 26 (3.4%) had more than 200 places. Bed occupancy was greater than 90% in three regions (Veneto, Umbria and Calabria).

33.3.2.1 Characteristics of RFs

Staff coverage was generally high. In fact, most RFs had 24-h staff coverage (n = 724; 96%). In many RFs, resident turn-over was very low. In 302 (40.1%) and 93 (12.3%) of the surveyed RFs there were no discharges and no admissions, respectively, in 2001. Only 141 (18.7%) RFs discharged more than nine patients in 2001. Most RFs were managed by locally run ($n = 220$; 29.2%) or religious ($n = 182$; 24.1%) institutions; the others were run by private, non-profit ($n = 158$; 21.0%) or private, for- profit organizations ($n = 131$; 17.4%); a very small percentage were directly managed by the Italian Health Service (IHS) ($n = 30$; 4.0%); the remaining RFs were of mixed management type. The majority of RFs relied on combined public–private funding ($n = 414$; 54.9%), whereas the percentage of facilities funded by the IHS ($n = 124$; 16.4%) was lower than those accommodating self-paying residents only ($n = 144$; 19.1%). In more than half of the RFs all residents contributed to expenses for their stay through pension deductions or other forms of payment. Only in 12.7% of surveyed facilities did no residents financially contribute to their expenses.

33.3.3 Characteristics of Patients

The sample is made up of 1,215 subjects resident in RFs. All were Caucasian, with a mean age of 83 years (SD 7.8), 73.1% (*n* 888) were female. In terms of age distribution, only one-sixth of the respondents were younger than 75. From the total numbers of patients, 74.5% of women and only 51% of men were in the class of 80 years and older. 27% of the women were older than 90 years. Education level was rather low: 27% were illiterate, 58% had attended 1–5 years of schooling, 7.9% 6–8 years and 7.1% 9 years and more. A significantly higher prevalence of widowhood was found in women compared with men (60.5% and 53.8%).

The MMSE was administered to 895 patients (73.7%). The remaining subjects were excluded because of behavioural disorders (1.5%), neuro-sensorial disorders (66.1%) or refusal to participate (21.4%). The mean MMSE score corrected by age and years of education was 18.5 (SD + 8.7). The NPI was filled out for 1,192 patients (98.1%, missing data for 23 residents and the overall mean NPI total score was 9.7 (SD ± 14.3). 628 patients (51.7%) had at least one clinically relevant item. Apathy, depression and irritability were the most common symptoms in the sample patients with clinically relevant symptoms, while the mean NPI total in the same group was higher for elation, hallucinations and delusions. The item exploring conflictual relationships (CRs) shows that 10.7% of the residents presented conflicts with other residents, 8.6% with staff, 7.9% (*n* 93) with the person sharing the room and 5.5% with relatives. However, among these residents, only 1.2% presented CRs for all the above-mentioned categories (staff, relatives and residents) and 2.6% with at least two categories. As regards the use of medication in the week preceding the interview, 19.8% of the residents received anxiolytics, while 12.8% received antipsychotics and the same proportion received antidepressants. Hypnotics were administered to 8.8% of the sample.

Three indicators summarizing the health status of social and health residential institution recipients across five Italian regions were calculated. The first indicator evaluated the number of diseases for each recipient; the second estimated the presence of severe illness. The third indicator was calculated taking into account the same diseases used by the National Household Survey on Health conditions, in order to compare the health status of residential institution recipients with the health status of the elderly living with family. The descriptive analyses showed that 45% of elderly recipients in residential institutions had at least three diseases and 75% had had at least one severe illness, painting a picture of multiple and chronic diseases. In contrast, it seems that the elderly living with family have a better health status. In fact, 17% had never had any illness and only 30% had had at least one disease compared with 52% of elderly residential recipients. The different health status between the elderly in residential institutions and those living with family is also caused by the different age distribution of the two groups. In particular, in residential institutions the percentage of people aged 80 years and over was 43% compared to only 10% of those living with family. The different age distribution is related to different gender distribution. Residential institutions had a 73% proportion of females, while that of women living with family was 58%. Moreover, family structure also contributes to the difference between the two groups: 34% of recipients in residential institutions were unmarried, 58% were widowed and more than 46% had lived alone before admission. In contrast to this, 57% of the elderly living with family were married, while only 27% had lived alone. In conclusion, it seems there is a higher likelihood of being institutionalized due to bad health status caused by multichronic diseases or a severe disease, and in the absence of an informal network able to support health-care activities.

The relationship between the third indicator (i.e. the comparison of levels of disease in institutionalised patients with those living with family), the class of MMSE, the sex and the age of each resident was investigated with Multivariate Correspondence Analysis. Furthermore, region and dimension of the RFs (in class) were included as supplementary variables. The analysis showed that the 52% of inertia is due to the first axis and 9% is due to the second axis. The presence of no severe illness, no cognitive problems and residing small structures seems to characterize the patients in Umbria and in Sardinia. The presence of at least one or two severe illnesses, severe or moderate cognitive problems, being at least 80 years old, female and in a medium or large structure characterizes the elderly patients in Sicily. In Veneto, mild cognitive problems in males aged between 70 and 79 years residing middle-sized structures are in the majority. Finally, Calabria is characterized by small structures and patients with three or more severe illnesses and it is distant from the other regions. Therefore the MCA analysis highlights the differences in illnesses, cognitive problems, age and gender distribution between the five regions considered. It also discriminates between the different dimensions of the RFs (Fig. 33.2).

In order to analyse the social aspect of residential living, we conducted a Multiple Correspondence Analysis on some variables investigating social activities. Some questions were on whether the patient was able to go outside alone, read or had any hobbies, stayed alone, worked in group activities, went to sleep later, participated in religious activities and was strictly religious. The MCA highlighted the fact that generally the elderly who lived in smaller RFs with a religious

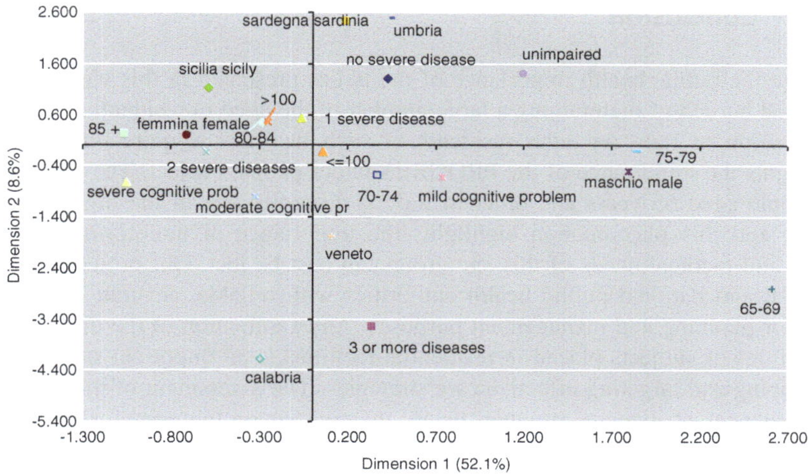

Fig. 33.2 Correspondence analysis map of illnesses index, cognitive problems, age and gender between the five regions and the different dimensions of the RFs

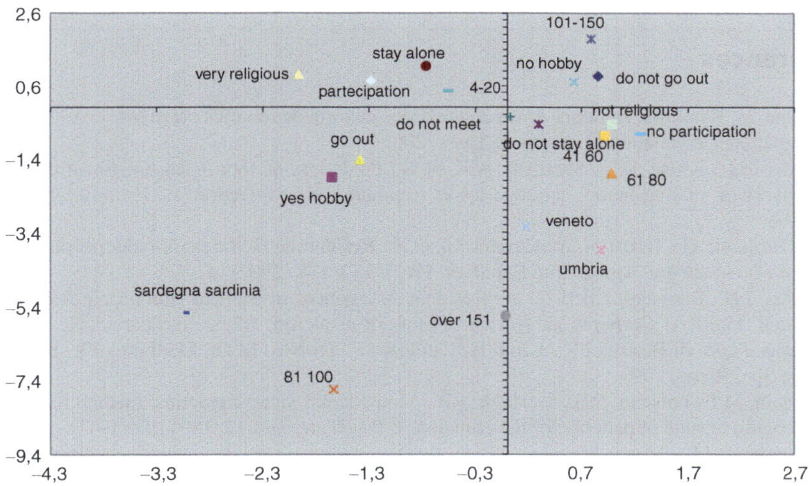

Fig. 33.3 Correspondence analysis map of social activities of elderly between the five regions and the different dimensions of the RFs

administration preferred being alone and found strength in their religious beliefs; people who lived in medium RFs with Health Authority administrations seemed to be more independent with some hobbies and participated in some activities. Finally, the elderly who lived in the biggest RFs with private administration were not independent and did not participate in any activities (Fig. 33.3).

33.4 Conclusion

Despite the public health importance of the issues presented in this work and the financial burden of maintaining a large number of the aged in residential facilities, information in Italy on older residents of such facilities is quite limited. This highlights the importance of the PROGRES-Older people Project. The percentage of people aged 65 years and older in Italy is currently one of the highest in the world, and this phenomenon highlights the importance of monitoring how the social and medical needs of this age group can best be met. Our project provides regional and national public health authorities with reliable, accurate data to be used for planning and management purposes. An examination of the diseases and disabilities of subjects placed in residential facilities is an important step towards developing and targeting effective care strategies. The assessment of the disabling consequences of illnesses provides useful information on the need for assistance, therapy and rehabilitation. The interventions required by the majority of subjects living in residential facilities are assistance in personal care and mobility, and in the management of behavioural disturbances. These requirements should determine the staffing profiles of residential facilities.

References

1. Cribier, F., Kych, A.: Parcours de fin de vie d'une cohorte de retraités parisiens, in XXV IUSSP International Population Conference, Tours, 2005
2. Binder, E.F., Kruse, R.L., Sherman, A.K. et al.: Predictors of short-term functional decline in survivors of nursing home- acquired lower respiratory tract infection. J. Gerontol. **58A**, 60–7 (2003)
3. De Girolamo, G., Tepestini, A., Cavrini, G., et al.: Residential facilities for older people in Italy: a five-region survey. Aging Clin. Exp. Res. **19**(2), 132–138 (2007)
4. Morris, J.N., Bernabei, F.B.E., et al.: Resident Assessment Instrument Home Care Assessment Manual. InterRAI Corporation, Anziano Ospite di residenza; Manuale d'istruzione. Edizione italiana a cura di Bernabei R., Landi F., Lattanzio F., Di Niro M.G., Manigrasso L. Ed. Pfizer Italia Spa. Roma, 1996
5. Folstein, M.F., Folstein, S.E., McHugh, P.R.: Mini-Mental State: a practical method for grading the cognitive state of patients for the clinician. J. Psychiatr. Res. **12**, 189–198 (1975)
6. Cummings, J.L., Mega, M., Gray, K. et al.: The neuropsychiatric inventory: comprehensive assessment of psychopathology in dementia. Neurology **44**, 2308–2314 (1994)
7. Giannakouris, K.: Population and social conditions. Eurostat, statistics in focus 72/2008

34 Dementia in the Elderly: Health Consequences on Household Members

V. Egidi, M.A. Salvatore, L. Gargiulo, L. Iannucci, G. Sebastiani, and A. Tinto

Abstract

The growing number of the oldest-old will cause an increase in the number of mentally ill elderly persons in the population, given that no positive evolution of senile dementia is expected in the near future. Living in the household is the best strategy to contain the pace of mental deterioration, to better manage the disease and to maintain as long as possible the vigilance of the demented person. But dementia is one of the most devastating impairments, both for the persons who are affected by it and for their entire family network and friends, and its impact is high on life and well-being for all persons living with the demented elderly and maximum for his/her caregiver. The aim of this work is to evaluate the impact of the presence of an elderly person with dementia on the perceived health of the co-living household members, using data from the Italian health interview survey carried out by the Italian National Institute of Statistics (Istat) in 2005.

Keywords

Aging • Dementia • Elderly • Household • Mortality and Health

34.1 Introduction

Dementia, and Alzheimer's disease, which is its most common form, is a complex disorder mainly affecting the elderly [17]. It is a chronic disease that leads to a progressive disability ultimately resulting in death. At present, in fact, there is no effective treatment against this disease.

V. Egidi (✉)
Rome University "Sapienza", Roma, Italy
e-mail: Viviana.Egidi@uniroma1.it

M.A. Salvatore · L. Gargiulo · L. Iannucci · G. Sebastiani · A. Tinto
Italian National Institute of Statistics (Istat), Italy

N. Torelli et al. (eds.), *Advances in Theoretical and Applied Statistics*,
Studies in Theoretical and Applied Statistics, DOI 10.1007/978-3-642-35588-2_34,

Among degenerative disorders, dementia is one of the most devastating, not only for the persons who are affected by it, but also for its effects on their entire family network and friends involving the cognitive abilities of the individual, his/her autonomy in the functions of daily life and behaviours. In the initial stages the sick person is able to remain independent and to live alone, while at the end of the process, when the sufferer is often dumb, unable to move and interact, confined to bed and incontinent, he/she is very likely to live in an institution [16]. The duration of the disease is rather long. It is estimated that from the first symptoms (loss of memory) to the death 10 years on average could elapse with a wide variation going from 3 to 20 years. It is also estimated that, after the age of 65, an increase of 5 years of age implies the doubling of the incidence of mental impairment [17]. Among people over 90 years the prevalence of dementia is estimated to be dramatically high, about 28% for men and 45% for women [8] and among the centenarian population it may range between 62% [20] and 100% [25]. As a result of the demographic ageing of the population, the number of prevalent cases for the whole European region is projected to grow from 7.1 million in 2000 to 16.2 million by 2050 (+2.3%) [26]. In the same period, the working age population having the main burden of economic and care support of demented old persons, will decrease so that the previous ratio of 69.4% persons of working age per one demented person will decrease to 21.1. Also in the favourable hypothesis of a delayed onset of dementia and a reduction in its disabling consequences, the impact on the expected increase would be minor [11].

The literature does not highlight any positive trend for the future evolution of the incidence of dementia, except for those that may be caused by a modification of cerebrovascular risk factors and stroke treatment [9].

All studies carried out so far agree that living in the household is the best strategy to contain the pace of mental deterioration to maintain as long as possible the vigilance of the demented person and reduce the high and growing health care costs [18,27], at least public ones, as the economic cost of caring for a demented patient at home paid by the household is also very high [24]. But this does not happen without consequences, often serious, on the physical and mental health of the household members and of the caregiver [3,19]. Moreover, the more and more frequent marital dissolution and the decreasing fertility rates indicate that there will be fewer spouses and fewer children of dementia patients available to take on the role of caregiver [2,22]. This evolution could have particularly heavy consequences in the near future when the wide cohorts born after the 2nd World War will enter into the higher risk ages [7].

The aim of this work is to evaluate the impact of the presence of an elderly person with dementia in the household on the perceived health of co-living household members, using data from the Italian Health Interview Survey carried out by Istat in 2005 in order to verify both the coherence of Italian data with international results and to evaluate the performance of a general health interview survey on households in reproducing results until now obtained by epidemiological surveys, frequently based on small sample sizes and selected populations. The large sample size and a questionnaire that allows for the study of a wide range of interactions between the presence of a demented old person in the household and socioeconomic and environmental factors may provide some additional element for effective health policies.

34.2 Data Sources and Definitions

The Italian health interview survey provides some information on the mental health of individuals and a wide range of information about the health status of the population living in the community and about the most important socio-demographic health correlates. The overall sample size is about 50.5 thousand households and 128 thousand persons. The sampling design is a two stage one, with stratification in the first stage based on population size of municipalities for each region. In the second stage a defined number of households is selected from each municipality and every member of the household is interviewed.

In the self-compiled part of the questionnaire, all interviewed persons (or their proxy respondents) are asked if they are affected by a set of chronic diseases and if they have been diagnosed by a doctor. One of the items of the list is Alzheimer's disease or other forms of senile dementia. In the questionnaire administered by interview, respondent is also asked if he/she is suffering from a disability (also not legally recognized) due to mental insufficiency. In this study the elderly person affected by dementia is, therefore, identified as a person who answered (directly or through a proxy) to be suffering from Alzheimer's disease or senile dementia, or who declared to be suffering from a permanent disability due to a severe mental impairment.

Using this definition and the survey occurrence of 2005 it is possible to estimate a prevalence of dementia of 3.7%, among people aged 65 years and more. A level that is significantly lower than that estimated for the age 65–84 years by the epidemiological survey ILSA (Italian Longitudinal Study on Aging) with reference to the years 1992–1993 (equal to 6.6%) [10]. Nevertheless, this study, that is one of the few examples of a health examination longitudinal survey carried out in Italy, adopted a survey methodology markedly different from that used for the Italian health interview survey. Among the most important differences are the survey technique (medical examination vs. personal interview) and the difference in the reference population (total persons living in Italy vs. persons living in private households). This latter one is a very important difference for the estimation of the mental impairment prevalence because of the exclusion, in the health interview survey, of people living in an institution presumably in poorer health. On the other hand, the ILSA survey refers to a sample which does not represent the population of the entire national territory, being carried out only in eight municipalities, and it adopts a different strategy for its units selection. The prevalence estimated by the Istat survey is also lower than that estimated by Ravaglia et al. [21] with reference to an Italian community resident in Conselice (Emilia-Romagna region): 2.6% for men and 8.5% for women aged 65 years and more. The same is true in comparison with estimates based on epidemiological surveys, carried out at an international level providing prevalence rates of around 5–6% with a high variability only partially explained by the different method used to detect the health condition (MRC [13] for England and Wales: 6.6%; Letenneur et al. [14] for Gironde (France): 6.5%; Lobo et al. [15] for Zaragoza (Spain): 7.4%; Zhang et al. [28] for China: 4.6%; Kurz et al. [12] for Belgium: 8.0%).[1]

It could be stressed, however, that with the aim of this study being to estimate, not the prevalence of dementia in the elderly population, but the impact of the presence of an elderly person with dementia on the household members' health, the exclusion of demented people living in institutions does not produce distorting effects. On the other hand, however, it is also possible that the health interview survey, underestimating the prevalence of demented older people living in households, selects relatively more severe cases. To address the possible distortion produced by this problem, we will investigate the impact of the dementia stage by using the demented old person's functional health, expressed by ADL limitations, as a proxy variable of the severity of the disease.

The outcome indicator selected was the poor perceived health as declared by the household member living with a demented old person (the demented old person excluded) answering "poor" or "very poor" to the question "How is your health in general?" (with the following possible answers: very good, good, fair, poor and very poor). The evaluation of the impact is made by controlling for the most important variables affecting the perception of health, such as age and gender, among demographic variables; living arrangement, education, housing condition, geographic area of residence and the demographic size of the municipality (less than 10,000 inhabitants, 10,001–50,000 inhabitants, more than 50,000 inhabitants), among social and territorial variables. The functional and objective health of respondents are controlled by considering disability (defined as the maximum level of difficulty in at least one activity describing motor abilities, sensory abilities and daily life activities) and multi-chronicity (defined as the presence of three or more chronic diseases). This last indicator was preferred over the traditional one (referring to the presence of at least one chronic disease) which has low variability and a weak association with perceived health.

34.3 Results

The Italian health interview survey 2005 allows for the identification of approximately 409 thousand people aged 65 years and over who declared themselves to suffer from dementia or mental impairment, according to the definition described in the previous paragraph. Of these, approximately 141 thousand are men and 268 thousand are women (slightly less than 2 women for each man are found in this condition). The disadvantage is surely due to greater female aging compared to men. However, a greater prevalence of mental impairment among women is also found when comparing the same age groups. The prevalence of dementia is 3.7% (3.0%

[1]For a recent meta-analysis of the prevalence of dementia in Europe, see Berr et al. [5].

Table 34.1 Proportion of persons cohabiting with an elderly affected by dementia on total population, by age and gender (* 1,000)

	Men			Women			Total		
Age	‰	IC 95%		‰	IC 95%		‰	IC 95%	
0–24	4.1	2.9	5.2	4.4	3.2	5.6	4.2	3.4	5.0
25–44	5.7	4.5	6.9	4.3	3.2	5.3	5.0	4.2	5.8
45–64	8.3	6.7	9.9	10.5	8.7	12.2	9.4	8.2	10.6
65–74	11.2	8.1	14.3	16.7	13.3	20.1	14.2	11.9	16.5
75+	21.5	16.4	26.6	17.9	14.4	21.5	19.3	16.4	22.2
Total	7.5	6.8	8.3	8.7	7.9	9.5	8.1	7.6	8.7

Source: Italian Health Interview Survey, 2005

for men and 4.1% for the women) for the total of the elderly population, with a steep age gradient: the percentage moves from less than 1% for people aged 65–69 years to more than 18% (12% among men and 21% among women) for those aged over 90 years.

Overall, we can identify 881 thousand individuals who on a daily basis face the mental health problems of elderly people, including both the elderly affected by dementia living alone or living with other family members (409 thousand), and the cohabiting relatives (472 thousand). Even considering that the available survey underestimates dementia diffusion among the elderly, the problem is noteworthy and it will be more and more important with the increase in longevity.

Excluding from the analysis the 111 thousand old people with mental problems living alone, 472 thousand relatives cohabiting with the remaining 298 thousand elderly with dementia were analyzed comparing them with persons not cohabiting with demented relatives. It can be estimated that around 1.8 cohabiting relatives, on average, share daily with each unhinged elderly his/her dreadful experience. This situation affects about 8 ‰ of the population, and it peeks to 19 ‰ (Table 34.1) for household members aged 75 years or over. The rate is higher among women in comparison to men; but among people aged 75 and over this difference no longer persists, because of the women's longer life expectancy, which allows them to outlive men, but does not protect them from sickness.

As expected, the mean age of household members living with the elderly with dementia is higher when compared to the population mean age (55 years versus 42 years); 39% are aged 65 years or more, and 21% are over 75; they are mainly women (121:100 on average) in all age groups, with the only exception being those aged 25–44 years (Table 34.2).

The consequences of dementia of the elderly people on the perceived health of their household's members were evaluated assuming a binomial distribution of the dichotomous response variable, and logistic regression models were used to model the observed data and to compare perceived health of people having and not having an elderly person with dementia as a cohabitant, controlling for confounding variables and dementia severity. Since several studies indicate that depression is a condition frequently involving poor perceived health [1], and that it

Table 34.2 Age structure of the household members of the elderly with dementia living in households. Italy 2005 (* 100)

	Men			Women			Total			F/M*
Age	%	IC 95 %		%	IC 95%		%	IC 95%		100
0–24	14.1	10.4	17.7	11.8	8.7	14.9	12.8	10.5	15.2	102.0
25–44	24.1	19.6	28.6	14.7	11.3	18.1	18.9	16.2	21.7	74.1
45–64	28.1	23.4	32.8	30.3	25.9	34.6	29.3	26.2	32.5	130.9
65–74	14.6	10.9	18.4	21.1	17.3	25.0	18.2	15.5	20.9	175.8
75+	19.1	15.0	23.3	22.1	18.2	26.0	20.8	17.9	23.6	140.5
Total	100.0			100.0			100.0			121.5

Source: Italian Health Interview Survey, 2005

could be both a cause of the bad perceived health and a consequence of the presence of a demented relative [6, 23] the analysis was performed both not assuming (Model 1) and assuming (Model 2) depression as perceived health correlate, in order to have indications on the possible mechanism producing the health perception of the demented elderly relatives. Among persons declaring to be suffering from depression, we selected only persons declaring that the disease was diagnosed by a doctor, in order to distinguish the chronic depression from sadness and feeling blue.

Because the households having an older component with dementia are a small proportion of the total (1.6%) we also evaluated the robustness of our results, performing the models on a subsample of households not having components affected by dementia (three times larger than the sample of households with an elderly person with dementia). Those results completely confirm our estimates, without significant improvements in the quality of the models.

The outcome variable of the models, shown in Table 34.3, is bad and very bad perceived health (labelled as "poor health") as assessed by respondents.

Results are in line with what is found in literature [1]: all confounding variables being equal, the presence in the household of an elderly person affected by severe dementia (the presence of difficulties in 6 ADL was considered as a proxy of severity of dementia) has a significant impact on the perceived health of household members (OR equal to 2.2 without controlling for depression). No significant impact is detected when the demented elderly is autonomous in his/her daily life functions. When chronic depression is considered among predictors of the poor perceived health, its OR is high, the model improves its goodness-of-fit, and the impact of the presence of an elderly person affected by severe dementia slightly decreases (OR from 2.2 to 1.9). This suggests that chronic depression takes over a limited part of the impact of the presence of a demented old person on the relative's perceived health, possibly acting as a intermediate condition of the poor perceived health. However a strong impact is also maintained among relatives not declaring to be depressed so that the mechanism of the poor perceived health could be independent from chronic depression. Further analysis showed that living with an elderly person affected by dementia has a significant impact on the mental component of health (measured through the SF12 index called MCS), compared to the physical component (PCS).

Table 34.3 Mutually adjusted odds ratios of "poor" perceived health declared by respondents (demented old persons excluded) for selected variables not assuming (Model 1) and assuming (Model 2) chronic depression as predictor. Health interview survey, Italy, 2005

	Model 1		Model 2	
Variable (reference value)	**Exp(B)**		**Exp(B)**	
Sex (*Male*)				
Female	1.38	***	1.21	***
Age(< 44)				
45–64	4.97	***	4.41	***
65–74	8.58	***	7.74	***
75+	8.69	***	8.15	***
Disability (*No*)				
Yes	12.50	***	11.95	***
Multi-chronicity (*No*)				
Yes	3.93	***	3.13	***
Education (*High*)				
Medium	1.50	***	1.48	***
Low	1.97	***	1.96	***
Housing condition (*Good*)				
Bad condition	1.69	***	1.63	***
Living arrangement (*Couple with children*)				
Couple	1.20	***	1.16	**
One parent family	1.19	**	1.14	*
Elderly living with his/her son's family	1.08	ns	1.07	ns
Other	1.17	*	1.16	ns
Geographic region (*North*)				
Centre	1.52	***	1.50	***
South	1.46	***	1.51	***
Size of municipality (Less than 10,000 inhabitants)				
10,001–50,000 inhabitants	1.14	**	1.13	**
More than 50,000 inhabitants	1.30	***	1.30	***
Presence of demented elderly (*No*)				
Yes, with ADL	2.20	***	1.92	***
Yes, without ADL	1.15	ns	1.04	ns
Presence of depression (*No*)				
Yes			4.84	***
Generalized R-squared	0.3262		0.3535	
AUC	0.862		0.879	
–2*Loglikelihood	29407.11		28361.46	
Degrees of freedom	19		20	

*** $p < 0.0001$ ** $p < 0.001$ * $p < 0.05$

Some interesting findings are obtained by analysing the interactions between the presence of a demented elderly and each one of the other variables considered in the previous models, included depression acting as determinant. In this model the presence of a demented elderly is not specified by functional health status to avoid too low frequencies. The large majority of the interaction effects are nonsignificant. However the interaction with the respondent's age indicates very high impacts for younger members of the family up to the age of 64 years (OR higher than 2), particularly high for people aged 45–64 years when different family and social roles may conflict with the need for care demanded by a cohabiting demented person (Table 34.4). The interaction with the household composition clarifies that the impact is higher for couples (with or without children), where more than 50% of demented elderly persons living with others live, and where the caregiver, almost always the spouse, can get no, or only limited, help from cohabiting relatives. The extended family with different generations living together, where the care giving, as well as emotional burden, can be shared among more persons, is the only structure associated with a positive impact on perceived health of the household members, although just below the threshold of statistical significance (p-value = 0.08) due to the limited number of cases (less than 30% of the demented elderly persons living with others live with their son's family). The last significant interactions underlie the well-known isolation condition that affects elderly people living in larger cities, especially when faced with illness. No significant interaction effect is detected with depression, confirming that the mechanism by which the presence of a demented elderly relative affects perceived health does not necessarily goes through chronic depression.

34.4 Discussion and Conclusions

Estimating the prevalence of diseases and their impact through interview surveys has several well-known limits [4]. It happens that respondents (or persons answering in their place through "proxy" interviews) do not necessarily know the right disease denomination, or they may not be keen to declare it, even having the correct knowledge of it. Alzheimer's disease and dementia in general makes no exception to this rule, in spite of scientific evidence that highlights dementia as having the lowest proportion of false negatives (clinically detected diseases, not declared by the respondent) and of false positives (diseases declared by the respondent, clinically not detected) [10].

Some other relevant limitations of the data used have to be highlighted. One of the first and most relevant data problem is that we are not able to identify who in the household concretely takes care of the demented person. There is no doubt that the negative impact on the relative's health status will be stronger as his/her involvement in care duties gets higher. Another limitation is that the data source used does not allow controlling for the stage of dementia, which would be an important element in the evaluation of the entity of the impact on the health of the cohabiting relatives, especially the caregiver, and in the analysis of the family's strategies to

Table 34.4 Mutually adjusted odds ratios of "poor" perceived health declared by respondents (demented old persons excluded) for selected variables (1). Interaction effects. Health interview survey, Italy, 2005

Variable (reference value)	Exp(B)	
Age* Demented elderly		
0–44* Demented elderly *(Yes vs. No)*	2.36	ns
45–64* Demented elderly *(Yes vs. No)*	2.52	***
65–74* Demented elderly *(Yes vs. No)*	1.19	ns
75+* Demented elderly *(Yes vs. No)*	0.94	ns
Living arrangement* Demented elderly		
Couple* Demented elderly *(Yes vs. No)*	2.14	**
Couple with children* Demented elderly	2.12	*
One-parent family* Demented elderly *(Yes vs. No)*	2.04	ns
Elderly living with his/her son's family * Demented elderly (*Yes vs. No*)	0.63	ns
Other* Demented elderly *(Yes vs. No)*	1.10	ns
Size of municipality* Demented elderly		
Less than 10,000 inhabitants (*Yes vs. No*)	1.22	ns
10,001–50,000 inhabitants (*Yes vs. No*)	1.75	*
More than 50,000 inhabitants (*Yes vs. No*)	1.94	***
Generalized R-squared	0.3539	
AUC	0.88	
–2* Loglikelihood	28344.61	
Degrees of freedom	28	

*** $p < 0.0001$ ** $p < 0.001$ * $p < 0.05$
(1) The model is adjusted for the variables sex, age, disability, multi-chronicity, education, housing condition, living arrangement, geographic region, size of municipality, presence of depression, presence of demented elderly (Yes/No)

cope with the disease. We tried to overcome this drawback by adopting the demented elderly person's functional limitations as proxy of the dementia severity showing that the impact on the relative's perceived health is higher when dementia affecting the elderly is severe. However other consequences of the progressive loss of self-awareness by the demented elderly could be interesting to investigate.

In any case the most important limitation is that we are considering cross-sectional data and, therefore, it is impossible to identify the order of the transitions from one health state to another and to establish causal relations between variables. The direction and the intensity of the evidenced relations, however, makes it unlikely that there are other underlying explanations.

Beyond all those limits, the results have clearly shown that a general health survey, based on a nationally representative sample, may give valuable information about the impact of dementia among the elderly on the perceived health of the cohabiting household members, being able to detect complex relations among relevant variables that cannot always be taken into account by more specific epidemiologic studies, such as geographic variables and urban settlement. All this information, together with a continuous monitoring of the spread of mental health problems among the elderly, is needed to design effective health policies able to cope with the social impact of dementia and help the informal network, having the charge of being the caregivers for the demented elderly. In this network the crucial role is played by the cohabiting household members (the "hidden victims" of dementia; [29]) who very often have to face alone and without adequate support, neither material nor psychological, the progressive and inexorable mental deterioration of a beloved person. In several countries, just in consideration of the social, but also therapeutic role, carried out by the household member of the elderly with dementia, structures and supports are put in place in order to put household members in the position of cohabiting with the disease without paying a high price for their health. In Italy the support of the territorial services to the families with mentally impaired persons is still completely insufficient and widely differentiated across the territory. It is extremely important for the future that formal nets of care, to effectively support the informal nets and households in this difficult task, are developed; moreover the availability of care structures in order to face the most severe stages of the disease should be improved.

References

1. Adams, K.B., Sanders, S.: Alzheimer's caregiver differences in experience of loss, grief reactions, and depressive symptoms across the stages of Alzheimer's disease: A mixed methods analysis. Dementia Int. J. Soc. Res. Prac. **3**(2), 195–210 (2004)
2. Albertini, M., Kohli, M., What childless older people give: is the generational link broken? Ageing Soc. **29**, 1261–1274 (2009)
3. Aminzadeh, F., Byszeski, A., Molnar, F.J., Eisner, M.: Emotional impact of dementia diagnosis: exploring persons with dementia and caregivers perspective. Aging Ment. Health **11**(3), 281–290 (2007)
4. Anstey, K.J., Burns, R.A., Birrell, C.L., Steel, D., Kiely, K.M., Luszcz, M.A.: Estimates of probable dementia prevalence from population-based surveys compared with dementia prevalence estimates based on meta-analyses. BMC Neurol. **10**, 62 (2010)
5. Berr, C., Wancata, J., Ritchie, K.: Prevalence of dementia in the elderly in Europe. Eur. Neuropsychopharmacol. **15**, 463–471 (2005)
6. Betts Adams, K., Sanders, S.: Alzheimer's caregiver differences in experience of loss, grief reactions and depressive symptoms across stage of disease. A mixed method analysis. Dementia **3**(2), 195–210 (2004)
7. Black, S.E., Gauthier, S., Danziel, W. et al.: Canadian Alzheimer's disease caregiver survey: baby boomer caregivers and burden of care. Int. J. Geriatr. Psychiatr **25**(8), 807–813 (2010)
8. Corrada, M.M., Brookmeyer, R., Berlau, D., Paganini-Hill, A., Kawas, C.H.: Prevalence of dementia after age 90. Results from the 90+ study. Neurology **71**, 337–343 (2008)
9. Feldman, H., Woodward, M.: The staging and assessment of moderate to severe Alzheimer disease. Neurology **65**, S10–7 (2005)
10. ILSA: Prevalence of chronic diseases in older Italians: comparing self-reported and clinical diagnosis. Int. J. Epidemiol. **26**(5), 995–1002 (1997)
11. Jagger, C., Matthews, R., Lindasay, J., Robinson, T., Croft, P., Brayne, C.: The effect of dementia trends and treatments on longevity and disability: a simulation model based on the MRC cognitive function and ageing study (MRC CFAS). Age Ageing. **38**, 319–325 (2009)
12. Kurz, X., Scuvée-Moreau, J., Salmon, E., Pepin, J.L., Ventura, M., Dresse, A.: Prevalence in Aged Patients Consulting in General Practice, vol. **56**(12), pp. 835–839. National Dementia Economic Study, Dementia in Belgium (2001)

13. MRC: CFAS cognitive function and dementia in six areas of England and Wales: The distribution of MMSE and prevalence of GMS organicity level in the MRC CFA study. The Medical Research Council Cognitive Function and Ageing Study (MRC CFAS). Psychol. Med. **28**, 319–335 (1998)
14. Letenneur, L., Dequae, L., Jacqmin, H., Nuissier, J., Decamps, A., Barberger-Gateau, P., Commenges, D., Dartigues, J.F.: Prevalence of Dementia in Gironde (France), Rev Epidemiol Sante Publique **41**(2), 139–45 (1993)
15. Lobo, A., Dewey, M.E., Copeland, J., Día, J.L., Saz, P.: The prevalence of dementia among elderly people living in Zaragoza and Liverpool. Psychol. Med. **22**(1), 239–243 (1992)
16. Matthews, F.E., Dening, T.: Prevalence of dementia in institutional care. Lancet **360**, 20 (2002)
17. Moise, P., Schwarzinger, M., Um, M.Y., and the Dementia Experts' Group: Dementia Care in 9 OECD Countries a Comparative Analysis, DELSA/ELSA/WD/HEA 4, OECD Health Working Papers 13, p. 109 (2004)
18. Olesem, J., Gustavsson, A., Svensson, M., Wittchen, H.U., Jönsson, B.: The economic cost of brain disorders in Europe. Eur. J. Neurol. **19**, 155–162 (2012)
19. Ory, M.G., Yee, J.L., Tennestedt, S.L., Schulz, R.: The Extent and Impact of Dementia Care: Unique Challenges Experienced by Family Caregivers,'in Handbook on Dementia Caregiving: Evidence-Based Interventions for Family Caregivers. Richard Schulz editor. New York, NY (2000)
20. Ravaglia, G., Forti, P., De Ronchi, D., Maioli, F., Nesi, B., Cucinotta, D., Bernardi, M.: Prevalence and severity of dementia among northern Italian centenarians, Neurology **53**(2), 416–418 (1999)
21. Ravaglia, G., Maioli, F., Sacchetti, L., Mariani, E., Nativio, V., Talerico, T., Vettori, C., Mancini, P.L.: Education, occupational and prevalence of dementia from the conselice study. Dement. Geriatr. Cogn. Disord. **14**, 90–100 (2002)
22. Riggs, J.A.: The health and long-term care policy challenges of Alzheimer's disease. Aging Ment. Health **5**(Supplement 1), 138–145(8) (2001)
23. Schulz, R., Martire, L.M.: Family caregiving of persons with dementia. Prevalence, health effects and support strategies.
24. Stommel, M., Collins, C.E., Given, B.A.: The costs of family contributions to care of persons with dementia. Gerontologist **34**(2), 199–205 (1994)
25. Thomassen, R., van Shaick, H.W., Blansjaar, B.A., Prevalence of dementia over age 100. Neurology **50**, 283–286 (1998)
26. Wancata, J., Musalek, M., Alexandrovicz, R., Krautgartner, M.: Number of dementia sufferers in Europe between the years 2000 and 2050. Eur. Psychiatr **18**, 306–313 (2003)
27. Wimo, A., Winblad, B., Jönsson, L.: The worldwide societal costs of dementia: Estimates for 2009. Alzheimer's Dementia **6**, 98–103 (2010)
28. Zhang, M.Y.: The prevalence study on dementia and Alzheimer's disease. Natl. Med. J. China **70**, 424–428 (1990)
29. Zarit, S.H., Orr, N.K., Zarit, J.M.: The Hidden Victims of Alzheimer's Disease: Families Under Stress. New York University, NY (1985)

Asset Ownership of the Elderly Across Europe: A Multilevel Latent Class Analysis to Segment Countries and Households

35

Omar Paccagnella and Roberta Varriale

Abstract

Wealth is a useful measure of the socio-economic status of the elderly, because it might reflect both accumulated socio-economic position and potential for current consumption. A growing number of papers have studied household portfolio in old age, both from a financial point of view (i.e. in the framework of the life-cycle model) and from a marketing perspective. In this chapter, we aim at providing new evidence on this issue both at the household and country level, by investigating similarities and differences in the ownership patterns of several financial and real assets among elderly in Europe. To do so, we exploit the richness of information provided by SHARE (Survey of Health, Ageing and Retirement in Europe), an international survey on ageing that collects detailed information on several aspects of the socio-economic condition of the European elderly. Given the hierarchical structure of the data, the econometric solution we adopt is a multilevel latent class analysis, which allows us to obtain simultaneously country and household segments.

Keywords

Ageing • Latent class analysis • Multilevel data analysis • Segmentation

O. Paccagnella (✉)
Department of Statistical Sciences, University of Padua, Via C. Battisti 241/243, 35121 Padova, Italy
e-mail: omar.paccagnella@unipd.it

R. Varriale
Istat - Italian National Statistical Institute, Via Cesare Balbo 16, 00184 Rome, Italy
e-mail: varriale@istat.it

N. Torelli et al. (eds.), *Advances in Theoretical and Applied Statistics*, Studies in Theoretical and Applied Statistics, DOI 10.1007/978-3-642-35588-2_35,

35.1 Introduction

International segmentation aims at structuring heterogeneity that exists among countries and units (individuals, firms, etc.) in order to identify relatively homogeneous groups of countries and/or units [16]. The ongoing political process of European unification, the introduction of the Euro currency and new developments in information and communication technology have contributed to create a potential large single market, the European Union. By treating the different member states as a unique market can lead to plan some attractive (because of economies of scale) unified marketing strategies. Nevertheless, the identification of substantial differences among countries (because of differences in economic structures, cultural identities or regulations) might support the use of multi-domestic or multi-regional policies.

The segmentation of the European market has been widely investigated over the last decade. An interesting review is provided by Lemmens et al. [9]. Mixed evidence raises from these studies, in terms of presence (or not) of cross-country segments, and the number and composition of such segments. However, most of these works investigate country and consumer segmentation sequentially, that is countries are screened at a first stage and grouping is usually based on macro characteristic analyses. Few studies simultaneously determine country segments and cross-national segments [1].

The identification of homogeneous segments can be helpful under different points of view. On the one hand, the revealed structure helps companies to develop and implement international marketing strategies [13]; on the other hand, the findings can support the researches on the household life-cycle effects within the financial market [5]. In this framework, it is of particular interest the study of portfolio holdings of the elderly. The elderly control, as a group, a substantial fraction of the household wealth; therefore their choices in investments or savings might affect the future course of the markets. However, despite their substantial wealth, the majority of elderly has restricted portfolio holdings and does not invest in risky financial assets [7]. Furthermore, there is evidence of a large heterogeneity in household portfolio ownership within and across European countries [3, 4]. The study of the portfolio choice made by old people is therefore important for policymakers, scientific researchers and marketing managers [11].

This chapter aims at shedding more light on the behaviour of households across Europe by investigating similarities and differences in the ownership patterns of many financial and real assets. To this aim, we use data from SHARE (Survey of Health, Ageing and Retirement in Europe). This is an international survey on ageing that collects detailed information on several aspects that characterize the socio-economic condition of the European elderly [2].

The statistical approach we use is the multilevel Latent Class (LC) analysis introduced by Vermunt [14]. As a LC model, one of its typical applications is clustering or constructing population typologies with observed variables. However, differently from traditional analyses for clustering, such as principal component or cluster analysis, the LC approach is a model-based solution that allows to obtain simultaneously country and household segments. As all multilevel models, this approach takes into account the dependencies between lower-level units resulting from the hierarchical data structure (households nested into countries). Hence, applications of multilevel LC models typically aim at simultaneously clustering individuals and groups: lower-level units are assumed to belong to lower-level LCs differing in the distribution of the observed responses; higher-level units are assumed to belong to higher-level LCs differing in the distribution of the lower-level LCs.

The remainder of this chapter is organized as follows. Section 35.2 briefly introduces the data analysed in this work, while Sect. 35.3 describes our econometric approach. Results of the empirical applications are reported and commented in Sect. 35.4. Section 35.5 ends the chapter, drawing some comments on the basis of our findings.

35.2 Data

In this chapter we use data from the 2006 to 2007 wave of the Survey of Health, Ageing and Retirement in Europe. SHARE is an interdisciplinary survey on ageing that is run every 2 years in a host of European countries. It collects extensive information on health, socio-economic status and family interactions of individuals aged 50 and over, using a computer-assisted personal interviewing (CAPI) program, supplemented by a self-completion paper and pencil questionnaire. In each country, a probability sample was selected through an appropriate national survey sampling design.

Even though data are collected at the individual level, for the purposes of our analysis we consider data at the household level. Within couples, the eligible reference person ("head") is chosen as the elder partner, or the female when partners have the same age.

Overall, our sample contains 23,238 households living in 14 countries (Sweden, Denmark, Ireland, Germany, The Netherlands, Belgium, France, Switzerland, Austria, Spain, Italy, Greece, Poland and Czech Republic). Heads are prevalently females (55 %) and 65 years old on average. Almost half of them have low or no education. More than 50 % are retired from work. About 65 % of the heads lives with a partner and the average household size is 2.12. The average number of living children (within or outside the household) is larger than 2. The financial and real asset information available in the dataset is the ownership of the following set of products: bank or postal accounts; bonds; stocks; mutual funds; individual retirement accounts (IRAs); life insurance; house; mortgage; other real estate (RE).

As shown in Table 35.1, there is evidence of large variations in the ownership across products and across countries. On the one hand, almost 75 % of the overall households own bank and postal accounts or the house; on the other hand, bonds and mutual funds are held by less than 12 % of the overall households. The ownership of risky assets shows the largest differences across European countries, i.e. the

Table 35.1 Percentages of the ownership of financial and real assets across European countries

Country	Accounts	Bonds	Stocks	Mutual funds	IRA	Life insurance	House	Other RE	Mortgage
Austria	88.4	3.0	6.9	7.9	8.7	22.8	60.0	10.6	8.4
Germany	94.0	17.8	14.1	14.0	13.6	31.0	59.6	12.1	15.0
Sweden	90.2	17.2	43.6	45.1	47.9	43.3	74.6	33.1	46.0
Netherlands	95.8	3.9	16.7	11.7	8.4	30.4	63.3	6.1	49.7
Spain	80.6	2.1	4.7	3.4	10.8	9.5	88.2	20.4	10.2
Italy	79.6	13.4	5.3	4.4	2.0	8.7	80.2	22.5	5.6
France	96.7	3.2	14.4	13.0	32.3	19.3	72.7	25.5	11.8
Denmark	94.2	19.7	41.7	15.7	37.9	29.8	73.3	21.3	49.0
Greece	44.0	0.9	2.7	0.6	1.1	3.5	85.7	39.1	6.4
Switzerland	94.0	23.9	27.1	22.0	27.3	22.6	56.6	22.5	46.8
Belgium	96.2	11.4	21.8	15.4	26.9	24.2	79.4	16.3	13.4
Czechia	54.7	1.3	3.7	2.9	36.1	15.6	74.0	18.0	4.4
Poland	25.2	0.7	1.3	0.8	1.9	35.1	72.1	6.1	1.4
Ireland	79.4	3.9	14.8	8.6	8.34	36.8	86.5	12.1	16.9
Total	78.7	8.6	15.7	11.9	19.6	23.1	73.9	19.9	19.8

percentage of holding mutual funds ranges from less than 1 % in Greece to more than 45 % in Sweden. The primary accommodation is owned by more than 80 % of the households in Mediterranean countries and Ireland. In Poland and Czech Republic more than 10 % of the households has none of the listed products, while in Sweden, France and Belgium this percentage falls to less than 2 %.

35.3 The Multilevel Latent Class Analysis

As introduced in the previous section, data are available for an international sample of households, ($i = 1, 2, \ldots, n_j$), belonging to a set of European countries ($j = 1, 2, \ldots, 14$). According to the hierarchical nature of the data, households represent lower-level units and countries higher-level units.

The multilevel Latent Class model assumes that households belong to one of K LCs (unobserved subpopulations) at the lower-level and countries to one of C LCs at the higher-level. The aim of the model is to classify, at the same time, households and countries in groups with some typical profile on the basis of the observed ownership of financial and real assets of the households.

The multilevel LC model proposed by Vermunt [14] consists of two basic equations, one representing the mixture model at the household level, and the other the mixture model at the country level. For each household i in country j, let y_{hij} denote the observed ownership of product h of household i in country j, where $y_{hij} = 1$ if household i from country j owns the financial product h, and $y_{hij} = 0$ otherwise. Vectors $\mathbf{y}_{ij}$ and $\mathbf{y}_j$ denote, respectively, the full vector of responses of household i of country j and the full set of responses of country j. At the household level, the model specifies the conditional probabilities of product ownership for household i of country j, given that country j belongs to LC c:

$$P(\mathbf{y}_{ij}|w_j = c) = \sum_{k=1}^{K} P(x_{ij} = k|w_j = c)P(\mathbf{y}_{ij}|x_{ij} = k) \quad (35.1)$$

$$= \sum_{k=1}^{K} P(x_{ij} = k|w_j = c)\prod_{h=1}^{H} P(y_{hij} = 1|x_{ij} = k), \quad (35.2)$$

where x_{ij} and w_j denote the variables representing the lower- and higher-level latent class memberships, and k ($k = 1, \ldots, K$) and c ($c = 1, \ldots, C$) denote a particular lower- and higher-level latent class.

As underlined by Bijmolt et al. [1], this represents a traditional mixture model, with the exception that the relative sizes of first-level LC, $P(x_{ij} = k|w_j = c)$, depend on higher-level LC membership. Within each LC at the household level, the ownership of different products is assumed to be independent, which is the basic assumption of most LC models and usually referred to as the local independence assumption [6].

The mixture model at the country level specifies the marginal probabilities of product ownership for a country j:

$$P(\mathbf{y}_j) = \sum_{c=1}^{C} P(w_j = c)P(\mathbf{y}_j|w_j = c) = \sum_{c=1}^{C} P(w_j = c)\prod_{i=1}^{n_j} P(\mathbf{y}_{ij}|w_j = c). \quad (35.3)$$

As typical in multilevel analysis, the observations of the n_j households in each country j are assumed to be independent of each other given the country LC membership.

Combining Eqs. (35.2) and (35.3), the final model is:

$$P(\mathbf{y}_j) = \sum_{c=1}^{C}\left[P(w_j = c)\prod_{i=1}^{n_j}\left[\sum_{k=1}^{K} P(x_{ij} = k|w_j = c)\prod_{h=1}^{H} P(y_{hij} = 1|x_{ij} = k)\right]\right]. \quad (35.4)$$

The probability of observing a particular response pattern for a country, $P(\mathbf{y}_j)$, is expressed by three components: (i) the probability that country j belongs to higher-level LC c, (ii) the probability that household i belongs to first-level LC k, given the country latent class membership, (iii) the probability that household i of country j owns the financial product h, given the household latent class membership [10].

Components (i) and (ii) are modelled through multinomial logit equations:

$$P(w_j = c) = \frac{\exp(\gamma_c)}{\sum_{r=1}^{C}\exp(\gamma_r)} \quad (35.5)$$

$$P(x_{ij} = k|w_j = c) = \frac{\exp(\gamma_{kc})}{\sum_{r=1}^{K}\exp(\gamma_{rc})}. \quad (35.6)$$

The third component is modelled through a logit equation:

$$P(y_{hij} = 1|x_{ij} = k) = \frac{\exp(\beta_{hk})}{1 + \exp(\beta_{hk})}. \quad (35.7)$$

Models with covariates affecting the class membership probability and/or the product ownership probability are obtained by adding the relevant covariate effects to the intercept term γ_c, γ_{kc} and β_{hk} in the linear predictor of the logit models.

35.4 Empirical Application

We run the multilevel LC model defined in Eq. (35.4) to the SHARE dataset described in Sect. 35.2 in order to classify countries and households in latent classes with some typical profiles. To choose the number of latent classes at the two levels of the analysis, we used the three step procedure proposed by Lukočienė et al. [10]. First, we determined the number of lower-level LCs ignoring the multilevel structure (i.e. assuming that $C = 1$); second, we fixed the number of lower-level LCs to the value of step 1 and determine the number of higher-level LCs; third, we fixed the number of higher-level LCs to the value of step 2 and re-determine the number of lower-level LCs. In particular, we used the BIC value computed with the number of lower-level units to decide about the number of LCs at the household level and we used the BIC value computed with the number of higher level units to decide about the number of LCs at the country level. Each step was performed using the Latent GOLD software [15].

The final model classifies the households in eight latent classes and the countries in six latent classes. Results at lower- and higher-level are reported in Tables 35.2 and 35.3, respectively. Table 35.2 also highlights the main demographic and economic characteristics of the households belonging to each LC. Since covariates have not been specified in the model, Table 35.2 reports percentages of the main characteristics on the basis of the ex-post assignment of each household to the LCs.

Table 35.2 shows that LC8 is the largest lower-level latent class and is characterized by households with high probability of owning bank accounts and the house, confirming the results obtained from the descriptive statistics in Sect. 35.2. The three largest LCs are LC 5, 7 and 8 (which jointly have the probability that household i belongs to one of them greater than 50 %) and are characterized by people with a low probability of holding any risky assets, such as stocks or mutual funds. Households with very large and diversified portfolios (in terms of probability of owning different assets) are in LC4. This LC is quite similar to LC3, where the probability of having the primary house and a mortgage is less important than LC4. This difference between the two classes might be related to the choice of long-time investments.

Looking at the main characteristics of the households belonging To the Five mentioned classes, it is interesting to note that, on the one hand, LCs characterized by the ownership of non-risky assets are mainly composed by females, low

Table 35.2 Household segments: LC size ($P(x_{ij} = k)$), asset ownership probabilities ($P(Y_{hij} = 1|x_{ij} = k)$) and main characteristics of LC components (computed after the household assignment to LC)

k	1	2	3	4	5	6	7	8
$P(x_{ij} = k)$	0.031	0.104	0.137	0.086	0.179	0.108	0.150	0.206
$P(Y_{hij} = 1\|x_{ij} = k)$								
Accounts	0.735	0.938	0.992	0.974	0.887	0.213	1	0.588
Bonds	0.020	0.037	0.313	0.244	0.031	0.001	0.053	0.030
Stocks	0.064	0.134	0.554	0.601	0.035	0.003	0.070	0.009
Mutual funds	0.049	0.089	0.434	0.44	0.041	0.001	0.052	0.003
Ira	0.996	0.159	0.443	0.664	0.065	0.004	0.154	0
Life insurance	0.337	0.381	0.383	0.526	0.149	0.266	0.199	0.018
House own	0.788	1	0.712	1	0.194	0.692	0.963	0.824
Other real estate	0.295	0.101	0.379	0.390	0.033	0.044	0.184	0.262
Mortgage	0.066	0.999	0	1	0	0.010	0	0.049
Household characteristics								
Female (%)	0.615	0.495	0.463	0.437	0.614	0.622	0.541	0.568
Age (mean)	59.6	60.3	65.0	60.8	68.8	65.8	67.5	66.9
Education (%)								
– Low	0.442	0.358	0.303	0.220	0.522	0.550	0.430	0.725
– Medium	0.367	0.310	0.311	0.329	0.308	0.340	0.319	0.171
– High	0.191	0.332	0.386	0.451	0.170	0.110	0.251	0.104
Partner (%)	0.697	0.793	0.728	0.842	0.433	0.584	0.651	0.641
Occupation (%)								
– Retired	0.469	0.317	0.537	0.316	0.593	0.655	0.615	0.508
– Worker	0.466	0.486	0.341	0.602	0.186	0.147	0.198	0.204
– Homemaker	0.009	0.110	0.059	0.023	0.098	0.048	0.132	0.232
– Other	0.056	0.087	0.063	0.059	0.123	0.150	0.055	0.056
Household size (mean)	2.29	2.31	2.02	2.21	1.65	2.55	1.97	2.29
Number of living Children (mean)	2.19	2.51	2.33	2.45	2.58	2.61	2.51	2.34

educated, particularly old (aged 67 years or more) as well as retired respondents. Many of these households are composed by singles: in two out of the three classes the average household size is lower than two. On the other hand, LCs characterized by the ownership of diversified portfolios are prevalently composed by males, highly educated individuals and workers. Few of these households are composed by singles, even though the average household sizes are not the largest among all LCs.

LCs 1 (the smallest class) and 6 are interesting because of the role of two specific products: high probability of having IRAs and low probability of having bank accounts, respectively. Since bank accounts are common in all countries, while IRAs are typical in a few countries (for instance, Nordic countries, Czech Republic, France, etc.), this finding might suggest the presence of country-specific features, probably related to national regulations to incentive the ownership of some specific assets to particular socio-economic groups of elderly. Indeed, LC1 is composed by

Table 35.3 Country segments: LC size ($P(w_j = c)$) and household segment probabilities ($P(x_{ij} = k|w_j = c)$)

c	1	2	3	4	5	6
$P(w_j = c)$	0.078	0.211	0.211	0.078	0.078	0.344
$P(x_{ij} = k\|w_j = c)$						
$k = 1$	0.008	0.002	0.008	0	0.343	0.005
$k = 2$	0.005	0.134	0.028	0.455	0.003	0.098
$k = 3$	0.015	0.284	0.048	0.047	0.015	0.177
$k = 4$	0	0.336	0.002	0.042	0.002	0.032
$k = 5$	0	0.228	0	0.434	0.022	0.276
$k = 6$	0.945	0	0	0	0.284	0.036
$k = 7$	0.026	0	0.096	0.022	0.025	0.359
$k = 8$	0	0.016	0.818	0	0.305	0.018

the youngest group of respondents, equally divided by workers and pensioners and with the highest prevalence of medium educated people among all latent classes. On the other hand, LC6 highlights the least proportion—among all LCs—of workers, as well as the highest percentages of females and low educated respondents. Even though the household head does not often live with a partner, LC6 shows the largest average household size, signalling the relevant presence of other relatives or persons within the household. This LC also highlights the largest number of living children (within or outside the household).

In the end, LC2 is characterized by households with high probability of owning a primary accommodation, a mortgage and bank accounts: this class basically includes "young" households (the "oldest old" people typically do not have mortgages), whose heads have a job. Similar to LC3, in LC2 respondents are equally distributed according to gender and education.

Table 35.3 shows the model results relative to the higher-level of the analysis: size ($P(w_j = c)$) and household segment probabilities ($P(x_{ij} = k|w_j = c)$) of the country latent classes. Using the empirical Bayes modal prediction [12], the countries have been assigned to one of the six latent classes. As represented in Fig. 35.1, we can segment the countries according to a geographical order.

Central Europe countries (plus Ireland) belong to the largest higher-level LC (higher-level LC6), and they are characterized by an interesting mix of household segments with non-risky (lower-level LC5 and LC7) and risky asset portfolio (lower-level LC3). According to the model, the richest countries (Sweden, Denmark and Switzerland) belong to the higher-level LC2, where the probability that households belong to lower-level LCs with risky asset portfolio (LC3 and LC4) is the highest; then, Mediterranean countries (higher-level LC3) are mainly characterized by household segments with a low probability of risky asset portfolio (lower-level LC8). Each of the remaining three LCs are composed by a single country: the Netherlands (higher-level LC4), Poland (higher-level LC1) and Czech Republic (higher-level LC5). At a first sight, this is a surprising result, in particular for the Eastern Europe countries. However, we should note that in each of these LCs, the

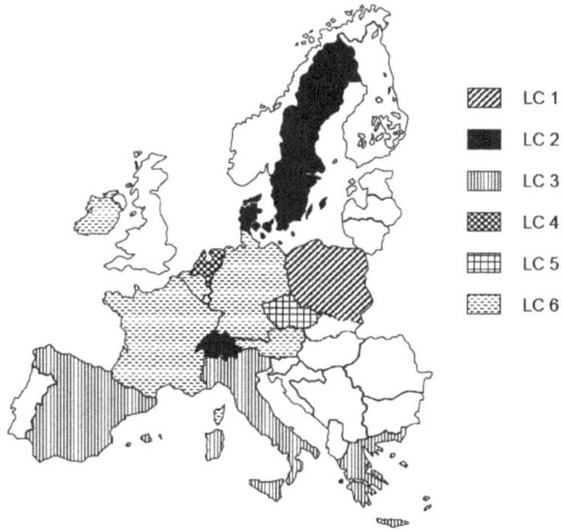

Fig. 35.1 Empirical Bayes modal prediction: country segments posterior membership

conditional probability of the household segments is the highest for the lower-level LCs 2, 6 and 1, respectively, that are just the three household segments described above as having some peculiar profiles.

Our results confirm the evidence of a substantial heterogeneity of portfolio holdings of the elderly within and across Europe [3, 4]. However, the innovative features of our approach (compared for instance to the probit modelling used in the aforementioned literature) allow us to highlight two other important findings: first, in each country we can recognize some groups of households whose portfolio characteristics are similar to those in other European countries (e.g. the richest households); second, when analysing household portfolios of the elderly in Europe, some countries are characterized by so similar patterns of asset ownership that can be grouped together (instead of specifying a full set of country-specific effects).

35.5 Conclusions

Hurd and Shoven [8] argue that wealth is a particularly useful measure of the socio-economic status of the elderly, because it might reflect both accumulated socio-economic position and potential for current consumption. People in older age are usually retired, so their income typically reflects their pensions. Consumption in later life can be then supported by spending down financial or real assets.

However, there is evidence from the Survey of Health, Ageing and Retirement in Europe of a huge cross-country variability on the number and the type (risky *vs* non-risky) of the asset ownership among the elderly.

For investigating similarities and differences in the ownership patterns of several financial and real assets among European elderly, we adopt a multilevel latent class solution that allows us to use the information collected from households to make inference both at the household and country level.

The overall result is that treating the different European member states as a unique market is difficult to support and most countries can be grouped in classes that follow a geographical division. Moreover, households can be divided in well-defined classes, where the largest have low probability of holding risky assets. Low educated, retired and very old individuals characterize the non-risky asset classes. Couples are more likely to hold risky assets than singles.

There is a broad space to future researches on these topics. On the one hand, to further investigate the role of the household characteristics in their choices of asset ownership, we aim at extending our model by adding first-level covariates and then comparing ex-post household profiles. On the other hand, since SHARE is a panel survey, there is data availability for a longitudinal investigation of the acquisition patterns of the assets, in the spirit of Paas et al. [11].

Acknowledgements The authors thank Dimitri Christelis and two anonymous referees for their helpful comments. This chapter uses data from SHARE wave 2 release 2.3.1, as of July 28, 2010. The SHARE data collection has been primarily funded by the European Commission through the 5th framework programme (project QLK6-CT-2001-00360 in the thematic programme Quality of Life), through the 6th framework programme (projects SHARE-I3, RII-CT-2006-062193, COMPARE, CIT5-CT-2005-028857, and SHARELIFE, CIT4-CT-2006-028812) and through the 7th framework programme (SHARE-PREP, 211909 and SHARE-LEAP, 227822). Additional funding from the U.S. National Institute on Aging (U01 AG09740-13S2, P01 AG005842, P01 AG08291, P30 AG12815, Y1-AG-4553-01 and OGHA 04-064, IAG BSR06-11, R21 AG025169) as well as from various national sources is gratefully acknowledged (see http://www.share-project.org for a full list of funding institutions).

References

1. Bijmolt, T.H.A., Paas, L.J., Vermunt, J.K.: Country and consumer segmentation: multi-level latent class analysis of financial product ownership. Int. J. Res. Marketing **21**, 323–340 (2004)
2. Börsch-Supan, A., Brugiavini, A., Jürges, H., Kapteyn, A., Mackenbach, J., Siegrist, J., Weber, G. (eds.): First Results from the Survey of Health, Ageing and Retirement in Europe (2004–2007). Starting the Longitudinal Dimension. MEA, Mannheim (2008)
3. Christelis, D., Georgarakos, D., Haliassos, M.: Economic integration and mature portfolios. CSEF Working Paper n. 194, Centre for Studies in Economics and Finance (CSEF), University of Naples (2008)
4. Christelis, D., Jappelli, T., Padula, M.: Cognitive abilities and portfolio choice. Eur. Econ. Rev. **54**, 18–38 (2010)
5. Guiso, L., Haliassos, M., Jappelli, T. (eds.): Household Portfolios. MIT, Cambridge (2002)
6. Hagenaars, J.A., McCutcheon, A.L.: Applied Latent Class Analysis Models. Cambridge University Press, Cambridge (2002)
7. Hurd, M.: Portfolio holdings of the elderly. In: Guiso, L., Haliassos, M., Jappelli, T. (eds.) Household Portfolios, pp. 431–472. MIT, Cambridge (2002)
8. Hurd, M., Shoven, J.B.: The economic status of the elderly. In: Bodie, Z., Shoven, J.B. (eds.) Financial Aspects of the United States Pension System, pp. 359–398. University of Chicago Press, Chicago (1983)
9. Lemmens, A., Croux, C., Dekimpe, M.G.: Consumer confidence in Europe: united in diversity? Int. J. Res. Marketing **24**, 113–127 (2007)
10. Lukočienė, O., Varriale, R., Vermunt, J.K.: The simultaneous decision(s) about the number of lower- and higher-level classes in multilevel latent class analysis. Sociol. Methodol. **40**, 247–283 (2010)

11. Paas, L.J., Bijmolt, T.H.A., Vermunt, J.K.: Acquisition patterns of financial products: a longitudinal investigation. J. Econ. Psychol. **28**, 229–241 (2007)
12. Skrondal, A., Rabe-Hesketh, S.: Generalized Latent Variable Modeling. Chapman and Hall, Boca Raton (2004)
13. Steenkamp, J.-B., ter Hofstede, F.: International market segmentation: issues and perspectives. Int. J. Res. Marketing **19**, 185–213 (2002)
14. Vermunt, J.K.: Multilevel latent class models. Sociol. Methodol. **33**, 213–239 (2003)
15. Vermunt, J.K., Magidson, J.: LG-Syntax Users Guide: Manual for Latent GOLD 4.5 Syntax Module. Statistical Innovations Inc., Belmont (2008)
16. Wedel, M., Kamakura, W.A.: Market Segmentation: Conceptual and Methodological Foundations, 2nd edn. Kluwer Academic, Dordrecht (2000)

The Longevity Pattern in Emilia Romagna, Italy: A Spatio-temporal Analysis

36

Giulia Roli, Rossella Miglio, Rosella Rettaroli, and Alessandra Samoggia

Abstract

In this chapter, we investigate the pattern of longevity in an Italian region at a municipality level in two different periods. Spatio-temporal modeling is used to tackle at both periods the random variations in the occurrence of long-lived individuals, due to the rareness of such events in small areas. This method allows to exploit the spatial proximity smoothing the observed data, as well as to control for the effects of a set of regressors. As a result, clusters of areas characterized by extreme indexes of longevity are well identified and the temporal evolution of the phenomenon can be depicted. A joint analysis of male and female longevity by cohort in the two periods is conducted specifying a set of hierarchical Bayesian models.

Keywords

Centenarian rate • Longevity • Space–time analysis

36.1 Introduction

In the last decades, the study of human longevity and its development has drawn the attention of researchers belonging to different fields of analysis. Various studies performed in different Italian regions showed the presence of specific areas where the prevalence of oldest-old people is higher than elsewhere. For instance, a definite geographical area in Sardinia is characterized by an exceptional male longevity [14], as well as a low female/male centenarian ratio; a significant negative correlation between surname abundance and index of longevity has been detected in Calabria

G. Roli (✉) · R. Miglio · R. Rettaroli · A. Samoggia
Department of Statistical Sciences, University of Bologna, Bologna, Italy
e-mail: g.roli@unibo.it; rossella.miglio@unibo.it; rosella.rettaroli@unibo.it; alessandra.samoggia@unibo.it

N. Torelli et al. (eds.), *Advances in Theoretical and Applied Statistics*, Studies in Theoretical and Applied Statistics, DOI 10.1007/978-3-642-35588-2_36,

where some isolated areas of male longevity present a high level of inbreeding [11]; in Emilia Romagna some longevity "clusters" have been identified and their persistence has been detected by comparing the results of different spatial scan statistic methods [12]. The explanatory analysis of disparities in the frequencies of the oldest-old population reminds of different genetic and environmental factors differently spread at a geographical level. In order to deepen these several aspects, one of the scientific approaches aims at mapping the geographical diffusion and the temporal evolution of extreme longevity. With this purpose, a recent statistical approach consists in the use of spatial and spatio-temporal models (see, e.g., [10]) which involve the geographical proximity and the time interaction of the areas resulting in a smoothing effect among the observations. As a result, more consistent estimates and inferences on quantities of interest are provided. Moreover, these methods allow to control for extra-variation among the units [5]. Indeed, when the geographical analysis is performed on a fine territorial scale and the phenomenon under study is characterized by a scarce number of units, the territorial distribution of the cases can be invalidated by random variations.

In this chapter, we use a Bayesian spatio-temporal model [1, 8] to manage both the geographical structure and the temporal dimension of the extreme longevity. In particular, we consider a modified version of Centenarian Rate [15] in two different 5-year periods (1995–1999 and 2005–2009), separately by municipality and gender. The small numbers at municipality level are properly tackled by the model, which further allows to embed into the analysis some areal features. The main purposes are to identify territorial groups of areas characterized by high or low levels of longevity and to investigate the development of these clusters by time. We consider a joint analysis of male and female indexes of longevity which involves a set of hierarchical models. Plausible shared and gender-specific risk components are tested by using goodness-of-fit and complexity criteria. As a result, the best suitable model is identified to be interpreted in its results. The chapter develops as follows. We first describe the regional area under study and corresponding data and indicators. Section 36.3 defines the statistical analysis. The model comparison and the results are in Sects. 36.4 and 36.5. The last section reports the main conclusions.

36.2 Area, Data, and Indicators

Emilia Romagna is a North-Eastern Italian region which shows one of the oldest age structures in Europe (with 22.5 % persons aged 65+, and 6.9 % persons aged 80+ in 2009). The region is characterized by a great geographical variability in terms of environmental context, social conditions, and economic resources. Emilia Romagna is split up into nine provinces and 341 municipalities. The different spread of longevity in the regional area is measured by a modified version of the Centenarian Rate (CR) [15]. This indicator, denoted by CR^{95+}, is obtained at a municipality level by dividing the count of people aged 95 and over, P^{95+}, by the number of 55–64 years old persons living in the same area 40 years earlier,

P^{55-64}, in order to avoid complications due to emigration.[1] In this framework, we assume that the individuals resident in an area 40 years before the time point t under study are the population exposed to the "risk" of becoming P^{95+}. To depict the temporal dimension of the phenomenon, we calculate the CR^{95+} separately by sex and municipality area centered on two 5-year periods: 1995–1999 and 2005–2009.[2] In practice, we draw the population exposed to the "risk" of becoming P^{95+} from the nearest available Italian censuses (i.e., in 1961 and 1971 for the two periods, respectively), with analogous considerations to identify the cohort age group.[3] Additional information on the features of the municipalities are collected. In detail, we consider the classification of areas with respect to the altimetry zone and population density. These are then combined to form the following groups: mountain, hill and density lower than 78 people per km^2, hill and density between 78 and 193, hill and density greater than 193, coastal hill, plain and density lower than 78, plain and density between 78 and 193, plain and density greater than 193. The sources of the data all refer to official statistics published by the Italian institutional agency for statistical data collection (ISTAT) and by the regional authorized agency of Emilia Romagna.

36.3 Methods

We develop a hierarchical Bayesian model to investigate the space–time pattern of "risk" of becoming P^{95+} among the 341 municipalities of Emilia Romagna by exploiting the adjacency and the interaction of the geographical areas. We associate a temporal dimension with the phenomenon, by considering the evolution of P^{95+} in the two periods 1995–1999 and 2005–2009. A joint analysis of male and female indexes of longevity is conducted. In detail, with respect to the i-th municipality, we assume the observable P^{95+} at each time t and gender k, denoted by y_{itk}, are Poisson distributed with means $p_{itk}\theta_{itk}$. In this formulation, p_{itk} represents the potential P^{95+} and θ_{itk} is the estimate of CR^{95+} in the required location, period, and gender. Then, we follow the conventional log-linear formulation on the rate θ_{itk} and we allow for the possibility of different components that additively contribute to explain the space–time distribution of these rates. We further control for the effects

[1]The CR has been shown to be an appropriate indicator of longevity as it takes into account for the effects of the work-related migration. Indeed, it is well known that in Italy, including the Emilia Romagna region, migration was very common in the recent past, especially for working ages population. Under this perspective, the CR removes the unknown influence of the migration process, which is assumed to be negligible only after the 60 years of age.

[2]A multiple-year aggregation of data is introduced to avoid random fluctuations due to specific years or cohorts.

[3]Since these censuses are referred to the second half of October 1961 and 1971, we conventionally consider the data as a proxy of population on 1 January 1962 and 1972, respectively. Therefore, for each period, area, and gender, the CR^{95+} is obtained as the ratio of the mean number of people reached age 95+ during the period to the count of individuals aged 60–69 at the censuses.

of some areal features by including the eight categories of altimetry and population density introduced above. As a result, the model for the logarithm of the rate θ_{itk} can be analytically expressed as follows:

$$\log(\theta_{itk}) = \alpha_{tk} + \beta_k x_i + u_{ik} + v_{ik} \tag{36.1}$$

where α_{tk} represents the time-varying and gender-specific intercept, x_i is the altimetry and population density group of the i-th municipality, β_k is the corresponding effect on the log rate by gender, u_{ik} and v_{ik} are the correlated and uncorrelated spatial heterogeneity by area and gender, respectively, which are both assumed to be constant in time. For the correlated spatial component u_{ik}, we assume a Gaussian conditionally autoregressive (CAR) model [1, 3], separately by sex:

$$p(u_{1k}, \ldots, u_{341k} | \tau_{uk}) \propto \exp\left\{-\frac{\tau_{uk}}{2} \sum_{i \neq j} w_{ij} (u_{ik} - u_{jk})^2\right\} \tag{36.2}$$

Although improper, the CAR prior leads to a posterior distribution which is proper, allowing the Bayesian inferences still proceed. The random effects v_{ik} which capture the region-wide heterogeneity are supposed to follow independent exchangeable normal priors by gender, that is

$$v_{ik} \overset{\text{iid}}{\sim} N\left(0, \frac{1}{\tau_{vk}}\right) \tag{36.3}$$

We specify independent, vague, and proper distributions for the other parameters (α_{tk} and β_k). At the third layer of the hierarchy we specify hyperprior distributions for the precision parameters τ_{uk} and τ_{vk} to be diffuse uninformative Gamma, which are also "fair," i.e., yield the proportion of the variability due to the spatial homogeneity to be $\approx\frac{1}{2}$ a priori [3]. Several specifications for $\mathbf{u}_k$, $\mathbf{v}_k$, and β_k are explored by considering various combinations of shared and sex-specific components (see Sect. 36.4). Note that we did not specify space–time interaction terms. We use the WinBUGS software to implement Markov chain Monte Carlo (MCMC) techniques. In order to compute the posterior estimates of the relative risks and, then, of the CRs^{95+}, we consider the last 10,000 iterations of 30,000 in total.

36.4 Model Comparison

Selecting a suitable model from a wide class of plausible models with a large number of random terms in a Bayesian setting is a difficult task. As discussed by Plummer [13], the use of the deviance information criterion (DIC) has practical limitations when the effective number of parameters in the model is much smaller than the number of independent observations. In disease mapping, this assumption does not hold and DIC under-penalizes more complex models. We consider an alternative deviance-based loss function, derived from the same decision-theoretic framework [13]. This criterion can be decomposed into the sum of the expected posterior deviance plus a penalty which is a measure of the degree of optimism. Interestingly, the degree of optimism is related to model complexity, and therefore any model choice criterion discounted for optimism can be viewed as a penalized criterion for model complexity:

$$\overline{D} + r_{\text{opt}} \tag{36.4}$$

The first term in the previous formula is the expected posterior deviance which is a Bayesian measure of goodness-of-fit. The penalty term is difficult to calculate and different proposed approximations could lead to important misspecifications [4, 13]. Here we consider an approximation formula of the optimism, that is:

$$r_{\text{opt}} \approx \sum_i \frac{p_{Di}}{1 - p_{Di}} \tag{36.5}$$

and

$$p_{Di} = \overline{D}_i - \hat{D}_i \tag{36.6}$$

where i denotes the generic observation, the first term in the subtraction is its contribution to the expected posterior deviance and the second term is the deviance contribution of the i-th observation having fixed the model parameters at their posterior expected values.

In this work, we consider and compare five alternative models with respect to the data under study. These are characterized by terms which are either common or not to sexes: model (1) no common terms ($\mathbf{u}_k$, $\mathbf{v}_k$, β_k); model (2) common covariate effects ($\mathbf{u}_k$, $\mathbf{v}_k$, β); model (3) common heterogeneity and covariate effects ($\mathbf{u}_k$, $\mathbf{v}$, β); model (4) common clustering and covariate effects ($\mathbf{u}$, $\mathbf{v}_k$, β); model (5) common clustering, heterogeneity and covariate effects ($\mathbf{u}$, $\mathbf{v}$, β).

36.5 Results and Discussion

In the whole region, we can observe a greater contribution of women to the longevity, with an average CR^{95+} ($\times$1,000) equal to 21.47 in 1995–1999 and 37.54 in 2005–2009. Conversely, men are characterized by lower values (6.39 in the first period and 10.68 in the second one). The increase of the phenomenon across time is thus larger again for females. The application of the hierarchical spatio-temporal model introduced above offers the opportunity to investigate different aspects of the longevity pattern, as well as control for variations due to random occurrences. The last feature is pursued through the so-called smoothing effect, which can be in practice highlighted by mapping and then comparing the observed with the estimated values of the CRs^{95+} for each areas. For both sexes, the model yields more

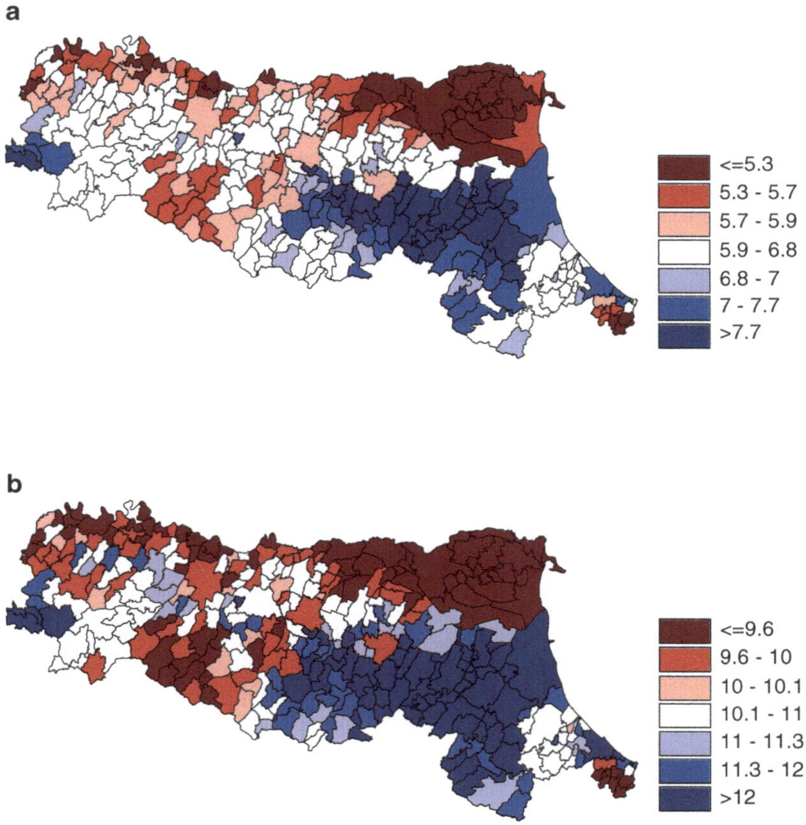

Fig. 36.1 Maps of CRs^{95+} estimated by the Bayesian spatio-temporal model. Men. (**a**) 1995–1999; (**b**) 2005–2009

homogeneous estimates according to the geographical proximity of areas (Figs. 36.1 and 36.2[4]). Conversely, the observed CRs^{95+} (not shown) are denoted by a set of spots, identifying locations with extremely different values from the other nearby areas.

As far as the model comparison is concerned, the specification with gender-specific covariate effects (i.e., model 1) shows a very bad index $\overline{D} + r_{\mathrm{opt}}$, which amounts to 5,555.08. A great improvement can be observed when the homogeneity component is supposed to be shared by sexes, with measures which decrease from 5,544.45 and 5,528.11 for models 2 and 3, respectively, to 5,501.27 and 5,510.58 for models 4 and 5. In the spirit of a goodness-of-fit and parsimony approach, the

[4] All the results we report for the CRs^{95+} refer to a number of 1,000 individuals exposed to "risk".

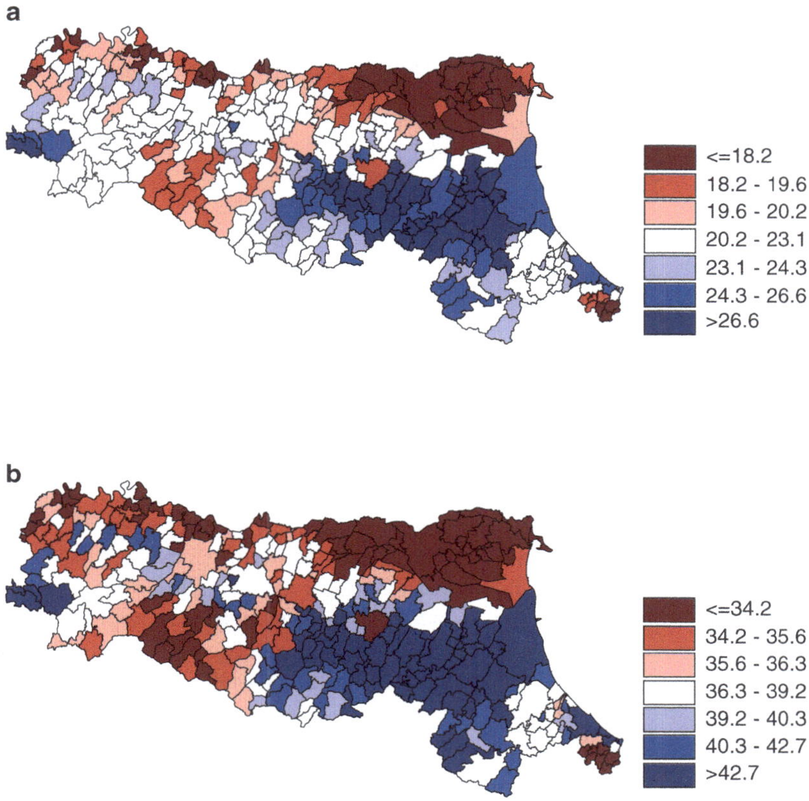

Fig. 36.2 Maps of CRs^{95+} estimated by the Bayesian spatio-temporal model. Women. (**a**) 1995–1999; (**b**) 2005–2009

best specification is then identified in model 4, where the unstructured spatial term keeps on varying between sexes. Results are thus referred to this model.

In order to evaluate the different territorial contributions to the CRs^{95+}, we first consider the ranking of the municipalities according to the decile distribution of the estimated values of CRs^{95+} in the first period, separately by sex, and then put together the four central classes. This grouping becomes the scale to map the result in the first period. For the second time observations, the amount of the regional rise in the CR^{95+} across the two periods is added to each class of the former classification. As a result, we control for the variation of each area in the value of CR^{95+} with respect to the overall regional time increase (Figs. 36.1 and 36.2). It is noteworthy a persistence of areas of lower and higher occurrences of P^{95+} across time, associated with a rise in the values of CRs^{95+}. In particular, mean and median values that are higher than the regional ones are shown in the municipalities belonging to the provinces of Ravenna and Forli-Cesena, spreading out in the

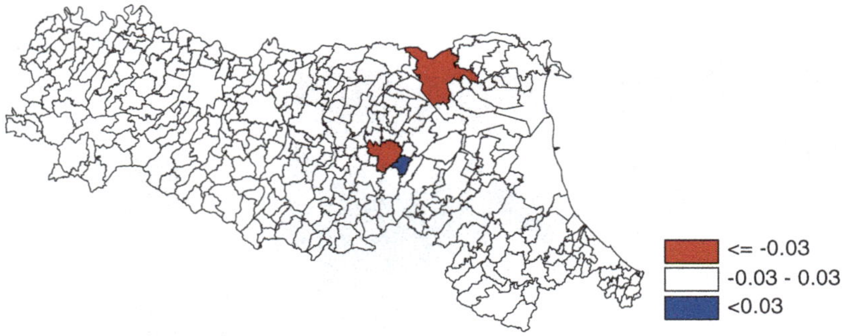

Fig. 36.3 Heterogeneity maps. Women

Adriatic coast, at one side, and in the Apennine municipalities of Bologna and Modena, on the other. Some areas of the province of Piacenza still stand out with high values of CR^{95+}. Conversely, the municipalities of the province of Ferrara are characterized by a lower longevity.[5] The widening of areas characterized by the lowest and the highest values of CR^{95+} in the second period shows that the smallest and largest increases of longevity concern the same municipalities identified in the first period and the adjacent ones. Low levels of longevity also appear for some groups of municipalities in the Apennines of Parma and Reggio Emilia, in the north-west of the region (province of Piacenza) and in the coastal hill of the province of Rimini. A less spatially structured component can be observed for females, where some sparse areas over the region also emerge for different values of CRs^{95+} with respect to adjacent ones (see also Fig. 36.3).

The contribution of the clustering and heterogeneity effects in terms of total territorial variability can be first evaluated through the estimates of the posterior variance of the two components **u** and **v**, given respectively by $\widehat{\text{Var}}(U|Y)$ and $\widehat{\text{Var}}(V|Y)$. For both sexes, the structured spatial variability seems to prevail. As a consequence, the global model representations for the CRs^{95+} are more influenced by the territorial clusters with relatively similar longevity risks, rather than by the heterogeneity effect. We estimated the 80 % for males and the 78 % for females of the total variability is due to clustering. Moreover, once the covariates are included into the model specification, they mainly contribute to explain the homogeneity across the municipalities. Since the unstructured variability due to the peculiarities of individual areas to the values of CR^{95+} is small, for men no areas appear to have noticeable values. For women, two municipalities spread out both with a stronger negative contribution of the residual heterogeneity component (Fig. 36.3). The first

[5]We cannot exclude that some fluctuations of CR^{95+} could depend on work-related migration movements. For example, Ferrara has been an emigration area for population in working ages and if we assume that emigrants have a better health profile, those who remain could be a selected subpopulation that unlikely could reach oldest-old ages.

Table 36.1 Effects of altimetry and population density on longevity

	Fitted rate ratios	90 % credibility Interval
Mountain	1	–
Hill and density <78	0.960	0.882–1.043
Hill and density 78–193	1.095	1.008–1.189
Hill and density >193	0.985	0.900–1.115
Coastal hill	0.801	0.639–1.008
Plain and density <78	0.779	0.677–0.894
Plain and density 78–193	0.930	0.849–1.021
Plain and density >193	0.997	0.919–1.085

one is Ferrara, which is part of a larger area identified by low values of CR^{95+}. The other municipality is Bologna, which already emerged as an individual area with low longevity as a result of its peculiarities significantly different from the other adjacent areas. Conversely, only one area (i.e., the municipality of S. Lazzaro) appears with a positive unstructured residual peculiarity. According to the model specification, the fitted rates for the intercepts in the two periods and sexes represent the level of probability of observing long-lived subjects in the mountain areas of Emilia Romagna, that is the reference group. An overall increase in these values can be observed over time (0.006–0.011 for men; 0.022–0.038 for women). Municipalities in different altimetry zones also classified with respect to the population densities all experienced rate ratios which are lower than 1, except for people living in hill areas with medium population density who have a significant higher rate of becoming long-lived than mountain residents (Table 36.1).

36.6 Conclusions

Statistical theory, several simulation studies, and a large number of applications all support the use of hierarchical Bayesian modeling for spatial and spatio-temporal data as a powerful method which allows to yield more consistent estimates on quantities of interest and inferences [8]. Indeed, the consequent smoothing effect in both spatial and temporal sense allows to control for the variation in the population size across the geographical areas [7] and for extra-variation among the units [5]. Moreover, the opportunity of including different levels of covariates allows to investigate some crucial effects for the study at hand and explain some parts of variability. A critical aspect of these methods is certainly represented by the sensitivity to the choice of the hyperpriors [2], which can be sometimes avoided by the use of an empirical-Bayes approach [6], rather than previous knowledge, or by making assumptions which are reasonable with the problem [1, 3]. In this application, a crucial point is represented by the choice of the Gamma priors for the precision parameters for τ_{uk} and τ_{vk}. Indeed, they control the strength of the smoothing and heterogeneity effect. Those components are shown to be quite sensitive to the prior specification, but the effect is mainly reflected only upon the variances of the parameters. Therefore, the posterior estimates are weakly affected.

The estimated values of the adopted longevity index allow to identify some groups of areas where the people reaching 95+ years of age are more likely to occur with a substantial persistence over time. As a result, the effect of some areal features and correlations can be jointly investigated. The use of a hierarchical Bayesian modeling for spatio-temporal data to analyze the longevity pattern in small areas represents a quite innovative application, which aims to confirm the usefulness of these methods also in socio-demographic research. The picture detected shows the presence of both quite large areas which are homogeneous with respect to the CR^{95+} values and isolated municipalities characterized by high levels of heterogeneity with respect to neighbor ones. As far as the effects of altitude and population density are concerned, plains are generally shown to be associated with a lower count of the oldest-old, as already detected in Campania and Sicily [9]. Moreover, hilly zones seem to be more favorable to longevity. Besides being influenced by spatial interactions and areal features, it is well known that the higher or lower presence of long-lived people across the areas largely depends on mortality features after 80 years [16, 17]. Therefore, a first development of the analysis should consist in the use of a space-cohort model where the area-specific levels of mortality for each group of age in the cohort are included as regressors. The results here obtained also call for more in-depth analysis aiming at revealing the specific reasons causing the extremely high or low levels of longevity that have been found in some areas, such as environmental effects, which are likely to influence the spatial distribution of longevity (e.g., pollution indicator, presence of incinerators, indexes of deprivation) or genetic factors.

References

1. Banerjee, S., Carlin, B.P., Gelfand, A.E.: Hierarchical Modeling and Analysis for Spatial Data. Chapman and Hall, London (2004)
2. Bernardinelli, L., Clayton, D.G., Montomoli, C.: Bayesian estimates of disease maps: how important are priors? Stat. Med. **14**, 2411–2431 (1995)
3. Best, N.G., Waller, L.A., Thomas, A., Conlon, E.M., Arnold, R.A.: Bayesian models for spatially correlated diseases and exposure data. In: Bernardo, J.M., et al. (eds) Bayesian Statistics, vol. 6. Oxford University Press, Oxford (1999)
4. Biggeri, A., Catelan, D., Dreassi, E.: The epidemic of lung cancer in Tuscany: a joint analysis 273 of male and female mortality by birth cohort. Spatial Spatio-Temporal Epidemiol. **1**(1), 31–40 (2009)
5. Breslow, N.E.: Extra-Poisson variation in log-linear models. Appl. Stat. **33**(1), 33–44 (1984)
6. Carlin, B., Louis, T.: Bayes and empirical Bayes methods for data analysis. Chapman and Hall/CRC, London (1998)
7. Clayton, D.G., Kaldor, J.: Empirical Bayes estimates of age-standardized relative risks for use in disease mapping. Biometrics **43**, 671–681 (1987)
8. Lawson, A.B.: Bayesian Disease Mapping. CRC press, New York (2009)
9. Lipsi, R.M.: Longevity in small areas and their socio-economic, demographic and environmental characteristics: a Hierarchical Bayesian approach. Poster Presented at XXVI IUSSP International Population Conference, Marrakech, 2009

10. Mollie, A.: Bayesian mapping of disease. In: Gilks, W., Richardson, S., Spiegelhalter, D.J. (eds.) Markov Chain Monte Carlo in Practice. Chapman and Hall, London (1994)
11. Montesanto, A., Passarino, G., Senatore, A., Carotenuto, L., De Benedictis, G.: Spatial analysis and surname analysis: complementary tools for shedding light on human longevity patterns. Ann. Hum. Genet. **72**, 253–260 (2008)
12. Miglio, R., Marino, M., Rettaroli, R., Samoggia, A.: Spatial analysis of longevity in a Northern Italian region. Paper Presented at the PAA Annual Conference, Detroit, 2009
13. Plummer, M.: Penalized loss functions for Bayesian model comparison. Biostatistics **9**(3), 523–39 (2008)
14. Poulain, M., Pes, G., Grasland, C., Carru, C., Ferrucci, L., Baggio, G., Franceschi, C., Deiana, L.: Identification of a geographic area characterized by estreme longevity in the Sardinia Island: the AKEA study. Exp. Gereontol. **39**, 1423–1429 (2004)
15. Robine, J.M., Caselli, G.: An unprecented increase in the number of centenarians. Genus **LXI**, 57–82 (2005)
16. Robine, J.M., Paccaud, F.: Nonagenarians and centenarians in Switzerland, 1860–2001: a demographic analysis. J. Epidemiol. Community Health **59**, 31–37 (2005)
17. Thatcher, R.A.: The demography of centenarians in England and Wales. Population **13**(1), 139–156 (2001)

Material Deprivation and Incidence of Lung Cancer: A Census Block Analysis

37

Laura Grisotto, Dolores Catelan, and Annibale Biggeri

Abstract

We study the relationship between incidence of lung cancer in males in the Tuscan region and material deprivation defined at census block level. We developed a bi-variate hierarchical Bayesian model to assess completeness of registration of incidence data, and we proposed a series of random effect hierarchical Bayesian models to estimate the degree of association with material deprivation. Model comparison is addressed by a modified Deviance Information Criterion. We estimated a percent increase in risk of lung cancer for an increase of one standard deviation of material deprivation at census block level of 3.36% vs 5.87% at municipality level. The random slope models reported a paradoxically negative effect of material deprivation at census block level in some areas. Spatially structured random intercept models behaved better and random slope models were penalized by their extra complexity.

Keywords

Hierarchical models • Lung cancer • Material deprivation • Random effect • Spatial data analysis

---

L. Grisotto (✉) · D. Catelan · A. Biggeri
Departement of Statistics, Informatics, Applications "G.Parenti", University of Florence, Viale Morgagni 59,
50134 Florence, Italy

Biostatistics Unit, ISPO Cancer Prevention and Research Institute, Via Cosimo il Vecchio 2, 50139 Florence, Italy
e-mail: grisotto@disia.unifi.it; catelan@disia.unifi.it; abiggeri@disia.unifi.it

N. Torelli et al. (eds.), *Advances in Theoretical and Applied Statistics*,
Studies in Theoretical and Applied Statistics, DOI 10.1007/978-3-642-35588-2_37,

37.1 Introduction

The relationship between socioeconomic factors and health has been studied in many circumstances, mainly on aggregated data. A strong association with mortality for lung cancer is reported in [1].

In the epidemiological literature material deprivation indexes defined at census block level were used as a proxy of individual socio-economic status (e.g. [2], and in Italy [3]). In fact census block is considered a small enough aggregation level to assume that individuals be homogeneously exposed and thus to avoid the problem of ecological bias [4].

The aim of this work is to study the relationship between incidence of lung cancer in males in the Tuscan region and material deprivation defined at the census block level.

We take advantage of individual incidence data are made available by the Tuscan Cancer Registry for the year 2004.

We developed a hierarchical Bayesian model to assess completeness of registration of incidence data, and we proposed a series of random effect hierarchical Bayesian models to estimate the association with material deprivation. We also addressed model selection issue.

37.2 Data

37.2.1 Incidence and Mortality Data

Incidence data come from the Tuscany Cancer Registry [5]. Individual records report information on disease code and demographic characteristic of the person including the address at diagnosis. For the present analysis we considered the incident cases of lung cancer in males in the year 2004 for the whole region (a total of 2,097 patient records).

Through the Iter.net system [6] 1,545 over 2,097 cases have been linked to census block indicator, because the Arezzo Province was not covered by the system. The corresponding number of analyzed municipalities is 248 (over the whole set of 287 Tuscan municipalities).

The expected number of cases for each census block was computed under indirect standardization applying a set of age specific (16 age classes, 0–4,. . . ,75 or more) reference rates (Tuscany 2004) to the population of each area.

To assess the completeness of registration ISTAT death certificates of lung cancer in males in the period 2003–2005 for the 248 municipalities for which incidence data are available were recovered from the Regional Mortality Register.

37.2.2 Material Deprivation Data

Data on socio-economic factors derived from individual records of the 2001 census were made available by the Tuscany Longitudinal Study (SLTo) [7, 8].

Following the current epidemiological literature we used the following variables to compute an indicator of socio-economic inequalities:
- Unemployment
- Low education (less than 6 years of schooling)
- House tenure
- Overcrowding

The material deprivation index has been computed as the sum of z-score of the proportion of people with the condition listed above for each census block. Details on the construction of a social deprivation index at census block and municipality level for Italy are in [9].

37.3 Methods

37.3.1 Bayesian Analysis of Completeness of Registration

To check for completeness of registration we conducted a Bayesian analysis on the mortality/incidence ratio (M/I) [10]. Let event counts Y_{lk} be distributed as Poisson ($\theta_{lk} \times \text{Pop}_k$), where θ_{lk} is the rate parameter, Pop_k the person years at risk, $l = 1, 2$ indexes death or incidence event, respectively, and k indexes the death. A log linear bivariate model was specified on the rate parameter:

$$\log \theta_{lk} = \alpha + \mu_k + \delta_k I_1$$

where α is the intercept, μ_k is an area-specific random term, I_1 an indicator variable which is 1 for mortality and 0 for incidence, δ_k is the parameter for the M/I ratio. We compared different prior specifications for the μ_k and δ_k parameters: spatially unstructured (heterogeneity) Gaussian(0, σ^2) (with uninformative Gamma hyperprior for σ^2) or spatially structured (clustering) Gaussian conditional autoregressive $c_k | c_{v \sim k} \sim Normal(\bar{c}_k, \lambda_c n_k)$ where $v \sim k$ denotes areas adjacent to the kth one, $\bar{c}_k$ is the mean over the set of adjacencies and n_k its cardinality [11].

The expected M/I ratio is one. In fact, when the median survival time is very short as in the case of lung cancer, the number of incident cases will be almost equal to the number of deaths. A value of the M/I ratio greater than one suggests incompleteness of registration of incident cases; an M/I ratio smaller than one suggests misclassification of prevalent cases as incident cases because of poor definition of the date of diagnosis [10].

37.3.2 Bayesian Analysis of the Association Between Mortality and Material Deprivation: Fixed-Effect Models

Let x_{jk} be the material deprivation index for the jth ($j = 1, \ldots, 23, 182$) census block in the kth ($k = 1, \ldots, 248$) municipality and $\bar{x}_k$ the mean deprivation for the

kth municipality. Let Y_{jk}, the number of incident cases, follow a Poisson distribution with mean $\theta_{jk} X E_{jk}$ where θ_{jk} is the Relative Risk and E_{jk} the expected number of cases under indirect standardization. We specified a series of log linear models for θ_{jk}.

37.3.2.1 Ecological Model at Census Block Level

$\log\ \theta_{jk} = {}^{0}\beta + {}^{1}\beta x_{jk}$, where ${}^{1}\beta$ is the ecological effect of the census block material deprivation value.

37.3.2.2 Ecological Model at Municipality Level

$\log\ \theta_{jk} = {}^{0}\beta + {}^{2}\beta \bar{x}_k$, where ${}^{2}\beta$ is the ecological effect of the average material deprivation at municipality level.

37.3.2.3 Cronbach Model

$\log\ \theta_{jk} = {}^{0}\beta + {}^{1*}\beta(x_{jk} - \bar{x}_k) + {}^{2}\beta(\bar{x}_k - \bar{x})$, where ${}^{1*}\beta$ is the effect of the average census block material deprivation within municipality and ${}^{2}\beta$ is the ecological effect of the average material deprivation at municipality level [12].

It is important to notice that it is erroneous to compare lower level coefficients (e.g. ${}^{1}\beta$ model Sect. 37.3.2.1) vs higher level coefficients (e.g. ${}^{2}\beta$ model Sect. 37.3.2.2). In fact, the lower level coefficient is ${}^{1}\beta = {}^{1*}\beta + {}^{2*}\beta \frac{\mathrm{Var}(\bar{x})}{\mathrm{Var}(x)}$ and the higher level coefficient is ${}^{2}\beta = {}^{1*}\beta + {}^{2*}\beta$ and it is not surprising to observe that ${}^{1}\beta > {}^{2}\beta$. A more useful comparison is ${}^{2}\beta - {}^{1*}\beta$ vs ${}^{1*}\beta$, which are easily obtained from the Cronbach model Sect. 37.3.2.3.

37.3.3 Random-Effect Models

The previous models do not consider random intercept, i.e. variability of the baseline risk among areas, or random slope i.e. varying material deprivation coefficient by municipality. We defined:

37.3.3.1 A Random Intercept Model

$\log\ \theta_{jk} = {}^{0}\beta_{\mathrm{k}} + {}^{1*}\beta(x_{jk} - \bar{x}_k) + {}^{2}\beta(\bar{x}_k - \bar{x})$ where ${}^{0}\beta_k \sim Normal(\mu_{{}^{0}\beta}, \sigma^2_{{}^{0}\beta})$.

37.3.3.2 A Random Intercept and Slope Model

$\log \theta_{jk} = {}^{0}\beta_k + {}^{1*}\beta_k(x_{jk} - \bar{x}_k) + {}^{2}\beta(\bar{x}_k - \bar{x})$ where $({}^{0}\beta_k, {}^{1*}\beta_k) \sim MVNormal$ $(\boldsymbol{\gamma}, \mathbf{T}^{-1})$, $\boldsymbol{\gamma} \sim MVNormal(\mathbf{a}_\gamma, \mathbf{b}_\gamma)$ and $\mathbf{T} \sim Wishart(\mathbf{a}_T, \mathbf{b}_T)$.

37.3.3.3 A Spatially Structured Random Intercept Model

We used the conditional autoregressive specification $\log\ \theta_{jk} = {}^{0}\beta_{\mathrm{k}} + {}^{1*}\beta(x_{jk} - \bar{x}_k) + {}^{2}\beta(\bar{x}_k - \bar{x})$ where ${}^{0}\beta_k |^{0}\ \beta_{v \sim k} \sim N\left({}^{0}\bar{\beta}_k, \lambda_{{}^{0}\beta}\ n_k\right)$, where $v \sim k$ indicates the adjacent areas to the kth municipality, ${}^{0}\bar{\beta}_k$ is the mean over the adjacencies, n_k their number and $\lambda_\beta \sim IGamma(a_\beta, b_\beta)$ is the precision parameter.

37.3.3.4 A Spatially Structured Random Intercept and a Random Slope Model

As before we used a conditional autoregressive specification for the intercept and a spatially unstructured prior for the random slope $\log\theta_{jk} = {}^{0}\beta_k + {}^{1*}\beta_k(x_{jk} - \bar{x}) + {}^{2}\beta(\bar{x}_k - \bar{x})$ where ${}^{0}\beta_k|{}^{0}\beta_{v\sim k} \sim Normal({}^{0}\bar{\beta}_k, \lambda_{{}^{0}\beta}\, n_k)$ and ${}^{1*}\beta_k \sim Normal(\mu_{{}^{1*}\beta}, \sigma^2_{{}^{1*}\beta})$.

Non-informative distribution specifications were assumed for all the hyperparameters.

37.3.4 Model Selection

Comparison of several competing hierarchical Bayesian models is a difficult task and requires special consideration. When comparing models which involve random terms of growing complexity usually the Deviance Information Criterion (DIC) is used [13]. DIC is defined as the sum of two components DIC $= \bar{D} + p_{\mathrm{D}}$; $\bar{D}$ is the posterior deviance expectation and summarizes the fit of the model, p_{D} is the expected deviance minus the deviance evaluated at the posterior expectations. It represents the "effective number" of model parameters and reflects the complexity of the model.

Plummer [14] discussed the validity of DIC when the number of independent observation is less than the number of parameters (as common in geographical studies). He proposed to adjust for the "degree of optimism" present in the DIC statistics when evaluating model complexity. A penalized loss function based on cross-validation or a simpler modified version of the DIC was suggested. Here we used the modified DIC.

In particular:

$$\mathrm{DIC}c = \bar{D} + r_{\mathrm{opt}}$$

being

$$r_{\mathrm{opt}} \approx \sum_{j=1}^{J} p_{\mathrm{D}j}/(1 - p_{\mathrm{D}j})$$

where j indexes census block as before, and r_{opt} denotes the residual optimism.

37.4 Results

Figure 37.1(Left) shows the spatial distribution of lung cancer standardized incidence ratios (SIR) at municipality level for the Tuscany region (excluding the Province of Arezzo) and Fig. 37.1(Right) shows the spatial distribution of material deprivation index at municipality level. Both maps are highly spatially structured with the west part of the region showing higher lung cancer incidence and higher level of material deprivation, which is suggestive of a direct association between the two variables.

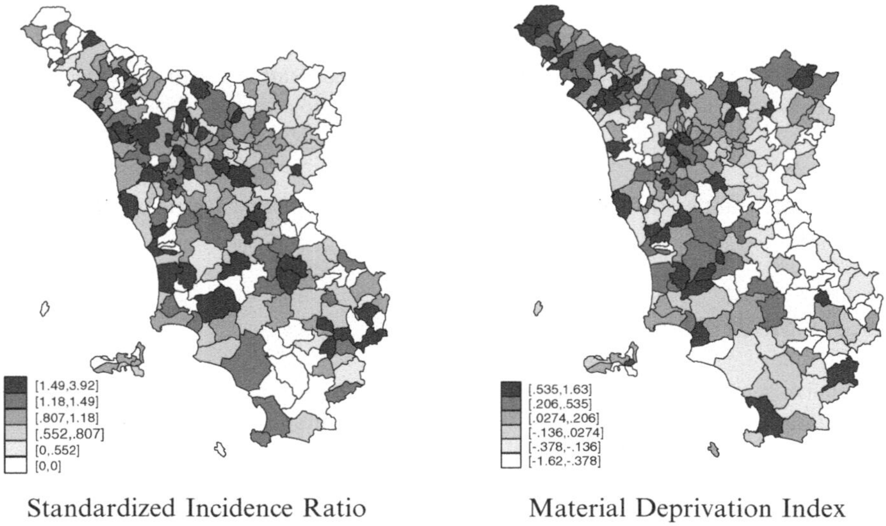

Fig. 37.1 (*Left*) Spatial distribution of the standardized incidence ratio (SIR, in quintiles). Lung cancer. Males, Tuscany (without Arezzo Province). Tuscany Cancer Registry, 2004. (*Right*) Spatial distribution of Material Deprivation index (MDI, in quintiles). Tuscany municipalities (without Arezzo Province), ISTAT 2001Census

37.4.1 Completeness of Cancer Registration

Before addressing the analysis of the association between socioeconomic factors and lung cancer incidence we must evaluate potential flaws in cancer registration which could bias the results. The mortality/incidence ratio for each municipality was estimated by means of a series of bi-variate hierarchical Bayesian models. Figure 37.2(Left) shows the predicted M/I ratios by the best fitting model with spatially structured terms. The distribution is slightly skewed. A slightly incompleteness of registration is seen in the north-west part of the region and a misclassification of prevalent cases in the south-eastern areas. However we cannot exclude selective survival by covariates such as socio-economic factors. Eventually the observed M/I ratio correlate with the material deprivation index.

37.4.2 Deprivation and Incidence of Lung Cancer

The estimate of the rate ratio (RR) for a unit increase of material deprivation at municipality level on incidence of lung cancer ranged from RR $= 1.16$ (Credibility interval CrI90% 1; 1.35) to RR $= 1.18$ (CrI90% 1.05; 1.33) depending on the hierarchical Bayesian random effect model considered. To compare the estimated

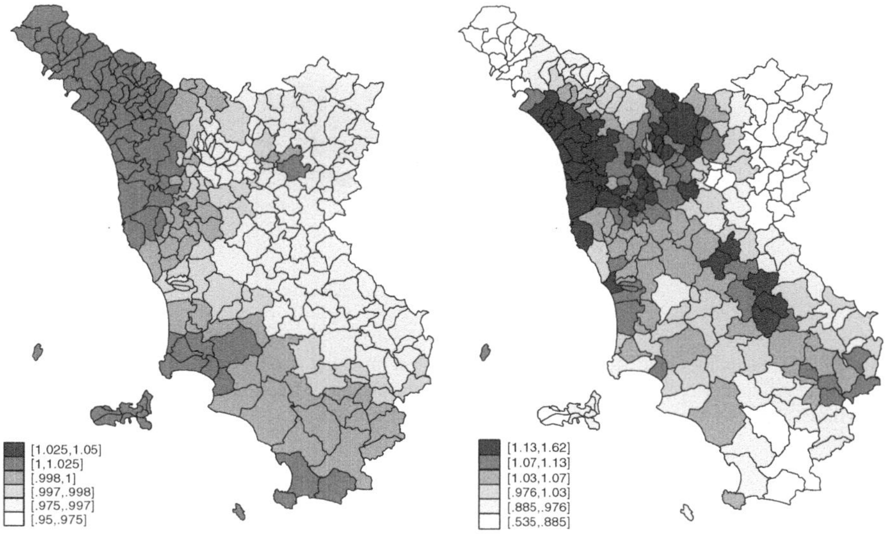

M/I ratio Baseline Relative Risk

Fig. 37.2 (*Left*) Spatial distribution of the estimated Lung Cancer Mortality/Incidence (M/I) ratio (in quintiles). Males, Tuscany (without Arezzo Province), Tuscany Cancer Registry, 2004. (*Right*) Spatial distribution of baseline relative risks from the spatially structured random intercept (see text). Lung Cancer, Males, Tuscany (without Arezzo Province), Tuscany Cancer Registry, 2004

Table 37.1 Indices of goodness of fit: $\bar{D}$, DIC, p_D, corrected DIC (DICc), r_{opt}, EDP

Model	$\bar{D}$	DIC	p_D	DIC*c*	r_{opt}	EDP
Random intercept	4,578	4,670	92	4,666	88	8,469
Random intercept and slope	4,580	4,680	100	4,684	104	8,466
Spatially structured random intercept	4,583	4,646	63	4,648	65	8,471
Spatially structured random intercept and random slope	4,570	4,663	93	4,667	97	8,471

rate ratios for material deprivation at census level to those at municipality level we express the rate ratio for an increase of one standard deviation and we report here for brevity only the best fitting model results (Table 37.1). The percent increase in risk for an increase of one standard deviation of material deprivation at census block level was 3.36% (CrI90%–1.44%; 8.40%) vs 5.87% (CrI90%–0.21%; 12.33%) at municipality level. Notice that introducing a spatially structured random intercept deflated the risk estimates from those under the fixed-effect Cronbach model - 3.87% (CrI90%–1.37%; 9.21%) and 6.72% (CrI90% 1.19%; 11.64%) respectively. The random slope models reported a paradoxically negative effect of material deprivation at census block level in some areas (Fig. 37.3). Table 37.1 reports

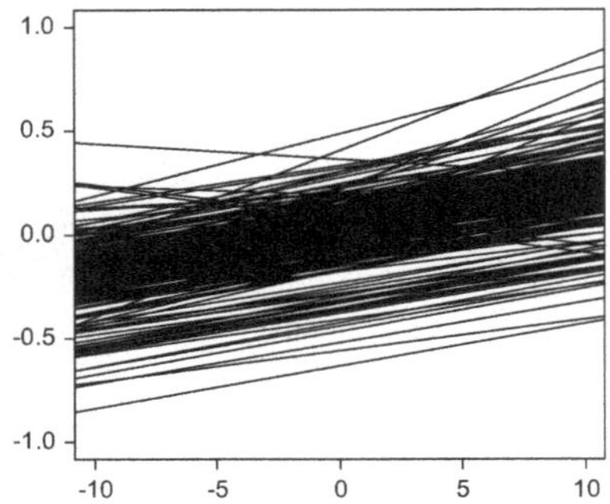

Fig. 37.3 Census block effect of material deprivation on Lung cancer mortality by municipality (see text: random slope model Sect. 37.3.3.4). Lung cancer. Males, Tuscany (without Arezzo Province), Tuscany Cancer Registry, 2004. *Y-axis*: log relative risk; *X-axis*: material deprivation (percent)

model comparison statistics. Despite a lower D-medio the spatially structured random intercept model behaves better in terms of DIC*c*. Random slope models are penalized by their extra complexity.

Figure 37.2(Right) shows the spatial distribution of random intercepts from the best fitting model. A lower risk in the north- western part of the region might by spurious as highlighted by the predicted distribution of the M/I ratio.

37.5 Discussion and Conclusions

In Italy there are few studies that compare results from analysis at the individual and the census block level [15]. In this paper we assumed that census block information may be used as a surrogate for the individual level. However since a contextual effect is plausible we expect the effect estimates be larger than those obtained having the complete information [16].

The results showed that municipality level has a larger impact than census block level even when we adjust for baseline risk variability. We must be cautious in interpreting this finding, since the ecological bias may menace the validity of this comparison [17]. We report confounding by location that is controlled by spatially structured random effects.

The percent increase in mortality for one standard deviation increase in material deprivation was estimated 3.36% ($\exp {}^{1*}\beta$) vs 2.43% ($\exp({}^{2}\beta - {}^{1*}\beta)$). The residual effect of the aggregate variable (2.43%) can be interpreted as contextual effect or ecological bias. It is outside the scope of this paper to address this issue in detail (see Fleischer and Diez Roux [18], Dowd [19]).

In conclusion, the Bayesian approach allows flexible random effect modelling. Future work will address a joint modelling of completeness of registration and disease risk.

Acknowledgements Italian Ministry of Education, University and Scientific Research Project PRIN20072S2HT8.

References

1. Van der Heyden, J.H., Schaap, M.M., Kunst, A.E., Esnaola, S., Borrell, C., Cox, B., Leinsalu, M., Stirbu, I., Kalediene, R., Deboosere, P., Mackenbach, J.P., Van Oyen, H.: Socioeconomic inequalities in lung cancer mortality in 16 European populations. Lung Canc. **63**, 322–330 (2009)
2. Krieger, N., Chen, J.T., Waterman, P.D., Soobader, M.J., Subramanian, S.V., Carson, R.: Geocoding and monitoring of US socioeconomic inequalities in mortality and cancer incidence: does the choice of area-based measure and geographic level matter? the Public Health Disparities Geocoding Project. Am. J. Epidemiol. **156**, 471–82 (2002)
3. Michelozzi, P., Perucci, C.A., Forastiere, F., Fusco, D., Ancona, C., Dell'Orco, V.: Inequality in health: socioeconomic differentials in mortality in Rome, 1990–95. J Epidemiol. Community Health. **53**, 687–93 (1999)
4. Morgenstern, H.: Ecologic studies. In: Rothmam, K.J. et al. (eds.) Modern Epidemiology, 3rd edn. Lippincott, Williams, & Wilkins, Boston (2008)
5. Regione Toscana: Registro Tumori della Regione Toscana. DGR 1200 12–12–2005 (2005) http://www.ittumori.it/ITA/chisiamo/registro-tumori.shtml
6. Regione Toscana: Progetto ITER-NET sistema regionale delle strade e indirizzi. DGR 44 30–01–2006 (2006) http://www.iternet.fi.eng.it/SigmaPortal2
7. Accetta, G., Grisotto, L., Terni, G., Biggeri, A.: Mortalità per condizione socio-economica. Studio Longitudinale Toscano 2001–2005. Informazioni Statistiche, Studi e Ricerche. Edizioni Regione Toscana. 1–116 (2007)
8. Regione Toscana: Studio Longitudinale Toscano. DGR 423 19–04–1999 (1999)
9. Caranci, N., Biggeri, A., Grisotto, L., Pacelli, B., Spadea, T., Costa, G.: L'indice di deprivazione italiano a livello di sezione di censimento: definizione, descrizione e associazione con la mortalità. Epidemiologiae Prevenzione. **34**, 167–176 (2010)
10. Parkin, D.M., Chen, V.W., Ferlay, J., Galceran, J., Storm, H., Whelan, S.: Comparability and quality control in cancer registration. IARC Technical Report, No.19, IARC, Lyon (1994)
11. Besag, J.: Spatial interaction and the statistical analysis of lattice systems. JRSS B **36**, 192–236 (1974)
12. Cronbach, L.J., Webb, N.: Between class and within class effects in a reported aptitude × treatment interaction: a reanalysis of a study by G.L. Anderson. J. Educ. Psychol. **67**, 717–724 (1975)
13. Spiegelhalter, D.J., Best, N., Carlin, B.P., van der Linde, A.: Bayesian measures of model complexity an fit (with discussion). J. R. Stat. Soc. B **64**, 583–639 (2002)
14. Plummer, M.: Penalized loss functions for Bayesian model comparison. Biostatistics **9**, 523–539 (2008)
15. Biggeri, A., Dreassi, E., Marchi, M.: A multilevel Bayesian model for contextual effect of material deprivation. Stat. Method Appl. **13**, 88–102 (2004)
16. Geronimus, A.T., Bound, J.: Use of Census-based aggregate variables to proxy for socioeconomic group: evidence from National samples. Am. J. Epidemiol. **148**(5), 475–486 (1998)
17. Geronimus, A.T., Bound, J., Neidert, L.J.: On the validity of using census geocode characteristics to proxy individual socioeconomic characteristics. J. Am. Stat. Assoc. **91**(434), 529–537 (1996)

18. Fleischer, N.L., Diez Roux, A.V.: Using directed acyclic graphs to guide analyses of neighbourhood health effects: an introduction. J. Epidemiol. Community Health. **62**, 842–846 (2008)
19. Dowd, B.E.: Separated at birth: statisticians, social scientists, and causality in health services research. Health Serv. Res. **46**, 397–420 (2011)

Mining Administrative Health Databases for Epidemiological Purposes: A Case Study on Acute Myocardial Infarctions Diagnoses

38

Francesca Ieva, Anna Maria Paganoni, and Piercesare Secchi

Abstract

We present a pilot data mining analysis on the subset of the Public Health Database (PHD) of Lombardia Region concerning hospital discharge data relative to Acute Myocardial Infarctions without ST segment elevation (NON-STEMI). The analysis is carried out using nonlinear semi-parametric and parametric mixed effects models, in order to detect different patterns of growth in the number of NON-STEMI diagnoses within the 30 largest clinical structures of Lombardia Region, along the time period 2000–2007. The analysis is a seminal example of statistical support to decision makers in clinical context, aimed at monitoring the diffusion of new procedures and the effects of health policy interventions.

Keywords

Biostatistics and bioinformatics • Data mining • Generalized linear mixed models • Health service research

38.1 Introduction

Recent years have witnessed a growing interest in the use of performance indicators in health care; they may measure some aspects of the health care process, clinical outcomes or epidemiological incidence and prevalence of diseases. In response, a sizeable literature has emerged questioning the right use of such indicators as a measure of quality of care, as well as stating more specific criticism of the statistical methods used to obtain estimates adjusted for patient case-mix.

F. Ieva (✉) · A.M. Paganoni · P. Secchi
MOX - Dipartimento di Matematica, Politecnico di Milano, via Bonardi 9, 20133 Milano, Italy
e-mail: francesca.ieva@mail.polimi.it; anna.paganoni@polimi.it; piercesare.secchi@polimi.it

N. Torelli et al. (eds.), *Advances in Theoretical and Applied Statistics*,
Studies in Theoretical and Applied Statistics, DOI 10.1007/978-3-642-35588-2_38,

Health care service scheduling is strictly connected with a deep knowledge of current health needs; with respect to this, a total novelty is represented by the potential offered by the statistical data mining of administrative data-banks for collecting clinical and epidemiological information. Indeed, there is an increasing agreement among epidemiologists on the validity of disease and intervention registries based on administrative databases [1, 4, 12]. In this work we focus on growth curves for the number of diagnoses of Acute Myocardial Infarction without ST segment elevation (NON-STEMI), and we explore the question as to whether they had a trend in the time interval 2000–2007. Indeed, clinical best practice maintains that there is no evidence for a greater incidence of NON-STEMI in this period; however, since the early 2000s a new diagnostic procedure—the *troponin* exam—has been introduced and this, by easing NON-STEMI detection, could have produced an increased number of positive diagnoses. Administrative data-banks can be used to check for growth in the number of NON-STEMI diagnoses along with the adoption by clinical institutions of new diagnostic procedures or devices, like the troponin exam.

In this work we will illustrate a pilot data mining case study on hospital discharges data for patients with NON-STEMI diagnosis; data come from the Lombardia Region Public Health Database (PHD), an ongoing collection of data used, up to now, only for administrative purposes. The study is part of the Strategic Program "Exploitation, integration and study of current and future health databases in Lombardia for Acute Myocardial Infarction" (AMI Project) funded by the Italian Ministry of Health and by "Direzione Generale Sanità—Regione Lombardia" and started in January 2009. The major objective of this project is the identification of new diagnostic, therapeutic, and organizational strategies to be applied to patients with acute coronary syndromes (ACS), in order to improve clinical outcomes. To achieve this goal Regione Lombardia authorized the extraction from the PHD database of data concerning patients with Acute Coronary Syndromes.

The statistical analysis is conducted along different phases. The visual evidence for growth in the number of NON-STEMI diagnoses is first questioned by fitting a semi-parametric mixed effect model, in order to capture the shape of growth curves and to test the significance of the grouping factor effect. The relevant features emerged with this first analysis are then modeled by means of parametric nonlinear models of decreasing complexity, which are easier to interpret and more suited to inferential purposes.

All the analyses have been performed with the R program [15]; the `mgcv` [18] package and the `nlme` package [14], respectively, for generalized additive mixed models and for nonlinear mixed effects models have been used.

38.2 Data Mining Discharge Data on Acute Myocardial Infarctions

In this section we describe a data mining study of the Hospital Discharge Database (*Database Ricoveri*), which is one of the three main databases belonging to the Star scheme [10] that composes the PHD of Lombardia Region. We focus on the numbers

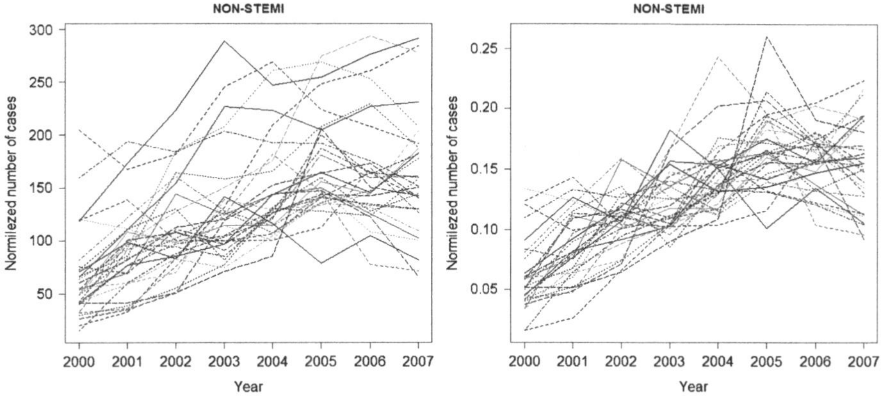

Fig. 38.1 *Left panel*: Number of AMI without ST-elevation diagnoses in the period 2000–2007 in the 30 largest clinical institutions of Lombardia Region. *Right panel*: Standardized number of AMI without ST-elevation diagnoses in the period 2000–2007 in the 30 largest clinical institutions of Lombardia Region. For each hospital the yearly number of diagnoses has been divided by the hospital total number of diagnoses in the time period 2000–2007

of hospital discharges with a diagnosis of NON-STEMI, grouped by hospital and relative to the 30 largest clinical institutions of Lombardia Region, during years 2000–2007. Detection of cases is performed according to the AHQR guidelines [5].

Figure 38.1, left panel, represents the number of acute myocardial infarction without ST-elevation (NON-STEMI) diagnoses, along the time period 2000–2007, for the 30 hospitals. The total number of diagnoses in the time period 2000–2007 has a considerable variability between institutions: in fact it ranges from a minimum value of 715 to a maximum of 1,872. This difference is due to the different exposure of different hospitals; indeed, exposure could be a confounding factor in a statistical analysis focused on the growth trend of the number NON-STEMI cases. Hence, in order to analyze comparable data, for each hospital the yearly number of diagnoses has been standardized by the hospital total number of diagnoses in the time period 2000–2007, thus adjusting for hospital exposure (see Fig. 38.1, right panel).

The high variability between hospitals and the structure of the data grouped by hospital motivate the use of mixed effects models [13] for the analysis of these longitudinal data. A first explorative analysis conducted by means of a simple linear mixed model, where the standardized number of NON-STEMI diagnoses appears as a linear function of time, with hospital as a grouping factor, shows a significant linear trend over time (the p-value of the test on the "year" fixed effect is less than 10^{-14}).

Since the use of a linear parametric model can be quite binding, a further enquire into the growth trend has been conducted by fitting a semi-parametric mixed effect model. Indeed, we set $\tilde{N}_{ij}$ to be the standardized number of NON-STEMI diagnoses for hospital $i = 1, \ldots, 30$ and year $j = 1, \ldots, 8$, where $j = 1$ is for year 2000

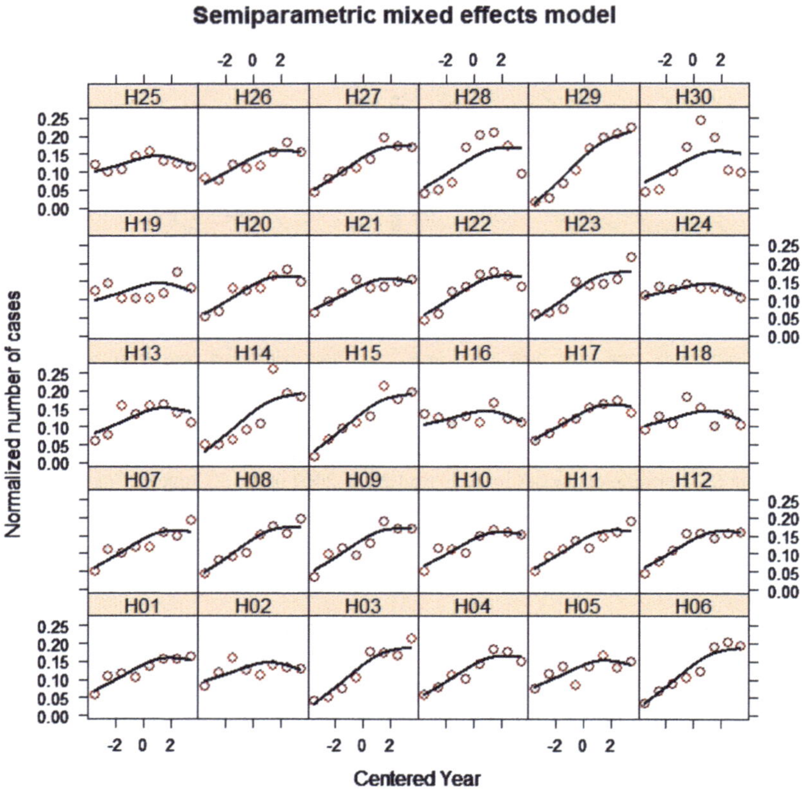

Fig. 38.2 Estimated growth curves through model (38.1) together with the original data

and $j = 8$ is for year 2007, and following [17], we fit the following mixed effects semi-parametric model with respect to time

$$\tilde{N}_{ij} = s(t_j) + b_{0i} + b_{1i}t_j + \varepsilon_{ij} \qquad i = 1, \ldots 30, \ j = 1, \ldots, 8, \tag{38.1}$$

where t_j is the centered time covariate (i.e. $t_0 = 2000 - 2003.5 = -3.5$, $t_1 = 2001 - 2003.5 = -2.5$ and so on), s is a common cubic regression spline, while b_{0i} and b_{1i} are i.i.d samples of the random variables $b_0 \sim \mathscr{N}(0, \sigma^2_{b_0})$ and $b_1 \sim \mathscr{N}(0, \sigma^2_{b_1})$ respectively, representing gaussian additive independent random effects, grouped by hospital. The quantities ε_{ij} are i.i.d. samples from the random variable $\varepsilon \sim \mathscr{N}(0, \sigma^2)$ representing residual error: ε, b_0 and b_1 are assumed to be independent. Estimates are obtained by maximization of restricted likelihood. Figure 38.2 shows the estimated growth curves together with the original data.

We fitted a semi-parametric mixed effects model in order to catch a common behavior in the growth of normalized number of NON-STEMI diagnoses in the years 2000–2007, smoothing data and taking into account overdispersion due to

Table 38.1 Fixed effects estimates and Anova table for model (38.2)

Fixed effects estimates:

	Value	Std. Error
Asym	0.1544	0.0026
Tmid	−2.7017	0.1368

Anova Table:

	numDF	denDF	F-value	p-value
Asym	1	209	5417.630	$< .0001$
Tmid	1	209	389.845	$< .0001$

the grouping factor. In fact, inspection of Fig. 38.2 suggests a common "S-shaped" growing pattern. Concerning the random effects, the estimated parameters are: $\hat{\sigma}_{b_0} = 2.702 * 10^{-07}$, $\hat{\sigma}_{b_1} = 0.00765$, and $\hat{\sigma} = 0.02297$. The negligible effect of the random variable b_0 suggests that the curves are in fact different only with respect to their growth rate. The greater effect of the random variable b_1 is conducive to a further analysis of these data by means of a model that captures the common growth trend while taking into account overdispersion in the growth rates. Indeed, the following (parametric) logistic mixed effects model accommodates for the "S-shaped" common growing pattern, pointed out by the nonparametric analysis, while enabling the testing of its significance:

$$\tilde{N}_{ij} = \frac{\text{Asym} + \alpha_i}{(1 + \exp(\text{Tmid} + \tau_i - t_j))} + \varepsilon_{ij}, \qquad i = 1, \ldots 30, \;\; j = 1, \ldots, 8, \tag{38.2}$$

where t_j is the centered time covariate, the fixed effects Asym and Tmid represent, respectively, the asymptote and the inflection point of the logistic curve, while α_i and τ_i are i.i.d samples of the random variables $\alpha \sim \mathcal{N}(0, \sigma_\alpha^2)$ and $\tau \sim \mathcal{N}(0, \sigma_\tau^2)$, respectively, representing gaussian additive random effects, grouped by hospital. The quantities ε_{ij} are i.i.d. samples from the random variable $\varepsilon \sim \mathcal{N}(0, \sigma^2)$ and they represent residual error. The two random effects α and τ are assumed to be independent, and independent of ε; all estimates are computed by restricted maximum likelihood. Table 38.1 shows that both fixed effects Asym and Tmid are significant.

Concerning the random effects, the estimated parameters are: $\hat{\sigma}_\alpha = 6.8183 * 10^{-07}$, $\hat{\sigma}_\tau = 0.4821$, and $\hat{\sigma} = 0.0287$. It is then confirmed that the variability of the additive random effect relative to the asymptote is negligible; thus α_i can be removed from model (38.2) without loss in model performance. On the contrary, the variability of the random effect relative to the inflection point is large and implies a very significant effect; this stimulates an interesting interpretation, since, in the logistic model, the inflection point indicates the time of maximum growth speed and this, in turn, is directly related to the timing of a growth speed significantly different from zero.

Inspection of the set of (estimated) random effects $\tau_i, i = 1, \ldots, 30$, related to the inflection point suggests a clustering structure that has been captured by partitioning the set in $k = 1, 2, \ldots,$ clusters by means of the Partitioning Around Medoids procedure (PAM, [11]), implemented with the Euclidean distance, denoted by d. A critical point is the choice of k, the number of groups: an helpful method is the computation of the average silhouette width, and the inspection of the silhouette plot of PAM. For each estimated τ_i, let A be the cluster to which τ_i has been assigned and compute $a(\tau_i)$, the average dissimilarity of τ_i to all other objects in A,

$$a(\tau_i) = \frac{1}{|A| - 1} \sum_{\tau_j \in A, \tau_j \neq \tau_i} d(\tau_j, \tau_i).$$

Now, if C is a cluster different from A, denote by

$$d(\tau_i, C) = \frac{1}{|C| - 1} \sum_{\tau_j \in C} d(\tau_j, \tau_i)$$

the average dissimilarity of τ_i from all objects in C and set $c(\tau_i)$ to be the smallest value of $d(\tau_i, C)$ when C is let to range over the set of all clusters different from A. The *silhouette value* $s(\tau_i)$ of τ_i is defined as

$$s(\tau_i) = \frac{c(\tau_i) - a(\tau_i)}{\max\{a(\tau_i), c(\tau_i)\}}.$$

Clearly $s(\tau_i)$ lies between -1 and 1; large values of $s(\tau_i)$ support the fact that the element τ_i is well classified in A. The entire silhouette plot, i.e., the plot of all $s(\tau_i)$, and the Average Silhouette Width, i.e., the average of all silhouette values, are qualitative indexes helpful to judge and compare the results obtained by PAM for different values of k [16].

By inspecting the silhouette plot, represented in Fig. 38.3, the presence of $k = 3$ clusters can be sustained. Indeed, for $k = 3$, the Average Silhouette Width is equal to 0.58 and, as a general rule, it can be asserted that a reasonable clustering structure has been found when the Average Silhouette Width is greater than 0.5. The medoids representative of the three clusters correspond to years $y_A = 2000$, $y_B = 2001$ and $y_C = 2002$. "Cluster A" denotes the institutions for which the estimated time of inflection point Tmid + τ_i in model (38.2) is closer to -3.1692, i.e., closer to year $y_A = 2000$. Analogously, "Cluster B" denotes the institutions for which the estimated time of inflection point is closer to -2.6839, i.e., closer to year $y_B = 2001$, and "Cluster C" denotes the institutions for which the estimated time of inflection point is closer to -2.3014, i.e., closer to year $y_C = 2002$.

In the left panel of Fig. 38.4, the curves estimated by model (38.2) are represented, one curve for each hospital, together with the real data; the right panel shows the estimated logistic growth curves. The thick red, black, and green curves represent the three benchmarks growth curves, i.e., medoids for cluster A, B, and C, respectively.

The particular interest in analyzing the clustering structure of the random effects related to the inflection points derives by the clinical surmise about their

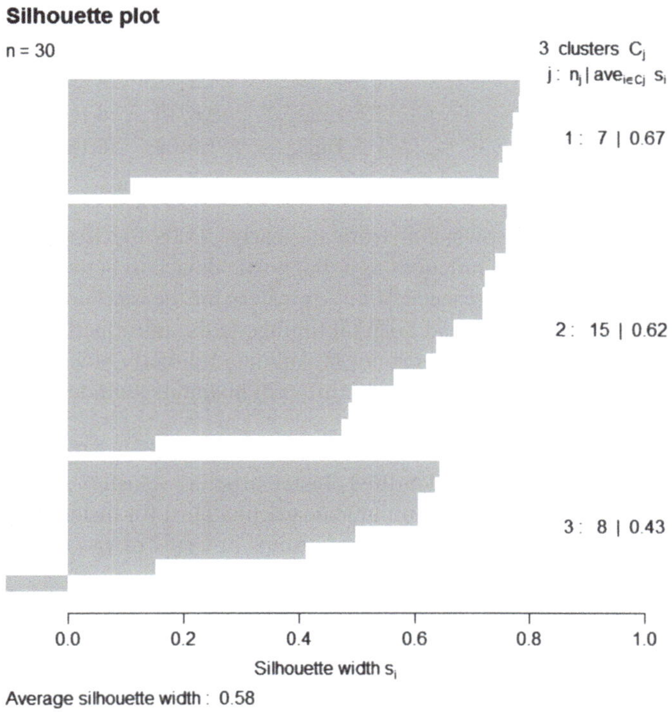

Fig. 38.3 Silhouette plot of PAM procedure on the estimated inflection points with $k = 3$ clusters

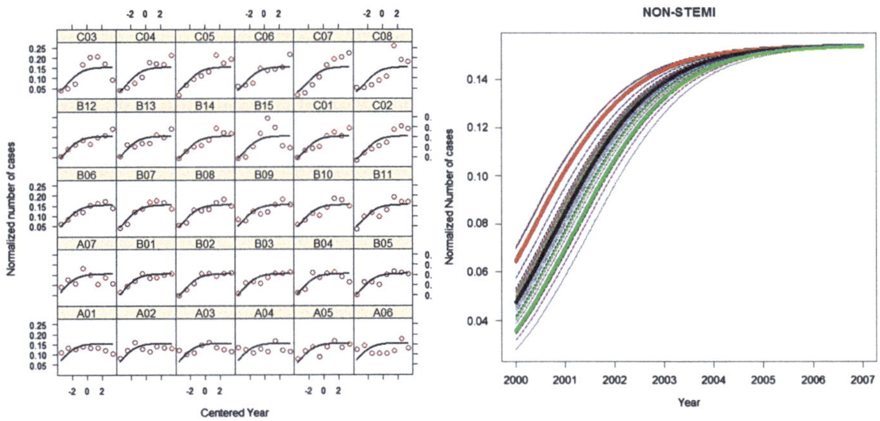

Fig. 38.4 Estimated logistic growth curves for different medical institutions

Table 38.2 Fixed effects estimates for model (38.3)

	Value	Std. Error	p-value
Asym	0.1540	0.0025	$< .0001$
Tmid_A	−3.9434	0.2383	$< .0001$
Tmid_B	−2.6719	0.1294	$< .0001$
Tmid_C	−1.9108	0.1637	$< .0001$

presence. Indeed, it is known that from the early 2000s the troponin exam has been introduced in hospital practices as a diagnostic device to better identify NON-STEMI events; hence, the presence of three clusters for the random effects τ_i could be a consequence of the different hospital timings in the introduction and adoption of this practice. This hypothesis cannot be validated directly since the timings of adoption of the troponin exam by the 30 different hospitals included in the analysis are not available.

The previous analysis suggests a simpler model with fixed effects only, where dummy variables represent the identified cluster structure (clusters A, B, or C). This model is easier to interpret and communicate to clinicians; for instance, it quantifies the statistical evidence of the existence of groups in terms of p-values reported in Table 38.2. The model is:

$$\tilde{N}_{ij} = \frac{\text{Asym}}{(1 + \exp(\text{Tmid}_A \cdot 1_{i \in A} + \text{Tmid}_B \cdot 1_{i \in B} + \text{Tmid}_C \cdot 1_{i \in C} - t_j))} + \varepsilon_{ij}, \tag{38.3}$$

where $i = 1, \ldots 30$, is the institution index, $j = 1, \ldots, 8$, is the year index, and ε is defined as before. Estimates for the effects of model (38.3) appear in Table 38.2; they are all significant. Notice that the fixed effects estimates reported in Table 38.2 are close to the values identifying the inflection points of the three medoids y_A, y_B and y_C generated by the analysis of model (38.2). Testing all possible contrasts between the three different fixed effects related to the inflection point always generates a p-value less than 10^{-4}; there is a strong evidence of different inflection points in the three groups. Diagnostic checks show that normality assumption of residuals can be sustained.

In conclusion, the statistical analysis advocates the presence of three groups of hospitals, possibly distinguished by different timings of introduction and adoption of the troponin test and supports the clinical tenet that in the time period 2000–2007 there has been an apparent increase in the normalized number of NON-STEMI diagnoses that is not due to a real increase in the disease incidence, but to a new diagnostic procedure adopted in hospitals along different timings.

38.3 Conclusions and Further Developments

The study presented in this chapter is a pilot example of an advanced statistical analysis performed on data drawn from a PHD. Administrative health care databases play today a central role in epidemiological evaluation of Lombardia health care system because of their widespread diffusion and low cost of information. Public health care regulatory organizations can assist decision makers in providing information based on available electronic health records, promoting the development and the implementation of the methodological tools suitable for the analysis of administrative databases and answering questions oriented to disease management. The aim of this kind of evaluation is to estimate adherence to best practice (in the setting of evidence-based medicine) and potential benefits and harms of specific health policies. Health care databases can be analyzed in order to calculate measures of quality of care (quality indicators); moreover the implementation of disease and intervention registries based on administrative databases could enable decision makers to monitor the diffusion of new procedures (as was in troponin exam adoption example) or the effects of health policy interventions. The unassailable benefit of the use of the PHD is the high data quality, and the real time data availability without costs increase. This innovative perspective was a paramount motivation for the Strategic Program "Exploitation, integration and study of current and future health databases in Lombardia for Acute Myocardial Infarction" (AMI Project). More details about the AMI Project and the planned analyses can be found in [2, 3, 6–9].

The case study illustrated in the previous section is an example of the potential offered by the statistical analysis of an administrative database for clinical and epidemiological purposes. Statistical analysis is conducted in different phases: an explorative analysis of data conducted by means of semi-parametric models to guide the study towards an appropriate parametric model, the fit of a suitable nonlinear parametric model with mixed effects estimated under the usual assumptions of random effects and residuals normality, the criticism of this model assumptions, offered by the presence of a clustering structure in the random effects, and thus the final improvement obtained through the introduction of appropriate dummy variables, taking into account the identified clustering structure, and leading to a simple and significant fixed effect logistic model.

Acknowledgements This work is a part of the Strategic Program "Exploitation, integration and study of current and future health databases in Lombardia for Acute Myocardial Infarction" supported by "Ministero del Lavoro, della Salute e delle Politiche Sociali" and by "Direzione Generale Sanità—Regione Lombardia". The authors wish to tank the Working Group for Cardiac Emergency in Milano and the Cardiology Society.

References

1. Balzi, D., Barchielli, A., Battistella, G., et al.: Stima della prevalenza della cardiopatia ischemica basata su dati sanitari correnti mediante un algoritmo comune in differenti aree italiane. Epidemiologia e Prevenzione **32**(3), 22–29 (2008)
2. Barbieri, P., Grieco, N., Ieva, F., Paganoni, A.M., Secchi, P.: Exploitation, integration and statistical analysis of Public Health Database and STEMI archive in Lombardia Region. "Complex data modeling and computationally intensive statistical methods", Contribution to Statistics, pp. 41–56. Springer, New York (2010)

3. Determinazioni in merito alla "Rete per il trattamento dei pazienti con Infarto Miocardico con tratto ST elevato(STEMI)": Decreto N° 10446, 15/10/2009, Direzione Generale Sanità - Regione Lombardia (2009)
4. Every, N.R., Frederick, P.D., Robinson, M. et al.: A comparison of the national registry of myocardial infarction with the cooperative cardiovascular project. J. Am. Coll. Cardiol. **33**(7), 1887–1894 (1999)
5. Glance, L.G., Osler, T.M., Mukamel, D.B., et al.: Impact of the present-on-admission indicator on hospital quality measurement experience with the Agency for Healthcare Research and Qualit (AHRQ) Inpatient Quality Indicators. Med. Care **46**(2), 112–119 (2008)
6. Grieco, N., Ieva, F., Paganoni, A.M.: Performance assessment using mixed effects models: a case study on coronary patient care. IMA J. Manag. Math. **23**(2), 117–131 (2012)
7. Ieva, F., Paganoni, A.M.: Statistical Analysis of an integrated Database concerning patients with Acute Coronary Syndromes, S.Co.2009 - Sixth conference - Proceedings, MAGGIOLI, Milano, 223–228 (2009)
8. Ieva, F., Paganoni, A.M.: Multilevel models for clinical registers concerning STEMI patients in a complex urban reality: a statistical analysis of MOMI^{2}2 survey. Comm. Appl. Ind. Math. **1**(1), 128–147 (2010)
9. Ieva, F.: Designing and mining a multicenter observational clinical registry concerning patients with acute coronary syndromes. In: Grieco, N., Paganoni, A.M., Marzegalli, M. (eds.) Identification and Development of New Diagnostic, Therapeutic and Organizational Strategies for Patients with Acute Coronary Syndromes. Springer (2013, to appear)
10. Inmon, W.H.: Building the Data Warehouse, 2nd edn. Wiley, New York (1996)
11. Kaufman, L., Rousseeuw, P.: Finding Groups in Data. Wiley Series in Probability and Mathematical Statistics. Wiley, New York (1990)
12. Manuel, D.G., Lim, J.J.Y., Tanuseputro, P. et al.: How many people have a myocardial infarction? Prevalence estimated using historical hospital data. BMC Publ. Health **7**, 174–89 (2007)
13. Pinheiro, J.C., Bates, D.M.: Mixed-Effects Models in S and S-Plus. Springer, New York (2000)
14. Pinheiro, J.C., Bates, D.M., DebRoy, S., Sarkar, D. and the R core team: nlme: Linear and Nonlinear Mixed Effects Models (2009)
15. R Development Core Team: R: A Language and Environment for Statistical Computing. R Foundation for Statistical Computing, Vienna, Austria, (2009). Available: http://www.R-project.org
16. Struyf, A., Hubert, M., Rousseeuw, P.J.: Clustering in an object-oriented environment. J. Stat. Software **1** (1996)
17. Wood, S.N.: Modelling and smoothing parameter estimation with multiple quadratic penalties. J. R. Stat. Soc. B **62**(2), 413–428 (2000)
18. Wood, S.N.: Generalized Additive Models: An Introduction with R. Chapman and Hall/CRC, Boca Raton, Florida (2006)

Part VI

Economic Statistics and Econometrics

Fractional Integration Models for Italian Electricity Zonal Prices

39

Angelica Gianfreda and Luigi Grossi

Abstract

In the last few years we have observed an increasing interest in deregulated electricity markets. Only few papers, to the authors' knowledge, have considered the Italian Electricity Spot market since it has been deregulated recently. This contribution is an investigation with emphasis on price dynamics accounting for technologies, market concentration, and congestions as well as extreme spiky behavior. We aim to understand how technologies, concentration, and congestions affect the zonal prices since all these combine to bring about the single national price (*prezzo unico d'acquisto*, PUN). Implementing Reg–ARFIMA–GARCH models, we draw policy indications based on the empirical evidence that technologies, concentration, and congestions do affect Italian electricity prices.

Keywords

ARFIMA models • Congestions • Electricity prices • Long memory • Market power

39.1 Introduction and Literature Review

Mean-reversion, seasonality, volatility clustering, inverse leverage effect, long memory, and extreme values (*spikes*) characterize the daily electricity price dynamics.

These stylized facts should be taken into account when modeling electricity prices [6]. In this chapter we propose to fit ARFIMA–GARCH to the seasonally adjusted series of Italian electricity prices. ARFIMA processes are able to model mean-reversion and residual seasonality, while volatility clustering is captured by

A. Gianfreda · L. Grossi (✉)
University of Verona, Department of Economics, Via dell'artigliere, 19, 37129 Verona, Italy
e-mail: angelica.gianfreda; luigi.grossi@univr.it

N. Torelli et al. (eds.), *Advances in Theoretical and Applied Statistics*,
Studies in Theoretical and Applied Statistics, DOI 10.1007/978-3-642-35588-2_39,

GARCH models. Long memory of the generating process of electricity prices has received small consideration in the literature. Frequently electricity prices are in fact transformed and returns are then modeled in order to deal with stationary time series [11]. Actually, when unit root tests are applied to electricity prices, the null is usually rejected, while the stationarity hypothesis is rejected as well. This is the case of the Italian electricity prices which present long memory features and can be modeled by fractional integrated processes (as in [14]), given that there is not unanimous consent on this issue and the fact that the Italian structure has been modified several times inducing structural breaks into the prices series. Indeed, it is well known that the presence of structural breaks can induce a correlation structure very similar to that observed for long memory processes [2]. Starting from this general framework, we propose a model in which different regressors are included.

Several explanatory variables have been proposed for this peculiar process to understand which factors drive the price dynamics. Traded volumes, as well as demand, reserve margin, volatility, and water supply have been explored, as for instance in [12]. The Italian market was previously investigated by [3], however, without considering none of previous factors. Here we propose to determine if technical factors (as technologies, congestions, and market power) do affect the zonal prices. We identify a *congestion* every time that prices of couples of contiguous zones are different, as suggested by [10]. Moreover, the behavior of generators can be heavily influenced by their geographical locations within zones and by transmission network capacity, see [8]. Therefore generation, congestions, and market power are strongly interdependent and crucial factors, see [5].

A generator has market power if it is able to raise the electricity price above marginal cost without experiencing a significant decline in demand. Previous studies focussed on this topic in the electricity generation sector relying on oligopoly theory, implementing simulation techniques to model the electricity generators' behavior. For a survey on models to detect market power see [4]. Traditionally, analysts and anti-trust regulators investigate market power issues using various measures of market concentration such as the popular *Hirschmann-Herfindahl* index (HHI), the *residual supply index* (RSI), and the *Lerner* index. Since there is not a consensus on which measure is the best indicator of market power for the electricity markets, because there is a number of factors to account for (transmission constraints are an example), we have decided to consider two structural indexes (the HHI and the RSI) provided by the Italian system operator (*Gestore del Mercato Elettrico*, GME).

Another issue which is addressed in this chapter is the effect of spikes in the time series of electricity prices. Extreme values or spikes are results of abnormally large variations in price caused by weather conditions, outages, or transmission failures, as pointed out in [16]. The influence of extreme observations is not generally considered in studying the generating processes of electricity time series, but the bias induced by outliers on model estimates is a very well-known problem (see [1] and cited references). We propose to treat daily spikes with median daily prices and consider network congestions as proposed by [10] recalling that the Italian market is segmented showing characteristics as those of Nord Pool because the most congested links identify aggregations of zones.

Summarizing our contribution, we propose a price dynamic modeling which takes into account simultaneously long memory, production technologies, concentration, and congestions. We also provide evidence that the special zonal structure of this market must be considered when modeling these prices since all series have weighted influence in determining the single national price.

The structure of the chapter is as follows. In Sect. 39.2 the Italian zonal electricity market is presented and issues related to production technologies and market power are discussed. The theoretical econometric background of the chapter is explained in Sect. 39.3 which also contains the main results of the empirical analysis. In Sect. 39.4 we conclude and discuss some further research issues related to the topic of this chapter.

39.2 The Italian Zonal Market: Structure, Technologies, Concentration, and Congestions

The Italian wholesale electricity market started its operations in April 2004 but became an Exchange only in 2005 registering an increasing in traded volumes from 73 TWh in 2004 to 232 TWh in 2008. It is important to emphasize that, since this market is comparatively young, there are continuous structural changes as for instance the abolition of the Calabria zone and its inclusion in the Southern zone starting from the beginning of 2009. Furthermore, the market has come to a steady phase only in the last years. Hence, our investigations refer only to a time period going from January 2007 to the end of 2008, see [9].

Italian electricity is produced by the following plants: thermal power plants only with coal, or with fuel oil or with natural gas; multi-fuel thermal power plants with oil and coal or with oil and natural gas; combined cycle gas turbines (CCGT); hydro power plants with pumped storage, with run of the river (fluent) or with reservoirs (modulation); gas turbine plants (GT); wind power plants; and finally other generation plants not included in the previous ones. These technologies, used in [7] to detect influences of generation sources on price and volatility dynamics, have been clustered into the following six types of the marginal technology index (MTI) distinguishing between oil, gas, and coal producing plants: *Coal* (all multi-fuel and thermal power plants with coal), *Thermal* (plants without coal), *Hydro*, *Wind* (renewables), *CC* represents combined cycles (CCGT and GT) and finally *Other* is meant for plants not included in the previous ones. In this work we have computed for every group of technology the number of hours (frequency) in which it has fixed the price over the corresponding zone and built a set of six dummies, one for each group, and then attributed one to the group with the maximum frequency over the day and zero to the others. Formally, let f_{rjt} be the number of hours for the r-th technology group used in zone j on day t. The dummy variable for the r-th group in zone j is then defined as $d_{rjt} = 1$ if $f_{rjt} = max_r(f_{rjt})$ and $d_{rjt} = 0$ otherwise.

Table 39.1 Frequencies (number of days) of technologies fixing the price over individual zones

	Coal	CC	Thermal	Wind	Hydro	Other
North	73	632	366	0	449	40
CNorth	122	462	702	0	218	26
CSouth	143	362	817	0	183	26
South	151	356	815	0	185	25
Calabria	188	351	810	0	156	28
Sicily	18	325	1,106	0	59	1
Sardinia	296	274	700	0	251	20

From the summary reported in Table 39.1, it is possible to exclude two technologies, *Wind* and *Other* in all zones, from our analysis since they have a low influence compared to the other sources.

Beside technologies, we want to analyze the effect of market power on wholesale prices. Hence we have considered the *Hirschmann–Herfindahl* index (HHI) and the *Residual Supply* index (RSI), whereas the *Lerner* index could be constructed on zonal basis once the marginal costs of technologies will be available.

The first index, HHI, measures the degree of concentration and dispersion of volumes sold (and/or offered[1]) by market participants for each hour and each zone. It is the sum of the shares of the volumes sold in the market by market participants as indicated in the following equation

$$HHI(j,h) = \sum_{i=1}^{N} [Q_i(j,h) * 100]^2 \qquad (39.1)$$

with

$$Q_i(j,h) = \frac{V_i(j,h)}{\sum_{i=1}^{N} V_i(j,h)} \qquad (39.2)$$

where $i = 1, \ldots, N$ are market participants, j represents the individual zones, h is the considered hour, and finally V_i are volumes sold by the i-th participant.

The range values of this index are 0 when there is *perfect competition* and 10,000 points in the case of *monopoly*. In the former case, the index simply shows that the quantity sold by the i-th participant is irrelevant with respect to the total amount of electricity sold by the other participants, and indeed in this situation the ratio $Q_i(j,h)$ tends to zero when the denominator is much more large (almost infinite) than (compared to) the numerator. The latter case of monopoly is identified by 10,000 which is given by 100^2 meaning that the ratio Q_i is equal to one, or in other words that the i-th participant is the only one selling electricity. From a preliminary

[1]The shares are defined by considering the volumes sold and/or offered (including those covered by Bilateral Contracts) by individual market participants aggregated on the basis of the group to which they belong.

Table 39.2 Percentages of HHI levels with respect to the employed sample of 35,064 h

	No concentration $HHI \leq 1{,}000$	Moderate Concentration $1{,}000 < HHI < 1{,}800$	Concentration $HHI \geq 1{,}800$
North	1,06	90,49	8,45
CNorth	1,08	0,97	97,95
CSouth	1,93	0,11	97,96
South	2,01	2,57	95,42
Calabria	2,05	0,00	97,95
Sicily	2,03	1,02	96,95
Sardinia	2,05	0,00	97,95

analysis of the Italian zonal HHI as provided by GME (Table 39.2), it is possible to state that in all Italian zones (apart from North) there is a poor competition on the generation side producing expectations on a direct relation between price and HHI, since when the latter increases then the price should increase as an effect of market concentration (or market power). Looking at time series of certain hours belonging to delivery periods off-peak 1, off-peak 2, and peak,[2] it is possible to see that there is a sensible shift in level in the HHI hourly series during the entire month of November 2008, and this is also reflected in the daily series (see Fig. 39.1 for Central North).[3] In that period we observed a shift in the HHI dynamics but similar behaviors can be seen neither in the quantities sold (not reported in both figures) nor in the Residual Supply Index (RSI).

Therefore, given this strange dynamics for the HHI index and considering in addition unexpected results,[4] we have decided to verify the impact of market power on electricity price dynamics using only the latter index, implicitly assuming that this index is able to detect that kind of exercise.

The second measure of concentration, that is the Residual Supply Index, measures the presence of residual market participants necessary to cover the total demand, thus the index measures the ex-post residuality. The hourly zonal RSI published by GME has the following formulation

$$RSI_i(j,h) = \sum_{l=1, l \neq i}^{N} S_l(j,h) - V_i(j,h) \tag{39.3}$$

[2]The delivery periods for the Italian market refer to the following groups of hours: *off peak 1* from 00.00 to 06.00 until the end of 2005 then from 2006 to 07.00; *peak* is from 07.00 (08.00 from 2006) to 22.00 (to 20.00 from 2006); *off peak 2* from 23.00 (or 21.00 from 2006) to 24.00.

[3]Similar dynamics are observed on other zones and are not reported for lack of space.

[4]Contrary on expectations from Table 39.2, HHI is found to be significant and positive only in CSouth.

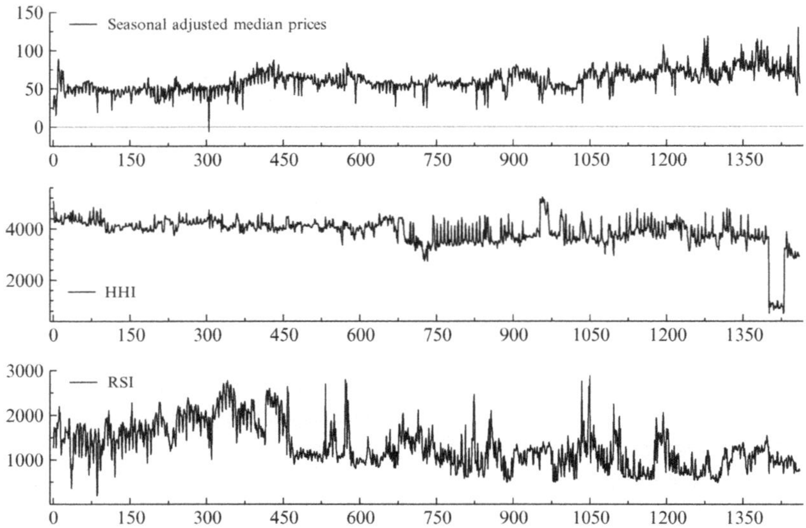

Fig. 39.1 Seasonal adjusted price series (*top panel*), HHI series (*middle panel*), and RSI series as provided by GME (*bottom panel*) for Central Northern zone

where $l, i = 1, \ldots, N$ are market participants, j represents the individual zones, h is the considered hour, and finally V_i are volumes sold by the i-th participant. This difference between the total supply and the sum of ith sellers' supply (or in other words the quantity offered by other market participants) represents the *non-contestable volumes*. Hence dividing this quantity by the total quantities sold in one zone at one particular hour, we determined the hourly and daily aggregated RSI_i. If the index is less than 1, then the ith firm is necessary to cover the demand and so it is a *pivotal* supplier in the market; if the index is greater or equal to 1, then the ith firm is not necessary and the market can be considered competitive, see [13] and [15]. We observed that all RSI_i adjusted for volumes are always less than one, apart for Calabria where the index is almost always equal to one. Therefore we expect that market power strongly affects the estimations implying a rise in electricity prices in all zones but Calabria.

The last issue we considered in this work is the influence of grid congestions on prices. In this chapter we only consider five physical national zones which are (until 2008): North, Central North (CNorth), Central South (CSouth), South, and Calabria (Calb). The two isles (Sardinia and Sicily) have not been considered because these are highly concentrated due to the limited connection to the peninsula. Electricity flows in both directions, and so a congestion occurs every time the transmission capacity is exceeded. We identify and define daily time series of frequencies of congestions every time we observe different zonal prices among contiguous zonal couples which are North–CNorth, CNorth–CSouth, CSouth–South, and South–Calb.

39.3 Model Specifications and Empirical Results

A preliminary empirical analysis of the Italian zonal market carried out using daily medians of prices has provided evidence of the presence of seasonality at daily level and a long memory autocorrelation structure, see [7].

Figure 39.2 represents the AutoCorrelation Function (ACF) and the Periodogram (estimated by the Fast Fourier Transformation) for seasonally adjusted prices collected in the Northern zone. Seasonal adjustment has been carried out by using a linear model with dummy regressors for days of the week and calendar effects (*CalEf*). The pattern of ACF is typical of a long memory process. Looking at the Periodogram, apart from peaks at low frequencies related to the trend-cycle, a peak can be clearly detected around 0.14 frequency, corresponding to a 7-day period. This means that the seasonal adjustment did not capture all the weekly seasonal dependence of the series. For this reason we decided to use the original series as dependent variable and explicitly model the seasonal pattern of the series. Long memory is usually captured by fractionally integrated processes. Taking into account the autocorrelation structure we could estimate ARFIMA models with seven autoregressive terms, that is a ARFIMA(7,0) or seven Moving Average terms, that is a ARFIMA(0,7). Another stylized fact that should not be neglected is the mean reversion of electricity prices which is usually estimated by a one-lag autoregressive term. For this reason we added one AR term also in the case of the ARFIMA model with seven Moving Average terms. Finally the two estimated ARFIMA models are: ARFIMA(7,0) and ARFIMA(1,7). Residual diagnostics (tables are not reported for lack of space) from these models show that ACF and PACF functions are inside the confidence regions, but the null hypothesis of homoscedasticity according to the Engle LM test cannot be accepted. For this reason we estimated ARFIMA–GARCH models. Now all residuals diagnostics lead to the acceptance of the model (see Table 39.3 for the ARFIMA(1,7)–GARCH(1,1) model). Comparing the information criteria (AIC and BIC), the best model seems to be the ARFIMA(1,7)–GARCH(1,1), which is the model we used as a basic one for testing the influence of exogenous explanatory variables on wholesale prices. To take into account the presence of many extreme values and consequent fat tails of the distribution of electricity prices, we estimated the models under the assumption of different distributions for residuals. The best performance has been obtained using a Student–t distribution with degrees of freedom jointly estimated with the other parameters of the model.

Hence, the effect of exogenous factors on wholesale prices has been measured implementing Reg–ARFIMA–GARCH models [12] with dummies for group of technologies, frequencies of congestions, and the market concentration index as explanatory variables.

The proposed models can be formalized as follows:

$$(1-L)^d(y_t-\mu_t) = \varepsilon_t + \theta_1\varepsilon_{t-1} + \ldots + \theta_q\varepsilon_{t-q} \qquad \varepsilon_t|I_{t-1} \sim t(0,\sigma_t^2) \qquad (39.4)$$

where

$$\sigma_t^2 = \omega + \alpha\varepsilon_{t-1}^2 + \beta\sigma_{t-1}^2 \qquad (39.5)$$

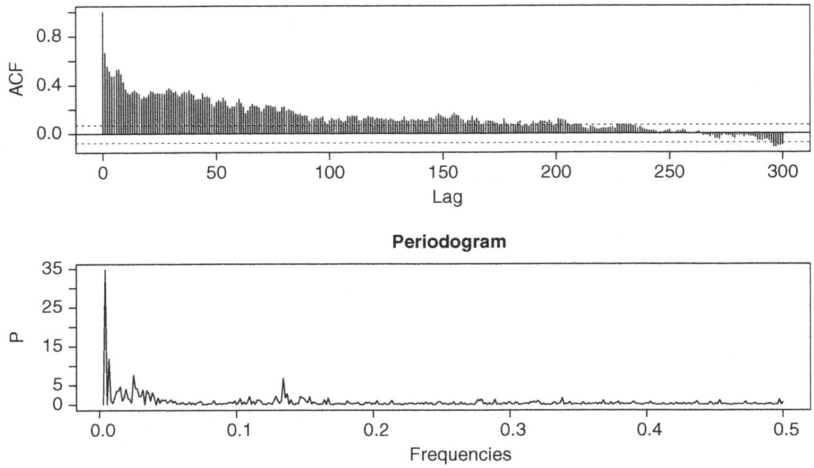

Fig. 39.2 AutoCorrelation Function (*upper panel*) and Periodogram (*lower panel*) of seasonally adjusted prices for North

Table 39.3 P-values of residuals diagnostic tests for ARFIMA(1,7)–GARCH(1,1) models estimated in five zones

	NORTH	CNORTH	CSOUTH	SOUTH	CALB
Q-Statistics on Standardized Residuals					
Q(15)	0.302	0.406	0.340	0.369	0.251
Q(20)	0.234	0.560	0.656	0.724	0.454
Q(30)	0.112	0.373	0.242	0.394	0.165
Q-Statistics on Squared Standardized Residuals					
Q(15)	0.581	0.915	0.704	0.627	0.749
Q(20)	0.640	0.972	0.718	0.609	0.775
Q(30)	0.635	0.552	0.042	0.715	0.929
Diagnostic test based on the news impact curve (EGARCH vs. GARCH)					
Sign Bias Test	0.542	0.594	0.459	0.274	0.535
Negative Size Bias Test	0.851	0.620	0.820	0.674	0.835
Positive Size Bias Test	0.712	0.780	0.719	0.682	0.512
Joint Test	0.749	0.891	0.720	0.479	0.618
LM Engle test					
ARCH 1–2 test	0.742	0.984	0.851	0.750	0.744
ARCH 1–5 test	0.390	0.990	0.362	0.294	0.342
ARCH 1–10 test	0.575	0.835	0.475	0.385	0.511

for $t = 1, \ldots, T$, y_t is the zonal median electricity price at time t, L is the lag operator defined by $Ly_t = y_{t-1}$, and $\mu_t = E(y_t|I_{t-1})$ is the mean equation conditioned to the set of information available at time $t-1$. The $\theta_j \varepsilon_{t-j}$ terms represent the moving average component of the price dynamics with coefficients

θ_j for $j = 1, \ldots, q$, indicating the number of moving average parameters. The following specification has been considered for the conditional mean function:

$$\mu_t = c + \nu_1 D_1 + \ldots + \nu_7 D_7 + \gamma CalEf + \phi_1 y_{t-1} + \lambda_1 Tech_t + \lambda_2 MarPow_t + \lambda_3 Cong_t \qquad (39.6)$$

where D_j with $j = 1, \ldots, 7$ are dummies for days of the week and ν_j are the corresponding coefficients; $CalEf$ is a dummy accounting for calender effects and γ is the corresponding coefficient; the $\phi_1 y_{t-1}$ term represents the autoregressive component of the price dynamics. *Tech*, *MarPow*, and *Cong* are respectively the dummy variables indicating the *technology group* determining the price, the index of *market power* and finally *Cong* are dummies for daily *congestion* events, λs are regression coefficients.

Table 39.4 shows the maximum likelihood estimates of Reg–ARFIMA–GARCH parameters applied to time series of daily median prices. Information criteria (Akaike Information Criteria and Bayesian Information Criteria) and log-likelihood values are reported in the final rows. Looking at estimates, we can draw the following conclusions:

1. *Calendar effects*, *seasonality*, *fractional integration*, and *volatility clustering* are important and salient features to take into account since the estimates of $CalEf$, days of the week, d, α and β are always significant. Moreover d is always significant and less than 0.5 for all zones as found previously in [7], hence confirming that these price processes have long memory.
2. The *autoregressive structure*, that is the ϕ_1 term, is found to be important to capture the stylized fact of mean-reversion of electricity prices. Whereas the inclusion of moving average terms has been used to obtain white noise residuals.
3. The employed *groups of technologies* determining the zonal prices are generally significant across zones. In details, looking at Table 39.4, *Combined Cycles* (CC) always reduce electricity zonal prices, whereas *Thermal* power (TNC) generally increases them.
4. *Concentration* (or market power) has been analyzed considering two indexes. Since we have found that the HHI has a controversial impact,[5] only results with the RSI has been reported. The RSI indicates competitive markets when it approaches (and is greater than) one, as in Calabria. And indeed, we have found that the exercise of market power, apart for Calabria and Central South, increases zonal prices given that estimated parameters are positive and significant.
5. *Congestions* are important only in North and CNorth. In the first case, it can be argued that when congestions affect the Northern zone, demand could be satisfied

[5]It is significant (with a negative sign) in some zones but it turns to be insignificant in some others, whereas it should always have a positive sign: when the HHI increases then the price increases as result of exercise of market power. These results are available on request.

Table 39.4 Reg–ARFIMA estimates (with significance levels in brackets) for Italian Electricity Zonal Prices (where ***, ** and * represent significance at 1%, 5% and 10%, respectively)

	NORTH		CNORTH		CSOUTH		SOUTH		CAL	
c	54.09	(***)	50.64	(***)	54.35	(***)	51.89	(***)	36.30	(***)
RSI	0.77	(***)	0.11	(**)	0.04		0.17	(**)	0.18	
Cong	−2.89	(***)	1.18	(**)	2.15		0.80		1.96	
CC	−1.37	(**)	−0.68		−1.44	(**)	−1.55	(**)	−1.69	(***)
TNC	1.30		1.38	(*)	1.84	(***)	1.74	(**)	2.16	(***)
CalEf	−12.55	(***)	−21.83	(***)	−22.32	(***)	−22.89	(***)	−21.81	(***)
Mon	8.22	(***)	20.14	(***)	20.17	(***)	19.80	(***)	18.99	(***)
Tue	11.11	(***)	21.27	(***)	21.14	(***)	20.89	(***)	20.41	(***)
Wed	10.84	(***)	21.07	(***)	21.17	(***)	20.96	(***)	20.58	(***)
Thu	10.11	(***)	20.51	(***)	20.41	(***)	20.18	(***)	19.89	(***)
Fri	8.76	(***)	19.86	(***)	19.29	(***)	18.89	(***)	18.86	(***)
Sat	2.16	(**)	5.59	(***)	5.90	(***)	6.44	(***)	6.09	(***)
d-Arfima	0.49	(***)	0.49	(***)	0.48	(***)	0.47	(***)	0.49	(***)
ϕ_1	0.19	(***)	0.22	(**)	0.19	(**)	0.13	(**)	0.18	(***)
θ_1	−0.26	(**)	−0.23	(**)	−0.16		−0.14		−0.16	
θ_2	0.00		−0.10	(**)	−0.14	(***)	−0.14	(***)	−0.12	(***)
θ_3	−0.02		−0.03		−0.05		−0.07		−0.08	(*)
θ_4	−0.01		0.00		0.00	(***)	0.01		0.02	
θ_5	−0.03		−0.01		0.01		−0.01		−0.01	
θ_6	0.04		0.07	(*)	0.03		0.03		0.02	
θ_7	0.22	(***)	0.22	(***)	0.24	(***)	0.24	(***)	0.22	(***)
ω	1.08		2.39	(*)	1.74	(*)	2.24	(*)	2.03	
α	0.06	(***)	0.12	(***)	0.10	(***)	0.12	(***)	0.10	(***)
β	0.92	(***)	0.86	(***)	0.89	(***)	0.86	(***)	0.88	(***)
Student(DF)	6.37	(***)	4.42	(***)	4.85	(***)	5.21	(***)	5.10	(***)
AIC	6.68		6.82		6.88		6.89		6.91	
BIC	6.84		6.98		7.04		7.04		7.07	
Log-likelihood	−2,491		−2468		−2416		−2492		−2501	

by imports. On the other hand, when congestions occur in CNorth electricity prices raise because of an excess of demand. Interestingly and not surprisingly, prices in CSouth, South, and Calabria are not influenced by congestions.[6]

[6]This could be due to the presence of limited production poles which only inject electricity into the system then providing the necessary supply: Brindisi in the Southern zone and Rossano in Calabria.

39.4 Conclusions

This chapter is an analysis of effects of technologies, concentration, and congestions on Italian Electricity zonal prices. According to the most recent contributions in the time series analysis applied to electricity prices for instance [10], we took into account the long memory feature of the generating stochastic process estimating a parameter of fractional integration, which turned out to lie very close to 0.5. A causal analysis in the framework of Reg–ARFIMA–GARCH models confirmed the significant impact of production technologies, market concentration, and congestions on these price dynamics. These results can be converted in tentative suggestions for policy indications to be followed when programming the medium–long-term energy policy in Italy.

Concluding, we have provided first insights on relationships between zonal electricity spot prices, technologies, concentration, and congestions finding parameters significant and coherent to our expectations according to the Italian market structure and mechanisms, especially about the degree of market power observed in Italy: the lack of market competition leads indeed to higher prices. In addition, we would like to emphasize that special attention should be spent on the construction of new transmission lines given that generators can serve only if there exists adequate transmission capacity, since the installation of new generating capacity is expected to produce even more and could lead to sudden bottleneck problems.

Acknowledgements We would like to thank the book's editors and two anonymous referees for their valuable comments on a previous version of the paper.

References

1. Battaglia, F., Orfei, L.: Outlier detection and estimation in nonlinear time series. J. Time Series Anal. **26**(1), 107–121 (2005)
2. Bisaglia, L., Gerolimetto, M.: Testing structural breaks vs. long memory with the Box-Pierce statistics: a Monte Carlo study. Stat. Method Appl. **18**, 543–553 (2009)
3. Bosco, B., Parisio, L., Pelagatti, M.: Deregulated wholesale electricity prices in Italy: An empirical analysis. Int. Adv. Econ. Res. **13**, 415–432 (2007)
4. Fridolfsson, S., Tangeras, T.: Market power in the Nordic electricity wholesale market: a survey of the empirical evidence. Energ Pol. **37**, 3681–3692 (2009)
5. Furió, D., Lucia, J.J.: Congestion management rules and trading strategies in the Spanish electricity market. Energ Econ. **31**, 48–60 (2009)
6. Gianfreda, A., Bunn, D.: Integration and shock transmissions across European electricity forward markets. Energ Econ. **32**(2), 278–291 (2010)
7. Gianfreda, A., Grossi, L.: Zonal Price Analysis of the Italian Wholesale Electricity Market. In: IEEE CNF Conference Proceedings of the European Energy Markets Conference (EEM09). Available on IEEE Xplore, 2009
8. Giulietti, M., Grossi, L., Waterson, M.: Price transmission in the UK electricity market: Was NETA beneficial? Energ Econ. **32**, 1165–1174 (2010)
9. GME: Technical Report 2008, available at http://www.mercatoelettrico.org
10. Haldrup, N., Nielsen, M.O.: A regime switching long memory model for electricity prices. J. Econometrics **135**(1–2), 349–376 (2006)

11. Karakatsani, N., Bunn, D.W.: Forecasting electricity prices: The impact of fundamentals and time-varying coefficients. Int. J. Forecast. **24**(4), 764–785 (2008)
12. Koopman, S.J., Ooms, M., Carnero, M.A.: Periodic seasonal Reg-ARFIMA-GARCH models for daily electricity spot prices. J. Am. Stat. Assoc. **102**(477), 16–27 (2007)
13. Manuhutu, C., Owen, A.D.: Gas-on-gas competition in Shangai. Energ Pol. **38**(5), 2101–2106 (2010)
14. Olbermann, B.P., Lopes, S.R.C., Reisen, V.A.: Invariance of the first difference in ARFIMA models. Comput. Stat. **21**, 445-461 (2006)
15. Rahimi, A.F., Sheffrin, A.Y.: Effective market monitoring in deregulated electricity markets. IEEE Trans. Power Syst. **18**(2), 486–493 (2003)
16. Weron, R.: Modeling and Forecasting Electricity Loads and Prices: A Statistical Approach. Wiley, New York (2006)

A Generalized Composite Index Based on the Non-Substitutability of Individual Indicators

40

Matteo Mazziotta and Adriano Pareto

Abstract

Composite indices for ranking territorial units are widely used by many international organizations in order to measure economic, social and environmental phenomena. The literature offers a great variety of aggregation methods, all with their pros and cons. In this chapter, we propose an alternative composite index which, starting from a linear aggregation, introduces penalties for the units with "unbalanced" values of the indicators. As an example of application, we consider the set of individual indicators of the Technology Achievement Index (TAI) and present a comparison between some traditional methods and the proposed index.

Keywords

Dimensionality reduction • Ranking

40.1 Introduction

Social and economic phenomena, such as development, poverty, wellness and malnourishment have been measured, in the past, principally from an unidimensional point of view, that is using only one variable. The more recent literature tends to consider these phenomena as multidimensional or complex since they are characterized by the combination of different variables. The measurement of complex phenomena is a difficult and dangerous operation since it requires simplifications that are inherently somewhat arbitrary, is always constrained by limited resources and time, inevitably involves competing and conflicting priorities, and rests on a foundation of values preferences that are typically resolved by pragmatic considerations, disciplinary biases and measurement traditions. Nevertheless, it is

M. Mazziotta (✉) · A. Pareto
Italian National Institute of Statistics, Rome, Italy
e-mail: mazziott@istat.it; pareto@istat.it

N. Torelli et al. (eds.), *Advances in Theoretical and Applied Statistics*,
Studies in Theoretical and Applied Statistics, DOI 10.1007/978-3-642-35588-2_40,

possible to consistently combine both the selection of variables representing the phenomenon and the choice of the "best" aggregation function in order not to lose much statistical information.

The aim of this chapter is to present a generalized composite index, denoted as MPI (Mazziotta-Pareto Index), suitable in the case in which the variables are "non-substitutable", i.e. they have all the same weight (importance) and a compensation among them is not allowed [7]. In Sect. 40.2, the steps to implement a composite index are presented; in Sect. 40.3, the MPI and its main properties are shown; finally, in Sect. 40.4, an empirical comparison among simple arithmetic mean, geometric mean, median and MPI is proposed.

40.2 Steps Towards the Synthesis of Indicators

In the scientific literature, there are many studies, by eminent authors, concerning the use of composite indices in order to measure complex, economic and social, phenomena about geographical areas. The main problems, in this approach, are the availability of the data, the choice of the more representative indicators and their treatment in order to allow comparisons (normalization). An other very important step is the definition of the synthesis measure because it is the tool used to "add" the normalized indicators; in this delicate phase, from a methodological point of view, the choices of the researcher assume a fundamental role. In fact, to define the aggregation function, there are many possible alternatives such as multivariate techniques, distance measures and linear or nonlinear functions.

The steps for constructing a composite index can be summarized as follows:

(a) Defining the phenomenon to be measured. The definition of the concept should give a clear sense of what is being measured by the composite index. It should refer to a theoretical framework, linking various sub-groups and underlying indicators.

(b) Selecting a group of individual indicators, usually expressed in different units of measurement. The strengths and weaknesses of a composite index reflect the quality of the underlying individual indicators. Ideally, indicators should be selected according to their relevance, analytical soundness, timeliness, accessibility, etc. [9]. Proxy measures can be used when the desired data are unavailable or when cross-country comparability is limited. Besides, there must not be redundancies and all the aspects of the phenomenon must be considered.

(c) Normalizing the individual indicators to make them comparable. Normalization is required prior to any data aggregation as the indicators in a data set often have different measurement units. Therefore, it is necessary to bring the indicators to the same standard, by transforming them into pure, dimensionless, numbers. There are various methods of normalization, such as ranking, re-scaling, standardization (or z-scores) and "distance" to a reference. Assigning the same "importance" to each indicator, it is indispensable to apply a transformation criterion that makes the indicators independent from both the unit of measurement and the variability, e.g. the standardization.

(d) Aggregating the normalized indicators by composite indices (mathematical functions). Different aggregation methods are possible. The most used are additive methods that range from summing up unit ranking in each indicator to aggregating weighted transformations of the original indicators. Multivariate techniques as Principal Component Analysis [3] and distance measures as Wroclaw Taxonomic Method [4] are also often used.

For this approach, obviously, there are several problems such as finding data, losing information and researcher arbitrariness for: (1) selection of indicators, (2) normalization, and (3) aggregation and weighting. In spite of these problems, how mentioned in the introduction, the advantages of this approach are clear and they can be summarized in: (a) unidimensional measurement of the phenomenon; (b) immediate availability; and (c) simplification of the geographical data analysis.

Many works and analysis have won over the critics and the scientific community concluded that it is impossible to obtain a "perfect" method where the results are universally efficient. On the contrary, the data and the specific targets of the work must, time by time, individuate the "best" method in terms of robustness, reliability and consistency of the solutions.

40.3 The Composite Index

Additive methods for constructing composite indices imply requirements and properties which are often not desirable or difficult to meet. They assume above all a full substitutability among the components of the index: a deficit in one dimension can be compensated by a surplus in another. However, a complete compensability among individual indicators is generally not acceptable and a "balanced" distribution of the values is required.

The proposed method wants to supply a composite measure of a set of indicators that are considered "non-substitutable", i.e. all the dimensions of the phenomenon must be "balanced" [5].

40.3.1 General Aspects

The MPI is designed in order to satisfy the following properties: (1) normalization of the indicators by a specific criterion that deletes both the unit of measurement and the variability effect [2]; (2) synthesis independent from an "ideal unit", since a set of "optimal values" is arbitrary, non-univocal and can vary with time [1]; (3) simplicity of computation; and (4) ease of interpretation.

These properties can be satisfied by the following approach.

It is known that distributions of different indicators, measured in different way, can be compared by the transformation in standardized deviations. Therefore, the individual indicators can be converted to a common scale with a mean of 100 and a standard deviation of 10: the obtained values will fall approximately in the range 70–130[1]. In this type of normalization the "ideal vector" is the set of mean values and it is easy to identify both the units that are above average (values greater than 100) and the units that are below average (values less than 100). Moreover, normalizing by standardized deviations allows to release the indicators from their variability and to assign them the same weight.

In such context, a penalty coefficient can be introduced that is function, for each unit, of the indicators' variability in relation to the mean value ("horizontal variability"): this variability is measured by the coefficient of variation. The proposed approach penalizes the score of each unit by adjusting the mean of the standardized values with a quantity that is directly proportional to the "horizontal variability". The purpose is to favour the units that, mean being equal, have a greater balance among the values of the indicators [8].[2]

The method provides a "robust" measure that is less "sensitive" to inclusion or exclusion of individual indicators [6].

The steps for computing MPI are the following.

1. Normalization

Let $\mathbf{X} = \{x_{ij}\}$ be the matrix with n rows (statistical units) and m columns (individual indicators) and let M_{x_i} and S_{x_j} denote the mean and the standard deviation of the j-th indicator:

$$\mathrm{M}_{x_j} = \frac{\sum_{i=1}^{n} x_{ij}}{n};\ \mathrm{S}_{x_j} = \sqrt{\frac{\sum_{i=1}^{n} (x_{ij} - \mathrm{M}_{x_j})^2}{n}}.$$

The standardized matrix $\mathbf{Z} = \{z_{ij}\}$ is defined as follows:

$$z_{ij} = 100 \pm \frac{(x_{ij} - \mathrm{M}_{x_j})}{\mathrm{S}_{x_j}} 10 \qquad (40.1)$$

where the sign $\pm$ is the "polarity" of the j-th indicator, i.e. the sign of the relation between the j-th indicator and the phenomenon to be measured (+ if the individual indicator represents a dimension considered positive and – if it represents a dimension considered negative).

2. Aggregation

Let cv_i be the coefficient of variation for the i-th unit:

$$\mathrm{cv}_i = \frac{\mathrm{S}_{z_i}}{\mathrm{M}_{z_i}}$$

[1] On the basis of Bienaymé-Cebycev theorem, the terms of the distribution within the range 70–130 are at least 89% of total terms.

[2] Note that the penalty can be viewed as the "price" to pay for having uneven achievement across dimensions.

where:

$$\mathrm{M}_{z_i} = \frac{\sum_{j=1}^{m} z_{ij}}{m};\ \mathrm{S}_{z_i} = \sqrt{\frac{\sum_{j=1}^{m} (z_{ij} - \mathrm{M}_{z_i})^2}{m}}.$$

Then, the generalized form[3] of MPI is given by:

$$\mathrm{MPI}_i^{+/-} = \mathrm{M}_{z_i}(1 \pm \mathrm{cv}_i^2) = \mathrm{M}_{z_i} \pm \mathrm{S}_{z_i}\mathrm{cv}_i$$

where the sign of the penalty (the product $\mathrm{S}_{zi}\mathrm{cv}_i$) depends on the kind of phenomenon to be measured.

If the composite index is "increasing" or "positive", i.e. increasing values of the index correspond to positive variations of the phenomenon (e.g. the socio-economic development), then MPI^- is used. Vice versa, if the composite index is "decreasing" or "negative", i.e. increasing values of the index correspond to negative variations of the phenomenon (e.g. the poverty), then MPI^+ is used.

In the first case, the penalty is subtracted and the index is less than the mean for unbalanced values; in the second one, the penalty is added and the index is greater than the mean for unbalanced values.

40.3.2 Computation and Properties

Table 40.1 shows an example of how to calculate the $\mathrm{MPI}^{+/-}$.

MPI^+ and MPI^- can also be written as follows:

$$\mathrm{MPI}_i^+ = \frac{\sum_{j=1}^{m} z_{ij}^2}{\sum_{j=1}^{m} z_{ij}} \quad \text{and} \quad \mathrm{MPI}_i^- = \frac{2}{m}\sum_{j=1}^{m} z_{ij} - \frac{\sum_{j=1}^{m} z_{ij}^2}{\sum_{j=1}^{m} z_{ij}}$$

where z_{ij} is given by (40.1). The MPI_i^+ is a concave function of the generic z_{ik} $(k = 1, \ldots, m)$, while the MPI_i^- is a convex function of z_{ik}.

Given the matrix $\mathbf{X} = \{x_{ij}\}$, the generalized index has the following properties:

i. The MPI^+ of the i-th unit is greater or equal than the MPI^- of the same unit, that is:

$$\mathrm{MPI}_i^+ \geq \mathrm{MPI}_i^-.$$

In particular, $\mathrm{MPI}_i^+ = \mathrm{MPI}_i^-$ if and only if $\mathrm{S}_{z_i} = 0$.

ii. The MPI^+ and the MPI^- of the i-th unit are linked by the relation:

$$\mathrm{MPI}_i^- = 2\mathrm{M}_{z_i} - \mathrm{MPI}_i^+.$$

[3]It is a generalized form since it includes "two indices in one".

Table 40.1 Example of calculating MPI

	Original indicators		Normalized indicators						
Unit	X1	X2	Z1	Z2	Mean	Std. dev.	CV	MPI$^+$	MPI$^-$
A	1	100	88.4	114.1	101.28	12.86	0.127	102.91	99.65
B	6	80	110.7	107.1	108.88	1.81	0.017	108.91	108.85
C	4	60	101.8	100.0	100.89	0.89	0.009	100.90	100.88
D	6	40	110.7	92.9	101.81	8.88	0.087	102.58	101.04
E	1	20	88.4	85.9	87.14	1.28	0.015	87.16	87.12
Mean	**3.6**	**60.0**	**100.0**	**100.0**					
Std. dev.	**2.2**	**28.3**	**10.0**	**10.0**					

iii. Given two units i and h, with $\mathrm{M}_{z_i} = \mathrm{M}_{z_h}$, we have:

$$\mathrm{MPI}_i^- > \mathrm{MPI}_h^- \quad \text{if and only if} \quad \mathrm{S}_{z_h} > \mathrm{S}_{z_i};$$
$$\mathrm{MPI}_i^+ > \mathrm{MPI}_h^+ \quad \text{if and only if} \quad \mathrm{S}_{z_i} > \mathrm{S}_{z_h}.$$

iv. Given two units i and h, with $\mathrm{M}_{z_i} > \mathrm{M}_{z_h}$, we have:

$$\mathrm{MPI}_i^- > \mathrm{MPI}_h^- \quad \text{if and only if} \quad \mathrm{M}_{z_i} - \mathrm{M}_{z_h} > \mathrm{S}_{z_i}\mathrm{cv}_i - \mathrm{S}_{z_h}\mathrm{cv}_h;$$
$$\mathrm{MPI}_i^+ > \mathrm{MPI}_h^+ \quad \text{if and only if} \quad \mathrm{M}_{z_i} - \mathrm{M}_{z_h} > \mathrm{S}_{z_h}\mathrm{cv}_h - \mathrm{S}_{z_i}\mathrm{cv}_i.$$

v. Let $x_j x_k$ be the linear correlation coefficient between the j-th and the k-th indicator; if $\mathrm{r}_{x_j x_k} = 1$, for each j and k with $j \neq k$, then:

$$\mathrm{MPI}_i^+ = \mathrm{MPI}_i^- = \mathrm{M}_{z_i}.$$

This result is due to the fact that, for the i-th unit, we have $z_{ij} = z_{ik}$ for $j \neq k$.

Property (v) is very interesting because it shows the relation between the behaviour of the MPI and the structure of the correlations among the individual indicators.

In general, the lower the correlation among the indicators, the higher the "horizontal variability" induced in each unit and the greater the difference between MPI and arithmetic mean.

Therefore, the MPI may be a useful tool to summarize uncorrelated variables, such as the principal components, in a "non-compensatory" point of view.

40.4 An Application to Real Data

In this section, an application of MPI to the variables of the Technology Achievement Index (TAI) is presented; see [9] for more details.

The TAI is a composite measure of technological progress that ranks countries on a comparative global scale [10].

Table 40.2 List of individual indicators of the Technology Achievement Index (TAI)

Indicator	Definition	Unit
PATENTS	Number of patents granted to residents, to reflect the current level of invention activities (1998)	Patents granted per 1,000,000 people
ROYALTIES	Receipts of royalty and license fees from abroad per capita, so as to reflect the stock of successful innovations of the past that are still useful and hence have market value (1999)	US $ per 1,000 people
INTERNET	Diffusion of the Internet, which is indispensable to participation in the network age (2000)	Internet hosts per 1,000 people
EXPORTS	Exports of high and medium technology products as a share of total goods exports (1999)	%
TELEPHONES	Number of telephone lines (mainline and cellular), which represents old innovation needed to use newer technologies and is also pervasive input to a multitude of human activities (1999)	Telephone lines per 1,000 people (log)
ELECTRICITY	Electricity consumption, which represents old innovation needed to use newer technologies and is also pervasive input to a multitude of human activities (1998)	kWh per capita (log)
SCHOOLING	Mean years of schooling (age 15 and above), which represents the basic education needed to develop cognitive skills (2000)	Years
UNIVERSITY	Gross enrolment ratio of tertiary students enrolled in science, mathematics and engineering, which reflects the human skills needed to create and absorb innovations (1995–1997)	%

In Table 40.2 is reported the indicators list of the TAI.

Note that the variables have all the same importance in order to represent the technological progress. So, we may assume that the variables of the TAI have the property of non-substitutability, or rather it is very important that there are no compensatory effects among the variables; they have the same weight and a balanced distribution of the values is desirable.

The data matrix is shown in Table 40.3. For the sake of simplicity, only 33 of the 72 original countries measured by the TAI are considered.

The aim of the application is to compare the MPI with some traditional methods in order to test the action of the penalty function. Indicators are normalized by (40.1).

The aggregation functions used are:

- Arithmetic Mean (AM)
- Geometric Mean (GM)
- Median (ME)

The AM is the most commonly used aggregation method: its popularity is due to its transparency and ease of interpretation. Nevertheless, it implies full compensability, whereby low values in some indicators can be compensated by high values of others. A better suited method, if the researcher wants some degree of non-compensability among individual indicators, is the GM, but it can be used only for sets of positive

Table 40.3 Raw data for the individual indicators of the TAI[a]

Country	PATENTS	ROYALTIES	INTERNET	EXPORTS	TELEPHONES	ELECTRICITY	SCHOOLING	UNIVERSITY
Argentina	8.0	0.5	8.7	19.0	2.5	3.3	8.8	12.0
Australia	75.0	18.2	125.9	16.2	16.2	3.9	10.9	25.3
Austria	165.0	14.8	84.2	50.3	3.0	3.8	8.4	13.6
Belgium	72.0	73.9	58.9	47.6	2.9	3.9	9.3	13.6
Bolivia	1.0	0.2	0.3	26.0	2.1	2.6	5.6	7.7
Brazil	2.0	0.8	7.2	32.9	2.4	3.3	4.9	3.4
Canada	31.0	38.6	108.0	48.7	2.9	4.2	11.6	14.2
China	1.0	0.1	0.1	39.0	2.1	2.9	6.4	3.2
Colombia	1.0	0.2	1.9	13.7	2.4	2.9	5.3	5.2
Czech Republic	28.0	4.2	25.0	51.7	2.8	3.7	9.5	8.2
Finland	187.0	125.6	200.2	50.7	3.1	4.2	10.0	27.4
France	205.0	33.6	36.4	58.9	3.0	3.8	7.9	12.6
Germany	235.0	36.8	41.2	64.2	2.9	3.8	10.2	14.4
Hungary	26.0	6.2	21.6	63.5	2.7	3.5	9.1	7.7
Ireland	106.0	110.3	48.6	53.6	3.0	3.7	9.4	12.3
Israel	74.0	43.6	43.2	45.0	3.0	3.7	9.6	11.0
Italy	13.0	9.8	30.4	51.0	3.0	3.7	7.2	13.0
Japan	994.0	64.6	49.0	80.0	3.0	3.9	9.5	10.0
Korea, Rep. of	779.0	9.8	4.8	66.7	3.0	3.7	10.8	23.2
Mexico	1.0	0.4	9.2	66.3	2.3	3.2	7.2	5.0
The Netherlands	189.0	151.2	136.0	50.9	3.0	3.8	9.4	9.5
New Zealand	103.0	13.0	146.7	15.4	2.9	3.9	11.7	13.1
Norway	103.0	20.2	193.6	19.0	3.1	4.4	11.9	11.2
Poland	30.0	0.6	11.4	36.2	2.6	3.4	9.8	6.6

(continued)

Table 40.3 (continued)

Country	PATENTS	ROYALTIES	INTERNET	EXPORTS	TELEPHONES	ELECTRICITY	SCHOOLING	UNIVERSITY
Portugal	6.0	2.7	17.7	40.7	3.0	3.5	5.9	12.0
Romania	71.0	0.2	2.7	25.3	2.4	3.2	9.5	7.2
Singapore	8.0	25.5	72.3	74.9	3.0	3.8	7.1	24.2
Slovakia	24.0	2.7	10.2	48.7	2.7	3.6	9.3	9.5
Slovenia	105.0	4.0	20.3	49.5	2.8	3.7	7.1	10.6
Spain	42.0	8.6	21.0	53.4	2.9	3.6	7.3	15.6
Sweden	271.0	156.6	125.8	59.7	3.1	4.1	11.4	15.3
UK	82.0	134.0	57.4	61.9	3.0	3.7	9.4	14.9
USA	289.0	130.0	179.1	66.2	3.0	4.1	12.0	13.9

[a]Only countries with no missing data.

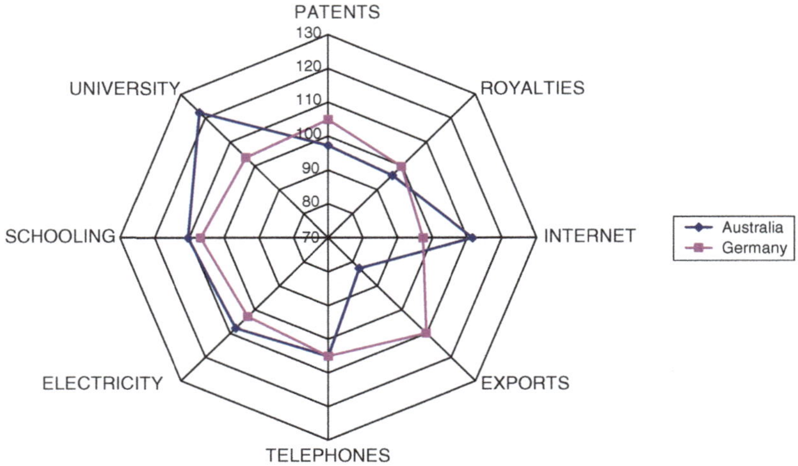

Fig. 40.1 Individual indicators of the TAI: a comparison of countries (normalized values)

values that are interpreted according to their product and not their sum. Moreover, the value of the GM is "biased" low. Both the AM and the GM are affected, in different way, by extreme values (*outliers*). On the contrary, the ME, i.e. the middle value of the variable in the ranking, is less sensitive to outliers.

In Table 40.4, a comparison between MPI with negative penalty (MPI^-) and the above methods is reported.

On the whole, the rankings of the countries (ordered by MPI^-) do not present relevant surprises from a geo-economic point of view, but this is not the aim of the work.

The most similar method to the MPI^- is the GM[4](mean absolute difference of rank of 0.24), the most different one is the ME (mean absolute difference of rank of 1.58). Comparing $MPI^-$and AM rankings, each country changes, on average, less than half position (mean absolute difference of rank of 0.48). The Spearman's rank correlation index is 0.995 so that there is not a great difference between AM and MPI^-.

The divergence between the two methods lies in the penalty function, as the MPI is a mean adjusted on the basis of the "horizontal variability" of standardized indicators. The greater differences between the rankings are in the middle part of the list where, probably, the countries have a higher variability of the indicators.

In Fig. 40.1 is displayed a comparison between the distributions of the normalized individual indicators of Australia and Germany.

[4]Note, however, that the GM is very different to MPI^+.

Table 40.4 TAI country rankings by different aggregation methods

	AM		GM		ME		MPI		Difference of rank		
Country	Value	Rank	Value	Rank	Value	Rank	Value	Rank	MPI-AM	MPI-GM	MPI-ME
Finland	112.7	1	112.3	1	111.5	1	112.0	1	0	0	0
USA	111.8	2	111.7	2	111.0	2	111.5	2	0	0	0
Sweden	111.4	3	111.2	3	111.0	3	111.1	3	0	0	0
Japan	109.5	4	108.8	4	105.5	6	107.8	4	0	0	2
Korea, Rep. of	107.8	5	107.1	5	107.9	4	106.4	5	0	0	1
The Netherlands	106.2	6	106.0	6	103.0	12	105.7	6	0	0	6
UK	105.3	8	105.2	7	103.5	11	105.0	7	−1	0	4
Canada	105.1	9	105.0	9	104.1	9	104.8	8	−1	1	1
Norway	105.8	7	105.0	8	104.9	7	104.3	9	2	1	2
Germany	103.7	12	103.7	10	104.3	8	103.6	10	−2	0	2
Singapore	104.0	11	103.5	11	103.7	10	103.1	11	0	0	1
Ireland	103.2	13	103.1	13	101.8	17	102.9	12	−1	1	5
Australia	104.1	10	103.5	12	106.4	5	102.9	13	3	1	8
Belgium	102.4	14	102.3	14	102.2	13	102.3	14	0	0	1
Austria	101.7	16	101.7	15	102.1	14	101.6	15	−1	0	1
France	101.4	17	101.4	17	102.0	15	101.3	16	−1	1	1
New Zealand	102.0	15	101.5	16	101.8	16	101.0	17	2	1	1
Israel	100.6	18	100.5	18	100.1	18	100.5	18	0	0	0
Spain	98.3	19	98.2	19	97.6	21	98.1	19	0	0	2
Italy	98.3	20	98.2	20	97.8	20	98.0	20	0	0	0
Czech Republic	97.6	21	97.6	21	96.8	22	97.5	21	0	0	1
Slovenia	97.3	22	97.2	22	97.9	19	97.1	22	0	0	3
Hungary	97.3	23	97.1	23	95.1	25	97.0	23	0	0	2

(continued)

Table 40.4 (continued)

	AM		GM		ME		MPI		Difference of rank		
Country	Value	Rank	Value	Rank	Value	Rank	Value	Rank	MPI-AM	MPI-GM	MPI-ME
Slovakia	96.6	24	96.6	24	95.7	23	96.5	24	0	0	1
Portugal	95.4	25	95.3	25	95.3	24	95.1	25	0	0	1
Poland	94.3	26	94.2	26	93.0	26	94.1	26	0	0	0
Argentina	92.8	27	92.7	27	92.1	27	92.6	27	0	0	0
Romania	92.0	29	91.9	29	91.0	29	91.7	28	−1	1	1
Mexico	92.3	28	92.0	28	91.6	28	91.6	29	1	1	1
Brazil	88.7	30	88.6	30	90.6	30	88.5	30	0	0	0
China	87.4	31	87.1	31	88.8	32	86.9	31	0	0	1
Colombia	86.8	32	86.7	32	86.7	33	86.6	32	0	0	1
Bolivia	86.0	33	85.6	33	89.3	31	85.3	33	0	0	2
Mean absolute difference of rank									**0.48**	**0.24**	**1.58**
Spearman's rank correlation index									**0.995**	**0.999**	**0.969**

Germany has a balanced distribution of the values (all, except INTERNET, fall approximately in the range 100–110), whereas Australia has an unbalanced distribution (UNIVERSITY has a very high value and EXPORT a low value).

Germany ranks 12th according to the AM, 10th according to the GM and MPI^{-} and 8th according to the ME.

On the other hand, Australia ranks 10th with the AM and 5th with the ME, while its ranking is much lower according to the other two methods (GM and MPI^{-}). In this case, the penalty function of the MPI penalizes the country causing a decrement of 3 positions respect to the AM, 1 position respect to the GM and a good 8 positions compared with the ME.

40.5 Concluding Remarks

Social and economic phenomena, such as development, poverty, wellness and malnourishment, are very difficult to measure and evaluate since they are characterized by a multiplicity of aspects or dimensions. The complex and multidimensional nature of these phenomena requires the definition of intermediate objectives whose achievement can be observed and measured by individual indicators. The idea of summarizing individual indicators into composite indices is not straightforward. It involves both theoretical and methodological assumptions which need to be assessed carefully to avoid producing results of dubious analytic rigour. The international literature on composite indices offers a wide variety of aggregation methods. The most used are additive methods, but they imply requirements and properties which are often not desirable or difficult to meet, such as a full substitutability among the components of the index.

To overcome this assumption, we proposed an alternative composite index, denoted as MPI (Mazziotta-Pareto Index). The MPI is independent from both the range and the "polarity" of the individual indicators and can be validly applied to different scientific contexts.

Acknowledgements This chapter is the result of the common work of the authors: in particular M. Mazziotta has written Sects. 40.1, 40.2, 40.3.2 and A. Pareto has written Sects. 40.3.1, 40.4 and 40.5.

References

1. Aureli Cutillo, E.: Lezioni di statistica sociale. Parte seconda, sintesi e graduatorie. CISU, Roma (1996)
2. Delvecchio, F.: Scale di misura e indicatori sociali. Cacucci Editore, Bari (1995)
3. Dunteman, G.H.: Principal Components Analysis. Sage Publications, Newbury Park (1989)
4. Harbison, F.H., Maruhnic, J., Resnick, J.R.: Quantitative Analyses of Modernization and Development. Princeton University Press, New Jersey (1970)
5. Mazziotta, M., Pareto, A.: Un indicatore sintetico di dotazione infrastrutturale: il metodo delle penalità per coefficiente di variazione. In: Lo sviluppo regionale nell'Unione Europea – Obiettivi, strategie, politiche. Atti della XXVIII Conferenza Italiana di Scienze Regionali. AISRe, Bolzano (2007)

6. Mazziotta, C., Mazziotta, M., Pareto, A., Vidoli, F.: La sintesi di indicatori territoriali di dotazione infrastrutturale: metodi di costruzione e procedure di ponderazione a confronto. Rev. Econ. Stat. Reg. Stud. 1 (2010)
7. Munda, G., Nardo, M.: Non-Compensatory Composite Indicators for Ranking Countries: A Defensible Setting. European Commission-JRC, Ispra (2005)
8. Palazzi, P.: Lo sviluppo come fenomeno multidimensionale. Confronto tral'ISU e un indice di sviluppo sostenibile. Moneta e Credito 227 (2004)
9. OECD: Handbook on Constructing Composite Indicators. Methodology and user guide. OECD Publications, Paris (2008)
10. United Nations: Human Development Report. Oxford University Press, New York (2001)

Evaluating the Efficiency of the Italian University Educational Processes through Frontier Production Methods

41

Luigi Biggeri, Tiziana Laureti, and Luca Secondi

Abstract

The university efficiency indicators, computed as partial indices following the input–output approach, are subject to criticism mainly concerning the *problem of comparability* of institutions due to the influence of several environmental factors. The aim of this study is to construct *comparable efficiency indicators* for the university educational processes by using frontier production methods which enable us to solve the problem. The indicators are computed for the Italian State Universities by using a new data set referring to the cohort of students enrolled in the academic year 2004/2005. A comparison with the corresponding partial indicators usually employed is presented.

Keywords

Educational process • Stochastic Frontier Models • University Efficiency Indicators

41.1 Introduction

University performance indicators (UPIs) are widely used in several OECD countries (including Italy) due to their key role concerning the distribution of public funds (core fund or targeted fund) for teaching and research among universities. Nevertheless, there has been a great deal of debate over the validity and reliability of the different partial indicators used, each measuring a different aspect of a

L. Biggeri
University of Florence, Florence, Italy
e-mail: biggeri@disia.unifi.it

T. Laureti (✉) · L. Secondi
University of Tuscia, Viterbo, Italy
e-mail: laureti@unitus.it; secondi@unitus.it

N. Torelli et al. (eds.), *Advances in Theoretical and Applied Statistics*, Studies in Theoretical and Applied Statistics, DOI 10.1007/978-3-642-35588-2_41,

university's performance [2]. The main criticism regarding UPIs concerns the problem of comparability of universities and faculties. This problem is particularly significant when partial efficiency indicators (PEIs) of the educational processes are considered since they refer to the production process which can be influenced by exogenous factors that characterize the context in which the universities operate.

The aim of this paper is threefold. Firstly, to suggest a new approach based on frontier production methods for constructing comparable efficiency indicators of the educational processes in order to avoid misleading comparisons of university performance. Secondly, to estimate comparable efficiency indicators of the educational processes of the Italian State Universities, referring to the entire production process of a graduate and the various sub-processes in which the formation of a graduate can be divided. For the first time in Italy, the estimation is carried out by using data from the Italian University Student Archive considering a specific cohort of students. The third aim is to analyse the differences between the estimated comparable technical efficiency indicators and the partial indicators used by the Ministry of Education, University and Research (MIUR) in order to verify if the use of the partial indicators for the allocation of funds could be biased.

41.2 From Partial Efficiency Indicators to Comparable Technical Efficiency Indicators

The usual practice for measuring the efficiency of the university educational process has been to find a set of PEIs (i.e. drop-out rate, percentage of credits achieved and graduation rate) constructed by referring to the input–output approach [3]. Although these indicators are useful, because they are simple to construct and easily understood, they do not provide a reliable base for comparing institutions with different features. Therefore PEIs must be adjusted to "contextual" factors, i.e. university differences in "quality" of students and course-related characteristics, in order to potentially avoid misleading comparisons of university performance.

In order to construct comparable technical efficiency indicators (CTEIs), which consider the various characteristics of universities, we need to refer to the production theory framework. Therefore, bearing in mind the requirements of the neoclassical production theory we can model the university educational process of the first level degree courses as a series of production processes: the entire educational process where freshmen are transformed into graduates and the three sub-processes represented by the results obtained by the freshmen at the end of each year of the degree course. These different processes can be regarded as processes of human capital formation. The university, through the training it delivers, transforms a cultural "raw" material (input) into a cultural "refined" material (output) by using a number of other inputs, such as teachers, textbooks and class rooms.

With the aim of measuring the technical efficiency of these processes the stochastic frontier approach [8, 9] is used to allow for the various "environmental" factors which can affect the results of the educational processes, such as the students' characteristics [1], the resources used and the socio-economic context.

In this way we can obtain the efficiency measures that represent CTEIs which enable us to make comparisons among the universities at *ceteris paribus* conditions.

We propose a heteroscedastic stochastic frontier model [5] for cross-sectional data as a general model to be used for describing the various processes. This model can be written in natural logs as:

$$\ln y_i = \ln f(\mathbf{x}_i; \boldsymbol{\beta}) + v_i - u_i \qquad (41.1)$$

where y_i is the output of production unit $i\,(i = 1, \ldots, n)$, $\mathbf{x}_i$ is a vector of s inputs, $f(\mathbf{x}_i; \boldsymbol{\beta})$ is the production function and $\boldsymbol{\beta}$ a vector of technological parameters to be estimated; the v_i components, which capture the effect of statistical noises, are assumed to be normal with zero mean and constant variance $v_i \sim i.i.d N(0, \sigma_v^2)$. Concerning the one-sided error terms u_i, which express technical inefficiency of each production unit, the scaled half-normal model was selected since it satisfies the scaling property that has several statistical advantages and attractive features.

The model specification was carried out by referring to literature on this subject and considering the results obtained from preliminary analyses. First of all, the production unit is represented by the faculty[1] since teaching activities are considered the most important aspect of human capital formation. The production technology is specified by the Cobb–Douglas production function as it is suitable for describing the attitude of the faculties in carrying out the formation process as well as estimating a parsimonious number of parameters.

In the university formation process framework the issue of distinguishing between heterogeneity in the production model and heterogeneity in the inefficiency model is particularly important. The problem of heterogeneity in the production model is due to different technologies, which may simultaneously coexist as the various faculties carry out their production in different environments. In order to deal with this problem and allow for a statistical test of the assumption of a common technology, we included the type of faculty in the production function [4]. To overcome the issue of heterogeneity in the inefficiency model, caused by the students' characteristics and other contextual factors, we assumed that the variance of the inefficiency terms is a function of a vector $\mathbf{z}_i$ of variables which are exogenous to the production process but influence its efficiency, such as the students' individual characteristics and the size and territorial location of the universities, that is $\sigma_{ui}^2 = \exp(\boldsymbol{\delta}'\mathbf{z}_i)$. Since the study subject can influence both the technology of production and the inefficiency level, we also included the degree subject, which represents the type of faculty according to the NUEC classification, in the model of technical inefficiency [4, 10].

[1]The faculty is the entity of a university that organizes teaching activities in a specific area (study subject). We are aware that there are various degree courses with different educational objectives within a faculty but it was not possible to consider them in our study due to lack of disaggregated data.

The estimates of the unknown parameters in (41.1) can be obtained by maximizing the generalized log likelihood function [7]. The individual technical efficiency (TE) of each faculty, expressed by $TE_i = \exp\{-\hat{E}(u_i|\varepsilon_i)\}$ is of particular interest in our study because it expresses a CTEI. An unbiased estimate of technical inefficiency $\hat{E}(u_i|\varepsilon_i)$ is obtained by adjusting for heteroscedasticity the formula of the conditional mean suggested by Jondrow et al. [6] when u is homoscedastic.

41.3 Data and Model Specification

We constructed a specific database by using data from the newly established Italian University Student Archive referring to the cohort of students enrolled in first level degree courses in the academic year 2004/2005 and following their progress up to the academic year 2007/2008. We also included data concerning the university facilities and equipment gathered yearly by the National University Evaluation Committee (NUEC) and the Ministry of University and Research (MIUR). The final dataset refers to the 60 Italian State Universities[2] and to the 452 faculties. The database contains a lot of valuable information, but most variables are only observed at faculty level while other variables are observed at university level.

Bearing in mind the aim of our paper we considered the three sub-processes mentioned above for which we specified three models by identifying the following outputs as semi-manufactured products that indicate the students' progress in their human capital formation:

1. Full Credit Equivalent Students in the first year (FCES1) who are the theoretical number of students (which is less than or equal to the number of freshmen) who would have obtained the 60 credits required.
2. Full Credit Equivalent Students at the end of the second year (FCES2) who are the theoretical number of students who would have obtained the 120 credits required.
3. Full Credit Equivalent Students at the end of the third year (FCES3), who are the theoretical number of students who would have obtained the 180 credits required.

We analysed the efficiency of the entire educational process by considering two different models identifying two different outputs as finished products:

1. Number of Graduates within institutional time (GT), equal to three years.
2. Number of Graduates who got their degree within a year after the institutional time[3] (G1L).

[2]After examining the reliability and quality of the data, two universities were excluded from the analysis.

[3]Although the number of graduates within the institutional time is the real output of the process, the indicator G1L can be considered as a second best result of the entire formation process bearing in mind the fact that some students take longer than three years to obtain their degrees.

We considered the following variables as inputs in the educational processes: (1) Freshmen who enrolled in the academic year 2004/2005; (2) Full and/or Associate Professors and/or researchers; (3) Seats in lecture halls; (4) Seats in laboratories; (5) Seats in libraries; (6) Books in libraries; (7) Subscriptions to journals in libraries; (8) Weekly opening hours of libraries. Among the environmental factors we specified the following characteristics of freshmen: (1) Percentage of female freshman; (2) Percentage of freshmen with best results of secondary school leaving examination (90–100/100); (3) Percentage of freshmen with Lyceum Diploma; (4) Percentage of freshmen aged over 25; (5) Percentage of freshmen from other areas (outside the University region). Moreover, we included variables on size (small, medium and big) and territorial location (North, Centre, South) of the universities, the Gross Domestic Product (GDP) of the province in which the university is located and the survival rate of the students enrolled in the academic year 2004/2005 in the model up to the second and third year.

It is interesting to note that a descriptive analysis of the PEIs, concerning credits achieved and graduation rates, shows that on average the Italian university educational process is characterized by a low degree of efficiency, with great variability among the faculties. For example, the average rate of credits achieved in the first year is 0.43, with a standard deviation of 0.137, and ranges from a minimum of 0.28 (Veterinary medicine) to a maximum of 0.60 (Medicine) while the average graduation rate within the institutional time is equal to 0.16, with a standard deviation of 0.146, and ranges from 0.06 (Law) to 0.46 (Medicine).

41.4 Estimation Results

Although we estimated five different models concerning the various educational processes (FCES1, FCES2, FCES3, GT, G1L), testing different specifications with different sets of variables, hereunder there are the results regarding the sub-process specified by the first year of the degree course (Tables 41.1 and 41.4) and the estimation results of the entire process[4] considering the number of graduates within institutional time (Tables 41.2 and 41.3) as the output.

Concerning the efficiency analysis of the first year of the degree course, which represents the most interesting sub-process to be evaluated due to the high drop-out rate observed (about 30% on average for the 2004/2005 cohort), a likelihood ratio test showed that there was no significant statistical heterogeneity in the technology of production (Table 41.1). On the contrary the hypothesis that faculties share the same technology was strongly rejected by the data in the case of the stochastic production model describing the entire formation process (Table 41.2).

Therefore in the first year of the degree course, the evaluation of technical efficiency can be performed by considering a common technology since the process

[4]The results of the models FCES2, FCES3 and G1L are available from the authors on request.

Table 41.1 FCES1 model: production function parameter estimates

Variable	Coefficient
Undergraduate enrolments	0.973 (0.023)***
Professors	0.057 (0.022)***
Seats in lecture hall	0.085 (0.003)***
Seats in laboratories	0.018 (0.026) **
Seats in libraries	0.019 (0.032)
Journal subscriptions	−0.019 (0.015)
Labs – office hours	0.182 (0.088)
Constant	−0.992 (0.403)***
Two-sided error variance	
$\text{Ln}\sigma_v^2$	−3.373 (0.217)***

Note: *Significant at the 10% level **Significant at the 5% level ***Significant at the 1% level

Table 41.2 GT model: production function parameter estimates

Variable	Coefficient
Undergraduate enrolments	0.797 (0.054)***
Professors	0.167 (0.056)***
Seats in lecture hall	0.180 (0.023)**
Seats in laboratories	0.078 (0.059)
Seats in libraries	0.019 (0.063)
Journal subscriptions	0.021 (0.026)
Labs – office hours	0.091 (0.186)
Degree subject (Reference group = Law)	
Agricultural sciences	−0.272 (0.224)
Architecture	0.564 (0.207)***
Economics	0.259 (0.161)
Pharmacology	−0.366 (0.220)*
Engineering	−0.155 (0.175)
Medicine	0.619 (0.194)***
Literature	0.108 (0.166)
Languages	0.373 (0.205)*
Psychology	0.437 (0.034)**
Veterinary medicine	−0.931 (0.260)***
Education	0.161 (0.200)
Biology and Mathematics	−0.285 (0.185)
Physical science and education	0.185 (0.283)
Political sciences	0.346 (0.175)**
Constant	−0.977 (0.875)
Two-sided error variance	
$\text{Ln}\sigma_v^2$	−3.317 (0.446)***

Note: *Significant at the 10% level **Significant at the 5% level ***Significant at the 1% level

Table 41.3 GT model: variance parameter estimates and marginal effects

Variable	Coefficient	Marginal effect on $E(u_i)$	Marginal effect On $Var(u_i)$
Degree subject			
(Ref. group = Law)			
Agricultural sciences	−0.777 (0.530)	−0.384	−0.567
Architecture	−0.608 (0.449)	−0.301	−0.444
Economics	−1.877 (0.416)***	−0.928	−1.371
Pharmacology	−1.494 (0.476)***	−0.738	−1.090
Engineering	−0.990 (0.570)*	−0.495	−0.723
Medicine	−3.806 (0.603)***	−1.877	−2.779
Literature	−1.762 (0.379)***	−0.871	−1.286
Languages	−1.625 (0.611)***	−0.802	−1.187
Psychology	−2.776 (0.750)***	−1.372	−2.027
Veterinary medicine	−1.727 (0.616)***	−0.854	1.261
Education	−1.843 (0.597)***	−0.911	−1.346
Biology and Mathematics	−1.588 (0.392)***	−0.785	−1.160
Physical science and education	−2.122 (0.723)***	−1.049	−1.549
Political sciences	−2.185 (0.469)***	−1.080	−1.595
Students' characteristics			
Percentage of female	0.805 (0.884)	0.398	0.587
Best freshmen	−4.413 (1.298)***	−2.181	−3.222
Freshmen with Lyceum Diploma	−1.122 (1.061)	−0.555	−0.819
Freshmen aged over 25	−0.518 (1.020)	−0.256	−0.378
Freshmen from other areas (outside of University region)	−0.679 (0.530)	−0.335	−0.496
University Dimension			
(Ref. group: Big University)			
Medium University	−0.463 (0.190)**	−0.229	−0.338
Small University	−0.663 (0.268)**	−0.328	−0.484
University Localization			
(Reference Group: North)			
Centre	1.033 (0.234)***	0.511	0.754
South	0.541 (0.400)	0.267	0.395
GDP per capita	−0.072 (0.28)**	−0.036	−0.053
Constant	4.794 (1.101)***		

Note: *Significant at the 10% level **Significant at the 5% level ***Significant at the 1% level

of achievement of credits in the first year may not greatly differ among the various groups of faculties.

The production function parameter estimates (Tables 41.1 and 41.2) show that the statistically significant inputs are the number of undergraduate enrolments, the number of seats in lecture halls, the number of professors and the number of seats

Table 41.4 FCES1 model: variance parameter estimates and marginal effects

Variable	Coefficient	Marginal effect on $E(u_i)$	Marginal effect on $Var(u_i)$
Degree Subject			
(Reference group = Law)			
Agricultural sciences	0.316 (0.540)	0.063	0.031
Architecture	−1.838 (0.514)	−0.369	−0.180
Economics	−0.463 (0.409)	−0.093	−0.045
Pharmacology	0.838 (0.470)**	0.168	0.082
Engineering	−1.327 (0.585)**	−0.266	−0.130
Medicine	−1.556 (0.538)***	−0.311	−0.152
Literature	0.253 (0.359)	0.052	0.025
Languages	−0.659 (0.597)	−0.133	−0.064
Psychology	0.319 (0.660)	0.064	0.031
Veterinary medicine	1.723 (0.534)***	0.344	0.168
Education	−0.120 (0.584)	−0.024	−0.012
Biology and Mathematics	0.610 (0.394)*	0.122	0.060
Physical science and education	−0.428 (0.698)	−0.086	−0.042
Political sciences	−0.735 (0.431)	−0.147	−0.072
Students' characteristics			
Percentage of Female	−2.853 (0.978)***	−0.572	−0.279
Best freshmen	0.355 (0.949)	0.071	0.035
Freshmen with Lyceum Diploma	−2.809 (1.043)***	−0.564	−0.275
Freshmen aged over 25	1.480 (1.063)	0.297	0.145
Freshmen from other areas (outside of University region)	−0.914 (0.587)	−0.183	−0.089
University Dimension			
(Ref. group: Big University)			
Medium University	−0.468 (0.209)**	−0.094	−0.046
Small University	−0.144 (0.305)	−0.029	−0.014
University Localization			
(Reference Group: North)			
Centre	0.087 (0.239)	0.017	0.008
South	−0.521 (0.452)	−0.105	−0.051
GDP per capita	−0.052 (0.031)*	−0.010	−0.005
Constant	2.702 (1.246)**		

Note: *Significant at the 10% level **Significant at the 5% level ***Significant at the 1% level

in laboratories (the latter only for the FCES1 sub-process). The variance parameter of the two-sided error term is statistically significant.

Considering the entire formation process, the coefficients related to the various faculties suggest that there is also significant variation in the levels of inefficiency across the faculties (Table 41.3). The degree subject and the students' characteristics play an important role in shaping technical efficiency. For example the negative significant coefficient of "Best Freshmen" makes the variance of the best freshmen lower than the variance of the other freshmen, *ceteris paribus*, thus showing a lower level of inefficiency of best freshmen.

It is interesting to note that the percentage of females and the percentage of students from Lyceum significantly influence the results obtained by the freshmen at the end of the first year of their degree courses (Table 41.4). Therefore faculties including a high percentage of students with these characteristics achieve a higher level of efficiency. Moreover, the size of the university and the economic indicators of the province are also statistically significant in both models.

In reference to the other formation processes, it is worth noting that in the FCES2 and FCES3 models the statistically significant inputs are identical to those found for FCES1 and GL models, respectively. The FCES3 model shows the importance of the survival rate of students after the first year course among the exogenous variables in the inefficient component u. Finally the results obtained from the G1L model are very similar to the GT model.

41.5 Comparable Efficiency Indicators *Versus* Usual Indicators

The most important result of our study is the estimation of the CTEIs obtained from the five models which take into account the different contexts characterizing each faculty. It is interesting to compare the ranking of faculties obtained by using CTEIs with the ranking based on the usual PEIs, which are obtained by dividing each of the five outputs by the number of freshmen enrolled in the academic year 2004/2005. This comparison enables us to verify if a faculty modifies its position in the classification according to the two efficiency indicators (Fig. 41.1). The comparison regarding the FCSE3 output is not reported in Fig. 41.1 since the scatter plot representation is similar to that of the graduation process.

Plots in Fig. 41.1 show the sensitivity of the rankings to the change of the efficiency measurement method. Each point in the four plots represents the position of a faculty according to the PEI and the CTEI indicators. Clearly, if the positions of the faculties in the two classifications had remained the same, the dots would be situated on the bisecting line. There is major stability in the ranking in the higher part of the classification situated at the bottom of the graphs and in the lower part situated in the upper right-hand corner. Thus, the most efficient and less efficient faculties tend to be stable regardless of the efficiency measurement method used. Yet there is much more dispersion in the central part of the distribution and we can note that, especially in case of GL1, there are sharp differences between the adjusted and the unadjusted UPIs and thus the use of the unadjusted indicators (PEIs) could result in misleading comparisons.

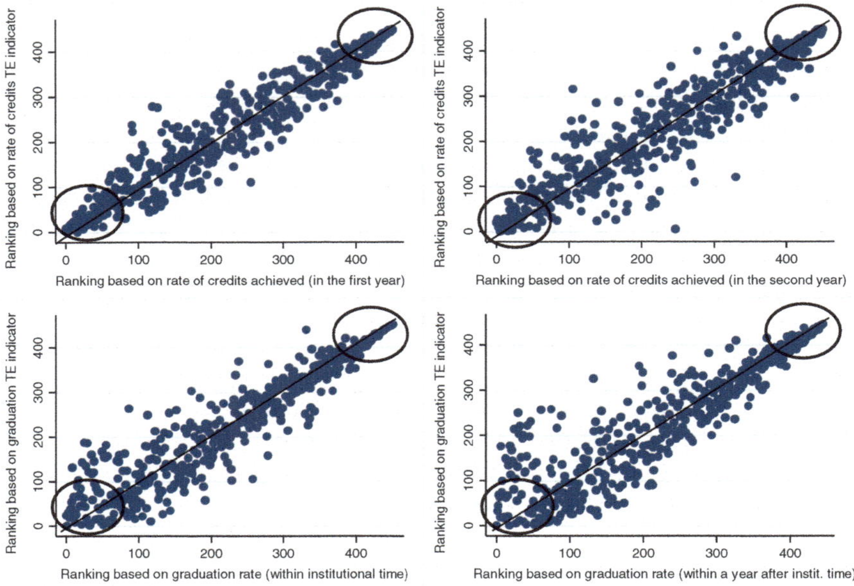

Fig. 41.1 Comparison of faculty rankings by usual PEIs vs. CTEIs indicators

41.6 Concluding Remarks

The models estimated in this study proved to be suitable for describing the educational production process in Italian State Universities. Moreover, the results obtained enable us to underline some important issues concerning the use of UPIs constructed as PEIs for the evaluation of the universities. CTEIs are more accurate for describing the level of efficiency among the various faculties in the educational production process since they take into consideration several contextual factors such as the available resources and the students' characteristics. The CTEIs are adjusted efficiency indicators which could be used by MIUR for allocating funds among universities.

References

1. Boero, G., Broccolini, C., Laureti, T., Naylor, R.: Velocità di progressione negli studi universitari: un confronto tra coorti pree post-riforma in *Performance accademica e tassi di abbandono: una analisi dei primi effetti della riforma universitaria*, a cura di Boero G. e Staffolani, S., Edizioni Cuec (2007)
2. European University Association (EUA): *A Reference System for Indicators and Evaluation Procedures*, by Tavenas T., Brussels (2004)

3. Goldstein, H., Spiegelhalter, D.J.: League tables and their limitations: statistical issues in comparisons of institutional performance (with discussion). J. R. Stat. Soc. A **159**, 385–443 (1996)
4. Greene, W.: The econometric approach to efficiency analysis. In: Fried, O.H., Lovell, C.A.K., Schimdt, S. (eds.) The Measurement of Productive Efficiency and Productivity Growth. Oxford University Press, Oxford (2008).
5. Hadri, K.: Estimation of a doubly heteroscedastic frontier cost function. J. Bus. Econ. Stat. **17**(3), 359–363 (1999)
6. Jondrow, J., Lovell, C.A.K., Materov, I.S., Schimdt, P.: On the estimation of technical inefficiency in the stochastic frontier production function model. J. Econometrics **19**, 233–238 (1982)
7. Kumbhakar, S.C., Lovell, C.A.K.: Stochastic Frontier Analysis, Cambridge Books. Cambridge University Press, Cambridge (2000)
8. Laureti, T.: Modelling exogenous variables in human capital formation through a heteroscedastic stochastic frontier. Int. Adv. Econ. Res. **14** (1), 76–89(2008)
9. Laureti, T., Secondi, L.: Constructing university performance indicators in Italy: a comparative approach. In: Proceedings of the Business and Economic Statistics Section, JSM, 2009, pp. 4542–4554
10. Stevens, P.A.: The Determinants of Economic Efficiency in English and Welsh Universities, National Institute of Economic and Social Research, Discussion Paper N.185 (2001)

Modeling and Forecasting Realized Range Volatility

42

Massimiliano Caporin and Gabriel G. Velo

Abstract

In this chapter, we estimate, model, and forecast Realized Range Volatility, a realized measure and estimator of the quadratic variation of financial prices. This quantity was early introduced in the literature and it is based on the high-low range observed at high frequency during the day. We consider the impact of the microstructure noise in high frequency data and correct our estimations, following a known procedure. Then, we model the Realized Range accounting for the well-known stylized effects present in financial data. We consider an HAR model with asymmetric effects with respect to the volatility and the return, and GARCH and GJR-GARCH specifications for the variance equation. Moreover, we consider a non-Gaussian distribution for the innovations. The analysis of the forecast performance during the different periods suggests that the introduction of asymmetric effects with respect to the returns and the volatility in the HAR model results in a significant improvement in the point forecasting accuracy.

Keywords

Econometrics • Forecasting methods • Long-memory • Realized range volatility • Realized volatility • Statistical analysis of financial data • Time series analysis • Volatility forecasting

JEL codes: C22, C52, C53, C58

M. Caporin · G.G. Velo (✉)
Department of Economics and Management "Marco Fanno", University of Padua, Italy
e-mail: massimiliano.caporin@unipd.it; gabriel.velo@unipd.it

N. Torelli et al. (eds.), *Advances in Theoretical and Applied Statistics*, Studies in Theoretical and Applied Statistics, DOI 10.1007/978-3-642-35588-2_42,

42.1 Introduction

In the last years, realized volatility measures, constructed from high frequency financial data and modeled with standard time series techniques, have shown to perform much better than traditional generalized autoregressive conditional heteroskedasticity (GARCH) and stochastic volatility models, when forecasting conditional second order moments. Most of the works that forecast volatility through realized measures, have concentrated on the *Realized Variance* (*RV*) introduced by Andersen et al. [1] and Barndorff-Nielsen and Shephard [4]. The *RV* is based on the continuous time price theory and it is defined as a function of the sum of squared intraday returns. The *RV* is a highly efficient and unbiased estimator of the quadratic variation and converges to it when the intraday measurement period goes to zero. Later on, Martens and van Dijk [13] and Christensen and Podolskij [6] introduced the *Realized Range Volatility* (*RRV*), another realized estimator consistent for the quadratic variation. The *RRV* is based on the difference between the minimum and maximum prices observed during a certain time interval. This new estimator tries to exploit the higher efficiency of the range relatively to that of the squared daily close-to-close return in the estimation of quadratic variation.

When dealing with high frequency financial market data, the asymptotic properties of the simple estimators are highly affected by the microstructure noise (noncontinuous trading, infrequent trade, bid ask bounce). As a result, an important part of the literature has presented different corrections to restore the efficiency of realized estimators for the volatility. These studies aimed at improving over the first generation of models, whose purpose was to construct estimates of realized variances by using series at a moderate frequency (see [2]). Some of the corrections presented to the *RV* are the Two Time Scale Estimator (TTSE), the subsampling method of [18] and the generalization introduced by Zhang [17]. Furthermore, kernel estimation was introduced by Hansen and Lunde [11], while [3] provide a generalization of this approach. Differently, Martens and van Dijk [13] proposed a correction for the *RRV* based on scaling the range with the daily range, while [7] presented another approach based on an adjustment by a constant which has to be estimated by simulation methods.

With the availability of new observable series for the volatility, many authors have applied traditional discrete time series models for their forecast (and implicitly for the forecast of returns volatility). Financial data are characterized by a series of well-known stylized facts. Being able to capture them will result in a more accurate prevision of our variable of interest. These stylized facts are also observable over realized variance series and require appropriate modeling strategies. The presence of long memory in volatility, documented in several studies, has been modeled through different specifications: Andersen et al. [2] introduced an ARFIMA model, and their forecasts for the RV generally dominate those obtained through GARCH models; Corsi [8] presented the Heterogenous autoregressive (HAR) model that reproduces the hyperbolic decay of the autocorrelation function by including the sums of *RV* over different horizons in order to capture the time strategies of the agents in the market. The second model has the advantage to be much simpler to estimate. Additionally, asymmetry, leverage effects, and fat tails should also be taken into account. Martens et al. [14] specified a flexible unrestricted high-order AR model. They also considered leverage effects, days of the week effects, and macroeconomic announcements. Differently, Corsi [8] presented the Heterogeneous Autoregressive model (HAR) and [9] introduced two important extensions specifying a GARCH component modeling the volatility of volatility and assuming non-Gaussian errors. Their results suggested an improvement in the accuracy in the point forecasting and a better density forecast.

In this work, we model and forecast volatility through the *Realized Range Volatility*. Our main objective is to study the prediction performance of the range as a proxy of the volatility. An accurate forecast of financial variability should have important implications in asset and derivative pricing, asset allocation, and risk management. In the first part of this chapter, we construct and analyze the realized range series, correct it from the microstructure noise following [13]. In the second part, we implement time series techniques to model and capture the stylized facts within the volatility equation to gain in forecasting accuracy. In details, we consider an HAR model, we introduce leverage effects with respect to the return and the volatility, and a GARCH and a GJR-GARCH specification, introduced by Bollerslev [5] and Glosten et al. [10], for the volatility of volatility. Furthermore, in order to capture the statistical feature of the residuals of our model, we also consider a Normal Inverse Gaussian (NIG) distribution.

The remainder of this chapter is structured as follows. In Sect. 42.2, we present the data and the correction procedure. In Sect. 42.3, we present the model and we discuss the results for the estimation and forecast in Sect. 42.4. Finally, Sect. 42.5 presents the results and futures steps.

42.2 Data and Correction Procedure

Under the assumption that there are no market frictions and there is continuous trading, Christensen and Podolskij [6] demonstrate that the *RRV* is five times more efficient than *RV*. In the reality, there are evidences against these assumptions and the realized estimators become inconsistent and unbiased, see for example [15]. Hence, a corrected version for the *RRV* should restore the efficiency of this estimator over the *RV*. In this chapter, we follow [13] that proposed a correction based on scaling the realized range with a ratio involving the daily range and the realized range over the previous trading days.

Basically, the *scaling* bias correction is not difficult to implement because it does not require the availability of tick by tick data. The idea of [13] is based on the fact that the *daily range* is almost not contaminated by market frictions. The simulation results of [13] confirm the theory that the range is more efficient than the *RV*, and in the presence of market frictions the scaling correction removes the bias and restores

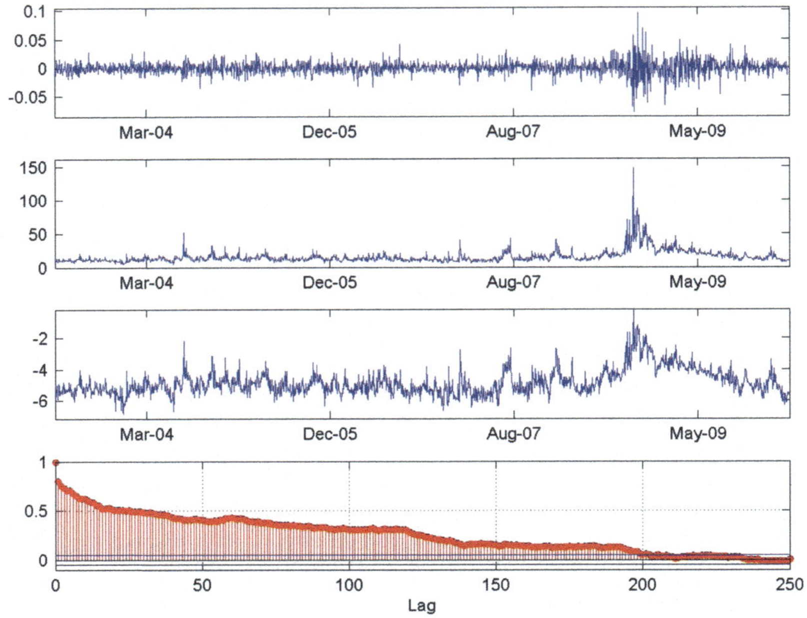

Fig. 42.1 Returns, annualized $RRV^{\Delta}_{\text{scaled}}$, log $RRV^{\Delta}_{\text{scaled}}$ and ACF for log $RRV^{\Delta}_{\text{scaled}}$

the efficiency of the Realized Range estimator over the Realized Volatility. The RRV^{Δ}_{t} is defined as

$$RRV^{\Delta}_{t} = \frac{1}{\lambda^2}\sum_{i=1}^{n}(\ln\ p^{\text{hg}}_{t,i} - \ln\ p^{\text{lo}}_{t,i})^2 \tag{42.1}$$

where $p^{\text{hg}}_{t,i}$ and $p^{\text{lo}}_{t,i}$ are the high and low prices observed in the ith interval of length Δ of an equidistant partition of day t, and λ is the scaling factor. Therefore, the *scaled* RRV^{Δ}_{t} is defined as:

$$RRV^{\Delta}_{\text{scaled},t} = \left(\frac{\sum_{l=1}^{q} RRV_{t-l}}{\sum_{l=1}^{q} RRV^{\Delta}_{t-l}}\right) RRV^{\Delta}_{t} \tag{42.2}$$

where $RRV_t \equiv RRV^{\Delta}_{t}$ (with $\Delta = 1$ day) is the daily range and q is the number of previous trading days used to compute the scaling factor. If the trading intensity and the spread do not change, q must be set as large as possible. However, in the reality only recent history should be taken into consideration.

Our database includes more than 7 years of 1 min high, low, open, and close prices for 16 stocks quoted on the NYSE. Because of space limitation, we concentrate on the analysis and present the detailed results for Procter and Gamble

Company. However, similar conclusions emerge from the other series. The original sample covers the period from January 2, 2003 to March 30, 2010, from 09:30 trough 16:00 for a total of 1,823 trading days. We construct the series for the range for the 1, 5, 30 min and daily sampling frequency (i.e., $\Delta = 1, 5, 30$ min or 1 day). We correct them using one, two, and three previous months (i.e., $q = 22, 44$, or 66). The results of the corrections show that, after scaling with different q, the mean volatility stabilized across the different sampling frequencies. We choose to sample every 5 min and to correct for the presence of noise with the 66 previous days, as in [13].

A statistical analysis of the return and volatility series confirms the presence of the stylized facts vastly documented in the literature. Figure 42.1 displays the returns, the annualized $RRV^{\Delta}_{\text{scaled}}$, the log $RRV^{\Delta}_{\text{scaled}}$, and the sample autocorrelation function for log $RRV^{\Delta}_{\text{scaled}}$. The distribution of the returns exhibits excess of kurtosis and volatility clustering. The plot of the annualized $RRV^{\Delta}_{\text{scaled}}$ also displays volatility clustering (volatility of volatility) while the long-memory pattern is observed in the hyperbolic and slow decay of the ACF for the log $RRV^{\Delta}_{\text{scaled}}$.

42.3 Models for the Observed Volatility Sequences

Different models have been presented to capture the stylized facts that financial series exhibit. Based on the statistical features briefly mentioned before, we consider the HAR model of Corsi [8] to capture the long-memory pattern. We also account for asymmetric effects with respect to the volatility and the returns. Moreover, following [9] we include a GARCH specification to account for heteroskedasticity in the observed volatility sequences and a standardized Normal Inverse Gaussian (NIG) distribution to deal with the observed skewness of the residuals. Finally, to account for asymmetric effects in the variance equation or volatility of the volatility we consider the GJR-GARCH specification.

We thus estimate the following model:

$$\begin{aligned} h_t &= \alpha + \delta_s I_s(h_{t-1})h_{t-1} + \beta_d h_{t-1} + \beta_w h_{(t-1:t-5)} + \beta_m h_{(t-1:t-22)} \\ &\quad + \gamma_R R_{t-1} + \gamma_{IR} I(R_{t-1}) R_{t-1} + \sqrt{\sigma_t}\epsilon_t \\ \sigma_t &= \omega + \beta_1 \sigma_{t-1} + \alpha_1 u_{t-1}^2 + \phi_1 u_{t-1}^2 I(u_{t-1}) \\ &\quad \epsilon_t | \Omega_{t-1} \sim d(0,1) \end{aligned} \tag{42.3}$$

where h_t is the log $RRV^{\Delta}_{\text{scaled},t}$, $h_{(t-1:t-j)}$ is the HAR component defined as

$$h_{(t-1:t-j)} = \frac{1}{j}\sum_{k=1}^{j} h_{t-k} \tag{42.4}$$

with $j = 5$ and 22 in order to capture the weekly and monthly components. $I_s(h_{t-1})$ is an indicator for $RRV^{\Delta}_{\text{scaled},t-1}$ bigger than the mean over $s = 5, 10, 22, 44$ and 66 previous days and the unconditional mean (um) up to $t - 1$. These variables capture the asymmetric effects with respect to the volatility. $R_t = \ln(p_t^{\text{cl}}/p_{t-1}^{\text{cl}})$ is the return, with p_t^{cl} the closing price for the day t and $I(R_{t-1})$ is an indicator for negative returns in $t - 1$, that captures the asymmetric effects with regard to the lagged return. $u_t = \sqrt{\sigma_t}\epsilon_t$ is the error term. The full specification for σ_t is a GJR-GARCH model, introduced by Glosten et al. [10], to account for the asymmetric effect in the volatility of the volatility, where $I(u_{t-1})$ is an indicator for $u_{t-1} < 0$.

Initially, we consider 21 alternative specifications for the mean equation, three possible variance equations and two different distributions for the innovations. Combining those three elements, we have a total of 126 models. After an in-sample selection, we decide to concentrate on six different specifications for the mean equation. The specifications for the mean are the AR(1) (*I*), the HAR model of Corsi [8] (*II*), the HAR model with symmetry effects with respect to the historical volatility and the returns and the lagged returns (*III*), the HAR with asymmetric effects with respect to the returns and lagged returns (*IV*), the HAR model with only asymmetric effects with respect to the returns (*V*) and finally the HAR model with symmetric effects for the past weekly volatility and the returns (*VI*). The last five models also include an AR(1) term. Finally, we have 36 models that are used to compute one-step-ahead rolling forecast for the volatility series and to assess the importance of the variable. Because of a space restriction, we only present estimation results for some of the considered models. The forecast analysis is carried out for different horizons.

42.4 Estimation and Forecast Results

First, we estimate the models for the entire sample from January 2003 to March 2010. The aim is to assess the impact and significance of our different variables in our models. Second, we compute one-day-ahead out-of-sample rolling forecast from January 3, 2006 to March 30, 2010 for a total of 1,067 periods. We estimate the models until December 30, 2005 and then we re-estimate each model at each recursion.To evaluate the performance, we compute the mean absolute error (*MAE*) and the mean square error (*MSE*). We compare the different performances of the models with the Diebold Mariano Test based on the *MAE* and on the *MSE*, that is a *robust* loss function in the sense of [16]. Besides, we consider the Model Confidence Set approach of [12] based on the same two loss functions.[1] Table 42.1 presents the results for the 2003–2010 estimation of some of the models, whereas Table 42.2 reports the forecast performance evaluation.

[1] In this version, we only present the results based on the *MSE* loss function. Similar results are obtained with the other loss function.

Table 42.1 Estimation results

	II Co-NO	III Co-NO	IV Co-NO	V Co-NO	VI Co-NO	II Gj-NI	III Gj-NI	IV Gj-NI	V Gj-NI	VI Gj-NI
α	-0.279^{a}	-0.493^{a}	-0.515^{a}	-0.472^{a}	-0.468^{a}	-0.263^{a}	-0.456^{a}	-0.449^{a}	-0.394^{a}	-0.393^{a}
	(0.069)	(0.091)	(0.084)	(0.078)	(0.078)	(0.076)	(0.097)	(0.084)	(0.078)	(0.078)
β_d	0.360^{a}	0.324^{a}	0.319^{a}	0.323^{a}	0.291^{a}	0.334^{a}	0.307^{a}	0.309^{a}	0.314^{a}	0.294^{a}
	(0.024)	(0.026)	(0.025)	(0.025)	(0.030)	(0.027)	(0.031)	(0.028)	(0.028)	(0.038)
β_w	0.460^{a}	0.468^{a}	0.468^{a}	0.469^{a}	0.505^{a}	0.462^{a}	0.465^{a}	0.466^{a}	0.466^{a}	0.487^{a}
	(0.042)	(0.042)	(0.042)	(0.042)	(0.048)	(0.042)	(0.041)	(0.041)	(0.041)	(0.051)
β_m	0.122^{a}	0.113^{a}	0.115^{a}	0.117^{a}	0.116^{a}	0.148^{a}	0.141^{a}	0.140^{a}	0.143^{a}	0.144^{a}
	(0.035)	(0.035)	(0.035)	(0.035)	(0.035)	(0.034)	(0.033)	(0.033)	(0.033)	(0.033)
δ_{um}	–	0.002	–	–	–	–	−0.000	–	–	–
		(0.005)					(0.004)			
δ_5	–	–	–	–	−0.007	–	–	–	–	−0.003
					(0.005)					(0.005)
γ_{RT}	–	1.532	1.588	–	–	–	2.025	2.009	–	–
		(1.671)	(1.667)				(1.612)	(1.609)		
γ_{IRT}	–	-11.79^{a}	-11.94^{a}	-9.656^{a}	-9.658^{a}	–	-10.90^{a}	-10.87^{a}	-7.918^{a}	-7.908^{a}
		(2.474)	(2.451)	(1.147)	(1.144)		(2.599)	(2.584)	(1.312)	(1.311)
ω	0.182^{a}	0.177^{a}	0.177^{a}	0.177^{a}	0.177^{a}	0.009^{a}	0.011^{a}	0.011^{a}	0.011^{a}	0.010^{b}
	(0.004)	(0.004)	(0.004)	(0.004)	(0.004)	(0.003)	(0.004)	(0.004)	(0.004)	(0.004)
β_1	–	–	–	–	–	0.898^{a}	0.884^{a}	0.884^{a}	0.887^{a}	0.892^{a}
						(0.029)	(0.035)	(0.035)	(0.034)	(0.032)
α_1	–	–	–	–	–	0.063^{a}	0.067^{a}	0.067^{a}	0.066^{a}	0.064^{a}
						(0.017)	(0.019)	(0.019)	(0.019)	(0.018)
ϕ_1	–	–	–	–	–	-0.036^{c}	-0.045^{c}	-0.045^{c}	-0.044^{c}	-0.041^{c}
						(0.022)	(0.024)	(0.024)	(0.024)	(0.023)
α_{NIG}	–	–	–	–	–	1.584^{a}	1.527^{a}	1.528^{a}	1.537^{a}	1.532^{a}
						(0.181)	(0.170)	(0.170)	(0.171)	(0.170)
β_{NIG}	–	–	–	–	–	0.429^{a}	0.367^{a}	0.368^{a}	0.370^{a}	0.367^{a}
						(0.112)	(0.103)	(0.103)	(0.103)	(0.103)
LLF	−985.03	−963.07	−963.17	−963.66	−962.90	−884.67	−868.95	−868.96	−869.87	−869.62
AIC	1980.0	1942.1	1940.3	1939.3	1939.8	1789.3	1763.9	1761.9	1761.7	1763.2
BIC	2007.3	1985.8	1978.5	1972.0	1978.0	1843.9	1834.8	1827.4	1821.7	1828.7
Lj_{30}	0.186	0.318	0.315	0.301	0.273	0.164	0.268	0.271	0.269	0.258
Lj_{40}	0.400	0.615	0.612	0.594	0.561	0.388	0.574	0.577	0.566	0.553
JB	0.001	0.001	0.001	0.001	0.001	–	–	–	–	–
KS	0.000	0.000	0.000	0.000	0.000	–	–	–	–	–
LL	0.001	0.001	0.001	0.001	0.001	–	–	–	–	–

Note: Estimation results for the whole sample from January 2003 to May 2010. We only present results for some of the models. Model *II* is an *HAR*, *VII* is an $HAR + I_{um}(h_{t-1})h_{t-1} + R_{t-1} + I(R_{t-1})R_{t-1}$, *VIII* is an $HAR + R_{t-1} + I(R_{t-1})R_{t-1}$, *IX* is an $HAR + I(R_{t-1})R_{t-1}$ and *XIV* is an $HAR + I_5(h_{t-1})h_{t-1} + I(R_{t-1})R_{t-1}$. They also include an AR(1) term. *NO* is Normal distribution and *NI* indicates Normal Inverse Gaussian distribution. *Co* is a constant variance specification, *Ga* is a *GARCH* and *Gj* is a *GJR* variance specification. LLF is the Log-likelihood function, *AIC* is the Akaike Information Criteria, and *BIC* is the Bayesian information criterion. LJ_{30} and LJ_{40} are the p-values for the Ljung Box test for 30 and 40 lags. *JB*, *KS*, and *LL* are the p-values for the Jarque–Bera test for Normality, the Kolmogorov–Smirnov test, and the Lilliefors test, respectively. Standard errors in bracket. "a," "b," and "c" indicate significance at the 1%, 5%, and 10%.

Table 42.2 Out-of-sample forecast evaluation: Diebold Mariando test and Model Confidence Set

	Full				Crisis			
		DM	MCS			DM	MCS	
	MSE	vs II	pv_R	pv_{SQ}	MSE	vs II	pv_R	pv_{SQ}
II-Co-NO	0.196	–	0.08	0.19^{a}	0.249	–	0.46^{b}	0.23^{a}
III-Co-NO	0.192	1.43	0.28^{b}	0.29^{b}	0.225	1.74^{c}	0.82^{b}	0.81^{b}
IV-Co-NO	0.191	1.60	0.35^{b}	0.40^{b}	0.224	1.79^{c}	1.00^{b}	1.00^{b}
V-Co-NO	0.191	1.83^{c}	1.00^{b}	1.00^{b}	0.225	1.75^{c}	0.82^{b}	0.79^{b}
VI-Co-NO	0.191	1.75^{c}	0.61^{b}	0.61^{b}	0.225	1.78^{c}	0.82^{b}	0.81^{b}
II-Ga-NO	0.200	–	0.08	0.09	0.265	–	0.30^{b}	0.10^{a}
III-Ga-NO	0.193	1.78^{c}	0.08	0.14^{a}	0.232	2.13^{b}	0.46^{b}	0.27^{b}
IV-Ga-NO	0.193	1.92^{c}	0.08	0.24^{a}	0.233	2.15^{b}	0.46^{b}	0.21^{a}
V-Ga-NO	0.192	2.08^{b}	0.22^{a}	0.29^{b}	0.234	2.11^{b}	0.31^{b}	0.15^{a}
VI-Ga-NO	0.193	2.01^{b}	0.08	0.27^{b}	0.233	2.16^{b}	0.46^{b}	0.29^{b}
II-Gj-NO	0.199	–	0.08	0.10^{a}	0.261	–	0.31^{b}	0.11^{a}
III-Gj-NO	0.192	1.74^{c}	0.08	0.27^{b}	0.227	2.10^{b}	0.82^{b}	0.52^{b}
IV-Gj-NO	0.192	1.84^{c}	0.22^{a}	0.29^{b}	0.228	2.09^{b}	0.82^{b}	0.52^{b}
V-Gj-NO	0.192	2.00^{b}	0.35^{b}	0.31^{b}	0.230	2.05^{b}	0.78^{b}	0.46^{b}
VI-Gj-NO	0.192	1.94^{c}	0.28^{b}	0.29^{b}	0.229	2.09^{b}	0.82^{b}	0.53^{b}
II-Co-NI	0.200	–	0.08	0.09	0.262	–	0.46^{b}	0.18^{a}
III-Co-NI	0.194	1.55	0.08	0.15^{a}	0.230	1.96^{b}	0.78^{b}	0.46^{b}
IV-Co-NI	0.193	1.70^{c}	0.08	0.23^{a}	0.229	1.98^{b}	0.81^{b}	0.48^{b}
V-Co-NI	0.192	2.01^{b}	0.28^{b}	0.29^{b}	0.231	2.00^{b}	0.68^{b}	0.38^{b}
VI-Co-NI	0.193	1.97^{b}	0.19^{a}	0.28^{b}	0.231	2.01^{b}	0.78^{b}	0.43^{b}
II-Ga-NI	0.200	–	0.08	0.08	0.265	–	0.31^{b}	0.12^{a}
III-Ga-NI	0.193	1.68^{c}	0.08	0.12^{a}	0.231	2.05^{b}	0.56^{b}	0.36^{b}
IV-Ga-NI	0.193	1.78^{c}	0.08	0.17^{a}	0.231	2.04^{b}	0.56^{b}	0.38^{b}
V-Ga-NI	0.193	2.05^{b}	0.08	0.27^{b}	0.233	2.05^{b}	0.56^{b}	0.31^{b}
VI-Ga-NI	0.193	2.04^{b}	0.08	0.27^{b}	0.232	2.09^{b}	0.68^{b}	0.39^{b}
II-Gj-NI	0.199	–	0.08	0.09	0.262	–	0.31^{b}	0.13^{a}
III-Gj-NI	0.193	1.68^{c}	0.08	0.25^{b}	0.228	2.06^{b}	0.82^{b}	0.62^{b}
IV-Gj-NI	0.192	1.80^{c}	0.19^{a}	0.28^{b}	0.227	2.07^{b}	0.82^{b}	0.72^{b}
V-Gj-NI	0.192	2.06^{b}	0.28^{b}	0.29^{b}	0.230	2.07^{b}	0.81^{b}	0.47^{b}
VI-Gj-NI	0.192	2.03^{b}	0.28^{b}	0.29^{b}	0.230	2.07^{b}	0.82^{b}	0.49^{b}

Note: Forecast performance evaluation for the full out-of-sample period *Full* (1,067 obs.) and the financial crisis period *Crisis* (200 obs.). Model *II* is an *HAR*, *VII* is an $HAR + I_{um}(h_{t-1})h_{t-1} + R_{t-1} + I(R_{t-1})R_{t-1}$, *VIII* is an $HAR + R_{t-1} + I(R_{t-1})R_{t-1}$, *IX* is an $HAR + I(R_{t-1})R_{t-1}$, and *XIV* is an $HAR + I_5(h_{t-1})h_{t-1} + I(R_{t-1})R_{t-1}$. They also include an AR(1) term. *NO* is Normal distribution and *NI* indicates Normal Inverse Gaussian distribution. *Co* is a constant variance specification, *Ga* is a *GARCH* and *Gj* is a *GJR* variance specification. *MSE* is the mean square error. *DM* is the Diebold Mariano test for equal predictive accuracy between two models based on the *MSE* loss function. *vs II* indicates a test for the row model versus the model *II* with same distribution and variance specification (indicated in the row model). Under Ho, both models have the same performance. T-statistic in the column. "a," "b," and "c" indicate significance at the 1%, 5%, and 10%. Positive T-statistic favors the row model. *MCS* is the Model Confidence Set, a procedure to determine the "best" models from a collection of models based on the *MSE* loss function and two t-statistics: pv_R is the p-value for the *range* deviation method and pv_{SQ} is the p-value for the *semi-quadratic* deviation method (see [12]). "a" and "b" denote that the model belongs to the 10% and 25% MCS.

Estimation results for the full sample period (2003–2010) suggest that HAR components are significant for the three variance specifications (constant variance, GARCH, and GJR-GARCH) and the two different distributions (Normal and NIG distribution). The asymmetric effects with respect to the return and the volatility improve the goodness of fit of the model. The first one is highly significant and it increases the volatility after a negative return. On the contrary, the asymmetric effects with respect to the volatility are not significant. The sign and significance of the coefficients in the mean equation remain stable for the different specifications in the variance equation. The inclusion of the GARCH or GJR specifications improves the fitting of the models. The models that best fit the series are the ones that include the HAR and leverage effects, with GARCH and GJR variances. Diagnostic tests for the residuals present different results. Only for the models that include the HAR components we cannot reject the null hypothesis of no serial correlation in the residual, implying a good performance. The Normality hypothesis for the residuals is rejected for all the models, which is an argument to introduce a non-Gaussian distribution. As we said, the estimated parameters of the mean equation for the models with NIG distribution are similar to the models with Normal distribution. The estimated parameters of the NIG distribution (α_{NIG} and β_{NIG}) capture the right skewness and excess of kurtosis displayed in the residuals. We analyze the results for the out-of-sample forecast in two different periods. In particular, we study the accuracy of our models for the full out-of-sample (1,067 observations) and during the financial crisis, from September 15, 2008 to July 30, 2009 (200 observations).

For the full out-of-sample forecast, the model that performs better, based on the MAE and MSE, is the HAR with lagged and asymmetry over the return, with constant variance and Normal distribution. Other models that include asymmetric effects with respect to the volatility over the five previous days and the unconditional mean perform similarly. Models with different specifications for the variance and distribution for the innovation perform as the models with constant variance. Although, GARCH and GJR improve the goodness of fitness in the estimation, they do not have impact in the forecast. The third column of Table 42.1 presents the t-statistics for the Diebold Mariano test for equal predictive accuracy between the HAR specification (II) and five different specifications for the mean equations[2] (III, IV, V, VI) with same variance and distribution assumptions. The results suggest that models with symmetric effects with respect to the volatility and the returns perform significantly better than the HAR models. This result is confirmed by the Model Confidence Set approach, p-values displayed in the third and fourth column of the same table, that recursively eliminates the models that worst perform starting from an initial set that includes the 36 models. For the range test statistic, the HAR model with different variance specifications and distributions for the innovations remains out of the set of best models. Similar results are obtained for the semi-quadratic test statistic.

---

[2]Results for the test between model I and II are not presented. As expected, HAR models perform significantly better than AR(1) model.

During the financial crisis, the model that performs better is the HAR with lagged return and asymmetric effect over the returns. The results of the Diebold Mariano Test, based on the *MSE* loss function, display evidence in favor of models with asymmetric effects with respect to the volatility and the returns. However, when considering the results of the Model Confidence Set approach, the set of best models includes the HAR component of Corsi [8] with different distributions and variance specifications.

42.5 Conclusions and Future Steps

In this chapter, we have modeled and forecasted price variation through the *Realized Range Volatility* introduced by Martens and van Dijk [13] and Christensen and Podolskij [6]. We have estimated the series for different sampling frequencies and corrected them with the scaling procedure of [13]. After the corrections, the volatility stabilizes across different sampling frequencies and scaling factors which suggest that the bias caused by the microstructure friction was removed, restoring the efficiency of the estimator. We have considered a model which approximates long memory, has asymmetric effects with respect to the return and the volatility in the mean equation, and includes GARCH and GJR-GARCH specifications for the variance equation (which models the volatility of the volatility). A non-Gaussian distribution was also considered for the innovations.

The results suggest that the HAR model with the asymmetric effects with respect to the volatility and returns is the one that better fit the data. The analysis of the forecast performances of the different models provides similar results for the two considered periods, the full sample and the financial crisis. The introduction of asymmetric effects numerically improves the point forecasting performance. Following the Diebold Mariano test and the Model Confidence Set approach, we find that the models that include asymmetries over the return and the volatility perform statistically better than the HAR for the different variance specifications and distributions for the innovations. As we expected, models with GARCH and GJR-GARCH specifications and different distributions for the innovations do not lead to more accurate point forecasts than models with constant variance.

Finally, the HAR with asymmetric effects with respect to the returns is able to capture most of the variability during the out-of-sample prevision. Then, in order to improve this performance, the introduction of financial and macroeconomic variables should be considered. Other future steps are the possible correction for jumps in the volatility series and an economic analysis of the performances of the models forecast.

Acknowledgements The authors wish to thank the participants to the Italian Statistical Society XLV Conference held in Padova in June 2010 for their helpful comments and suggestions.

References

1. Andersen, T.G., Bollerslev, T., Diebold, F.X., Labys, P.: The distribution of realized exchange rate volatility. J. Am. Stat. Assoc. **96**(453), 42–55 (2001)
2. Andersen, T.G., Bollerslev, T., Diebold, F.X., Labys, P.: Modeling and forecasting realized volatility. Econometrica **71**(2), 579–625 (2003)
3. Barndorff-Nielsen, O.E., Hansen, P.R., Lunde, A., Shephard, N.: Designing realized kernels to measure the ex post variation of equity prices in the presence of noise. Econometrica **76**(6), 1481–1536 (2008)
4. Barndorff-Nielsen, O.E., Shephard, N.: Econometric analysis of realized volatility and its use in estimating stochastic volatility models. J. R. Stat. Soc. B (Stat. Meth.) **64**(2), 253–280 (2002)
5. Bollerslev, T.: Generalized autoregressive conditional heteroskedasticity. J. Econometrics **31**(3), 307–327 (1986)
6. Christensen, K., Podolskij, M.: Realized range-based estimation of integrated variance. J. Econometrics **141**(2), 323–349 (2007)
7. Christensen, K., Podolskij, M., Vetter, M.: Bias-correcting the realized range-based variance in the presence of market microstructure noise. Finance Stochast. **13**(2), 239–268 (2009)
8. Corsi, F.: A simple approximate long-memory model of realized volatility. J. Financ. Econometrics **7**(2), 174–196 (2009)
9. Corsi, F., Mittnik, S., Pigorsch, C., Pigorsch, U.: The volatility of realized volatility. Economet. Rev. **27**(1–3), 46–78 (2008)
10. Glosten, L.R., Jagannathan, R., Runkle, D.E.: On the relation between the expected value and the volatility of the nominal excess return on stocks. J. Finance **48**(5), 1779–1801 (1993)
11. Hansen, P.R., Lunde, A.: Realized variance and market microstructure noise. J. Bus. Econ. Stat. **24**(2), 127–161 (2006)
12. Hansen, P.R., Lunde, A., Nason, J.M.: The model confidence set. Econometrica **79**(2), 453–497 (2011)
13. Martens, M., van Dijk, D.: Measuring volatility with the realized range. J. Econometrics **138**(1), 181–207 (2007)
14. Martens, M., van Dijk, D., de Pooter, M.: Forecasting s&p 500 volatility: Long memory, level shifts, leverage effects, day-of-the-week seasonality, and macroeconomic announcements. Int. J. Forecast. **25**(2), 282–303 (2009)
15. McAleer, M., Medeiros, M.C.: Realized volatility: a review. Economet. Rev. **27**(1), 10 (2008)
16. Patton, A.J.: Volatility forecast comparison using imperfect volatility proxies. J. Econometrics **160**(1), 246–256 (2011)
17. Zhang, L.: Efficient estimation of stochastic volatility using noisy observations: a multi-scale approach. Bernoulli **12**(6), 1019–1043 (2006)
18. Zhang, L., Mykland, P.A., Ait-Sahalia, Y.: A tale of two time scales. J. Am. Stat. Assoc. **100**(472), 1394–1411 (2005)

Clusters and Equivalence Scales

43

Gustavo De Santis and Mauro Maltagliati

Abstract

Equivalence scales (S) are difficult to estimate: even apparently solid microeconomic foundations do not necessarily lead to consistent results. We contend that this depends on "style" effects: households with the same economic resources and identical "needs" (e.g. same number of members) may spend differently, following unobservable inclinations ("style" or "taste"). We submit that these style effects must be kept under control if one wants to obtain unbiased estimates of S. One way of doing this is to create clusters of households, with different resources (income), different demographic characteristics (number of members) but similar "economic profile", in terms of both standard of living and "style". Cluster-specific scales, and the general S that derives from their average, prove defensible on theoretical grounds and are empirically reasonable and consistent.

Keywords

Cluster analysis • Equivalence scales • Measuring poverty and inequality

43.1 Equivalence Scales: A Dead End?

As Muellbauer and van de Ven [9:1] put it, "despite a considerable research effort…, almost every aspect of equivalence scale specification remains controversial". Indeed, the difficulties seem to be of both micro-economic and statistical nature.

Micro-economic theory suggests that equivalence scales S cannot be derived from the observation of empirical consumption patterns, because of their implicit compensation nature. Consider, for instance, the case when the size of the family

G. De Santis (✉) · M. Maltagliati
Department of Statistics, University of Florence, Florence, Italy
e-mail: desantis@disia.unifi.it; mmaltag@disia.unifi.it

N. Torelli et al. (eds.), *Advances in Theoretical and Applied Statistics*,
Studies in Theoretical and Applied Statistics, DOI 10.1007/978-3-642-35588-2_43,

increases because of the birth of a child, welcome by his/her parents. By definition, happy parents cannot be worse off, but the equivalence scale will nonetheless indicate the extra money they need to be "compensated" for the presence of the child. With varying emphasis, this objection can be found in virtually all authors (see, e.g., [8]) since when Pollak and Wales [11] first raised the issue. We submit, instead, that the idea is defensible, as long as one keeps it confined to a strictly economic ground: parents can be both happier and poorer, and equivalence scales try to estimate precisely if they are poorer, by how much.

Estimation methods are either mono- or multi-equational. In the former case, one variable (e.g. food share, as in the Engel method) is assumed to vary monotonically with (and thus represent) the standard of living of the household. Two households, say H_1 and H_2, where this indicator reaches the same value are thus assumed to enjoy the same standard of living, and, if this assumption holds, the ratio of their incomes (Y_2/Y_1) provides the desired equivalence index. Mono-equational models are easy to estimate, and, among them, one in particular, Engel's method, generally leads to reasonable results. This is probably the main reason why the "Carbonaro scale" (based on the Engel method) is generally used in Italy, by both Istat [5] and the Government (with the ISEE scale, a variant of Carbonaro's one— https://servizi.inps.it/servizi/isee/default.htm).

Mono-equational models, unfortunately, are scarcely defensible on a theoretical plan: whatever indicator one chooses (e.g. food share) there is always the possibility that two households at the same level of economic well-being will behave differently, that is, spend a different amount of money on that good (food), and thus incorrectly appear to enjoy different standards of living. This may happen in part because of price effects (consumers tend to purchase less of what is, or becomes, comparatively more expensive), and in part because of a set of other reasons which, in this paper, we will call "taste" or "style" for the sake of simplicity.

In order to see more clearly how this could happen, let us assume that the object of the research is the estimate of child costs (or, which is the same, the estimate of the equivalence scales for households with and without children) and that the selected indicator is food share (Engel's method). The result may be biased basically for two, non-alternative reasons:

(a) Households with preference for (and therefore greater presence of) children are selected in several ways, including some that directly affect our indicator of economic well-being. Imagine for instance that these households are (or, after the birth of their child, become) more indoor-oriented and, consequently, tend to consume more food at home. They will incorrectly appear poorer than they actually are, and this apparent lack of economic resources will normally be associated with, if not attributed to, the presence of children.
(b) A child differs from his or her parents: he/she may consume proportionally more food than other items (e.g. travel, culture and apparel). Overall, the household will shift its consumption pattern towards food, which, here too, will make households with children appear poorer than they actually are.

With complete demand systems and multi-equational models the problem becomes less visible but is still present, and it emerges clearly whenever one looks deeper into the data: equivalence scales that abruptly change from one year to the next without any apparent reason or that prove too sensitive to patently minor details (e.g. number or type of expenditure items considered), suggest that the estimates that one obtains are not robust. And methods specifically aimed at limiting this bias do not necessarily perform better than others (see, e.g., [2, 3].

This depends probably on two main causes. One is that there is an intrinsic multicollinearity between the object of interest (the equivalence scale) and style effects: to estimate the two at the same time proves difficult [4]. The other is that individual (household) consumption data are collected in so short a lapse of time and are therefore so variable, and so full of zeros, that even the best models can explain only a tiny fraction of the observed variance (see, e.g., [1]). Aggregating observations, or expenditure shares, or both, is an option, of course, but then other questions arise: how to proceed? When to stop? Besides, as mentioned before, aggregation is not neutral: it may affect the results very deeply [4], and, to the best of our knowledge, this major shortcoming has not been overcome yet. Are equivalence scales at a dead end?

43.2 A Possible Solution: Clusters

Both theoretical and practical issues (multicollinearity and "proper" aggregation of expenditure items) may find a solution in clustering observations in such a way that each cluster contains households that "behave similarly" in economic terms and can be assumed to enjoy more or less the same standard of living and also have a comparable lifestyle.

Let us imagine that each household H can be characterized by the following set of variables: $\mathbf{E}$(Economic indicators), Y (income) and N (Number of members), so that

$$H = (\mathbf{E}_h, Y_h, N_h) \tag{43.1}$$

For any given household typology, each economic indicator in $\mathbf{E}_h$ will reflect both its level of affluence (some households are richer than others) and its lifestyle (e.g. some households prefer the car and others the motorbike, if they can't afford to buy both).

Indicators of lifestyle tend to "sum to zero" for any given level of affluence. This is mathematically true if one refers to budgets shares, where styles are, by definition, deviations from the mean [3]. In other cases, this is no longer true in mathematical terms, but the notion tends to preserve its validity in a more general sense. In order to see why this happens and also how the system works, consider the following, extremely simplified, case. Imagine that we work only with 1- and 2-person households, which we want to classify on the basis of only two dichotomous active variables: having a car (yes/no) and having a motorbike (yes/no). We can form four clusters ($K = 4$), depending on whether the observed households own both car and motorbike, only car, only motorbike, or none of these. In each of these four groups we find households of both types, with one and two members. We *assume* that, within each cluster, households share a similar economic profile, in terms of both affluence and lifestyle, and are therefore comparable, which is an essential prerequisite for the construction of equivalence shares. Note that income Y is *not* an active variable and that we do not use it in clustering. Note, also, that we do not need to rank households on the basis of their standard of living (who is worse and who is better off)—but, of course we expect (and, in our empirical applications, we successfully checked for) consistency: households with both car and motorbike should be richer that others, both among 1- and 2-person households.

Next, we calculate the average income of households belonging to each cluster k with, respectively, 1 and 2 members: $Y_{1,k}$ and $Y_{2,k}$. We expect that $Y_{2,k} > Y_{1,k}$ (because two persons need more resources than one to reach any given standard of living) and since, by assumption, the standard of living is the same within each cluster, we can estimate $K(=4)$ cluster-specific equivalence scales $S_{2,k} = Y_{2,k}/Y_{1,k}(k = 1, 2, \ldots, 4)$. We average the four results, and what we obtain is an estimate of the "true" equivalence scale $S_2 = Y_2/Y_1$.

In practice, we refer to more household dimensions ($n = 1$–5), we use more active variables (14), we form many more clusters $k(1 - 200)$ and we try several clustering methods (see below), but the basic procedure is as just outlined. All estimates are weighted, both in the calculation of cluster-specific average incomes $Y_{n,k}$ and in the estimate of equivalence scales S_n

$$S_n = \sum_k S_{n,k} w_{n,k} \tag{43.2}$$

where weights $w_{n,k}\left(=\sqrt{H_{n,k} \cdot H_{1,k}}/\sum_k \sqrt{H_{n,k} \cdot H_{1,k}}\right)$ keep into account the number of households both in the numerator and in the denominator. Finally, as shown in the next section, we can also compute confidence interval, both for cluster-specific estimates $S_{n,k}(n = \text{household size})$ and for the general estimate S_n.

43.3 Estimating the Variance of Our Equivalence Scale

The equivalence scales that we compute are random variables, and we can estimate their variance. Within each cluster, we get $S_{n,k} = Y_{n,k}/Y_{1,k}$, where Y is income, k is the cluster and n is the household dimension ($n = 1$ for the reference household). $Y_{n,k}$ and $Y_{1,k}$ are sample estimates of averages, the variances of which $[Var(Y_{n,k}), Var(Y_{1,k})]$ can be estimated [6, p. 187].

$$Var(S_{n,k}) = \frac{1}{(Y_{1,k})^2}\left\{Var(Y_{n,k}) + S_{n,k}^2\left[Var(Y_{1,k})\right]\right\} \tag{43.3}$$

Since we estimate our equivalence scale S_n by averaging a series of independent cluster-specific estimates, $S_{n,k}$, each with its own weight w_k (given by Istat, for each household), the variance of S_n can be estimated as follows [6, p. 206]

$$Var(S_n) = \sum_k w_{n,k}^2 \cdot Var(S_{n,k}) \tag{43.4}$$

43.4 An Empirical Application to Italian Data on Consumption

We work on data that derive from Istat's consumption survey for the years 2003–2008 (6 years), after inflating monetary values so as to make them comparable. We consider households made up of 1–5 persons, dropping larger households, because they are rare, and because at this preliminary stage we want to make sure that our method works correctly on "standard" cases. To this end, we also decided to discard outliers, i.e. households that, given their dimension and their budget, spend proportionally too much or too little on each expenditure typology (of which we formed eight: food, housing, apparel, etc. the acceptable range is centred on the median, and spans 4 interquartile intervals, more and less—that is eight interquartile intervals, overall). This has two purposes: first, there may be errors in these cases, and it is better to get rid of them. Second, we look for households of various dimensions that are comparable in economic terms—and these outliers, by definition, differ from virtually every other household in the survey. This leaves us with about 128,000 "acceptable" households—but we also checked that our estimates practically do not change if we keep all the original households (about 148,000—results not shown here). Note that we do not have information on household's income Y: we will instead use their total expenditure X, as a proxy.

For each household there are, in principle, very many indicators apt to describe its affluence, either directly (the indicator increases with total spending X—e.g. probability of having heating in one's home) or inversely (the indicator decreases with X—e.g. food share). In practice, we checked about a hundred of them, but we eventually retained only 14 of them (Table 43.1): the others did not prove consistent, which means that they did not vary monotonically with X, for each household dimension, in the relevant range of X.

The two "ancillary" variables that appear in Table 43.1 are needed to form subgroups on which other active variables (luxuries) can be identified, because only non-vegetarian households spend on proteins, and it is only among oil consumers that the better off, consuming prevalently or exclusively olive oil, can be identified. On the basis of these 14 economic indicators $\mathbf{E}_h$, we form clusters.

In a clustering process, several potentially important choices are arbitrary: type of clustering (fuzzy, disjoint, hierarchical,...), what criterion to use (Centroid, Ward,...), whether to preliminary standardize the variables, and how many groups to form. In the following, we show only a subset of our results (with hierarchical,

Table 43.1 Active variables

Category	Label	#	Variable	Filters	Notes
Luxuries	%Lux	1	Share of luxuries on total expenses		(1)
	Ktchn	2	Kitchen as an independent room (Dummy: yes/no)		
	Garage	3	Garage (Dummy: yes/no)		
	Motor	4	Motorbike (Dummy: yes/no)		
	Washr	5	Washer (Dummy: yes/no)		
	Dishw	6	Dishwasher (Dummy: yes/no)		
	Hclnrs	7	Home-cleaning machines (Dummy: yes/no)		
	Cond	8	Air conditioning (Dummy: yes/no)		
	Tel	9	Telephone (Dummy: yes/no)		
	%Rchprt	10	Share of rich proteins on total proteins	var. 13	(2)
	5Olive	11	Share of olive oil out of total oil purchased	var. 14	(3)
Necessities	%Food	12	Share of food on total expenditure		
Ancillaries	Vgtm	13	Vegetarian (Dummy: yes/no)		
	Oilcnsmr	14	Oil consumer (Dummy: yes/no)		

Notes. (1) Includes a long list of expenditure items ranging from men's apparel to domestic help; from sportswear to pets; from private lessons to jewelry; etc. (2) This variable applies only to non- vegetarian households (var. 13). "Rich" are the proteins that derive from the purchase of meat (except chicken) or fish. Total protein expenditure includes "rich" protein, plus chicken, milk and eggs. (3) This variable applies only to oil-consuming households (var. 14). Besides olive oil, households can also purchase the less valuable seed oil and pomace oil

Ward method, on non-standardized variables, with different number of clusters), but other outputs, not presented here, differ only marginally: the method seems to be fairly robust.

43.5 Main Results

Our main results are as follows. First, the equivalence scale that we estimate increases monotonically with N (household size), which was not granted a priori, since this is a non-parametric model, and each equivalence factor, for each household dimension, forms independently of each other. Second, our equivalence scales depend in part on the criterion and, within that, on the number of clusters that we use (see Fig. 43.1). This is unfortunate, of course: but the good news is that the differences are not huge, or, at least, not with Ward (and with k-means—not shown here), while with other hierarchical methods this is not always granted (not shown here). As the number of clusters increases, the scales flatten, but they eventually converge towards a unique and apparently stable value, which is tempting to interpret as the "true" value of the scale. Incidentally, this is also the value towards which other clustering methods converge, too, although less rapidly (not shown here).

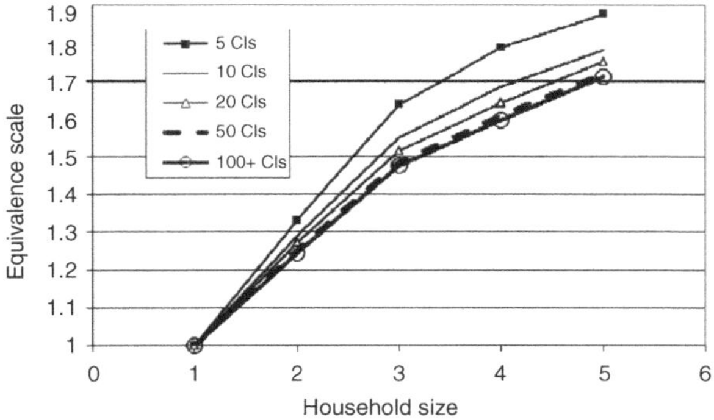

Fig. 43.1 Equivalence scales obtained with clustering, and a varying number of clusters (Households with 1–5 members). Source: Authors' elaborations on Istat data

Table 43.2 Estimated equivalence scale (S) and confidence intervals (95%; Italy 2003–2008)

	HhldS with 100 Clusters (Ward method)			Other selected S's	
Hhld Size	Average	Min (95%)	Max (95%)	Carbonaro	OECD Sq.Root
1	1	1	1	1	1
2	1.263	1.253	1.272	1.667	1.414
3	1.526	1.515	1.536	2.222	1.732
4	1.672	1.658	1.686	2.716	2.000
5	1.786	1.770	1.803	3.169	2.236

Source: Authors' elaborations on Istat data

Because of the very large number of observations that we have, and because of clustering, which dramatically reduces the variance of each estimate, the 95% confidence intervals that we can construct around our estimated equivalence scales are all very small (Table 43.2).

Our equivalence scale is considerably flatter than those that are normally used in Italy and worldwide (Fig. 43.2). It is flatter that Carbonaro's scale, which is normally considered to be too steep, as all Engel-based scales, but it is also somewhat flatter than the OECD square-root scale (square root of household size).

How variable are our cluster-specific estimates? This is a crucial question, but the answer is not easy, because it depends, among other things, on the clustering method, on the number of clusters, on the variables used for clustering, on the number of years and therefore on the number of households within each cluster, etc. An illustrative result is shown in Fig. 43.3, which refers to the estimates of $S_{3,k}$ (but the same holds for other household dimensions): cluster-specific estimates $S_{3,k}$ appear to be rather disperse, but outliers have very few observations and therefore scarce weight (not shown in the figure). Besides, there is at least one important systematic influence: equivalence scales appear to be lower at high income levels. Indeed, if the standard of reference is a one-person households with 1,000 Euros

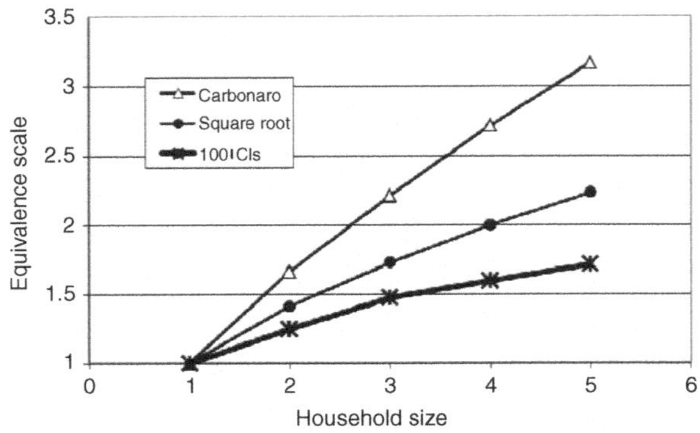

Fig. 43.2 Equivalence scales with Carbonaro's method, square-root of members (OECD) and our own (100 clusters; Households with 1–5 members). Source: Authors' elaborations on Istat data and [5]

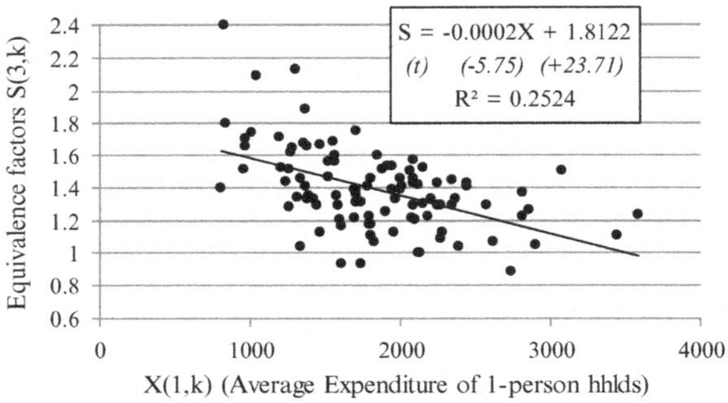

Fig. 43.3 Estimates of cluster-specific equivalence factors $S_{3,k}(= X_{3,k}/X_{1,k})$ at different utility levels, Italy 2003–2008. Source: Authors' elaborations on Istat data. 100 clusters; Ward method. Regression: unweighted

(2,008 values) of total monthly expenditure, the income-adjusted equivalence scale turns out to be very close to the OECD square root one (details not shown here).

This finding suggests, among other things, that the so-called *IB* (or Independence of Base) property does not hold [7]. Assuming that equivalence scales do not vary with the level of utility (i.e. income or total expenditure of the reference household) may be tempting and makes parametric estimation considerably easier, but can only be accepted as a first approximation: extra members are more expensive for poorer households—arguably, because of fixed costs.

43.6 Discussion

The method that we propose in this paper for the estimation of equivalence scales has, in our view, several merits. In the first place, to the best of our knowledge, it is completely original. Besides, our results, although based on non-parametric methods, are all consistent with a priori expectations: for instance, they increase with household size, but less then proportionally as N increases, they indicate that marginal costs are proportionally more relevant to poorer households, etc.

Not surprisingly, our estimates vary with the clustering method and, within that, with the number of clusters, but, at least with Ward (and with k-means), not too much, and not at all once the number of clusters is sufficiently high (100 or more). Our results are also robust to the choice of partially alternative sets of active variables (those used for clustering—not shown here), although we should like to insist that active variables must be chosen carefully, and must all be "well-behaved", i.e. they must vary consistently with household resources, separately for each household typology.

Our confidence intervals are extremely narrow. With fewer observations, of course, results may differ, but thanks to the efficiency of clustering, the increase is not dramatic, as our first tests suggest (not shown here). Besides, our method can be applied also to alternative databases, which, thus far, could not be exploited for the estimation of equivalence scales. Take, for instance, the Bank of Italy SHIW—Survey on Household Income and Wealth (http://www.bancaditalia.it/statistiche/indcamp/bilfait); or all the surveys of the same kind that converge into the LIS—Luxemburg Income study (http://www.lisdatacenter.org/). They do not detail household expenditure, and this is an essential ingredient in all the "traditional" methods for the estimation of equivalence scales. Our method, instead, only demands that a few well-behaved indicators of economic affluence be present in the database (e.g. current accounts, life insurances, and jewelry). Indeed, our preliminary estimates on the SHIW database show not only that estimating equivalence scales is possible, but also that they prove consistent with those estimated, in the same period, on Istat consumption data. Once again, to the best of our knowledge, the mere possibility of such a comparison is an absolute novelty in this field of study. Precisely because of its simplicity and flexibility, the method appears apt to several extensions: for instance, the study of the *evolution* of equivalence scales (or child costs) over time, possibly with the use of pseudo panels [10, 12].

Theoretically speaking, the method is more solid than it may appear at first sight: it is a conscious attempt at circumventing the thorny, but underestimated problems that derive from style effects—i.e. households of comparable standards of living who, nonetheless, choose to spend their money in different ways.

But this method has its limits, too. For instance, it relies on the implicit assumption that certain indicators (e.g. having a motorbike) have the same economic meaning for small and large households, or in different parts of the country. It may become difficult to handle when households differ not only because of their size but also because of other characteristics (e.g. the sex and age of their members). It mixes flow data (expenditure) with stock data (durables), which may not be fully justifiable.

Obviously, the method is still in its very early phase: it needs to be tested on several more databases, for several years, with different indicators of affluence, before it can qualify as a viable alternative for the estimation of equivalence scales. But the preliminary results seem to us to be encouraging.

Acknowledgements Financial support from the Italian MIUR is gratefully acknowledged. We thank Andrea Giommi (Un. of Florence) for his advice on the estimation of the variance of equivalence scales.

References

1. Bollino, C.A., Perali, F., Rossi, N.: Linear household technologies. J. Appl. Econ. **15**, 275–287 (2000)
2. De Santis, G., Maltagliati, M.: De gustibus non est disputandum? A new approach to the estimation of equivalence scales, Statistical Methods and Applications. J. Ital. Stat. Soc. **10**(1–3), 211–236 (2001)
3. De Santis, G., Maltagliati, M.: Equivalence scales: a fresh look at an old problem. In: Dagum, C., Ferrari, G. (eds.) Theory and Empirical Evidence, Household Behavior, Equivalence Scales and Well-being, pp. 29–53. Physica Verlag, New York/Berlin (2003)
4. De Santis, G., Seghieri, C.: How robust are equivalence scales? Proceedings of the 43rd SIS Sci. Meet. (Torino, 14–16 June), pp. 133–136, Padova, Cleup (2006)
5. Istat: La povertà in Italia nel 2009. Rome, (2010) (http://www.istat.it/salastampa/comunicati/in_calendario/povita/20100715_00/testointegrale20100715.pdf)
6. Kish, L.: Survey Sampling. Wiley, New York (1965)
7. Lewbel, A.: Household equivalence scales and welfare comparisons. J. Pub. Econ. **39**, 377–391 (1989)
8. Lewbel, A., Pendakur, K.: Equivalence Scales. Entry for The New Palgrave Dictionary of Economics, 2nd edn. Boston College and Simon Fraser University, Boston (2006), http://www.sfu.ca/~pendakur/palequiv.pdf
9. Muellbauer, J., van de Ven, J.: Estimating Equivalence Scales for Tax and Benefits Systems, NIESR (Natl. Inst. Econ. Soc. Res.) discussion papers 229, (2004), http://www.niesr.ac.uk/pubs/dps/dp229.pdf
10. Pashardes, P.: Contemporaneous and intertemporal child costs: Equivalent expenditure vs. equivalent income scales. J. Public Econ. **45**, 191–213 (1991)
11. Pollak, R.A., Wales, T.J.: Welfare comparisons and equivalence scales. Am. Econ. Rev. Pap Proc. **69**, 216–221 (1979)
12. Veerbek, M., Nijman, T.E.: Can cohort data be treated as genuine panel data? Empir. Econ. **17**, 9–23 (1992)

The Determinants of Income Dynamics 44

Gustavo De Santis and Giambattista Salinari

Abstract

Models of income distribution more or less succeed in linking the current *level* of household (or individual) income to household (or individual) characteristics. However, they are typically far less satisfactory in explaining income *dynamics*. Gibrat's model proves helpful in highlighting the predominant role of randomness in the short run (here, 2–4 years), and this explains why other systematic influences are difficult to identify. One empirical regularity that does emerge, however, is that small incomes tend to increase more, and with more variability, than large ones. The traditional version of Gibrat's model does not incorporate this peculiarity, but this shortcoming can be overcome with a relatively minor modification of the original model.

Keywords

Income dynamics • Poverty and inequality • Statistical modelling

44.1 Household Income Dynamics and the Problem of Heteroscedasticity

The analysis of income dynamics is frequently carried out with regressions and OLS estimators, both of which assume homoscedasticity. But results can be biased if this condition is violated, and this, unfortunately, is virtually always the case with income dynamics.

G. De Santis (✉)
Department of Statistics, University of Florence, Italy
e-mail: desantis@disia.unifi.it

G. Salinari
University of Sassari, Piazza D'Armi, 17, 07100 Sassari, Province of Sassari, Italy
e-mail: gsalinari@uniss.it

N. Torelli et al. (eds.), *Advances in Theoretical and Applied Statistics*,
Studies in Theoretical and Applied Statistics, DOI 10.1007/978-3-642-35588-2_44,

At least two systematic violations of homoscedasticity have been observed. First, small incomes tend to increase more, and with more variability, than large ones [8]. Second, there are occupational categories whose incomes follow markedly different paths in terms of both average rates of increase and variability. Consider, for instance, two employed persons, a dependent and an independent worker. In any given period, the income growth rate of the independent worker tends to be not only higher but also more variable. Therefore, in order to determine, for instance, which of the two individuals is more likely to fall in poverty in any given time interval, both aspects (average and variance of growth rates) must be modelled.

And this is precisely what this paper sets out to do: present a model of income dynamics where heteroscedasticity is explicitly taken into account and can be (partly) explained by covariates. In so doing, we will also highlight the great variability of income trajectories—greater, indeed, than it is normally acknowledged in literature.

Already at the individual level, income trajectories are affected by very many variables: education, age, gender, industry, SES (Socio-Economic Status), changes in the labour market, economic cycles, etc. At the household level, which is what this paper refers to, the number of these influences increases exponentially, if one takes interactions into account, and soon becomes very difficult, if not impossible, to handle.

In order to simplify the problem, we break it in two: (A) we first model income trajectories in a theoretical, perfectly homogeneous population; and (B) later on, we complicate the model, by explicitly considering heterogeneity. This procedure permits us to disentangle the general features of income dynamics, common to all trajectories (point A), from those that are contingent on households characteristics (point B).

An empirical application to household incomes and their dynamics in the period 1998–2006 suggests that our model can keep both dimensions under control (the general process and the influence of specific situations, in terms of both average and variability of growth rates), and, both theoretically and empirically, compares favourably with other more conventional models of income dynamics.

44.2 Gibrat's Model: Traditional and Modified Version

Our model of income dynamics in a homogeneous population (see [8]), derives from [2] seminal paper, where incomes are assumed to evolve in a multiplicative form. In Gibrat's formulation, for every individual, income at time $t + k$ (y_{t+k}) can be obtained by multiplying income at time t (y_t) by a series of random shocks R_j occurring in between

$$y_{t+k} = y_t \prod_{j=1}^{k} R_j \tag{44.1}$$

which, in logarithms, becomes

$$\log(y_{t+k}) = \log(y_t) + \sum_{i=1}^{k} R_i \tag{44.2}$$

The multiplicative form implies that income is always strictly positive: depending on how income is defined and on the length of the period under consideration, this limitation may occasionally force researchers to drop a few rather extreme observations (but not in our application). In all cases, this does not constitute a major obstacle to the use of a model that has repeatedly proved very effective in the description of income dynamics.

Gibrat's model implies four main theoretical predictions, all of them relevant and empirically testable (see, e.g., [4–6, 8, 9]):

1. Log-growth is independent of log-income
2. The distribution of log-incomes approximates the normal
3. The variance of log-income increases linearly with k
4. The covariance of log-income increases with t, but decreases with k.

Empirical analyses, reported, for instance, by Hart [1, 3], or by Salinari and De Santis [7, 8], show that implications (44.2)–(44.4) hold, but implication (44.1) does not, because incomes that start from low levels tend to increase more and with more variability than the model predicts.

We contend that this divergence between theoretical expectations and empirical results depends on a sort of mechanical reason. In order to see this more clearly, it is convenient to transform Gibrat's model in its additive version:

$$y_{t+1} = y_t + y_t(R-1) = y_t + y_t r \tag{44.3}$$

The growth rate $r(= R\text{–}1)$ can be thought of as the difference between what might be defined as the "gain rate" g and the "loss rate" l, so that Eq. (44.3) can be rewritten as:

$$y_{t+1} = y_t + G - L \tag{44.4}$$

where both $G(= g \cdot y_t)$ and $L(= l \cdot y_t)$ are random, non-negative variables. Losses L cannot be greater than $y_t + G$, otherwise income at time $t + 1$ would be negative. Therefore, L is a truncated random variable, the probability density function of which ranges between 0 and $y_t + G$. The impact of truncation is obviously greater for small than for large incomes, so that the growth rate of small incomes tends to be positively affected.

But why is it also more variable? Under our homogeneity assumption, gains and losses can be imagined to occur to individuals as in a Bernoulli experiment, where the probability of getting a "unit" of loss (or gain) is proportional to individual income. In this case, the gains and losses experienced by households distribute as a Poisson, the mean and variance of which increase linearly with income. As a consequence, the variance of the growth rate $V[r] = V[g\text{–}l]$ decreases as income increases.

A possible way of improving over Gibrat's model is to use a truncated normal distribution (T) to model income variations ($\Delta y = y_{t+1} - y_t$). Let these variations be normally distributed with mean and variance proportional to y_t. Since variations cannot be smaller than $-y_t$ (otherwise y_{t+1} would be negative), we restrict the domain of the probability distribution to the interval $(-y_t, +\infty)$. The probability density function $f(\cdot)$ of the relevant truncated random variable T is therefore:

$$f(\Delta y | y_t, \alpha, \beta) = \frac{\phi\left(\frac{\Delta y - a y_t}{b y_t}\right)}{1 - \Phi\left(\frac{-y_t - a y_t}{b y_t}\right)} \tag{44.5}$$

where a and b are two parameters defining the mean ($a y_t$) and the variance ($b y_t$) of the normal distribution, $\phi(\cdot)$ is the probability density function of the standard normal variable, and $\Phi(\cdot)$ its cumulative function.

We can now define Gibrat's "modified model" as:

$$y_{t+1} = y_t + T \tag{44.6}$$

where the two parameters a and b can be estimated numerically, for instance with maximum likelihood. Figure 44.1 shows a comparison between the actual and the expected (44.5) income variations registered for 5,195 Italian households in a series of 2-year intervals (1998–2000, 2000–2002,..., 2004–2006), conditional on the income class to which households belonged at the beginning of each period. Note that the variability of absolute income variations increases with income and that our model curves approximate reality rather well, especially if one considers that no covariate has been taken into consideration yet (homogeneity assumption).

The model of Eq. (44.6) can be extended to the case of heterogeneous populations, where households differ by their socio-economic characteristics. Let $\mathbf{x}_{t,i} = (1, x_{1,i}, \ldots, x_{p,i})$ be a time-dependent vector of covariates for household i, while $\boldsymbol{\alpha} = (\alpha_0, \alpha_1, \ldots, \alpha_p)$ and $\boldsymbol{\beta} = (\beta_0, \beta_1, \ldots, \beta_p)$ are two vectors of parameters describing the effects produced by these covariates, respectively, on the mean and on the variance of the distribution of the variations. Model (44.5) can now be generalized as follows:

$$\begin{cases} a_i = \exp(\boldsymbol{\alpha}' \mathbf{x}_{t,i}) \\ b_i = \exp(\boldsymbol{\beta}' \mathbf{x}_{t,i}) \end{cases} \tag{44.7}$$

44.3 SHIW Data

For the empirical part of our study, we use micro data taken from the Survey on Household Income and Wealth (SHIW). The Bank of Italy carries out this survey every other year, on about 8,000 households (about 20,000 individuals): it

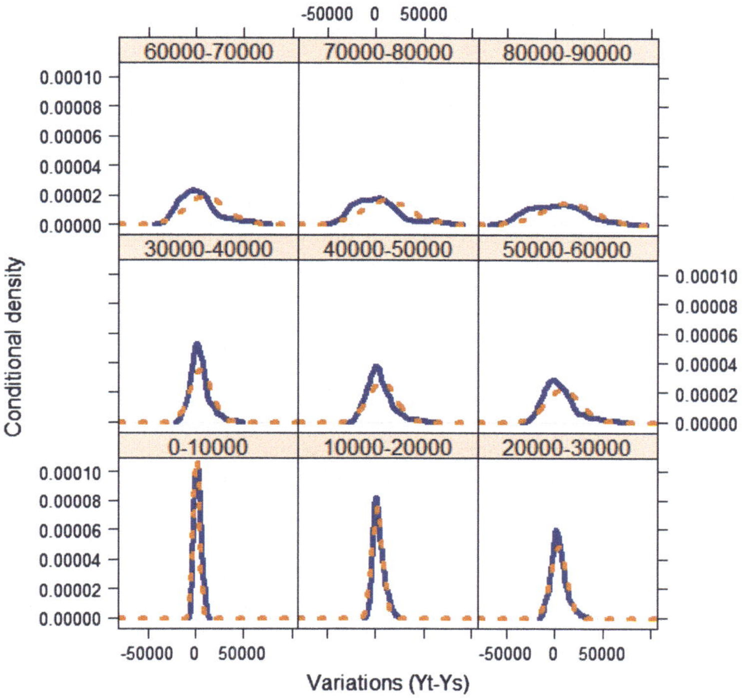

Fig. 44.1 Probability density distribution of conditional income variations (Italy 1998–2006). The variations (Δy) are observed in the periods 1998–2000,..., 2004–2006 for 5,195 Italian households initially belonging to different income classes (0–10,000,..., 80,000–90,000 Euros). The *solid line* indicates the empirical densities estimated with Gaussian kernel. The *dashed lines* indicate the theoretical distribution generated by Eq. (44.5) (with $a = 0.16$ and $b = 0.32$)

is described in detail on the website of the Bank,[1] from where elementary data can be downloaded, and, most importantly, it contains a panel part that we will exploit in this application.

We work on five rounds of the survey (years 1998, 2000, 2002, 2004 and 2006), and we study the change in the total net household income over a period of either 2 years (1998–2000; 2000–2002;...) or 4 years (1998–2002; 2000–2004;...). We work on a subset of the panels, selected on the basis of two criteria: the demographic structure of the household must not change in the period under examination (2 years or 4 years, depending on the application; see further in the text) and, at the beginning of the period, the reference person must not be older than 64 years and must be in the labour market. After dropping a few outliers (fewer than 1%, with abnormally high or low growth rates), we are left with about 5,100 biannual transitions and 2,400 quadrennial transitions, approximately equally distributed among the various time intervals considered.

[1] http://www.bancaditalia.it/statistiche/indcamp/bilfait.

44.4 A Regression Analysis of Income Dynamics

Before applying our modified version of Gibrat's model, let us take a step backward and consider the results produced by an ordinary regression analysis of household income and growth rates, which we will later compare to our own.

We are interested in modeling income dynamics, using "independent" explanatory variables. Actually, most of the variables that we use are not truly exogenous (as we will briefly discuss below) but our interest here is not really about how to "explain" income: merely to show that our approach (modified Gibrat's model) works better than other, more traditional ones, using the same set of explanatory variables.

In the traditional version, the dependent variable is the growth rate (Table 44.1, columns 2 and 3); we, instead, model transition probabilities (not reported here; but see, e.g. [8]) which result, among other things, in growth rates (Table 44.2). In both cases, we use a set of "independent" variables, of two broad types:

(a) *Demographic*: number of children $(0, 1, 2, 3+)$, region (North, Center, South), age, and educational attainment of the reference person ($1 =$ illiterate or primary school; $2 =$ secondary education; $3 =$ tertiary education; $4 =$ university degree or PhD).

(b) *Labour force*: occupation of the reference person (dependent, self-employed, unemployed, other), period.

As for the traditional approach, although we are basically interested in dynamics (growth rates), we also model income levels. In the latter case, we use:

$$\log(y_t) = \boldsymbol{\beta}^{\mathrm{T}}\mathbf{x}_{t,i} + \varepsilon_{t,i} \qquad (44.8)$$

while in the former (growth rates—both for 2- and 4-year periods), we use:

$$\log\left(\frac{y_{t+1,i}}{y_{t,i}}\right) = \boldsymbol{\beta}^{\mathrm{T}}\mathbf{x}_{t,i} + \varepsilon_{t,i} \qquad (44.9)$$

Our results are summarized in Table 44.1. In all the cases, the baseline is a household living in the north of Italy, with one child and whose reference person is dependent worker, aged 45, with an average level of education. The static regression of income (first column) confirms what is generally known: all the covariates we use affect income in the expected direction: incomes are higher for better educated and older (but younger than 65 years) reference persons, while living in the south, or being a dependent worker (or, worse still, unemployed), depresses income. Finally, income increases with the number of children. In all of these cases, the interpretation must be particularly cautious, because reverse causation, selection and unobserved heterogeneity surely play a major role, but, as mentioned before, we will not attempt to investigate these issues here; we are simply comparing models of income dynamics, and we are using this regression on income levels as an entry point, whose utility will appear shortly. This simple model "explains" about 43% of the variance in incomes.

Table 44.1 Regression analysis of income and income growth rates

Dependent variable	Income		Growth rate over 2 years		Growth rate over 4 years	
Intercept	10.223	***	0.087	***	0.167	***
Children 0	−0.089	***	−0.025		−0.056	
Children 2	0.077	***	0.010		0.034	
Children 3	0.069	**	0.040	*	0.051	
Age (Stand.)	0.140	***	−0.014	*	−0.011	
Education 1	−0.254	***	0.052	**	0.095	**
Education 3	0.240	***	0.009		0.008	
Education 4	0.562	***	0.023		0.006	
Self-Empl.	0.064	***	0.008		0.037	
Unempl.	−0.616	***	0.231	***	0.403	***
Other	−0.160	***	0.036	.	0.084	**
Period 2	0.067	***	−0.011		0.012	
Period 3	0.110	***	0.002		−0.010	
Period 4	0.168	***	−0.023		/	
Center It.	−0.124	***	0.017		0.030	
South It.	−0.476	***	−0.008		−0.028	
R^2	0.43		0.01		0.03	

Significance level: *** = 1%;** = 5%;* = 10%. Periods: 1 = from 1998 (ref.); 2 = from 2000; 3 = from 2002; 4 = from 2004. Education: 1 = low; 2 = medium (ref.); 3 = high; 4 = very high. For the reasons of comparability, R^2 for the growth rates regressions have been calculated as var[exp(fitted.values)]/var(actual.growth.rates)

If we now turn to the analysis of dynamics, i.e. of 2-year growth rates (second column) we find that only a few of our covariates exert a measurable effect, and the explained variance is much lower—merely 1%. The only variables that seem to matter are those associated with a low income at the start (which favours a more rapid increase in the period): for instance being unemployed, or with a low education.

Basically the same happens over a longer time span (4 years, third column), with a very slight improvement in the overall goodness of fit (R^2, from 1% to 3%). In short, while our covariates are associated with income levels, they do not seem to be associated with growth rates. How is this possible?

In order to answer this question, we note that:

(a) Income growth is highly erratic
(b) Growth is a cumulative process: very small systematic differences, even if they go unnoticed in the short run, may result in large differences in the long run
(c) Income trajectories can (and indeed do seem to) depend on their starting point
(d) What we observe here are net incomes, which are also influenced by the effects of the fiscal policy (progressive taxation, subsidies, etc.)

Table 44.2 Analysis of income rates of growth through the modified Gibrat's model

	Variation after 2 years				Variation after 4 years			
Coeff.	Estim. α		Estim. β		Estim. α		Estim. β	
Intercept	−1.498	*	−0.599	*	−0.934	*	−0.266	*
Children 0	0.114	*	−0.010		−0.260	*	−0.002	
Children 2	−0.105		0.028		−0.067		−0.150	*
Children 3	0.025		0.075		0.081		−0.058	
Age (Stand.)	−0.083	*	−0.009		0.059		0.013	
Education 1	−0.063		0.233	*	0.150		0.357	*
Education 3	−0.249	*	−0.121	*	−0.155	*	−0.089	
Education 4	−0.232	*	−0.057		−1.160	*	−0.597	*
Self-Empl.	0.203	*	0.295	*	0.201	*	0.289	*
Unempl.	0.518	*	0.347	*	0.568	*	0.486	*
Other	0.047		0.023		0.294	*	0.080	
Period 2	−0.071		−0.306	*	0.045		−0.216	*
Period 3	0.011		−0.111	*	0.004		−0.214	*
Period 4	0.150	*	−0.121		\		\	
Center It.	−0.011		−0.077		0.093		0.091	*
South It.	0.118		0.196	*	0.142	*	0.050	
R^2	0.02				0.05			

* Significant at 10%
R^2 calculated as var(fitted.growth-rates)/var(actual.growth-rates)

As for the first two points, consider, for example, "Self-employment" in Table 44.1 (2nd and 3rd column): the coefficients are positive (income increases more rapidly for this category), but not significant. However, the mean income of the self-employed is significantly higher than that of the dependent workers. Besides, over a 4-year period, the effect of self-employment is slightly more significant than it is over a 2-year period, and this is a pattern that holds for most of our variables.

Unfortunately, this explanation cannot be simply extended to all our covariates. Take "Education 1", for instance: this has a positive and significant effect on both the biannual and quadrennial growth rates, but the income of the poorly educated workers is significantly smaller than that of highly educated ones. Our interpretation is that this apparent contradiction depends on the different starting income of the two groups: highly educated workers are richer at the beginning (they typically come from richer families, and their entry wage is higher), but they tend to lose some of their initial advantage as time goes by.

44.5 Income Dynamics from Gibrat's Perspective

Let us now apply our modified Gibrat's model to the same data set. Our starting values are the estimates of the coefficients obtained in the previous regression (Table 44.1, columns 2 and 3), and the confidence intervals for the coefficients have been estimated through bootstrapping (500 repetitions). The results of our analysis are summarized in Table 44.2, where the $\boldsymbol{\alpha}$ coefficients represent the effect produced by the covariates on the average increment, and, apart from the intercept, are directly comparable with those estimated in the regression of growth rates. The $\boldsymbol{\beta}$ coefficients, instead, represent the effects of the covariates on the variance of these rates of growth, and for these there is no corresponding value in the preceding table.

The $\boldsymbol{\alpha}$ parameters of Table 44.2 tell basically the same story as in Table 44.1: variations in income are, on average, more strongly positive for the unemployed and for the self-employed; the evolution is relatively worse for the well educated, and the effect of age is not so clear. Note, however, that the variance explained by the model, although still very low, improves slightly and, more importantly, the cases when the $\boldsymbol{\alpha}$ parameters are significant (if only at the 10% level) are now considerably more numerous than in Table 44.1. For example, self-employment exerts a significant positive effect in all the cases of Table 44.2, while this effect is weak or insignificant in the estimation of Table 44.1. We advance two tentative explanations for this improvement: (1) our modified Gibrat's model corrects the bias produced by what we referred to as the "differential dynamics of income" (small incomes increase faster and with a greater variability than others); (2) an optimization process based on likelihood is less influenced by the presence of outliers.

A distinctive feature of our modified Gibrat's model is that it allows us to measure the effects produced by household's characteristics also on the variability of income, and not only on its average. Our results suggest that low education, unemployment and self-employment all contribute to an increase in the variability of income variations in the subsequent period (of either 2 or 4 years). Note that periods 2 and 3 of both analyses (2 years and 4 years increments) are characterized by a significant lower variability than the baseline period (period 1). This may depend on the macroeconomic situation: the period 2002–2005 witnessed a slowdown of economic growth in Italy, with an average annual growth rate of the GDP of only about 0.6%. And in times of economic recession, the variability of income variations typically shrinks.

44.6 Conclusions

Our modified Gibrat's model seems to describe income dynamics better than other models from at least three different points of view:

1. We can theoretically justify, and model, why smaller incomes increase more, and with more variability, than others.
2. In our estimation, we circumvent the problem of heteroscedasticity, which is explicitly modelled. In our case, the bias turns out to be relatively modest (α coefficients are similar in Tables 44.1 and 44.2), but this need not be always the case.
3. Finally, and perhaps most importantly, our modified Gibrat's model permits us to model the *variance* of income dynamics, and to measure the impact of covariates

on this variance. This has several advantages: for instance, it allows researchers to better identify the population subgroups who are more exposed to the risk of poverty in any given time interval.

Acknowledgements Financial support from the Italian MIUR and from the EU is gratefully acknowledged.

References

1. Fields, G.S., Duval Hernández, R.D., Freije Rodríguez, S.F., Sánchez Puerta, M.L.: Income Mobility in Latin America, mimeo (2006). http://www.cid.harvard.edu/Economia/Mexico06%20Files/fieldsetal%20102306.pdf
2. Gibrat, R.: Les inégalités économiques: applications aux inégalité des richesses, à la concentration des entreprises, aux populations des villes, aux statistiques de familles, etc., d'une loi nouvelle, la loi de l'effet proportionnel, Librerie du Recueil Sirey, Paris (1931)
3. Hart, P.E.: The comparative statics and dynamics of income distributions. J. R. Stat. Soc. A Gen. **139**(1), 108–125 (1976)
4. Kalecki, M.: On the Gibrat distribution. Econometrica **13**(2), 161–170 (1945)
5. Mitzenmacher, M.: A brief history of generative models for power law and lognormal distributions. Internet Math. **1**(2), 226–251 (2003)
6. Neal, D., Rosen, S.: Theories of the distribution of earnings. In: Atkinson, A.B., e Bourguignon, F. (eds.) Handbook of Income Distribution, vol. 1, Chapter 7, pp. 379–427. Elsevier, London (2000)
7. Salinari, G., De Santis, G,: On the Evolution of Household Income, Luxemb. Income Study Work. Pap. Ser., No. 488, July (2008)
8. Salinari, G., De Santis, G.: The evolution and distribution of income in a life-cycle perspective. Riv. Ital. di Econ. Demogr. e Stat. **63**(1–2), 257–274 (2009)
9. Shorrocks, A.F.: Income mobility and the markov assumption. Econ. J. **86**(343), 566–578 (1976)

45 Benchmarking and Movement Preservation: Evidences from Real-Life and Simulated Series

Tommaso Di Fonzo and Marco Marini

Abstract

The benchmarking problem arises when time series data for the same target variable are measured at different frequencies with different level of accuracy, and there is the need to remove discrepancies between annual benchmarks and corresponding sums of the sub-annual values. Two widely used benchmarking procedures are the modified Denton *Proportionate First Differences* (PFD) and the Causey and Trager *Growth Rates Preservation* (GRP) techniques. In the literature it is often claimed that the PFD procedure produces results very close to those obtained through the GRP procedure. In this chapter we study the conditions under which this result holds, by looking at an artificial and a real-life economic series, and by means of a simulation exercise.

Keywords

Benchmarking • Causey and Trager GRP • Combining data from different sources • Movement preservation • Modified Denton PFD

T. Di Fonzo (✉)
Department of Statistical Sciences, University of Padua, Via C. Battisti 241, Padova, Italy
e-mail: difonzo@stat.unipd.it

M. Marini
International Monetary Fund (IMF), Statistics Department, Washington, DC, USA
e-mail: MMarini@imf.org

The views expressed herein are those of the authors and should not be attributed to the IMF, its Executive Board, or its management

N. Torelli et al. (eds.), *Advances in Theoretical and Applied Statistics*,
Studies in Theoretical and Applied Statistics, DOI 10.1007/978-3-642-35588-2_45,

45.1 Introduction

Benchmarking monthly and quarterly series to annual series is a common practice in many National Statistical Institutes. The benchmarking problem arises when time series data for the same target variable are measured at different frequencies with different level of accuracy, and there is the need to remove discrepancies between annual benchmarks and corresponding sums of the sub-annual values. The most widely used benchmarking procedures are the modified Denton *Proportionate First Differences* (PFD) technique [4, 6], and the [3] *Growth Rates Preservation* (GRP) procedure (see also Trager [11], and Bozik and Otto [2]). The PFD procedure looks for benchmarked estimates aimed at minimizing the sum of squared proportional differences between the target and the unbenchmarked values, and is characterized by an explicit benchmarking formula involving simple matrix operations. The GRP technique is a nonlinear procedure based on a "true" movement preservation principle, according to which the sum of squared differences between the growth rates of the target and of the unbenchmarked series is minimized. As in the literature [1, 4, 5] it is often claimed that the PFD procedure produces results very close to those obtained through the GRP procedure, in this chapter we study the conditions under which this result holds. We do that by showing how the two procedures work in practice, by looking at an artificial and a real-life economic series. Then a simulation exercise is performed in order to appreciate the impact on the benchmarked series of the variance of the observational error and of possible "steps" in the annual benchmarks.

The chapter is organized as follows. In Sect. 45.2 the two benchmarking procedures are described, and the way they take into account a "movement preservation principle" is discussed. In Sect. 45.3 the artificial time series of Denton [6] and a quarterly preliminary series of the EU Quarterly Sector Accounts [7] are benchmarked to their annual counterparts, using both modified Denton PFD and Causey and Trager GRP benchmarking procedures, and the results are discussed. In Sect. 45.4 we design a simulation exercise in order to analyze the distinctive features of the two procedures.

45.2 Two Benchmarking Procedures

Let Y_T, $T = 1, \ldots, N$, and p_t, $t = 1, \ldots, n$, be, respectively, the (say annual) totals and the (say quarterly) preliminary values of an unknown quarterly target variable y_t. The preliminary values being not in line with the annual benchmarks, i.e., $\sum_{t \in T} p_t \neq Y_T$, $T = 1, \ldots, N$, we look for benchmarked estimates y_t^b such that $\sum_{t \in T} y_t^b = Y_T$.

As Bozik and Otto [2, p. 2] stress, "Just forcing a series to sum to its benchmark totals does not make a unique benchmark series." Some characteristic of the original series p_t should be considered in addition, in order to get benchmarked estimates "as close as possible" to the preliminary values. In an economic time series framework, the preservation of the temporal dynamics (however defined) of the preliminary series is often a major interest of the practitioner. Thus in what follows we consider two procedures designed to preserve at the best the movement of the series p_t: modified Denton PFD and Causey and Trager GRP.[1]

Denton [6] proposed a benchmarking procedure grounded on the *Proportionate First Differences* between the target and the original series. Cholette [4] slightly modified the result of Denton, in order to correctly deal with the starting conditions of the problem. The PFD benchmarked estimates are thus obtained as the solution to the constrained quadratic minimization problem

$$\min_{y_t} \sum_{t=2}^{n} \left(\frac{y_t}{p_t} - \frac{y_{t-1}}{p_{t-1}} \right)^2 \qquad \text{subject to} \quad \sum_{t \in T} y_t = Y_T, \quad T = 1, \ldots, N. \quad (45.1)$$

In matrix notation, denoting $\mathbf{p}$ and $\mathbf{Y}$ the $(n \times 1)$ and $(N \times 1)$, respectively, vectors of preliminary and benchmark values, the PFD benchmarked series is contained in the $(n \times 1)$ vector $\mathbf{y}^{\text{PFD}}$ solution of the linear system [4, p. 40]

$$\begin{bmatrix} \mathbf{Q} & \mathbf{C}' \\ \mathbf{C} & \mathbf{0} \end{bmatrix} \begin{bmatrix} \mathbf{y}^{\text{PFD}} \\ \lambda \end{bmatrix} = \begin{bmatrix} \mathbf{Qp} \\ \mathbf{Y} \end{bmatrix}, \qquad (45.2)$$

where λ is a $(N \times 1)$ vector of Lagrange multipliers, $\mathbf{Q} = \mathbf{P}^{-1} \Delta_n' \Delta_n \mathbf{P}^{-1}$, $\mathbf{P} = \text{diag}(\mathbf{p})$, $\mathbf{C}$ is a $(N \times n)$ temporal aggregation matrix converting quarterly values in their annual sums, and Δ_n is the $((n-1) \times n)$ first differences matrix.

Notice that $\Delta_n' \Delta_n$ has rank $n - 1$, so $\mathbf{Q}$ is singular. However, provided no preliminary value is equal to zero, the coefficient matrix of system (45.2) has full rank (see the Appendix). After a bit of algebra, the solution of the linear system (45.2) can be written as

$$\begin{bmatrix} \mathbf{y}^{\text{PFD}} \\ \lambda \end{bmatrix} = \begin{bmatrix} \mathbf{P} & \mathbf{0} \\ \mathbf{0} & \mathbf{I}_N \end{bmatrix} \begin{bmatrix} \Delta_n' \Delta_n & \mathbf{PC}' \\ \mathbf{CP} & \mathbf{0} \end{bmatrix}^{-1} \begin{bmatrix} \mathbf{0} \\ \mathbf{Y} \end{bmatrix}. \qquad (45.3)$$

Causey and Trager [3] consider a different quadratic minimization problem, in which the criterion to be minimized is explicitly related to the growth rate, which is a natural measure of the movement of a time series:

$$\min_{y_t} \sum_{t=2}^{n} \left(\frac{y_t}{y_{t-1}} - \frac{p_t}{p_{t-1}} \right)^2 \qquad \text{subject to} \quad \sum_{t \in T} y_t = Y_T, \quad T = 1, \ldots, N. \quad (45.4)$$

[1]Empirical comparisons between the Cholette–Dagum regression-based benchmarking approach, which can be seen [5] as a generalization of the seminal contribution by Denton [6], and the Causey and Trager approach, are shown in Titova et al. [10].

Looking at the criterion to be minimized in (45.4), it clearly appears that, differently from (45.1), it is grounded on an "ideal" movement preservation principle, "formulated as an explicit preservation of the period-to-period rate of change" of the preliminary series [1, p. 100].

It should be noted that while problem (45.1) has linear first-order conditions for a minimum, and thus gives rise to an explicit solution as shown in (45.3), the minimization problem in (45.4) is inherently nonlinear. Trager [11, see Bozik and Otto [2]] suggests to use a technique based on the steepest descent method,[2] using $\mathbf{y}^{\text{PFD}}$ as starting values, in order to calculate the benchmarked estimates y_t^{GRP}, $t = 1, \ldots, n$, solution to problem (45.4).

We employ the Interior Point method of the Optimization Toolbox of MATLAB® (version 2009b). It consists of an iterative procedure that solves a sequence of approximate unconstrained minimization problems by standard (quadratic) nonlinear programming methods. In each iteration the procedure exploits the exact gradient vector and hessian matrix of the Lagrangian function [8], which enables to make informed decisions regarding directions of search and step length. This fact makes the procedure feasible and robust, in terms of reduced numbers of iterations required for the convergence, as far as of quality of the found minimum.

It is interesting to go deep into the relationship between the criteria optimized by the two alternative procedures. Let

$$C_{\text{PFD}} = \sum_{t=2}^{n} \left(\frac{y_t}{p_t} - \frac{y_{t-1}}{p_{t-1}} \right)^2 \quad \text{and} \quad C_{\text{GRP}} = \sum_{t=2}^{n} \left(\frac{y_t}{y_{t-1}} - \frac{p_t}{p_{t-1}} \right)^2$$

be the objective functions of the PFD and GRP benchmarking procedures, respectively. We can write (U.S. Census Bureau [12], p. 96):

$$C_{\text{PFD}} = \sum_{t=2}^{n} \left[\frac{y_{t-1}}{p_t} \left(\frac{y_t}{y_{t-1}} - \frac{p_t}{p_{t-1}} \right) \right]^2. \tag{45.5}$$

Expression (45.5) makes clear the relationship between C_{PFD} and C_{GRP}. The term in parentheses is the difference between the growth rates of the target and the preliminary series, namely the addendum of C_{GRP}. In C_{PFD} these terms are weighted by the ratio between the target series at $t-1$ and the preliminary series at t. When these ratios are relatively stable over time, which is the case when the "benchmark-to-indicator ratio" [1]

$$\frac{Y_T}{\sum_{t \in T} p_t}, \qquad T = 1, \ldots, N$$

[2]For a recent survey on this issue, see Di Fonzo and Marini [8].

is a smooth series, C_{PFD} and C_{GRP} are very close to each other. On the contrary, when the ratios (y_{t-1}/p_t) behave differently each term in the summation is over-(under-)weighted according to the specific relationship between target and preliminary series in that period. For example, sudden breaks in the movements of y_{t-1}/p_t might arise in case of large differences between the annual benchmarks and the annually aggregated preliminary series. The situation is rather similar to the one described by Dagum and Cholette [5, p. 121], when the risk of producing negative benchmarked values is discussed: "These situations occur when the benchmarks dramatically change from one year to the next, while the sub-annual series changed very little in comparison; or, when the benchmarks change little, while the annual sums of the sub-annual series change dramatically."

Keeping in mind this relationship, we move to investigate on the differences between the PFD and the GRP benchmarking solutions in simulated and real-life cases.

45.3 Evidences from Artificial and Real Time Series

In this section we apply both the PFD and GRP benchmarking procedures to two illustrative examples, in order to show to what extent the former solution can be used effectively to approximate the "ideal" movement preservation criterion based on growth rates. We consider also a distance measure between the growth rates of the preliminary and target series given by the absolute, rather than the squared, value of their difference. The results are thus evaluated looking at the two ratios

$$r_\alpha = \left(\frac{\sum_{t=2}^{n} \left| \frac{y_t^{\text{GRP}}}{y_{t-1}^{\text{GRP}}} - \frac{p_t}{p_{t-1}} \right|^\alpha}{\sum_{t=2}^{n} \left| \frac{y_t^{\text{PFD}}}{y_{t-1}^{\text{PFD}}} - \frac{p_t}{p_{t-1}} \right|^\alpha} \right)^{\frac{1}{\alpha}} \qquad \alpha = 1, 2. \tag{45.6}$$

When $\alpha = 2$, this index is simply the square root of the ratio between the Causey and Trager "Growth Rate Preservation" criteria computed from the two solutions. Obviously, we expect that the GRP technique always reaches a lower value of the chosen criterion than PFD, and thus the ratio r_2 should be never larger than 1. Put in other words, r_2 is the ratio between the Root Mean Squared Adjustments to the preliminary growth rates produced by the Causey and Trager GRP and the Denton PFD benchmarking procedures. On the other hand, r_1 can be seen as the ratio between the Mean Absolute Adjustments: sometimes this index can be larger than 1, thus indicating a better performance of Denton PFD when the size of the corrections to the preliminary growth rates is measured according to an absolute rather than a squared form.

The first example we consider is the artificial preliminary series used in the seminal paper of Denton [6]. It consists of a 5-year artificial quarterly series, with a fixed seasonal pattern invariant from year to year. The values are 50, 100, 150, and 100 in the four quarters, for a total yearly amount of 400. The annual

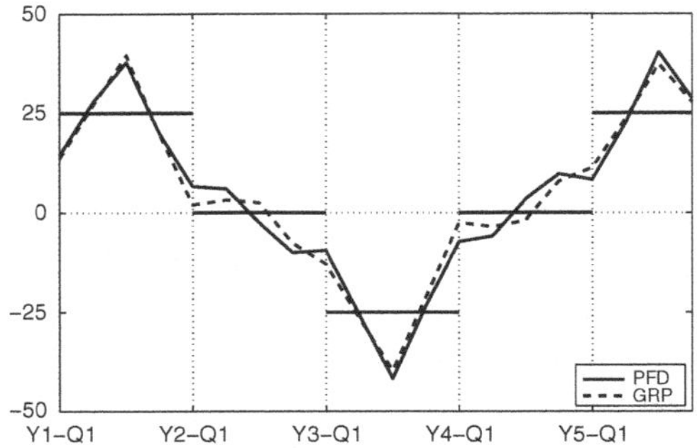

Fig. 45.1 Adjustments to the artificial series produced by the PFD and GRP procedures

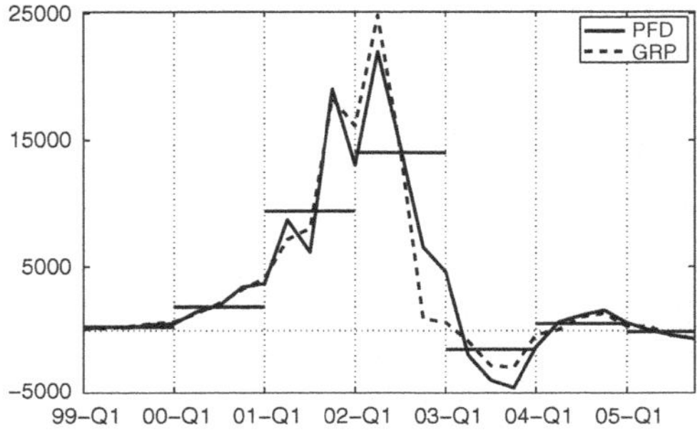

Fig. 45.2 Adjustments to the real-life series produced by the PFD and GRP procedures

benchmarks are assumed to be 500, 400, 300, 400, and 500 in the 5 successive years. The corresponding discrepancies (i.e., the differences between the known benchmarks and the annual sums of the preliminary series) are therefore 100, 0, −100, 0, and 100, respectively. As expected, the minimum C_{GRP} is achieved by the GRP procedure (0.04412 against 0.14428 of PFD, with $r_2 = 0.553$). The GRP procedure shows better results as regards the movement preservation, also when the distance between the preliminary and the target growth rates is measured by the absolute differences ($r_1 = 0.539$). Figure 45.1 shows the adjustments to the levels of the original series in the two cases. The horizontal lines in each year denote the (average) annual discrepancy to be distributed.

The second example is a real-life economic series coming from the European Quarterly Sector Accounts (EU-QSA). The EU-QSA system has been dealt with by Di Fonzo and Marini [7] in a *reconciliation* exercise, where several time series have to be adjusted in order to be in line with both temporal and contemporaneous known aggregates [5]. In this chapter we consider the series "Other Property Income" of the Financial Corporation sector, showing a considerable amount of temporal discrepancies. Figure 45.2 shows the large discrepancy in 2002, when the original series accounts for just 65% of the annual target. From 2003 onwards the discrepancies are much more contained. This is a typical practical situation where the preservation of the original growth rates can be better guaranteed by the GRP procedure ($r_2 = 0.579$ and $r_1 = 0.615$). The quarterly adjustments in the two cases are also displayed in Fig. 45.2. The differences are large in the years with large discrepancies (2001–2002), but they are also remarkable in 2003, when the discrepancy is limited. In this case the smoother distribution produced by the GRP procedure is clearly visible.

45.4 The Simulation Exercise

By means of this experiment we wish to shed light on the conditions under which the PFD benchmarking procedure produces results "close" to the GRP technique in terms of differences between the growth rates of the benchmarked and preliminary series. We consider quarterly series covering a period of 7 years ($n = 28$).

Let $\theta_t = \theta_{t-1} + \varepsilon_t$ be a random walk process, where ε_t is a Gaussian white noise with unit variance ($\sigma_\varepsilon^2 = 1$) and $\theta_0 = \varepsilon_0$. The target series of the exercise, y_t, is derived as

$$y_t = \theta_t^* + \mu_t, \qquad t = 1, \ldots, n$$

with $\theta_t^* = 100 + \theta_t$, where the constant term 100 is large enough to prevent negative values, and μ_t is given by

$$\mu_t = \begin{cases} \mu & t = 9, \ldots, 16 \\ -\mu & t = 17, \ldots, 24 \\ 0 & \text{elsewhere} \end{cases} .$$

The preliminary series p_t is related to y_t as follows:

$$p_t = \theta_t^* + e_t, \qquad t = 1, \ldots, n$$

where e_t is a Gaussian white noise with variance σ_e^2. It is clear that preliminary and target series are different for the effects of μ_t and e_t. The former is introduced in the model for y_t in order to simulate yearly biases of the preliminary series. The first control parameter of the experiment is thus μ. When $\mu > 0$, the target series contains a positive drift from p_t in years 3 and 4, followed by a negative step (of the same amount) in years 5 and 6. We set $\mu = 0, 15, 30, 45, 60$. The second control parameter is σ_e, the standard deviation of the innovation process e_t. The larger this parameter is, the larger the observational error in the preliminary series will be. We set $\sigma_e = 5, 10, 15, 20, 25$.

We drew two sets of 1,000 n-dimensional vectors as $N(0, 1)$. One set is used to simulate ε_t; the other is used to derive the innovation e_t according to the five levels of σ_e. By using the five values of μ, we achieved 1,000 experiments for each of the 25 combinations. For each combination, we computed summary statistics on the ratios r_1 and r_2 obtained over the 1,000 experiments.

Table 45.1 Median of r_2 for different values of σ_e and μ (across 1,000 experiments)

	μ				
σ_e	0	15	30	45	60
5	0.992	0.984	0.959	0.903	0.763
10	0.967	0.961	0.935	0.879	0.739
15	0.928	0.922	0.895	0.839	0.702
20	0.870	0.866	0.841	0.785	0.655
25	0.800	0.795	0.771	0.718	0.598

Table 45.2 Maximum of r_2 for different values of σ_e and μ (across 1,000 experiments)

	μ				
σ_e	0	15	30	45	60
5	0.999	0.995	0.982	0.946	0.870
10	0.996	0.993	0.978	0.950	0.888
15	0.992	0.987	0.977	0.946	0.903
20	0.987	0.982	0.973	0.941	0.907
25	0.980	0.977	0.965	0.943	0.885

Tables 45.1 and 45.2 show, respectively, the median and the maximum values[3] of r_2 under different values of σ_e (rows) and μ (columns).

According to Bozik and Otto [2], we used the series benchmarked *via* modified Denton PFD as starting values of the GRP procedure, and this turned out to be a good choice[4]: as one would expect, from Table 45.2 it appears that in all cases the GRP procedure improves on the modified PFD starting values and reaches a lower value of the criterion.

However, from Table 45.1 we observe that the PFD procedure provides very similar results to GRP when discrepancies are small and unsystematic (median $r_2 \geq 0.9$ when $\sigma_e \leq 10$ and $\mu \leq 15$). The reduction is stronger as both σ_e and μ increase.

These results are confirmed by r_1, whose median and maximum values are shown in Tables 45.3 and 45.4, respectively, with an important remark: from Table 45.4 we observe that if the absolute differences between preliminary and target growth rates are considered, and when the bias is either absent or small ($\mu \leq 15$), there are cases where Denton PFD gives benchmarked estimates whose dynamics is "closer" to the preliminary series than Causey and Trager GRP does.[5]

[3]The median is more representative than the mean in the case of atypical values. We also calculated mean, standard deviation, minimum, and range of r_1 and r_2, available on request from the authors.

[4]When the preliminary series were used as starting values, in 50 out of 25,000 cases (0.2%) the GRP procedure produced benchmarked series with $r_2 > 1$.

[5]The index r_1 is greater than one for 488 out of 25,000 series (1,95%). The highest number of cases with $r_1 > 1$ (270) is observed for $(\sigma_e, \mu) = (5, 0)$, followed by 101 cases for $(\sigma_e, \mu) = (10, 0)$. The remaining cases are: 50 for $(\sigma_e, \mu) = (15, 0)$, 15 for $(\sigma_e, \mu) = (20, 0)$, 4 for $(\sigma_e, \mu) = (25, 0)$, 20 for $(\sigma_e, \mu) = (5, 15)$, 15 for $(\sigma_e, \mu) = (10, 15)$, 8 for $(\sigma_e, \mu) = (15, 15)$, 2 for $(\sigma_e, \mu) = (20, 15)$, and 3 for $(\sigma_e, \mu) = (25, 15)$.

Table 45.3 Median of r_1 for different values of σ_e and μ (across 1,000 experiments)

	μ				
σ_e	0	15	30	45	60
5	0.993	0.977	0.941	0.880	0.748
10	0.969	0.956	0.922	0.860	0.725
15	0.928	0.917	0.884	0.825	0.693
20	0.873	0.864	0.832	0.773	0.650
25	0.805	0.795	0.765	0.710	0.597

Table 45.4 Maximum of r_1 for different values of σ_e and μ (across 1,000 experiments)

	μ				
σ_e	0	15	30	45	60
5	1.035	1.015	0.997	0.936	0.841
10	1.053	1.013	0.997	0.964	0.871
15	1.047	1.017	0.982	0.952	0.896
20	1.028	1.022	0.982	0.941	0.903
25	1.013	1.057	0.970	0.929	0.874

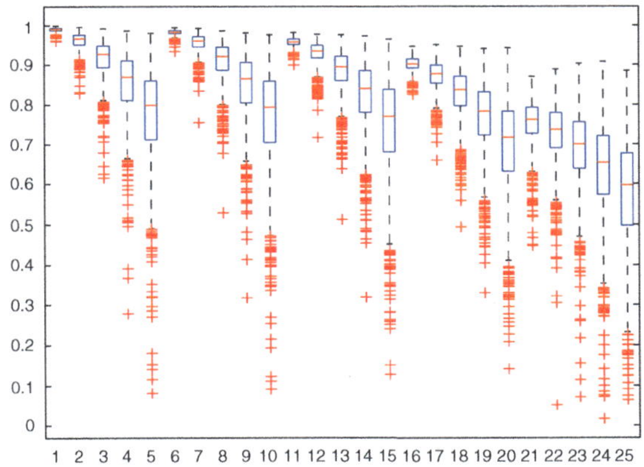

Fig. 45.3 Boxplots of r_2

A visual, more comprehensive overview of the results of the simulation experiment is given by the boxplots of r_2 (Fig. 45.3), and r_1 (Fig. 45.4). The 25 boxplots in each figure are ordered (from left to right) so that the first group of five boxplots corresponds to $\sigma_e = 5$ and $\mu = 0, 15, 30, 45, 60$, respectively, the second group to $\sigma_e = 10$ and $\mu = 0, 15, 30, 45, 60$, and so on.

From these evidences, and for this dataset, we conclude that the modified Denton PFD benchmarking procedure can be viewed as a sensible approximation of the Causey and Trager benchmarking procedure when the variability of the preliminary series and/or its bias are low with respect to the target variable. When this is not the case (high variability and/or large bias), the quality of the approximation clearly worsens. In addition, as regards the "movement preservation," we have found that generally the Causey and Trager GRP benchmarking procedure gives better

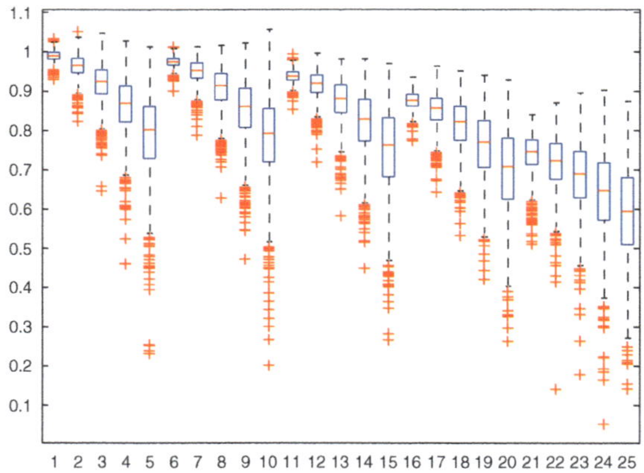

Fig. 45.4 Boxplots of r_1

performances as compared to Denton PFD. This turned out to be always the case (as one would expect) when the comparison criterion is the one optimized by the Causey and Trager procedure, and in the very largest amount (more than 98%) of the 25,000 series of our simulation experiment, when the distance between the growth rates of the preliminary and the benchmarked series is measured by the absolute difference.

Appendix: Non-singularity of the Coefficient Matrix of System (45.2)

Luenberger [9, p. 424] shows that a unique solution to the problem

$$\min_{\mathbf{y}} \mathbf{y}'\mathbf{Q}\mathbf{y} \qquad \text{subject to} \quad \mathbf{C}\mathbf{y} = \mathbf{Y}$$

exists if the matrix $\mathbf{C}$ is of full rank, and the matrix $\mathbf{Q}$ is positive definite on the null space of matrix $\mathbf{C}$: $\mathscr{N}(\mathbf{C}) = \{\mathbf{x} \in \mathbb{R}^n : \mathbf{C}\mathbf{x} = \mathbf{0}\}$.

Let us consider the matrix $\mathbf{Q} = \mathbf{P}^{-1}\Delta_n'\Delta_n\mathbf{P}^{-1}$, and let the vector $\mathbf{y}$ belong to $\mathscr{N}(\mathbf{C})$. We assume $p_t \neq 0$, $t = 1, \ldots, n$ (otherwise the objective function is not defined), and $\mathbf{C}\mathbf{p} \neq \mathbf{Y}$, which corresponds to exclude the trivial solution $\mathbf{y}^* = \mathbf{p}$, valid when there is no benchmarking problem. Given that

$$\mathbf{y}'\mathbf{Q}\mathbf{y} = \sum_{t=2}^{n} \left(\frac{y_t}{p_t} - \frac{y_{t-1}}{p_{t-1}} \right)^2,$$

it is immediately recognized that the expression above is strictly positive, which means that matrix $\mathbf{Q} = \mathbf{P}^{-1}\Delta_n'\Delta_n\mathbf{P}^{-1}$ is positive definite on the null space spanned by the columns of matrix $\mathbf{C}$, and thus the coefficient matrix of system (45.2) is nonsingular.

References

1. Bloem, A., Dippelsman, R., Mæhle, N.: Quarterly National Accounts Manual. Concepts, Data Sources, and Compilation. International Monetary Fund, Washington D.C. (2001)
2. Bozik, J.E., Otto, M.C.: Benchmarking: Evaluating methods that preserve month-to-month changes. Bureau of the Census - Statistical Research Division, RR-88/07 (1988) http://www.census.gov/srd/papers/pdf/rr88-07.pdf
3. Causey, B., Trager, M.L.: Derivation of Solution to the Benchmarking Problem: Trend Revision. Unpublished research notes, U.S. Census Bureau, Washington D.C. (1981). Available as an appendix in Bozik and Otto (1988)
4. Cholette, P.A.: Adjusting sub-annual series to yearly benchmarks. Surv. Meth. **10**, 35–49 (1984)
5. Dagum, E.B., Cholette, P.A.: Benchmarking, Temporal Distribution, and Reconciliation Methods for Time Series. Springer, New York (2006)
6. Denton, F.T.: Adjustment of monthly or quarterly series to annual totals: An approach based on quadratic minimization. JASA **333**, 99–102 (1971)
7. Di Fonzo, T., Marini, M.: Simultaneous and two-step reconciliation of systems of time series: methodological and practical issues. JRSS C (Appl. Stat.) **60**, 143–164 (2011a)
8. Di Fonzo, T., Marini, M.: A Newton's Method for Benchmarking Time Series According to a Growth Rates Preservation Principle. Department of Statistical Sciences, University of Padua, Working Paper Series **17** (2011b). http://www.stat.unipd.it/ricerca/fulltext?wp=432
9. Luenberger, D.G.: Linear and Nonlinear Programming. Addison-Wesley, Reading (1984)
10. Titova, N, Findley, D., Monsell, B.C.: Comparing the Causey-Trager method to the multiplicative Cholette-Dagum regression-based method of benchmarking sub-annual data to annual benchmarks. In: *JSM Proceedings*, Business and Economic Statistics Section. American Statistical Association, Alexandria, VA (2010)
11. Trager, M.L.: Derivation of Solution to the Benchmarking Problem: Relative Revision. Unpublished research notes, U.S. Census Bureau, Washington D.C. (1982) Available as an appendix in Bozik and Otto (1988)
12. U.S. Census Bureau: X-12-ARIMA Reference manual, Version 0.3. U.S. Census Bureau, U.S. Department of Commerce, Washington D.C. (2009). http://www.census.gov/srd/www/x12a/

Cumulation of Poverty Measures to Meet New Policy Needs

46

Vijay Verma, Francesca Gagliardi, and Caterina Ferretti

Abstract

Reliable indicators of poverty and social exclusion are an essential monitoring tool. Policy research and application increasingly require statistics *disaggregated to lower levels and smaller subpopulations*. This paper addresses some statistical aspects relating to improving the sampling precision of such indicators for subnational regions, in particular through the cumulation of data.

Keywords

Longitudinal data analysis • Measuring poverty and inequality • Sample design and estimation • Variance components models.

46.1 Context and Scope

Reliable indicators of poverty and social exclusion are an essential monitoring tool. In the EU-wide context, these indicators are most useful when they are comparable across countries and over time for monitoring trends. Furthermore, policy research and application increasingly require statistics *disaggregated to lower levels and smaller subpopulations*. Direct, one-time estimates from surveys designed primarily to meet national needs tend to be insufficiently precise for meeting these new policy needs. This is particularly true in the domain of poverty and social exclusion, the

V. Verma (✉) · F. Gagliardi
Università di Siena, Province of Siena, Italy
e-mail: verma@unisi.it; gagliardi10@unisi.it

C. Ferretti
Università di Firenze, Florence, Italy
e-mail: ferretti@ds.unifi.it

N. Torelli et al. (eds.), *Advances in Theoretical and Applied Statistics*, Studies in Theoretical and Applied Statistics, DOI 10.1007/978-3-642-35588-2_46,

monitoring of which requires complex distributional statistics—statistics necessarily based on intensive and relatively small-scale surveys of households and persons.

This paper addresses some statistical aspects relating to improving the sampling precision of such indicators for subnational regions in EU countries [8], in particular through the cumulation of data over rounds of regularly repeated national surveys [9]. The reference data for this purpose are based on EU Statistics on Income and Living Conditions (EU-SILC), which is the major source of comparative statistics on income and living conditions in Europe. EU-SILC covers data and data sources of various types: cross-sectional and longitudinal; household-level and person-level; on income and social conditions; and from registers and interview surveys depending on the country. A standard integrated design has been adopted by nearly all EU countries. It involves a rotational panel in which a new sample of households and persons is introduced each year to replace one quarter of the existing sample. Persons enumerated in each new sample are followed up in the survey for 4 years. The design yields each year a cross-sectional sample, as well as longitudinal samples of various durations. Two types of measures can be so constructed at the regional level by aggregating information on individual elementary units: *average measures* such as totals, means, rates and proportions constructed by aggregating or averaging individual values and *distributional measures*, such as measures of variation or dispersion among households and persons in the region. Average measures are often more easily constructed or are available from alternative sources. Distributional measures tend to be more complex and are less readily available from sources other than complex surveys; at the same time, such measures are more pertinent to the analysis of poverty and social exclusion. An important point to note is that, more than at the national level, many measures of averages can also serve as indicators of disparity and deprivation when seen in the regional context: the dispersion of regional means is of direct relevance in the identification of geographical disparity. Survey data such as from EU-SILC can be used in different forms and manners to construct regional indicators.

1. Direct estimation from survey data—in the same way as done normally at the national level—provided that the regional sample sizes are adequate for the purpose.
2. Constructing alternative (but with a substantively similar meaning) indicators which utilise the available survey data more intensively.
3. Cumulation of data over survey waves to increase the precision of the direct estimates.
4. Using survey data in conjunction with data from other (especially administrative) sources—which are larger in size but less detailed in content than survey data—in order to produce improved estimates using small area estimation (SAE) techniques.
5. Going altogether beyond the survey by exploiting administrative and other sources.

46.2 Cumulation Over Waves in a Rotational Panel Design

46.2.1 Illustration from EU-SILC Survey

One of the most important regular social surveys in the EU is the Statistics on Income and Living Conditions survey (EU-SILC). The EU-SILC was launched starting from 2003 in some countries; it covered 27 EU and EFTA countries by 2005, and all 30 by 2008. In each country it involves an annual survey with a rotational panel design. Its content is comprehensive, focusing on income, poverty and living conditions.

EU-SILC involves comprehensiveness in the substantive dimension (coverage of different topics), in space (coverage of different countries) and in time (regular waves or rounds). In EU-SILC most countries use the standard rotational household panel design as shown below. The survey is annual, and each panel stays in the survey for four consecutive years.

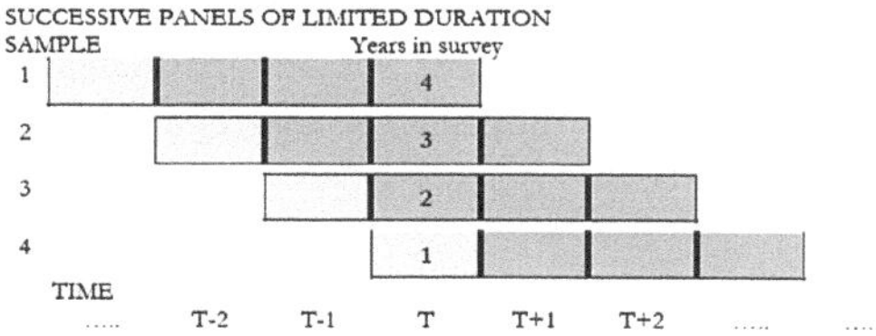

46.2.2 Pooling of Data Versus Pooling of Estimates

When two or more data sources contain—for the same type of units such as households or persons—a set of variables measured in a comparable way, then the information may be pooled either (a) by *combining estimates* from the different sources or (b) by *pooling data* at the micro level. Technical details and relative efficiencies of the procedures depend on the situation. The two approaches may give numerically identical results, or the one or the other may provide more accurate estimates; in certain cases, only one of the two approaches may be appropriate or feasible in any case.

Consider, for instance, the common case of pooling results across countries in a multi-country survey programme such as EU-SILC or EU-LFS. For linear statistics such as totals, pooling individual country estimates say ϕ_i with some appropriate weights P_i gives the same result as pooling data at the micro level with unit weights w_{ij} rescaled as $w'_{ij} = w_{ij} \cdot (P_i / \sum_j w_{ij})$ where i is the country and j is the final unit (e.g. individuals). For ratios of the form $\phi_i = \sum_j w_{ij} \cdot v_{ij} / \sum_j w_{ij} \cdot u_{ij}$, the two forms give very similar but not identical results, corresponding, respectively, to the "separate" and "combined" types of ratio estimate.

This paper is concerned with a different but equally common type of problem, namely pooling of different sources pertaining to the same population or largely overlapping and similar populations. Estimates from samples from the same population are most efficiently pooled with weights in proportion to their variances (meaning, with similar designs, in direct proportion to their sample sizes). Alternatively, the samples may be pooled at the micro level, with unit weights inversely proportion to their probabilities of appearing in *any* of the samples. This latter procedure may be more efficient (e.g., [4]), but be impossible to apply as it requires information, for every unit in the *pooled sample*, on its probability of selection into *each of the samples* irrespective of whether or not the unit appears in the particular sample [10]. Another serious difficulty in pooling samples is that, in the presence of complex sampling designs, the structure of the resulting pooled sample can become too complex or even unknown to permit proper variance estimation. In any case, different waves of a survey like EU-SILC or EU-LFS do not correspond to exactly the same population. The problem is akin to that of combining samples selected from multiple frames, for which it has been noted that micro level pooling is generally *not* the most efficient method [3].

For the above reasons, pooling of wave-specific estimates rather than of micro data sets is generally the appropriate approach to aggregation over time from surveys such as EU-SILC and EU-LFS.

46.3 Gain in Precision from Cumulation Over Survey Waves

Consider that for each wave, a person's poverty status is determined based on the income distribution of that wave separately, and the proportion poor at each wave is computed. These proportions are then averaged over a number of consecutive waves. The issue is to quantify the gain in sampling precision from such pooling, given that data from different waves of a rotational panel are highly correlated. Variance for the pooled estimators can be estimated on the following lines, using, for instance, the Jackknife Repeated Replication (JRR) procedure (see Sect. 46.4). The total sample of interest is formed by the union of all the cross-sectional samples being compared or aggregated. Using as basis for the common structure of this total sample, a set of JRR replications is defined in the usual way. Each replication is formed such that when a unit is to be excluded in its construction, it is excluded simultaneously from every wave where the unit appears. For each replication, the required measure is constructed for each of the cross-sectional samples involved, and these measures are used to obtain the required averaged measure for the replication, from which variance is then estimated in the usual way [1].

In terms of the quantities defined above, rows (1)–(5) of Table 46.1 are as follows.

Standard error of average HCR over 2 years (assuming independent samples)	$(1) = 1/2 \cdot (V_1 + V_2)^{1/2}$
Factor by which standard error is increased due to positive correlation between waves	$(2) = \cdot(1 + b.\,(n/n_H))^{1/2}$
Standard error of average HCR over 2 years (given correlated samples)	$(3) = (1) \cdot (2) = \cdot (V)^{1/2}$
Average standard error over a single year	$(4) = [(V_1)^{1/2} + (V_2)^{1/2}]/2$
Average gain in precision (variance reduction, or increase in effective sample size, over a single year sample)	$(5) = 1 - ((3)/(4))^2$

In place of the full JRR application, it is more illuminating to provide here the following simplified procedure for quantifying the gain in precision from averaging over waves of the rotational panel. It illustrates the statistical mechanism of how the gain is achieved. Indicating by p_j and p'_j the (1, 0) indicators of poverty of individual j over the two adjacent waves, we have the following for the population variances:

$$\text{var}(p_j) = \sum (p_j - p)^2 = p \cdot (1 - p) = v; \text{similarly, } \text{var}(p'_j) = p' \cdot (1 - p') = v'$$

$$\text{cov}(p_j, p'_j) = \sum (p_j - p) \cdot (p'_j - p') = a - p \cdot p' = c_1, \text{say,}$$

where "a" is the persistent poverty rate over the 2 years. For the simple case where the two waves completely overlap and $p' = p$, variance v_A for the averaged measure is: $v_A = \frac{v}{2} \cdot (1 + b)$, with correlation $b = (\frac{c_1}{v}) = (\frac{a-p^2}{p-p^2})$. The correlation between two periods is expected to decline as the two become more widely separated. Consider, for example, the case when the correlation between two points k waves apart can be approximated as $(c_k/v) = (c_1/v)^k$. In a set of K periods there are $(K - k)$ pairs exactly k periods apart, $k = 1$ to $(K - 1)$. It follows that variance v_K of an average over K periods relates to variance v of the estimate from a single wave as:

$$f_c = \left(\frac{v_k}{v}\right) = \frac{1}{K} \cdot \left(1 + 2 \cdot \sum_{k=1}^{k-1} \cdot \left(\frac{K-k}{K}\right) \cdot \left(\frac{c_1}{v}\right)^k\right)$$

where a, the persistent poverty between pairs of adjacent waves, and p, the cross-sectional poverty rate, are averages over the waves involved. For application to pairs of waves in EU-SILC, it is necessary to allow for variations in cross-sectional sample sizes and partial overlaps. The result is:

$$V = [(V_1 + V_2)/4] \cdot (1 + b \cdot (n/n_H))$$

where V_1 and V_2 are the sampling variances, b the correlation coefficient over the two cross-sections, n is the overlap between the cross-sectional samples and n_H is the harmonic mean of their sample sizes n_1 and n_2.

Table 46.1 Gain from cumulation over two waves: cross-sectional and persistent poverty rates. Poland EU-SILC 2005–2006

Sample			n	%se*		Mean	HCR: poverty line	
Base	Poverty rate	Est	persons	actual		income	National	Regional
CS-2006	HCR 2006	19.1	45,122	0.51	(1)	0.42	0.34	0.40
CS-2005	HCR 2005	20.6	49,044	0.45	(2)	1.31	1.18	1.18
LG 05–06	HCR 2006	18.5	32,820		(3)	0.55	0.40	0.47
LG 05–06	HCR 2005	20.2	32,820		(4)	0.60	0.48	0.56
LG 05–06	Persistent '05–06	12.5	32,820		(5)	14%	30%	30%

The methodology described above was applied to the 2005–2006 cross-sectional and longitudinal EU-SILC samples for Poland. Table 46.1 shows some results at the national level. Averaging the poverty rate (head count ratio, HCR) over two waves leads to a variance of this averaged estimator that is 30% less than the variance of the HCR estimated from just a single wave.

46.3.1 Reduction from Averaging Over Rounds in a Rotational Design

Consider a rotational sample in which each unit stays in the sample for n consecutive periods, with the required estimate being the average over Q consecutive periods, such as $Q = 4$ for 4-year average. The case $n = 1$ corresponds simply to independent samples each quarter. In the general case, the total sample involved in the estimation consists of $(n + Q - 1)$ independent subsamples. With some simplifying but reasonable assumptions, it can be proved [9] that the variance of the pooled estimate is approximately

$$V_a^2 = \left(\frac{V^2}{n \cdot Q}\right) \cdot \left\{ m_1 \cdot [m_2 - (m_1 - 1)] \cdot [1 + f(m_1)] \right.$$

$$\left. +2 \sum_{m=1}^{m_1-1} m \cdot [1 + f(m)] \right\} /(n \cdot Q) = \left(\frac{V^2}{n \cdot Q}\right) \cdot F(R)$$

$$f(m) = \frac{2}{m} \cdot \left\{(m-1) \cdot R + (m-2) \cdot R^2 + \ldots + R^{m-1}\right\}$$

The first factor is the variance which would be obtained in the absence of correlations between waves; $F(R)$ is the increase over that as a result of correlations.

46.4 Variance and Design Effects

Comparisons and cumulation over correlated cross-sections, with which this paper is concerned, add another layer of complexity.

JRR provides a versatile and straightforward technique for variance estimation in these situations. It is one of the classes of variance estimation methods based on comparisons among replications generated through repeated re-sampling of the same parent sample. Once the set of replications has been appropriately defined for any complex design, the same variance estimation algorithm can be applied to a statistic of any complexity. We have extended and applied this method for estimating variances for subpopulations (including regions and other geographical domains), longitudinal measures such as persistent poverty rates and measures of net changes and averages over cross-sections in the rotational panel design of EU-SILC [5]. Appropriate coding of the sample structure, in the survey micro-data and accompanying documentation, is an essential requirement in order to compute sampling errors taking into account the actual sample design. Lack of information on the sample structure in survey data files is a long-standing and persistent problem in survey work, and unfortunately affects EU-SILC as well. Indeed, *the major problem in computing sampling errors for EU-SILC is the lack of sufficient information for this purpose in the micro-data available to researchers.* We have developed approximate procedures in order to overcome these limitations at least partially and used them to produce useful estimates of sampling errors [7]. Use has been made of these results in this paper, but it is not possible here to go into detail concerning them.

A most useful concept for the computation, analysis and interpretation of sampling errors concerns "design effect" [2]. Design effect is the ratio of the variance (v) under the given sample design to the variance (v_0) under a simple random sample of the same size: $d^2 = v/v_0$, $d = se/se_0$. Proceeding from estimates of sampling error to estimates of design effects is essential for understanding the patterns of variation in and the determinants of magnitude of the error, for smoothing and extrapolating the results of computations, and for evaluating the performance of the sampling design.

Analysis of design effects into components is also needed in order to understand from where inefficiencies of the sample arise to identify patterns of variation, and through that to extend the results to other statistics, designs and situations. And most importantly, with JRR (and other replication methods) the total design effect can only be estimated by estimating (some of) its components separately [6]. In applications for EU-SILC, there is in addition a most important and special reason for decomposing the total design effect into its components. Because of the limited information on sample structure included in the micro-data available to researchers, direct and complete computation of variances cannot be done in many cases. Decomposition of variances and design effects identifies more "portable" components, which may be more easily imputed (carried over) from a situation where they can be computed with the given information, to another situation where such direct computations are not possible. On this basis valid estimates of variances can be produced for a wider range of statistics, thus at least partly overcoming the problem due to lack of information on sample structure. We may decompose total variance v (for the actual design) into the components or factors as $v = v_0.d^2 = v_0.(d_W.d_H.d_D.d_X)^2$, where d_W is the effect of sample weights, d_H of clustering of individual persons into households, d_D of clustering of households into dwellings and d_X that of other complexities of the design, mainly clustering and stratification. All factors other than d_X do not involve clusters or strata, but depend only on individual elements (households, persons etc.), and the sample weight associated with each such element in the sample. Parameter d_W depends on variability of sample weights and also on the correlation between the weights and the variable being estimated; d_H is determined by the number of and correlation among relevant individuals in the household, and similarly d_D by the number of households per dwelling in a sample of the latter. By contrast, factor d_X represents the effect on sampling error of various complexities of the design such as multiple stages and stratification. Hence unlike other components, d_X requires information on the sample structure linking elementary units to higher stage units and strata. This effect can be estimated as follows using the JRR procedures. We compute variance under two assumptions about structure of the design: variance v under the actual design, and v_R computed by assuming the design to be (weighted) simple random sampling of the ultimate units (addresses, households, persons as the case may be). This can be estimated from a "randomised sample" created from the actual sample by completely disregarding its structure other than the weights attached to individual elements. This gives $(d_x)^2 = (v/v_R)$, with $v_R = v_0.(d_W.d_H.d_D)^2$.

Table 46.2 gives standard error, design effect and components of design effect for the cross-sectional 2006 EU-SILC sample for Poland. The sample was a two-stage stratified sample of dwellings containing 45,122 individual persons. With "%se" (3rd and last column) we mean: for mean statistics, e.g. equivalised disposable income—standard error expressed as percentage of the mean value; for proportions and rates (e.g. poverty rates)—standard error given as absolute percent points. Terms (%se actual) and (%se SRS) relate, respectively, to the variances v and v_0 in the text. Parameter d_D cannot be estimated separately because of lack of information, but its effect is small and is, in any case, already incorporated into overall design effect d.

Table 46.2 gives poverty rates defined with respect to two different "levels" of poverty line: country level and NUTS1 level. By this we mean the population level to which the income distribution is pooled for the purpose of defining the poverty line. Conventionally poverty rates are defined in terms of the country poverty line (as 60% of the national median income). The income distribution is considered at the country level, in relation to which a poverty line is defined and the number (and proportion) of poor computed. It is also useful to consider poverty lines at other levels. Especially useful for constructing regional indicators is the use of regional poverty lines, i.e. a poverty line defined for each region based only on the income distribution within that region. The numbers of poor persons identified with these lines can then be used to estimate regional poverty rates. They can also be aggregated upwards to give an alternative national poverty rate—but which still

Table 46.2 Estimation of variance and design effects at the national level. Cross-sectional sample. Poland EU-SILC 2006

	Est.	%se actual	Design effect d_X	d_W	d_H	d	%se SRS
(1) Mean equivalised disposable income	3,704	0.57	0.94	1.22	1.74	1.99	0.29
(2) HCR—"head count" or poverty rate, using national poverty line	19.1	0.51	1.02	1.09	1.74	1.94	0.26
(3) HCR—"head count" or poverty rate, using regional (NUTS1) poverty line	19.0	0.61	1.05	1.09	1.74	1.99	0.30

remains based on the regional poverty lines. So defined, the poverty measures are not affected by disparities in the mean levels of income among the regions. The measures are therefore more purely relative.

46.5 Illustrative Applications of Cumulation at the Regional Level

Table 46.3 shows results for the estimation of variance and design effect for the cross-sectional 2006 and 2005 Poland datasets. The results at national level for the three measures considered have been already presented in previous sections. Here we present the results at NUTS1 regional level. All the values, except "%se SRS" and d_X, are computed at regional level in the same manner as the national level. All factors other than d_X do not involve clusters or strata but essentially depend only on individual elements and the associated sample weights. Hence normally they are well estimated, even for quite small regions. Factor $d_{X(G)}$ for a region (G) may be estimated in relation to $d_{X(C)}$ estimated at the country (C) level on the following lines. For large regions, each with a large enough number of PSUs (say over 25 or 30), we may estimate the variance and hence $d_{X(G)}$ directly at the regional level. Sometimes a region involves an SRS of elements, even if the national sample is multi-stage in other parts; here obviously, $d_{X(G)} = 1$. If the sample design in the region is the same or very similar to that for the country as a whole—which is quite often the case—we can take $d_{X(G)} = d_{X(C)}$. It is common that the main difference between the regional and the total samples is the average cluster size (b). In this case we use $d^2_{X(G)} = 1 + (d^2_{X(C)} - 1) \cdot b_{(G)}/b_{(C)}$. The last-mentioned model concerns the effect of clustering and hence is meaningful only if $d_{X(C)} \geq 1$, which is often but not always the case in actual computations. Values smaller than 1.0 may arise when the effect of stratification is stronger than that of clustering, when units within clusters are negatively correlated (which is rare, but not impossible), or simply as a result of random variability in the empirical results. In any case, if $d_{X(C)} < 1$, the above equation should be replaced by $d_{X(G)} = d_{X(C)}$.

Table 46.3 Estimation of variance and design effects at the regional (NUTS1) level. Full cross-sectional dataset

	2006 Est.	n persons	%se* SRS	d_X	d	%se* actual	2005 Est.	n persons	%se* actual
Mean equivalised disposable income									
Poland	3,704	45,122	0.29	0.94	1.99	0.57	3,040	49,044	0.62
PL1	4,236	8,728	0.65	0.94	2.06	1.34	3,455	9,871	1.32
PL2	3,889	9,273	0.63	0.94	1.78	1.13	3,143	10,181	1.22
PL3	3,162	9,079	0.64	0.94	2.00	1.28	2,618	9,674	1.32
PL4	3,530	6,912	0.73	0.94	1.90	1.39	2,977	7,195	1.84
PL5	3,906	4,538	0.90	0.94	1.96	1.77	3,164	5,066	1.85
PL6	3,419	6,592	0.75	0.94	1.90	1.43	2,816	7,057	1.58
At-risk-of-poverty rate, national poverty line									
Poland	19.1	45,122	0.26	1.02	1.94	0.51	20.6	49,044	0.45
PL1	17.1	8,728	0.57	1.02	1.85	1.06	19.1	9,871	0.92
PL2	14.7	9,273	0.52	1.02	1.86	0.97	16.4	10,181	0.87
PL3	25.2	9,079	0.64	1.02	2.09	1.34	25.2	9,674	1.13
PL4	18.7	6,912	0.66	1.02	1.98	1.32	20.2	7,195	1.19
PL5	18.6	4,538	0.82	1.02	1.91	1.56	20.2	5,066	1.43
PL6	21.4	6,592	0.71	1.02	1.95	1.40	23.7	7,057	1.26
At-risk-of-poverty rate, regional poverty lines									
Poland	19.0	45,122	0.30	1.05	1.99	0.61	20.5	49,044	0.51
PL1	19.8	8,728	0.70	1.04	1.90	1.34	20.9	9,871	1.07
PL2	18.5	9,273	0.67	1.04	1.91	1.27	19.0	10,181	1.05
PL3	18.6	9,079	0.68	1.06	2.14	1.45	20.8	9,674	1.21
PL4	17.5	6,912	0.76	1.05	2.04	1.54	20.1	7,195	1.35
PL5	20.9	4,538	1.00	1.04	1.97	1.96	22.2	5,066	1.68
PL6	19.1	6,592	0.80	1.05	2.00	1.60	21.3	7,057	1.37

The quantity (%se* SRS) can be directly computed at the regional level as was done for the national level in Table 46.2. However, very good approximation can be usually obtained very simply without involving JRR computations of variance. The following model has been used in Table 46.3. For means (such as equivalised income) over very similar populations, assumption of a constant coefficient of variation is reasonable. The region-to-country ratio of relative standard errors (expressed as percentage of the mean value as in Table 46.3) under simple random sampling is inversely proportional to the square-root of their respective sample sizes: $(\%se^{*}SRS)^2_{(G)} = (\%se^{*}SRS)^2_{(C)} \cdot (n_{(C)}/n_{(G)})$. For proportions ($p$, with $q = 100 - p$), with standard error expressed in absolute percent points as in Table 46.3, we can take: $(\%se^{*}SRS)^2_{(G)} = (\%se^{*}SRS)^2_{(C)} \cdot (\frac{p_{(G)} \cdot q_{(G)}}{p_{(C)} \cdot q_{(C)}}) \cdot (n_{(C)}/n_{(G)})$. A poverty rate may be treated as proportions for the purpose of applying the above. We see from Table 46.3 that the (%se*actual) at regional level is generally, for all the three measures, 2–3 times larger than that at the national level.

Table 46.4 Gain in precision from averaging over correlated samples. Poland NUTS1 regions

	Country	PL1	PL2	PL3	PL4	PL5	PL6
Mean equivalised income							
(1)	0.42	0.94	0.83	0.92	1.15	1.28	1.07
(2)	1.31	1.33	1.30	1.31	1.27	1.32	1.32
(3)	0.55	1.26	1.08	1.20	1.47	1.70	1.41
(4)	0.60	1.33	1.17	1.30	1.62	1.81	1.51
(5)	14%	11%	15%	14%	18%	12%	12%
HCR national poverty line							
(1)	0.34	0.70	0.65	0.88	0.89	1.06	0.94
(2)	1.18	1.18	1.17	1.18	1.18	1.17	1.19
(3)	0.40	0.83	0.76	1.03	1.05	1.23	1.12
(4)	0.48	0.99	0.92	1.24	1.26	1.50	1.33
(5)	30%	29%	31%	30%	30%	32%	29%
HCR regional poverty line							
(1)	0.40	0.86	0.83	0.94	1.03	1.29	1.05
(2)	1.18	1.18	1.18	1.17	1.18	1.17	1.18
(3)	0.47	1.02	0.98	1.10	1.21	1.51	1.24
(4)	0.56	1.21	1.16	1.33	1.45	1.82	1.49
(5)	30%	29%	29%	31%	30%	31%	31%

Rows (1)–(5) have been defined in Table 46.1

Regional HCR estimates based on the national poverty line are quite different from those based on the regional ones. Also, while individual regional estimates of HCR using the regional poverty line are quite close to the national estimate (19.0 for 2006), the ones using the national poverty line are more variable (from 14.7 to 25.2 for 2006).

From Table 46.4 below it can be seen that generally for the HCR measures, both for country and for NUTS1 level poverty lines, cumulating the estimates over two waves leads to a reduction of 30% in variance compared to that for a single wave. This reduction of the variance is smaller for mean equivalised income due to a higher correlation between incomes for the 2 years—generally the coefficient of correlation of the equivalised income between waves exceeds 0.70.

References

1. Betti, G., Gagliardi, F., Nandi, T.: Jackknife variance estimation of differences and averages of poverty measures. Working Paper no° 68/2007, DMQ, Università di Siena (2007)
2. Kish, L.: Methods for design effects. J. Offic. Stat. **11**, 55–77 (1995)
3. Lohr, S.L., Rao, J.N.K.: Inference from dual frame surveys. J. Am. Stat. Assoc. **95**, 271–280 (2000)
4. O'Muircheataigh, C., Pedlow, S.: Combining samples vs. cumulating cases: a comparison of two weighting strategies in NLS97. In: American Statistical Association Proceedings of the Joint Statistical Meetings, pp. 2557–2562 (2002)

5. Verma, V., Betti, G.: Cross-sectional and Longitudinal Measures of Poverty and Inequality: Variance Estimation using Jackknife Repeated Replication. In: Conference 2007 'Statistics under one Umbrella', Bielefeld University (2007)
6. Verma, V., Betti, G.: Taylor linearization sampling errors and design effects for poverty measures and other complex statistics. J. Appl. Stat. (2010), on-line first
7. Verma, V., Betti, G., Gagliardi, F.: An assessment of survey errors in EU-SILC, Eurostat Methodologies and Working Papers, Eurostat, Luxembourg (2010)
8. Verma, V., Betti, G., Natilli, M., Lemmi, A.: Indicators of social exclusion and poverty in Europe's regions. Working Paper no° 59/2006, DMQ, Università di Siena (2006)
9. Verma, V., Gagliardi, F., Ferretti, C.: On pooling of data and measures. Working Paper no° 84/2009, DMQ, Università di Siena (2009)
10. Wells, J.E.: Oversampling through households or other clusters: comparison of methods for weighting the oversample elements. Aust. New Zeal. J. Stat. **40**, 269–277 (1998)

Long Memory in Integrated and Realized Variance

47

Eduardo Rossi and Paolo Santucci de Magistris

Abstract
The long memory properties of the integrated and realized volatility are investigated under the assumption that the instantaneous volatility is driven by a fractional Brownian motion. The equality of their long memory degrees is proved in the ideal situation when prices are observed continuously. In this case, the spectral density of the integrated and realized volatility coincide.

Keywords
Fractional Brownian Motion • Long memory • Measurement error • Realized volatility

47.1 Introduction

An overwhelming empirical evidence supports the hypothesis that the volatility of financial returns is a long-memory process, see e.g. [4, 7, 12, 14, 15]. More recently [1–3, 20], and [23] report evidence of long memory in the realized variance series, RV. As a consequence of long memory, realized volatility series are clustered and highly predictable, given its past realizations. Therefore, it is interesting to explore a possible explanation of long memory in volatility, in order to understand its dynamic behavior. As a matter of fact, no theoretical justification for the presence of long

E. Rossi (✉)
Dipartimento Economia politica e metodi quantitativi, University of Pavia, Via San Felice 5, 27100 Pavia, Italy
e-mail: erossi@eco.unipv

P.S. de Magistris
School of Economics and Management, Aarhus University, Bartholins Alle 10, 1322 University Park, DK-8000 Aarhus C, Denmark
e-mail: psantucci@creates.au.dk CREATES, Denmark

N. Torelli et al. (eds.), *Advances in Theoretical and Applied Statistics*,
Studies in Theoretical and Applied Statistics, DOI 10.1007/978-3-642-35588-2_47,

memory in integrated variance (or integrated volatility, IV henceforth) is given from a continuous time perspective. A possible explanation of long memory in RV is aggregation. Hurvich and Soulier [17], referring to the result in [22], show that RV, given by the sum of n squared returns, has long memory. A recent paper by [18] shows that the presence of long memory in the RV can be the outcome of the aggregation of a finite number of short-memory series and it depends on the choice of the sampling scheme. However, the theory of RV, as an ex-post estimator of IV, lies in the *continuous time* framework, while [18]'s proof is based on the aggregation of discretely sampled squared returns. Moreover, Hurvich and Soulier [17] do not mention the effect of the discretization of the trajectories of the underlying continuous time process.

In this chapter we focus on the dynamic properties of ex-post estimators of IV, such as RV, when we assume that the trajectories of the instantaneous volatility, $\sigma^2(t)$, are generated by a fractional Brownian motion of order d, see [6] and [10, 11], encompassing the regular Brownian motion. First, we demonstrate (see Proposition 1) that IV has the same fractional integration order of $\sigma^2(t)$, that is d. Second, following Proposition 2, which features the discretization error, we are able to characterize the spectral density of RV for a given sampling frequency. We show that, letting the sampling interval to zero, the spectral density of RV coincides with that of IV, so that they are characterized by the same long memory degree.

The chapter is organized as follows. In Sect. 47.2, we prove that the IV has the same long memory degree of the instantaneous volatility, when the latter is driven by a fractional Brownian motion. In Sect. 47.3, we characterize the properties of the discretization error and the spectral density of RV. Finally, Sect. 47.4 concludes.

47.2 Long Memory in Integrated Variance

Let us assume that the logarithm of the instantaneous volatility be driven by a fractional Ornstein–Uhlenbeck process, with zero long run mean, as in [11]:

$$\mathrm{d}\ln\sigma^2(t) = -k\ln\sigma^2(t)\mathrm{d}t + \gamma\mathrm{d}W_d(t) \tag{47.1}$$

where $k > 0$ is the drift parameter, while $\gamma > 0$ is the volatility parameter and $W_d(t)$ is the fractional Brownian motion (fBm). The literature on long memory processes in econometrics distinguishes between type I and type II fractional Brownian motion. These processes have been carefully examined and contrasted by Marinucci and Robinson [19] and Davidson and Hashimzade [13]. When considered as real continuous processes on the unit interval, they can be defined respectively by

$$B_d(t) = \frac{1}{\Gamma(1+d)}\int_0^t (t-s)^d\mathrm{d}W(s) + \int_{-\infty}^0 [(t-s)^d - (-s)^d]\mathrm{d}W(s) \tag{47.2}$$

and

$$W_d(t) = \frac{1}{\Gamma(1+d)} \int_0^t (t-s)^d \mathrm{d}W(s). \tag{47.3}$$

where $\Gamma(\cdot)$ is the gamma function, d is the long memory parameter, and $W(\cdot)$ is the Brownian motion. In the type II case, the second term in (47.2) is omitted, it is the truncated version of the general fBm, see [10, 11] and [19]. As shown by Marinucci and Robinson [19], the increments of (47.2) are stationary, whereas those of (47.3) are not. When $d = 0$, both definitions of fBm collapse into the usual Brownian motion.

The solution of (47.1) can be written as $\ln \sigma^2(t) = \int_0^t e^{-k(t-s)} \gamma \, \mathrm{d}W_d(s)$. The process $\ln \sigma^2(t)$ has long memory of order d, with $0 < d < 1/2$, when $k_\infty = \frac{\mathrm{d}\gamma}{k}$ is a nonzero constant, see [9]. This makes clear that the long memory feature of process in (47.1) is directly linked to the characteristics of the drift term. Indeed, when $k = 0$, i.e., the mean reversion is zero, the condition on k_∞ is no more satisfied. When $k > 0$, the volatility process $\sigma^2(t)$ is asymptotically equivalent (in quadratic mean) to the stationary process (see [11]) :

$$\tilde{\sigma}^2(t) = \exp\left(\int_{-\infty}^t e^{-k(t-u)} \gamma \, \mathrm{d}W_d(u) \right), \quad k > 0 \quad 0 < d < \frac{1}{2}. \tag{47.4}$$

where the solution is expressed using type II fBm. Comte and Renault [11] prove that the spectral density, $f_{\tilde{\sigma}^2}(\lambda)$, of the process $\tilde{\sigma}^2(t)$, is equal to $c\lambda^{-2d}$ for $\lambda \to 0$ where c is a positive constant, so that the volatility process inherits the long-memory property induced by the fBm.

Proposition 1. *Given the process in* (47.1) *for the logarithm of the instantaneous volatility, then* $\lim_{\lambda \to 0} \lambda^{2d} f_{\mathrm{IV}}(\lambda) = c \in \mathbb{R}_+$ *where* $f_{\mathrm{IV}}(\lambda)$ *is the spectral density of* $IV = \int_{t-1}^t \tilde{\sigma}^2(u) \mathrm{d}u$.

This means that the *IV* process has the same degree of fractional integration as the instantaneous volatility and is covariance stationary when $k > 0$.[1] Given this result, in the next section, we will provide a rationale for the empirical evidence of long memory in the ex-post realized measures of *IV*.

47.3 The Discretization Error

Let $P(t)$ be the price of an asset, where its logarithm, $p(t)$, follows the stochastic differential equation:

$$\mathrm{d}p(t) = m(t)\mathrm{d}t + \sigma(t)\mathrm{d}W(t) \tag{47.5}$$

[1] Since $\sigma^2(t)$ is asymptotically equivalent to $\tilde{\sigma}^2(t)$, then $\int_{t-1}^t \tilde{\sigma}^2(s)\mathrm{d}s$ and $\int_{t-1}^t \sigma^2(s)\mathrm{d}s$ asymptotically coincide.

where $W(t)$ is a standard Brownian motion and $m(t)$ is locally bounded and predictable. $\sigma(t)$ is assumed to be independent of $W(t)$ and càdlàg, see [5]. Meddahi [21] provide both qualitative and quantitative measures of the precision of measuring IV by RV for a given frequency. The parameter of interest is the IV

$$IV = \int_0^1 \sigma^2(u)\mathrm{d}u. \tag{47.6}$$

Meddahi [21] assumes that the underlying data generating process is a continuous time, continuous sample-path model. While RV^Δ converges to IV_t when $\Delta \to 0$, the difference may be not negligible for a given Δ. Consider an equidistant partition $0 = t_0 < t_1 < \ldots < t_n = 1$, where $t_i = i/n$, and $\Delta = 1/n$, that is the interval is normalized to have unit length. Define $r_{i\Delta,\Delta} = p_{i\Delta,\Delta} - p_{(i-1)\Delta,\Delta}$, the RV at sampling frequency n is

$$RV^\Delta = \sum_{i=1}^{n} r^2_{i\Delta,\Delta}. \tag{47.7}$$

From the theory of stochastic integration, when $n \to \infty$, then $RV^\Delta \xrightarrow{p} IV$. Following [21] we can decompose the difference between RV and IV, for a given Δ as

$$\begin{aligned} RV_t^\Delta &= \sum_{i=1}^{n}\left[\int_{t-1+(i-1)\Delta}^{t-1+i\Delta} \sigma^2(u)\mathrm{d}u\right] + \sum_{i=1}^{n} u_{t-1+i\Delta,\Delta} \\ &= IV_t + u_t^\Delta. \end{aligned} \tag{47.8}$$

with

$$\begin{aligned} u_{t-1+i\Delta,\Delta} &= (\mu_{t-1+i\Delta,\Delta})^2 + 2(\mu_{t-1+i\Delta,\Delta}\varepsilon_{t-1+i\Delta,\Delta}) \\ &\quad +2\int_{t-1+(i-1)\Delta}^{t-1+i\Delta}\left(\int_{t-1+(i-1)\Delta}^{u} \sigma(s)\mathrm{d}W(s)\right)\sigma(u)\mathrm{d}W_u \end{aligned} \tag{47.9}$$

where

$$\mu_{t-1+i\Delta,\Delta} \equiv \int_{t-1+(i-1)\Delta}^{t-1+i\Delta} m(u)\mathrm{d}u$$

$$\varepsilon_{t-1+i\Delta,\Delta} \equiv \int_{t-1+(i-1)\Delta}^{t-1+i\Delta} \sigma(u)\mathrm{d}W(u)$$

As in [5, p. 257], we assume that the drift in the price equation is null, $m(t) = 0$, and there is no leverage effect, so that, for finite $\Delta > 0$, the error term can be written as

$$u_{t-1+i\Delta,\Delta} = \sigma^2_{t,i,\Delta}\left(z^2_{t,i} - 1\right) \tag{47.10}$$

where $z_{t,i}$ is $iiN(0,1)$ and it is independent of $\sigma^2_{t,i,\Delta}$. According to [5, p. 253] this result is valid for any stationary instantaneous volatility process with locally square integrable sample paths and independent of the standard Brownian motion in the price equation. The process in Eq. (47.1) fulfills these conditions.

Note that $\sigma^2_{t,i,\Delta} = \int_{t-1+(i-1)\Delta}^{t-1+i\Delta} \sigma^2(s)\mathrm{d}s$, is the *IV* over the i-*th* subinterval of length Δ. Therefore,

$$u_t^{\Delta} = \sum_{i=1}^{n} \sigma^2_{t,i,\Delta}\left(z^2_{t,i} - 1\right). \tag{47.11}$$

It is clear from the result in Proposition 1 that $\sigma^2_{t,i,\Delta}$ has long memory. Given the representation of the measurement error in Eq. (47.10), we are able to characterize the error term and the spectral density of RV^{Δ}, when the *IV* has long memory. In particular,

Proposition 2. *Let Δ be an integer and consider the processes $p(t)$, RV^{Δ} and u_t^{Δ} defined respectively in* (47.5) (47.7) *and* (47.11). *Then:*

(i) u_t^{Δ} *is dynamically uncorrelated;*
(ii) The error term is uncorrelated with IV_t*;*
(iii) The variance of the error term is $2\Delta^{-1}E\left[(\sigma^2_{t,i,\Delta})^2\right]$*;*
(iv) $Cov(RV_t^{\Delta}, RV_{t-h}^{\Delta}) = Cov(IV_t, IV_{t-h}) \quad \forall h \neq 0.$

Given (*iv*) we can characterize the spectrum of RV, which includes the variance of u_t^{Δ}, say

$$f_{RV^{\Delta}}(\lambda) = \frac{1}{2\pi}\left\{Var(RV^{\Delta}) + 2\sum_{j=1}^{\infty}\left[Cov(IV_t, IV_{t-j})\cos(\lambda j)\right]\right\}$$

$$= \frac{1}{2\pi}\left\{Var(IV) + Var(u_t^{\Delta}) + 2\sum_{j=1}^{\infty}\left[Cov(IV_t, IV_{t-j})\cos(\lambda j)\right]\right\}$$

Therefore, given the properties of u_t^{Δ}, the spectral density of RV_t^{Δ} is

$$f_{RV^{\Delta}}(\lambda) = f_{\mathrm{IV}}(\lambda) + f_{u^{\Delta}}(\lambda) \tag{47.12}$$

where $f_{u^{\Delta}}(\lambda) = \frac{2\Delta^{-1}E\left[(\sigma^2_{i,\Delta})^2\right]}{2\pi}$. As showed by [5], the variance of u_t^{Δ} converges to zero as $\Delta \to 0$, so that $f_{u^{\Delta}}(\lambda) \to 0$. In fact, in the ideal situation where prices are recorded continuously and without measurement errors, since $\lim_{\Delta\to 0} RV^{\Delta} = IV$, it is evident that

$$\lim_{\Delta\to 0} f_{RV^{\Delta}}(\lambda) = f_{\mathrm{IV}}(\lambda)$$

where $f_{RV^{\Delta}}$ is the spectral density of the RV. Therefore

$$\lim_{\lambda \to 0} \left[\lim_{\Delta \to 0} \lambda^{2d} f_{RV^{\Delta}}(\lambda) \right] = \lim_{\lambda \to 0} \lambda^{2d} f_{\text{IV}}(\lambda) \qquad (47.13)$$

so that RV is characterized by the same degree of long memory of IV when the instantaneous volatility is generated according to (47.1). Differently from [18], we are able to demonstrate the presence of long memory in RV from a continuous time perspective, where the process of the instantaneous volatility is driven by a fractional Brownian motion of order d. It is also interesting to note that the additive noise term has a spectral density that depends on the variance of $\sigma^2_{i,t,\Delta}$, and on length of the intradaily interval, Δ. A possible extension would be to consider also the presence of microstructure noise, such that the measurement error does not vanish when $\Delta \to 0$ and it enters in the spectral density of RV.

From a practical point of view, the presence of the measurement error, due to the discretization effect and microstructure noise, poses problems from the point of view of the inference on the long memory parameter, d, as noted among others by Hurvich et al. [16] in a discrete time stochastic volatility context. Therefore, the investigation of the finite sample performances of the estimates of the long memory parameter of IV represents a promising aspect that will be addressed in future research.

47.4 Conclusions

A stylized fact is that RV has long memory. In this chapter, we investigate the dynamic properties and the source of the long-range dependence of RV. When the instantaneous volatility is driven by a fractional Brownian motion the IV is characterized by long-range dependence. As a consequence, the RV inherits this property in the ideal situation where prices are observed continuously and without microstructure noise, and the spectral densities of IV and RV coincide. In this chapter we also focus on the dynamic properties of ex-post estimators of IV, such as RV, when we assume that the trajectories of the instantaneous volatility, $\sigma^2(t)$, are generated by a fractional Brownian motion of order d. First, we demonstrate that IV has the same fractional integration order of $\sigma^2(t)$, that is d. It is therefore natural that realized measures of volatility have the same integration order of IV in the ideal situation where the price is recorded continuously and without market microstructure noise. In this case, we know that RV is a consistent estimator of IV, but also the spectral densities converge to the spectral density of IV. We also characterize the measurement error term, due to the discretization error, when the IV has long memory. These theoretical results represent a starting point for a refined inference technique on the long memory parameter of the IV, based on realized proxies. In particular, future research on this topic will study not only the effect of microstructure noise and leverage on the dynamic behavior of RV, but also the finite sample performance of the semi-parametric estimators of the long memory parameter of the IV, when a measurement error contaminates the observed price.

References

1. Andersen, T.G., Bollerslev, T., Diebold, F.X., Ebens, H.: The distribution of stock return volatility. J. Financ. Econ. **61**, 43–76 (2001)
2. Andersen, T.G., Bollerslev, T., Diebold, F.X., Labys, P.: The distribution of exchange rate volatility. J. Am. Stat. Assoc. **96**, 42–55 (2001)
3. Andersen, T.G., Bollerslev, T., Diebold, F.X., Labys, P.: Modeling and forecasting realized volatility. Econometrica **71**, 579–625 (2003)
4. Baillie, R.T.: Long memory processes and fractional integration in econometrics. J. Econometrics **73**(1), 5–59 (1996)
5. Barndorff-Nielsen, O., Shephard, N.: Econometric analysis of realized volatility and its use in estimating stochastic volatility models. J. R. Stat. Soc. B **64**, 253–280 (2002)
6. Beran, J.: Statistics for Long-Memory Processes. New York: Chapman & Hall (1994)
7. Bollerslev, T., Mikkelsen, H.O.: Modeling and pricing long memory in stock market volatility. J. Econometrics **73**, 151–184 (1996)
8. Chambers, M.J.: The estimation of continuous parameter long-memory time series models. Economet. Theory **12**, 374–390 (1996)
9. Comte, F.: Simulation and estimation of long memory continuous time models. J. Time Anal. **17**, 19–36 (1996)
10. Comte, F., Renault, E.: Long-memory continuous-time models. J. Econometrics **73**, 101–149 (1996)
11. Comte, F., Renault, E.: Long-memory in continuous-time stochastic volatility models. Math. Finance **8**, 291–323 (1998)
12. Dacorogna, M., Muller, U.A., Nagler, R.J., Olsen, R.B., Pictet, O.V.: A geographical model for the daily and weekly seasonal volatility in the foreign exchange market. J. Int. Money Finance **12**, 413–438 (1993)
13. Davidson, J., Hashimzade, N.: Type I and type II fractional Brownian motions: A reconsideration. Comput. Stat. Data Anal. **53**(6), 2089–2106 (2009)
14. Ding, Z., Granger, C., Engle, R.: A long memory property of stock market returns and a new model. J. Empir. Finance **1**, 83–106 (1993)
15. Granger, C., Ding, Z.: Modeling volatility persistence of speculative returns. J. Econometrics **73**, 185–215 (1996)
16. Hurvich, C.M., Moulines, E., Soulier, P.: Estimating long memory in volatility. Econometrica **73**(4), 1283–1328 (2005)
17. Hurvich, C.M., Soulier, P.: Stochastic Volatility Models with Long Memory," in Handbook of Financial Time Series. Springer, New York (2009)
18. Lieberman, O., Phillips, P.: Refined inference on long memory in realized volatility. Economet. Rev. **27**(27), 254–267 (2008)
19. Marinucci, D., Robinson, P.M.: Alternative forms of fractional Brownian motion. J. Stat. Plann. Infer. **80**, 111–122 (1999)
20. Martens, M., Van Dijk, D., De Pooter, M.: Forecasting S&P 500 volatility: long memory, level shifts, leverage effects, day-of-the-week seasonality, and macroeconomic announcements. Int. J. Forecast. **25**, 282–303 (2009)
21. Meddahi, N.: A theoretical comparison between integrated and realized volatility. J. Appl. Economet. **17**, 475–508 (2002)
22. Robinson, P.M.: The memory of stochastic volatility models. J. Econometrics **2**, 195–218 (2001)
23. Rossi, E., Santucci de Magistris, P.: Long Memory and Tail dependence in Trading Volume and Volatility, Discussion paper, CREATES (2009)

Proofs

Proof of Proposition 1

We know from [11] that $\lim_{\lambda \to 0} \lambda^{2d} f_{\tilde{\sigma}^2}(\lambda) = c \in \mathbb{R}_+$. Given that $IV = \int_0^1 \tilde{\sigma}^2(s)\mathrm{d}s$. Following [8, p. 388] we express the integral operator in the IV definition as a simple filter that has transfer function

$$T(\lambda) = \int_0^1 \mathrm{e}^{-\mathrm{i}\lambda u}\mathrm{d}u = \frac{1}{(-\mathrm{i}\lambda)}[\mathrm{e}^{-\mathrm{i}\lambda} - 1].$$

Therefore the spectral density of IV is given by

$$f_{\mathrm{IV}}(\lambda) = |T(\lambda)|^2 f_{\tilde{\sigma}^2}(\lambda).$$

The limit of the spectral density of IV for $\lambda \to 0$ is

$$\lim_{\lambda \to 0} f_{\mathrm{IV}}(\lambda) = \lim_{\lambda \to 0}[|T(\lambda)|^2 f_{\tilde{\sigma}^2}(\lambda)]$$

Since $|T(\lambda)|^2 = \frac{2(1-\cos(\lambda))}{|\lambda|^2}$ and $(1 - \cos(\lambda)) = |\lambda|^2/2 + O(\lambda^3)$, then $\lim_{\lambda \to 0} |T(\lambda)|^2 = 1$, thus

$$\lim_{\lambda \to 0} \lambda^{2d} f_{\mathrm{IV}}(\lambda) = \lim_{\lambda \to 0} \lambda^{2d} f_{\tilde{\sigma}^2}(\lambda) = c$$

that is IV has the same degree of long memory of $\tilde{\sigma}^2(t)$.

Proof of Proposition 2

(i)

$$\begin{aligned}
Cov(u_{t,i}^{\Delta}, u_{t,i+h}^{\Delta}) &= Cov(u_{t,i}^{\Delta}, u_{t,i+h}^{\Delta}) \\
&= E\left(\sigma_{t,i,\Delta}^2(z_{t,i}^2 - 1)\sigma_{t,i+h,\Delta}^2(z_{t,i+h}^2 - 1)\right) \\
&= E\left(\sigma_{t,i,\Delta}^2\sigma_{t,i+h,\Delta}^2\right) E\left((z_{t,i}^2 - 1)(z_{t,i+h}^2 - 1)\right) \\
&= E\left(\sigma_{t,i,\Delta}^2\sigma_{t,i+h,\Delta}^2\right) \cdot 0 \\
&= 0 \quad \forall h > 0.
\end{aligned}$$

(ii)

$$Cov(IV_t, u_t^{\Delta}) = Cov\left(IV_t, \sum_{i=1}^{n} \sigma_{t,i,\Delta}^2 \left(z_{t,i}^2 - 1\right)\right)$$
$$= E\left(IV_t \cdot \sum_{i=1}^{n} \sigma_{t,i,\Delta}^2 \left(z_{t,i}^2 - 1\right)\right)$$
$$-E\left(IV_t\right) \cdot E\left(\sum_{i=1}^{n} \sigma_{t,i,\Delta}^2 \left(z_{t,i}^2 - 1\right)\right)$$

Given that $z_{i,t}$ is i.i.d. $N(0,1)$, then

$$Cov\left(IV_t, u_t^{\Delta}\right) = E\left(IV_t \cdot \sum_{i=1}^{n} \sigma_{t,i,\Delta}^2 \left(z_{t,i}^2 - 1\right)\right)$$

given that $IV_t = \sum_{i=1}^{n} \sigma_{t,i,\Delta}^2$, then given the independence between $z_{t,i}$ and $\sigma_{t,i,\Delta}^2$, we have that $E\left(IV_t \cdot \sum_{i=1}^{n} \sigma_{t,i,\Delta}^2 \left(z_{t,i}^2 - 1\right)\right) = 0$. So, despite the presence of long memory in IV, the error term is uncorrelated with IV_t.

(iii) Given that $E\left(u_t^{\Delta}\right) = 0$, then

$$Var\left[u_t^{\Delta}\right] = E\left[(u_t^{\Delta})^2\right] = E\left[\left(\sum_{i=1}^{n} \sigma_{t,i,\Delta}^2 (z_{t,i}^2 - 1)\right)^2\right]$$
$$= E\left[\sum_{i=1}^{n}\sum_{j=1}^{n} \sigma_{t,i,\Delta}^2 (z_{t,i}^2 - 1)\sigma_{t,j,\Delta}^2 (z_{t,j}^2 - 1)\right].$$

Note that, for a generic cross product,

$$E\left[\sigma_{t,i,\Delta}^2 (z_{t,i}^2 - 1)\sigma_{t,j,\Delta}^2 (z_{t,j}^2 - 1)\right] = E\left[\left(\sigma_{t,i,\Delta}^2 \sigma_{t,j,\Delta}^2\right)\right] E\left[(z_{t,i}^2 - 1)(z_{t,j}^2 - 1)\right]$$

Given that $z_{i,t} \perp z_{j,t}$, then $E\left(\sigma_{t,i,\Delta}^2 (z_{t,i}^2 - 1)\sigma_{t,j,\Delta}^2 (z_{t,j}^2 - 1)\right) = 0$ and all the cross products are equal to 0. Therefore

$$E\left[(u_t^{\Delta})^2\right] = \sum_{i=1}^{n} E\left[\left(\sigma_{t,i,\Delta}^2 (z_{t,i}^2 - 1)\right)^2\right]$$
$$= \sum_{i=1}^{n} E\left[(\sigma_{t,i,\Delta}^2)^2\right] \cdot E\left[\left(z_{t,i}^2 - 1\right)^2\right]$$
$$= 2\Delta^{-1} \cdot E\left[(\sigma_{t,i,\Delta}^2)^2\right].$$

Since the interval are equally spaced, then $E\left[(\sigma^2_{t,i,\Delta})^2\right] = Var(\sigma^2_{t,i,\Delta}) + E(\sigma^2_{t,i,\Delta})^2 = Var(\sigma^2_{t,j,\Delta}) + E(\sigma^2_{t,j,\Delta})^2 = E\left[(\sigma^2_{s,j,\Delta})^2\right]$ $\forall s \neq t$ and $\forall i \neq j$.

(iv) The autocovariance function of RV_t^Δ is

$$\begin{aligned}
Cov\left(RV_t^\Delta, RV_{t+h}^\Delta\right) &= Cov\left(IV_t + u_t^\Delta, IV_{t+h} + u_{t+h}^\Delta\right) \\
&= E\left[\left(IV_t + u_t^\Delta\right)\left(IV_{t+h} + u_{t+h}^\Delta\right)\right] \\
&\quad -E\left(IV_t + u_t^\Delta\right) \cdot E\left(IV_{t+h} + u_{t+h}^\Delta\right) \\
&= E\left(IV_t \cdot IV_{t+h}\right) + E\left(u_t^\Delta \cdot u_{t+h}^\Delta\right) + E\left(IV_t \cdot u_{t+h}^\Delta\right) \\
&\quad +E\left(IV_{t+h}^\Delta \cdot u_t\right) - E\left(IV_t\right) \cdot E\left(IV_{t+h}\right) \\
&= E\left(IV_t \cdot IV_{t+h}\right) - E\left(IV_t\right) \cdot E\left(IV_{t+h}\right) \\
&= Cov(IV_t, IV_{t+h}) \quad \forall h \neq 0.
\end{aligned}$$

The manufacturer's authorised representative in the EU is Springer Nature Customer Service Centre GmbH, Europaplatz 3, 69115 Heidelberg, Germany. If you have any concerns regarding our products, please contact ProductSafety@springernature.com

Printed and bound by CPI Group (UK) Ltd, Croydon, CR0 4YY
25/09/2026
02222884-0015